Birkhäuser

Studies in Universal Logic

This series is devoted to the universal approach to logic and the development of a general theory of logics. It covers topics such as global set-ups for fundamental theorems of logic and frameworks for the study of logics, in particular logical matrices, Kripke structures, combination of logics, categorical logic, abstract proof theory, consequence operators, and algebraic logic. It includes also books with historical and philosophical discussions about the nature and scope of logic. Three types of books will appear in the series: graduate textbooks, research monographs, and volumes with contributed papers.

Jean-Yves Béziau

Editor

Universal Logic:
An Anthology

From Paul Hertz to Dov Gabbay

 Birkhäuser

Editor
Jean-Yves Béziau
Institute of Philosophy
Federal University of Rio de Janeiro
Rio de Janeiro, RJ, Brazil

ISBN 978-3-0346-0144-3   e-ISBN 978-3-0346-0145-0
DOI 10.1007/978-3-0346-0145-0
Springer Basel Dordrecht Heidelberg London New York

Library of Congress Control Number: 2012931945

2010 Mathematical Subject Classification: 03B22, 03B20, 03B45, 03B53, 03A05, 03B50, 03B62, 03G30, 03F03, 03G27, 03C95

Printed on acid-free paper

Springer Basel AG is part of Springer Science+Business Media (www.birkhauser-science.com)

Preface of an Anthology of Universal Logic From Paul Hertz to Dov Gabbay[1]

This book is a retrospective on universal logic in the 20th century. It gathers papers and book extracts in the spirit of universal logic from 1922 to 1996. Each of the 15 items is presented by a specialist explaining its origin, import and impact, supported by a bibliography of correlated works. Some of the pieces presented here, such as "Remarques sur les notions fondamentales de la méthodologie des mathématiques" by Alfred Tarski, are for the first time translated into English.

Universal logic is a general study of logical structures. The idea is to go beyond particular logical systems to clarify fundamental concepts of logic and to construct general proofs. This methodology is useful to understand the power and limit of a particular given system. Lindström's theorem is typically a result in this direction: it provides a characterization of first-order logic. Roughly speaking, Lindström's theorem states that first-order logic is the strongest logic having both the compactness property and the Löwenheim–Skolem property (see details in Part 10). Such a theorem is concerned not only with first-order logic but with other nearby possible logics. One has to understand what these other possible logics are and be able to compare them with first-order logic. In short: one has to consider a class of logics and relations between them. Lindström's theorem is a result in favor of first-order logic, but to claim the superiority of this logic one must have the general perspective of an eagle's eye. Moreover Lindström's theorem favors first-order logic within a limited galaxy of possible logics. At a more universal level, things change. One may want to generalize Lindström's theorem to other galaxies, such as the galaxy of modal logics (about such generalization see e.g. [5]). In order to do so, we need a clear understanding of what Lindström's theorem exactly depends on.

Comparison of logics is a central feature of universal logic. The question of *translation* of a logic into another one is directly connected to it. This topic is especially treated in Part 4. Gödel has shown that it is possible to translate intuitionistic logic into a system of modal logic and more surprisingly classical logic into intuitionistic logic, a surprising result since in some sense intuitionistic logic is strictly weaker than classical logic (among other things, the excluded middle holds in the latter but not in the former). There are other cases of such a paradoxical situation, e.g. the occasion of a logic weaker than another logic which can however be translated into its weaker sister (see [7] and [24]). This is

[1]The author is supported by a grant of the Brazilian research council (CNPq). Thank you to Arnold Koslow and Arthur de Vallauris Buchsbaum.

a phenomenon similar to Galileo's paradox showing that we may have a one-to-one correspondence between a set and one of its proper subsets. Galileo's paradox is cleared up by showing that there are two concepts corresponding to two contiguous but distinct notions. Clarification of concepts is also a device to solve a translation's paradox. We must have a good definition of what the *strength* of a logic is, understanding that there is not only one way to compare logics, in particular that there are different non-equivalent ways to translate one logic into another. It is important to point out that we cannot solve this problem just by straightforwardly importing concepts from other part of mathematics, thinking they will do the job in the logic realm. Intuitionistic logic is not a *sublogic* of classical logic in the same sense that rational arithmetic is a *subalgebra* of real arithmetic. One has to avoid the famous sufism of Nasruddin: a man at night looking for his key, not where he lost it, but under the light of a not so nearby lamp-post.

Another paradox appears when *combining* classical logic with intuitionistic logic. They may collapse into the same logic, contrary to the expectation of the theory of combination of logics: to get the smaller conservative extension of both logics preserving their own idiosyncrasies (see [35] and [14]). An opposite paradox in combination of logics is the copulation paradox: instead of having less, we have more: e.g. by putting together the logic of conjunction with the logic of disjunction, we may get distributivity (see [8] and [9]). To avoid these paradoxes we must develop a good theory of combination of logics and to do so we must find the right concepts. With these paradoxes the logician is confronted with some particular cases that must be taken into account and analyzed to build a nice abstract theory. So in some sense logic is an empirical science, in the sense that the logician is facing some objective phenomena that cannot be dropped, whatever their private reality is. As it is common in the history of mathematics, first particular cases are studied and then the level of abstraction rises. This is typically what has been happening in the theory of combination of logics. First logicians were combining modal logics, and then they started to develop a general theory of combination of logics, in particular Dov Gabbay with his pivotal concept of fibring (see Part 15).

But the abstraction rise is not necessarily progressive, there are also some radical jumps into abstraction. In logic we can find such jumps in the work of Paul Hertz on *Satzsysteme* (Part 1) and of Alfred Tarski on the notion of a *consequence operator* (Part 3). What is primary in these theories are not the notions of logical operators or logical constants (connectives and quantifiers) but a more fundamental notion: a relation of consequence defined on undetermined abstract objects that can be propositions of any science but also data, facts, events. Probably Hertz and Tarski did not directly think of all possible interpretations of such abstract objects. When performing jumps into abstraction, we cannot foresee the true depth and breadth of the realm which is being opened. In universal logic, consequence is the central concept. But this consequence relation is neither syntactical (proof-theoretical), nor semantical (model-theoretical). We are beyond the dichotomy syntax/semantics (proof theory/model theory). This level of abstraction is the highest vertex of an upward pointing triangle with syntax and semantics as base angles. It is the crucial point of the completeness theorem; by reaching it we are led to its trivialization, following Wójcicki's way of speaking. In the original work by Hertz and Tarski this is not so clear, and one may be confused by the fact that Hertz's name is rather connected with proof theory due to its influence on Gentzen's work, and that Tarski's name is rather connected with semantics due to his work on truth and model-theory. But Hertz's original work is not so proof-theoretical, as shown by its later development in a structuralist perspective by Arnold Koslow [25]. It is also worth recalling that in his first paper Gentzen stays at Hertz's abstract level, proving an abstract completeness theorem (this is discussed in Part 14). Concerning Tarski, it is important to clearly distinguish his work on consequence operators from his work on model theory although there is a connection between the two explained in Dana Scott's paper presented in Part 12.

When we consider a logic as a structure with the consequence relation as the central concept, a relation which can be defined in many different ways but that can also be considered independently of any particular specification, we can say that we are at the level of *abstract logic*. The terminology "abstract logic" was much used by Roman Suszko. In the 1950s Suszko developed his work with Jerzy Łoś defining the consequence operator over an absolutely free algebra, leading to the notion of a structural consequence operator, in the spirit of Lindenbaum representation theorem of logical matrix theory (Part 7). At the end of the 1960s Suszko, along with Donald Brown and Stephen Bloom, came back to a more abstract setting that he explicitly called "abstract logic", considering a consequence relation over an undetermined abstract algebra (Part 11), a level of generalization not as high as the one of Tarski's first framework of consequence operator, but higher than the one of abstract model theory where the expression "abstract logic" is sometimes also used. Suszko and other people in Eastern Europe had the idea that logic was part of universal algebra, a mathematical trend highly popular in the East. There was an assimilation of abstract logic with universal algebra, connected with a broader assimilation, that of universal algebra with mathematics as a general theory of structures. In this context the differences between Boolean algebras, lattices and any mathematical structures is just a question of level of abstraction. There may be confusion in this mixture, such as when many years ago the word "structure" was used as synonymous with "lattice" by Glivenko (see [20] and [17]).

Universal logic can be defined as a general study of logical structures in the same way that universal algebra is a general study of algebraic structures. The word "universal" in "universal logic" is used according to this analogy: as in universal algebra, universal logic is not a universal system, but rather a universal systematization. Universal algebra is not one algebraic system encompassing everything, but a bunch of global concepts allowing us to unify the treatment of the multiplicity of algebraic structures. These concepts were mainly put forward by Garrett Birkhoff in the 1930s (see [10] and [11]). The central concept is the concept of abstract algebra defined by Birkhoff just as a set with a family of operators. The spirit of universality is the same in universal logic and universal algebra, but these two fields are different because a logic structure is not necessarily an abstract algebra. Reduction of logic to algebra can be developed through algebraization of logic, which can mean both the reduction of logical structures to algebraic structures and the application of algebraic methods to logic. Generally the former is seen as a first step towards the latter. But although it is interesting to make a connection between logic and algebra, there is no good reason to think that logic reduces to algebra. There are indeed logical structures that cannot be algebraized (see [6]). One may also apply other mathematical methods and tools to develop logic, for example topology. The initial Tarskian concept of consequence operator is in fact closer to topology than algebra. But the very idea of universal logic is that logical structures are different from other mathematical structures and that more generally logic is different from other parts of mathematics.

To have a deeper understanding of this, we have to think at the level of a general theory of mathematical structures. One may think of category theory. But category theory will not, right at the start, clarify what a logical structure is. And to apply category theory to logic in the perspective of an alternative foundation of mathematics, as it was done by Lawvere [26], is not the same as developing a study of categories of logical structures [2]. In this latter case we consider logical structures encompassing classical and non-classical logics, working with a pre-categorial vision of the concept of logical structures based on the reality of the huge variety of logical systems. A famous categorization of logical structures is due to the late Joseph Goguen, originator of the concept of *institution* (see Part 13). What is fascinating is that the origin of this trend of general abstract nonsense is computer science, a very concrete and applied science. Goguen described the situation as follows: "The enormous and still growing diversity of logics used in computer science presents a formidable challenge. One approach to bringing some order to this chaos is to formalize the notion of *a logic* and then systematically study general properties of logics using this formalization, including the representation, implementation, and translation of logics. This is the purpose of the theory of institutions, as developed and applied in a literature that now has hundreds of papers." [21]

To understand what a logical structure is, it is worth having a look at Bourbaki's monumental work. Bourbaki was the first to develop a general theory of mathematical structures. To avoid misunderstandings, one must distinguish Bourbaki's informal theory expressed in his 1948 paper *L'Architecture des mathématiques* [12] and the formalization of it presented in a chapter of the 1954 book *Théorie des ensembles* entitled *Structures* [13]. The defect of this dated set-theoretical formalization cannot be used as a fatal argument against the informal theory. More than anything it is important to remember that Bourbaki was the first to develop a general theory of mathematical structures and to consider *morphism* as a central concept of mathematics (see [4] and [16]). From the bourbachic viewpoint it is clear that mathematical structures do not reduce to algebraic structures. According to Bourbaki there are three distinct classes of *mother structures* from which we can reconstruct mathematics, mixing them thus generating *cross structures*. These fundamental classes of structures are structures of order, algebraic structures and topological structures. Bourbaki was not excluding the existence of other mother structures, as recalled by Jean Porte, who tried to develop logic in the bourbachic spirit—his PhD advisor was René de Possel, one of the founding members of Bourbaki. Porte, in his very interesting book *Recherches sur la théorie générale des systèmes formels et des systèmes connectifs* published in 1965, studies different classes of logical structures taking into account and including the recent developments of the Polish school on consequence operators and logical matrix theory (see Part 9).

Porte clearly states that he is not doing *metamathematics*, explaining that the logical systems he is studying are not exclusively describing mathematical reasoning. For him logic is mathematical, but not necessarily about mathematics. The expression "mathematical logic" is highly ambiguous because it can mean the *logic of mathematics* or a *mathematical study of logic* (this ambiguity was already noted by Zermelo in 1908, see [30, p. 320]). First-order logic can be seen as a combination of the two. But these two orientations may be quite different and they indeed were different in the history of modern logic. They can be distinguished in a broad outline using the opposition between the *Boolean way* and the *Fregean way*. Boole was using mathematics to understand the laws of thought, and these are not only concerned with mathematical thinking. Boole had a general perspective on reasoning, as Aristotle had: syllogistic is about any kind of reasoning. But syllogistic is neither mathematical in a Boolean sense—it does not use mathematics to describe reasoning—nor in a Fregean sense—it does not give an accurate description of mathematical reasoning. Frege's main interest was to describe reasoning of arithmetics, for doing that he was not using mathematics, but some two-dimensional graphism with a cryptic name: *Begriffsschrift*. Such ideography is not more mathematical than musical notation. A similar tendency—the use of a non-mathematical technique to describe mathematical reasoning—can be found with Peano's pasigraphy and with Whitehead's and Russell's *Principia Mathematica*. Jean van Heijenoort puts these tendencies in the same basket: according to him the three of them have the feature of a *lingua characteristica* (see [23]). We can consider first-order logic as a mix: rules of syntax (construction of the language) and rules of proofs (proof theory) are generally closer to a *lingua characteristica* orientation, but mathematics is extensively used for the semantics developed in model theory. This mix was described by Chang and Keisler in their classical book [15], by the equation *model theory = logic + universal algebra*. In this equation "logic" may be interpreted as logic syntax and "universal algebra" as a class of mathematical structures (Chang and Keisler are here under the influence of the reduction we were mentioning: *structures ⊆ algebras*).

A mix also appears at another level qualified as *mathematics of metamathematics* by the Polish duo Rasiowa/Sikorski [33]. This can be considered as synonymous to *algebraic logic*. Such use of mathematics is different from the use of mathematics at the semantic level. This can be understood through the example of classical propositional logic: its semantics is the Boolean algebra on {0, 1} but classical logic can directly be considered as a Boolean algebra by factoring it. These are two different methodologies, the first is mainly due to Post [32] and the second, more than 10 years later, to Tarski (not to Lindenbaum as often erroneously stated). This is the second methodology which is now usually qualified as algebraization of logic. Paul Halmos has generalized this methodology to first-order logic popularizing the expression "algebraic logic" (see [22]), but this terminology was introduced first by Curry and with a different meaning more connected with universal logic.

Haskell Curry was the last PhD student of Hilbert and tried to systematically develop the formalist approach, both at the philosophical and metamathematical levels. But finally, as pointed out by Seldin [34], his approach would be better qualified as structuralist. In the present anthology Seldin translates and comments extensive extracts of Curry's monograph *Leçons de logique algébrique* (Part 6). This book is not very well known and has never been translated into English. It was published in 1952, eleven years earlier as a more famous book by Curry, *Foundations of mathematical logic* [19]. As indicated by the title of this later book, Curry is interested in the foundations of logic, not in the foundations of mathematics. This difference may not be immediately caught because: 1) someone may not pay attention to the way the three notions foundations/mathematics/logic are combined, understanding Curry's title as synonymous with logical foundations of mathematics; 2) if we consider that mathematical logic is mainly concerned with mathematical reasoning, foundations of mathematical logic has to do with foundations of mathematics. But Curry, like Porte and the Poles, is not interested only in mathematical reasoning, he has also interest in many systems of logic describing reasoning concerning other fields, like quantum logic. His 1952 monograph is a study of many different systems of logic and to do so he develops a general framework, using some mathematical tools, in particular mathematical structures that are more or less algebraic structures.

In the 1950s Curry was the first to pay attention to the work of Saul Kripke. Kripke as a teenager wrote to him and they maintained a correspondence during a couple of years (see

[19, p. 240, 243, 250, 306]). Curry had much interest in modal logic, however, due to his formalist background he had been working more on its proof theory. But Curry was open minded, interested in all aspects of logic, having an impressive knowledge of what was going on in logical research, as we can see through the excellent bibliography of [19]. It is worth emphasizing that Kripke also had broad interests, he was working in many different logical systems other than modal logic, as we can see through Curry's book, such as multiple conclusion sequent systems for intuitionistic logic (p. 306). Kripke was considering logic in a general perspective but probably he was not aware that by developing a semantic framework for modal logics, he would provide a universal tool whose applications go far beyond the field of modal logic. Modal logic is a field of research that is directly connected to universal logic: there are many different systems of modal logic and it has been natural to develop a general theory of the class of modal logics. This theory gives good indication of how universal logic can be developed, giving hints for the study of other classes of logics and for systematization of logical structures in general. For example it is obviously useful to find the general formulation of the completeness theorem beyond all variations of Kripke structures, or the general techniques for combining them. Kripke structures can also be used to deal with many systems of logics other than modal systems: intuitionistic logic, relevant logic, paraconsistent logic. Moreover they are a powerful tool for making links between propositional logics and higher-order logic, in particular reducing fragments of first-order logic into propositional modal systems, work especially developed by Johan van Benthem (see e.g. [1]) who presents Kripke's work in Part 8. We may introduce the expression *Kripke logics* to name the class of logics that can be defined using Kripke structures. This class does not reduce to modal logic. And vice versa the class of modal logics is not included in the class of Kripke logics.

It is important to recall that the first semantics for modal logic is due to Łukasiewicz, it is a three-valued matrix semantics [28]. Later on Łukasiewicz developed a four-valued matrix semantics [29]. Matrix semantics does not reduce to modal logic. The tendency in fact, due to the awkwardness of Łukasiewicz's systems and Dugundji's negative result about characterization of S5 by finite matrices, is to consider that we have here two disjoint classes of logics: on the one hand the class of logics definable using logical matrices that can be called *truth-functional logics* (rather than many-valued logics, see [31]), on the other hand the class of modal logics. This does not mean that the class of truth-functional logics and Kripke logics are disjoint. For example classical propositional logic is a truth-functional logic and also a Kripke logic, even if the Kripke semantics for it is rather trivial (one may also argue that the truth-functional semantics of classical propositional logic is trivial compared to much more complex truth-functional semantics). Łukasiewicz used matrix semantics in a philosophical perspective, but Tarski saw it rather as a universal tool and it was developed as such in the Polish school, in particular by Lindenbaum. The work of Lindenbaum on logical matrices has been published mainly through Łos's monograph [27]. But logical matrices is not a pure Polish product, it was developed also by Emil Post presenting bivalent matrices for classical propositional logic [32] and immediately generalizing the technique (Post was born in Poland, but grew up in the USA) and by Paul Bernays. In Part 2 Bernays's work is presented: using many-valued matrices, not to develop a particular non-classical logical system as Łukasiewicz did, but to analyze the independence of axioms for classical propositional logic.

As pointed out by Suszko [36], it is possible to provide a bivalent semantics for Łukasiewicz's logic L3. This apparent paradox is cleared up when we know that this bivalent semantics is not truth-functional. A general theory of non truth-functional bivalent semantics has been developed by Newton da Costa and his school, the *theory of bivaluations* (see Part 14). We can call *da Costa logics* logics that can be defined by non truth-functional bivalent semantics. Da Costa's idea was that non truth-functional bivalent semantics is a universal tool in the sense that any logic can be defined using it; in other words, the class of da Costa logics is the universal class of logics. This is a contestable claim because we can argue that some structures that can rightly be called logics are not definable with this tool. In fact it is doubtful that there exists any tool that can be used to define all logics. There is no such logic wand. But anyway the theory of bivaluations is very interesting for at least two reasons. Firstly it breaks the illusion of the completeness theorem as a magical result connecting two worlds, on the one hand a world made of strings of symbols, on the other hand a world of flesh and blood models. In the theory of bivaluations, a model is just a set of formulas. Secondly the theory of bivaluations is used to study many non-classical logics. In fact da Costa was led to this theory trying to provide semantics for his systems of paraconsistent logic [18]. As it is known, da Costa is the main promoter of paraconsistent logic, and contrary to other people like Asenjo [3] or Jaśkowski developing paraconsistent logic in a rather Sufist way, using respectively the lamp-post light of matrix semantics and modal logic, da Costa invented a new tool having broad applications, being a new light for the universal dimension of logic. Using da Costa's theory of bivaluations it is possible to develop *paranormal logics*, logics which are both paraconsistent and paracomplete. It definitively shows that the foundations of logics lay far beyond some principles such as the principle of non-contradiction or the excluded middle. At the abstract level everything is possible, concepts are elaborated that can be used to develop tools that can be applied to many concrete situations.

References

1. Andréka, H., van Benthem, J., Németi, I.: Back and forth between modal logic and classical logic. Bull. Interest Group Pure Appl. Log. **3**, 685–720 (1995)
2. Arndt, P., Freire, R.A., Luciano, O.O., Mariano, H.L.: A global glance on categories in logic. Log. Univers. **1**, 3–39 (2007)
3. Asenjo, F.: A calculus of antinomies. Notre Dame J. Form. Log. **7**, 103–105 (1966)
4. Beaulieu, L.: Nicolas Bourbaki: History and Legend, 1934–1956. Springer, New York (2008)
5. van Benthem, J.: A new modal Lindström theorem. Log. Univers. **1**, 125–138 (2007)
6. Béziau, J.-Y.: Logic may be simple. Log. Log. Philos. **5**, 129–147 (1997)
7. Béziau, J.-Y.: Classical negation can be expressed by one of its halves. Log. J. Interest Group Pure Appl. Log. **7**, 145–151 (1999)
8. Béziau, J.-Y.: A paradox in the combination of logics. In: Carnielli, W.A. et al. (eds.) Proceedings of Comblog'04, pp. 87–92. IST, Lisbon (2000)
9. Béziau, J.-Y., Coniglio, M.E.: To distribute or not to distribute. Log. J. Interest Group Pure Appl. Log. **19**, 566–583 (2011)
10. Birkhoff, G.: Universal algebra. In: Comptes Rendus du Premier Congrès Canadien de Mathématiques, pp. 310–326. University of Toronto Press, Toronto (1946)
11. Birkhoff, G.: Universal algebra. In: Rota, G.-C., Oliveira, J.S. (eds.) Selected Papers on Algebra and Topology by Garrett Birkhoff, pp. 111–115. Birkhäuser, Basel (1987)
12. Bourbaki, N.: The architecture of mathematics. Am. Math. Mon. **57**, 221–232 (1950) (Originally published in French in 1948)
13. Bourbaki, N.: Theory of Sets. Addison-Wesley, Reading (1968)
14. Caleiro, C.: From fibring to cryptofibring. A solution to the collapsing problem. Log. Univers. **1**, 71–92 (2007)

15. Chang, C.C., Keisler, H.J.: Model Theory. North-Holland, Amsterdam (1973)
16. Corry, L.: Nicolas Bourbaki and the concept of mathematical structure. Synthese **92**, 315–349 (1992)
17. Corry, L.: Modern Algebra and the Rise of Mathematical Structures. Birkhäuser, Basel (2004) (1st edn. (1998))
18. da Costa, N.C.A., Alves, E.H.: A semantical analysis of the Calculi C_n. Notre Dame J. Form. Log. **18**, 621–630 (1977)
19. Curry, H.B.: Foundations of Mathematical Logic, p. 1963. McGraw Hill, New York (1963)
20. Glivenko, V.: Théorie générale des structures. Hermann, Paris (1938)
21. Goguen, J.: Institutions, http://cseweb.ucsd.edu/~goguen/projs/inst.html (2006)
22. Halmos, P.: The basic concepts of algebraic logic. Am. Math. Mon. **63**, 363–397 (1956)
23. van Heijenoort, J.: Logic as calculus and logic as language. Synthese **17**, 324–330 (1967)
24. Humberstone, L.: Béziau's translation paradox. Theoria **71**, 128–181 (2008)
25. Koslow, A.: A Structuralist Theory of Logic. Cambridge University Press, New York (1992)
26. Lawvere, B.: The category of categories as a foundation for mathematics. In: Proceedings of the Conference on Categorical Algebra, La Jolla 1965, pp. 1–20. Springer, New York (1966)
27. Łoś, J.: O matrycach logicznych, Pr. Wroc. Tow. Nauk., B **19**, 1949 (1949)
28. Łukasiewicz, J.: O logice trójwartościowej. Ruch Filoz. **5**, 170–171 (1920)
29. Łukasiewicz, J.: A system of modal logic. J. Comput. Syst. **1**, 111–149 (1953)
30. Mancosu, P., Zach, R., Badesa, C.: The development of mathematical logic from Russell to Tarski 1900–1935. In: Haaparanta, L. (ed.) The Development of Modern Logic, pp. 318–470. Oxford University Press, Oxford (2009)
31. Marcos, J.: What is a non-truth-functional logic? Stud. Log. **92**, 215–240 (2009)
32. Post, E.: Introduction to a general theory of elementary propositions. Am. J. Math. **13**, 163–185 (1921)
33. Rasiowa, H., Sikorski, R.: The Mathematics of Metamathematics, Polish Academy of Science, Warsaw (1963)
34. Seldin, J.P.: Curry's formalism as structuralism. Log. Univers. **5**, 91–100 (2011)
35. Sernadas, C., Rasga, J., Carnielli, W.A.: Modulated fibring and the collapsing problem. J. Symb. Log. **67**, 1541–1569 (2002)
36. Suszko, R.: Remarks on Łukasiewicz's three-valued logic. Bull. Sect. Log. **4**, 87–90 (1975)

University of Brazil   Jean-Yves Béziau
Rio de Janeiro
January 15, 2011

Contents

Contributors and Translators

Jean-Yves Béziau Institute of Philosophy, Federal University of Rio de Janeiro, Rio de Janeiro, Brazil

C. Caleiro SQIG – Institute of Telecommunication, Department of Mathematics – Instituto Superior Técnico, TU Lisbon, Lisbon, Portugal

Walter Carnielli Centre for Logic, Epistemology and the History of Science – CLE and Department of Philosophy – UNICAMP, Campinas SP, Brazil

Itala M. Loffredo D'Ottaviano Centre for Logic, Epistemology and the History of Science – CLE, Department of Philosophy, UNICAMP, Campinas, SP, Brazil

Răzvan Diaconescu Institute of Mathematics "Simion Stoilow" of the Romanian Academy, Bucharest, Romania

Hércules de Araújo Feitosa Department of Mathematics, São Paulo State University – UNESP, Bauru, SP, Brazil

Marcel Guillaume Clermont-Ferrand, France   .

Lloyd Humberstone Monash University, Victoria, Australia

Ramon Jansana Department of Logic, History and Philosophy of Science, University of Barcelona, Barcelona, Spain

Javier Legris University of Buenos Aires and CEF/CONICET, Buenos Aires, Argentina

Mathieu Marion Chaire de recherche du Canada en philosophie de la logique et des mathématiques, Département de philosophie, Université du Québec à Montréal, Montréal, Québec, Canada

Robert Purdy Toronto, Canada

Jonathan P. Seldin Department of Mathematics and Computer Science, University of Lethbridge, Lethbridge, Alberta, Canada

A. Sernadas SQIG – Institute of Telecommunication, Department of Mathematics – Instituto Superior Técnico, TU Lisbon, Lisbon, Portugal

Jouko Väänänen Department of Mathematics and Statistics, University of Helsinki, Helsinki, Finland; Institute for Logic, Language and Computation, University of Amsterdam, Amsterdam, The Netherlands

Johan van Benthem Institute for Logic, Language and Computation, University of Amsterdam, Amsterdam, The Netherlands; Department of Philosophy, Stanford University, Stanford, CA, USA

Richard Zach Department of Philosophy, University of Calgary, Alberta, Canada

Jan Zygmunt Chair of Logic and Methodology of Science, University of Wrocław, Wrocław, Poland

Part 1
Paul Hertz (1922)

Paul Hertz and the Origins of Structural Reasoning

Javier Legris

Abstract The aim of this note is to introduce and summarize Paul Hertz's contributions in his paper "On axiomatic systems for arbitrary systems of sentences" placing it in its historical context. In it the first ideas concerning structural reasoning are to be found. Moreover, it influenced strongly Gentzen's sequent calculus. The analysis of the formal structure of proofs was one of Hertz's most important achievements and it can be regarded also as an anticipation of a "theory of proofs" in the current sense. In this note Hertz's philosophical ideas concerning the nature of logic will be also sketched.

Keywords Structural reasoning · History of symbolic logic · Proof theory · Logical consequence

Mathematics Subject Classification (2000) 03-03 · 03B47 · 03F03 · 03F07 · 03A05

The first ideas concerning what is now called *structural reasoning* (in the sense, for example, of Koslow's "structuralist logic", see [21]) and *structural proof-theory* (see, v.g., [22]) are to be found in the paper "On axiomatic systems for arbitrary systems of sentences" written by Paul Hertz at the beginning of the 20s of the last century. In the paper the author introduced his key concept of *system of sentences* [*Satzsysteme*] and obtained some results that should count as proof theoretic. The paper was originally published in German in *Mathematische Annalen* and a second part appeared in the same journal some years later [14]. Apart from them, Hertz wrote also further papers on the subject in philosophical journals in order to provide a more general understanding of his *Satzsysteme* ([15] is a good introduction to the whole system). The very idea of sequent calculi and of structural rules was first conceived by Hertz. This fact was explicitly acknowledged by Gentzen himself who devoted his first published paper to Hertz's systems [8]. The analysis of the formal structure of proofs was one of Hertz's most important achievements and it can be regarded as an anticipation of General Proof Theory in the current sense. It must be noticed that Hertz's contributions were made when the "metamathematical revolution" had not been completely accomplished, so that many metalogical notions had not been introduced yet. It can be asserted that Hertz's systems played the role of a bridge between traditional formal logic and Gentzen's logical work.

Paul Hertz (1881–1940) studied Physics and Mathematics in Göttingen, where he earned his PhD in 1904. He taught in Heidelberg until 1912. In this period, he contributed essentially to statistical mechanics and thermodynamics. According to Bernays, typical of his research was its conceptual and methodological clarity (see [5]). In 1913 he moved to Göttingen again, where he—under the influence of David Hilbert's program—became interested in Epistemology and Methodology of Science. He edited together with Moritz Schlick the work of Hermann von Helmholtz on Philosophy of Science (von Helmholtz 1921). In some of these writings, problems concerning the philosophy and methodology of mathematics were discussed, specially in relation with the evaluation of non-Euclidean geometry.[1]

In 1921 he received an appointment to teach on "Methoden der exakten Naturwissenschaften" (Methodology of exact natural sciences) in Göttingen as *ausserordentlicher Professor* (associated professor). In those years he discussed his work with Paul Bernays and also took part in philosophical discussions with members of the Vienna Circle and the Berlin Group (see [10, p. 76]). In 1933 his *venia legendi* was withdrawn due to the racial laws of the Nazi-regime. He received then a research grant from the American Rescue Committee to work at the University of Geneva and—from 1936 on—at the German University of Prague. In 1939 Hertz emigrated to the USA, where he died in 1940 (for further references see [5]). In this time he published on causality and philosophy of logic in the Journal *Erkenntnis* (see [16, 18]).

As a matter of fact, the logical work of Hertz and its influence on Gentzen remained rather unnoticed by logicians. However, Paul Bernays had a highly positive appraisal of Hertz's contributions. Alonzo Church listed in his *Bibliography* almost all of Hertz's logical writings (see [6, p. 187]), and also Haskell B. Curry pointed out the importance of Hertz's contributions in a historical note in his *Foundations of Mathematical Logic* [7, pp. 246 ff.].[2]

Hertz's main contributions to logic could be summarized as follows:

1. The notion of sequent, anticipating Gentzen's sequent calculus.
2. The development of methods for constructing *minimal* axiom systems.
3. An Analysis of the *structure* of proofs.
4. The idea of a system of deduction consisting exclusively of *structural* deductions.

Paper [11] devotes to 1. and 2. More precisely, he achieved (a) some formal results concerning minimal axiom systems in order to reduce their complexity, (b) the introduction of special symbols ("ideal elements") in order to reduce the complexity of axioms. In the following, a general overview of all these contributions is given.

---

[1]In the edition the following papers of von Helmholtz were included "Über den Ursprung und die Bedeutung der geometrischen Axiome", "Über die Tatsachen, die der Geometrie zugrunde liegen", "Zahlen und Messen" and "Die Tatsachen in der Wahrnehmung".

[2]In his biographical article on Hertz, Bernays wrote "Diese Untersuchungen [of Hertz] sind Vorläufer verschiedener neuerer Forschungen zur mathematischen Logik und Axiomatik, insbesondere hat G. Gentzens Sequenzenkalkul von den H.schen Betrachtungen über Satzsysteme seinen Ausgang genommen" [5, p. 712], and in a paper on sequent calculi, Bernays asserted: "... in der Hertz'schen Theorie der Satzsysteme ein gewiss bei weitem noch nicht hinsichtlich der möglichen Fragestellungen und Erkenntnisse ausgeschöpftes Forschungs-gebiet der Axiomatik und Logik vorliegt." [4, p. 5, footnote]. Haskell Curry in a historical note of his textbook on mathematical logic stated: "For the present context it is worthwhile to point out that Gentzen was apparently influenced by Hertz [...] This throws some light on the role of "Schnitt" in the Gentzensystem." [7, pp. 246 f.]. Vittorio Michele Abrusci wrote an introductory paper on Hertz's logical work [1]. Further information on Hertz can be found in the Nachlaß of Paul Bernays (ETH Zürich), the Nachlaß of David Hilbert (Göttingen), biographical notes (not published) by Adriaan Rezus (Nijmegen), and above all in Hertz's Nachlaß located at Archives for Scientific Philosophy, University of Pittsburgh.

Hertz's original interest in logic, as it is shown in the paper reproduced here, was largely focused on the formal properties of axiom systems and his goal consisted in developing reduction methods for axiomatic systems from which some sort of "minimal" and independent system could be obtained, that is, a system where proofs should be as elemental as possible. Taking into account the situation of different axiomatic systems for the same theory, he explicitly thought about the possibility of developing reduction procedures to achieve some sort of irreducible ("normal") axiomatic system (see [11, p. 246]). In doing so, he initiated an investigation of the general properties of axiomatic systems. In this respect, he referred to "the notion of axiom in general" [14, p. 427]. His research led him to the idea of *Satzsystem* (system of sentences), undoubtedly his main contribution to mathematical logic.

In Hertz's sense, a "sentence" [Ger. *Satz*], related to a "basic domain" of elements, is an expression of the forms

$$(1)\ a \longrightarrow b$$
$$(2)\ a_1, a_2, \ldots, a_n \longrightarrow b$$

where $a_1, a_2, \ldots, a_n$ are called the *antecedens* of the sentence and b its *succedens* [11, §1.1, p. 249]. Sentences of the form (1) are called "sentences of first degree" (*lineal* sentences in later papers), i.e., they have only one antecedens. Apart from this, he distinguished in his paper from 1929 between sentences with free individual variables (generally understood as universally quantified) and sentences with constants which he called respectively 'macrosentences' (*Makrosätze*) and 'microsentences' (*Mikrosätze*).

Hertz proposed different interpretations for these sentences, depending on the nature of the basic domain (events or predicates). Basically, we can find the following interpretations:

(i) elements as events (*Ereignisse*, see [14, p. 459]);
(ii) elements as predicates: If the predicates $a_1, a_2, \ldots, a_n$ satisfy an object, then b satisfies it too (see [13, p. 273]);
(iii) sentences as formal implications "in the sense of Russell" (see [11, p. 247 note 1]), that is, they are to be understood as general valid implications between sentences.

According to this, the very symbol $\longrightarrow$ is alternately interpreted either as a logical relation (between events, predicates or sentences) or as a logical constant. Thus, he conceived the idea of sentences as symbolic structures in an abstract way, that is, as a "complex" of elements (see [11, §1.1]).

The first reaction to Hertz's ideas was to see them as a contribution to positive logic (i.e. propositional logic restricted only to conjunction, disjunction and material implication), which was analyzed by Bernays and others in Göttingen and for which its decidability had already been proved (see [2, p. 11]). In the *Grundlagen der Mathematik* by Hilbert and Bernays, Hertz is mentioned related to this field of research (see [20, p. 68 n. 1]).

With systems of this kind, Hertz aimed to solve the problem of the deductive closure of formal theories, which was discussed at that time. Usually this problem consists in determining for a set of sentences that everything that can be proved in the theory belongs to that set. In other words, it consisted in assuring that every theorem of the theory is in the set. This problem led him to the idea of "closed systems (sets) of sentences" (*geschlossene Satzsysteme*, [11, §1.5]).[3]

In the introduction of this paper, Hertz describes intuitively the problem of obtaining a minimal axiom system by means of graphs. The basic elements of the sentences are represented by points and arrows connecting these points correspond to sentences. Then, it is shown how the introduction of ideal elements reduces the number of arrows (i.e. sentences, [11, p. 248]). I will go back to this subject.

After this first presentation, Hertz established a minimal set of rules for closed systems of sentences, which constituted an antecedens of the structural rules in Gentzen's sequent calculus. He considered the case of an "inference system" (*Schlusssystem*) consisting in proofs for a system of proposition through these rules. Following the characterization of closed systems of sentences, they must contain what Hertz called "tautological sentences", of the form

$$a \longrightarrow a$$

which are "essentially logical" sentences.

Appart from this, Hertz realized that the closure of the system depends on both the rules of the system and the proof procedures in it. In all the papers he wrote on the subject in the 20's he presented two basic inference rules, which not only constituted the minimal set of rules for building closed systems but also played the role of basic principles defining logical inference. The first rule was a generalization of transitivity of the arrow and was called by Hertz with the traditional name of *syllogism* (*Syllogismus*). It had the following structure:

$$a_{11}, a_{12}, \ldots, a_{1n} \longrightarrow b_1$$
$$a_{21}, a_{22}, \ldots, a_{2n} \longrightarrow b_2$$
$$\ldots$$
$$a_{m1}, a_{m2}, \ldots, a_{mn} \longrightarrow b_m$$
$$\frac{a_{11}, \ldots, a_{1n}, a_{21}, \ldots, a_{2n}, a_{m1}, \ldots, a_{mn}, b_1, \ldots, b_m \longrightarrow c}{a_{11}, \ldots, a_{1n}, a_{21}, \ldots, a_{2n}, a_{m1}, \ldots, a_{mn} \longrightarrow c}$$

Hertz conceived this rule as a generalization of the *modus Barbara* of the Aristotelian syllogistic (if we think the rule as applied to *Makrosätze*, that is those sentences having free variables, which should be understood as universally quantified), and for him it played a decisive role in the characterization of logic. It constituted the basis of the whole deductive logic (see his later paper of 1935). Moreover, Hertz himself considered his systems of sentences as "the old theory about the chained inferences according to the *modus Barbara*" [15, p. 178]. He understood 'chain inferences' (*Kettenschlüsse*) as the successive application of the syllogism rule in a system of sentences as some kind of sorites.

The second rule is the rule of 'immediate inference' (*ummitelbarer Schluß*) having the following form (see [14, p. 463]):

$$\frac{a_1, a_2, \ldots, a_n \longrightarrow b}{a^1, a^2, \ldots, a^m, a_1, a_2, \ldots, a_n \longrightarrow b}$$

---

[3]It was also handled explicitly by Fritz London in his doctoral thesis *Über die Bedingungen der Möglichkeit einer deduktiven Theorie*, published in 1923, and later by Alfred Tarski (see [24, p. 70]), but on a completely different background than Hertz.

This rule permits the introduction of every antecedens in a proposition stating also a kind of monotonicity principle for deductive inferences, another distinctive feature of logic. When this rule is applied to tautological sentences, the result are logical sentences, having the form

$$a_1, a_2, \ldots, a_n, a \longrightarrow a$$

and called *trivial sentences* by Hertz.

If we restrict ourselves to sequents with only one succedens, the similarities between Hertz's rules of syllogism and immediate inference, on the one hand, and the cut rule and thinning of Gentzen's sequent calculus, on the other hand are obvious. Moreover, Hertz's tautological sentences correspond to the logical axioms in Gentzen's sequent calculus. In his paper of 1933 on Hertz, Gentzen formulated a version of *Satzsysteme*, which constituted an intermediate system between Hertz's system and the sequent calculus (see [8]). This system contained tautological sentences, a version of immediate inference—called *Verdünnung* (thinning) by Gentzen—and the following simplified version of syllogism:

$$\frac{L \longrightarrow u \qquad M\, u \longrightarrow v}{L\, M \longrightarrow v}$$

which Gentzen called 'cut' (*Schnitt*).

To show that his rules provided closed deductive systems, Hertz had to analyze the structure of the proofs generated by them. Thus, he devised a whole proof theory for systems of sentences. He developed a specific terminology for the description of proofs (see the second part of [11]). With the expression 'chained-inference' (*Kettenschluss*) he referred to orderings in the applications of rules, and he called 'inference systems' the derivations resulting from tautologies by applying the rules of syllogism and immediate inference. These inference systems have always a tree-structure (see also [14, p. 467]).

In investigating the structure of proofs, Hertz identified two "proof-methods" for systems of sentences with lineal sentences, which constitute two different normal forms for proofs. Following the tradition in formal logic and the usage in logic textbooks of that time, Hertz called these two forms 'Aristotelian' and 'Goclenian'—the latter after the German logician Rudolph Göckel (Rodolphus Goclenius, 1547–1628), who had considered categorical syllogisms of an analogous form. An example of an Aristotelian normal form is the following

$$\frac{\dfrac{a \longrightarrow b \qquad b \longrightarrow m}{a \longrightarrow m} \qquad m \longrightarrow c}{a \longrightarrow c}$$

This normal form is characterized by the fact that the right premiss of each application of the syllogism rule is an axiom (or "upper proposition", see, e.g., [14, pp. 473 ff.]). On the contrary, in the Goclenian normal form the left premiss of each application is an axiom, that is

$$\frac{a \longrightarrow b \qquad \dfrac{b \longrightarrow m \qquad m \longrightarrow c}{b \longrightarrow c}}{a \longrightarrow c}$$

That is, the last upper proposition in an Aristotelian normal form is first in a Goclenian one, and viceversa: the last upper proposition in a Goclenian normal form is an Aristotelian one.

Hertz devised a procedure for transforming Goclenian normal form into an Aristotelian one (see [14, p. 473]). This procedure should be a decision method related to these normal forms (see [14, Section 3]). A reconstruction of these procedures with technical details can be found in [23].

To some extent, the tarskian usual definition of logical consequence by means of the three basic conditions of reflexivity, monotonicity and transitivity was anticipated by Hertz's conception. These similarities were acknowledged by Tarski (see [24, p. 62 fn.]).[4] Notwithstanding, Hertz has mainly and decisively influenced Gentzen's proof-theoretical view on logical consequence.

Hertz considered his rules as the "essence of logic", as he stated in his contribution to a conference on mathematical logic that took place later in Geneva in 1934. According to Hertz, the rule of syllogismus specially disclosed a process that could be summarized as the *search* for a middle-term, designated as *interpositum* by him, in order to apply this rule (see [17, pp. 249 ff.]). The assertion of an implication $S \longrightarrow A$ implies the existence of an interpositum C, so that $S \longrightarrow C$ and $C \longrightarrow A$ can be asserted, and this process goes on until the subject finds basic sentences of some kind. In this process lies the origin of logic. Here, a new formulation of Aristotle's main ideas on the nature of syllogism, and its methodological function, can be found. These ideas were expressed by the *inventio medii*, the search for a middle term to construct a syllogism in modus barbara (see Aristotle *An. Pr.* A26 43 a 16–24). However, Hertz emphasized the epistemological aspects. Originally, Hertz regarded deduction as a kind of mental process. His epistemology book, *Über das Denken*, made reference to logical processes in knowledge (see [12, p. 121]).

It is interesting to notice that this process is not itself a deductive procedure. It can be identified with what is generally called a *regressive method*, a method for finding the grounds or justification of a proposition (which follows then from them). This regressive method is opposed to the *progressive method*, which proceeds from the grounds to the proposition in a synthetic way. Deduction should be a kind of such method. The notion of regressive method was discussed in Göttingen at that time within Hilbert's research group, and it should be an essential part of the axiomatic method, its analytical part, consisting in the differentiation between what counts as axiom and what as theorem in an axiomatic system, that is, it consists in establishing the axioms of the system, according to statements of Hilbert (see [19]). The other part of the axiomatic method would be the progressive part consisting in deducing theorems from the axioms.

In his philosophical papers and in some letters to Paul Bernays, Hertz made reference in several opportunities to *synthetic* moments or aspects in logic and it should be understood in the sense of a combinatorial or progressive process, such as Bernays stated as well in his philosophical paper on Hilbert's proof-theory [3].

In paper [11], Hertz introduced the notion of *ideal elements* ("ideale Elemente") in an axiomatic system to the effect that the number of axioms in the system can be reduced. The expression "ideal element" was in vogue at that time and it came to be an important notion in Hilbert's program, where a distinction between a real and an ideal part in mathematics is made. In the case of Hertz, it can be said that he intended to grasp only some of the purely *formal* aspects of this notion, excluding semantic properties. These

---

[4]Tarski wrote: "The discussion in Hertz, P. (27) has some points of contact with the present exposition." [24, p. 62 fn 1].

formal aspects consisted in its connecting real elements and in its reductive power. He argued that these ideal elements (terms or formulas) serve as a way to connect real elements. Moreover, these ideal elements provide "an accurate representation of sentences with real elements", "sentences with ideal elements have no meaning in themselves (*Bedeutung an sich*). Only sentences including exclusively real elements have meaning" [11, p. 249].

Hertz gave the following example of a system with the elements a, b, c, d and e constituting its domain and having the following axioms (on the left), and producing the following simplified system through the ideal element i (on the right):

$$
\begin{array}{ccc}
a \longrightarrow d & & \\
a \longrightarrow e & & a \longrightarrow i \\
b \longrightarrow d & >>>> & b \longrightarrow i \\
b \longrightarrow e & & c \longrightarrow i \\
c \longrightarrow d & & i \longrightarrow d \\
c \longrightarrow e & & i \longrightarrow e
\end{array}
$$

(see [11, p. 248]).

Now, Hertz suggested that *molecular* sentences could also be seen as ideal sentences:

"Through introduction of ideal sentences, we avoid the use of the words 'or' in the antecedens, and the use of the word 'and' in succedens." [11, p. 262].

This idea can be easily illustrated through the following system

$$
\begin{array}{c}
A_1 \longrightarrow B_1 \\
A_1 \longrightarrow B_2 \\
A_2 \longrightarrow B_1 \\
A_2 \longrightarrow B_2
\end{array}
$$

which can be reduced alternatively to the systems

$$
\begin{array}{c}
A_1 \longrightarrow B_1 \,\&\, B_2 \\
A_2 \longrightarrow B_1 \,\&\, B_2
\end{array}
$$

or

$$
\begin{array}{c}
A_1 \vee A_2 \longrightarrow B_1 \\
A_1 \vee A_2 \longrightarrow B_2
\end{array}
$$

and finally to the system with the only proposition

$$
\alpha \longrightarrow \beta
$$

where $B1 \,\&\, B2$ is the ideal element α and $A1 \vee A2$ is the ideal element β. So, the introduction of molecular sentences in systems of sentences serves the purpose of reducing the number of axioms.

It seems reasonable to connect these different aspects of finding ideal elements, searching an interpositum, the regressive method and, finally, the existence of synthetic processes in deduction, considering them as the key notions in an epistemological foundation of logic as it was pursued by Hertz.

References

1. Abrusci, V.M.: Paul Hertz's logical works contents and relevance. In: Abrusci, V.M., Casari, E., Mugnai, M. (eds.) Atti del Convegno Internazionale di Storia della Logica, San Gimignano, 4–8 dicembre 1982, pp. 369–374. Clueb, Bologna (1983)
2. Bernays, P.: Probleme der theoretischen Logik. Unterrichtsblätter für Mathematik und Naturwisschenschaften **33**, 369–377 (1927)
3. Bernays, P.: Die Philosophie der Mathematik und die Hilbertsche Beweistheorie. Blätter für Deutsche Philosophie **4** (1930–1931), 326–367 (1930)
4. Bernays, P.: Betrachtungen zum Sequenzen-Kalkül. In: Tymieniecka, A.-T. (ed.) Contributions to Logic and Methodology in Honor of J.M. Bochenski, pp. 1–44. North-Holland, Amsterdam (1965)
5. Bernays, P.: Paul Hertz. In: Neue deutsche Biographie. Herausgegeben von der historischen Kommission bei der bayerischen Akademie der Wissenschaften, vol. 8, pp. 711 f. Ducker & Humblot, Berlin (1969)
6. Church, A.: A Bibliography of Symbolic Logic. The Journal of Symbolic Logic **1**, 121–218 (1936)
7. Curry, H.B.: Foundations of Mathematical Logic. Dover, New York (1977)
8. Gentzen, G.: Über die Existenz unabhängiger Axiomensystem zu unendlichen Satzsystemen. Mathematische Annalen **107**, 329–350 (1933)
9. Gentzen, G.: Untersuchungen über das logische Schließen. Mathematische Zeitschrift **39**, 176–210 and 405–431 (1935)
10. Haller, R.: Neopositivismus. Eine historische Einführung in die Philosophie des Wiener Kreises. Wissenschaftliche Buchgesellschaft, Darmstadt (1993)
11. Hertz, P.: Über Axiomensysteme für beliebige Satzsysteme. I. Teil. Sätze ersten Grades. Mathematische Annalen **87**, 246–269 (1922)
12. Hertz, P.: Über das Denken. Springer, Berlin (1923)
13. Hertz, P.: Reichen die üblichen syllogistischen Regeln für das Schließen in der positiven Logik elementarer Sätze aus? Annalen der Philosophie **7**, 272–277 (1928)
14. Hertz, P.: Über Axiomensysteme für beliebige Satzsysteme. Mathematische Annalen **101**, 457–514 (1929a)
15. Hertz, P.: Über Axiomensysteme beliebiger Satzsysteme. Annalen der Philosophie **8**, 178–204 (1929b)
16. Hertz, P.: Von Wesen des Logischen, insbesondere der Bedeutung des modus Barbara. Erkenntnis **2**, 369–392 (1931)
17. Hertz, P.: Über das Wesen der Logik und der logischen Urteilsformen. Abhandlungen der Fries'schen Schule. Neue Folge **6**, 227–272 (1935a)
18. Hertz, P.: Sprache und Logik. Erkenntnis **7**, 309–324 (1939)
19. Hilbert, D.: In: Rowe, D.E. (ed.) Natur und mathematisches Erkennen. Vorlesungen, gehalten 1919–1920 in Göttingen. Nach einer Ausarbeitung von Paul Bernays. Birkhäuser, Basel (1992)
20. Hilbert, D., Bernays, P.: Grundlagen der Mathematik, vol. I. Springer, Berlin (1934)
21. Koslow, A.: Structuralist Logic: Implications, Inferences and Consequences. Logica Universalis **1**, 167–181 (2007)
22. Negri, S., von Plato, J.: Structural Proof Theory. Cambridge University Press, Cambridge (2001)
23. Schroeder-Heister, P.: Resolution and the origins of structural reasoning: Early proof-theoretic ideas of Hertz and Gentzen. Bulletin of Symbolic Logic **8**(2), 246–265 (2002)
24. Tarski, A.: Fundamentale Begriffe der Methodologie der deduktiven Wissenschaften I. Monatshefte für Mathematik und Physik **37**, 361–404 (1930). English trans. in Tarski, A.: Logic, Semantics, Metamathematics. Trans. by Woodger, J.H. 2nd ed. Hackett, Indianapolis (1983)

J. Legris (✉)
University of Buenos Aires and CEF/CONICET, Buenos Aires, Argentina
e-mail: jlegris@retina.ar

On Axiomatic Systems for Arbitrary Systems of Sentences

Part I
Sentences of the First Degree
(On Axiomatic Systems of the Smallest Number
of Sentences and the Concept of the Ideal Element)

Paul Hertz

1 Introduction

Whenever a system of sentences is recognized to be valid, it is often not necessary to convey each and every sentence to memory; it is sufficient to choose some of them from which the rest can follow. Such sentences, as it is generally known, are called axioms. The choice of these axioms is to a certain degree arbitrary. One can ask, however, if the property of a system of sentences to have several axiom systems is interconnected with other remarkable properties, and if there are systematic approaches to find, as the case may be, that axiomatic system which contains the least possible number of sentences. In the following some thoughts shall be communicated, which might be useful as a pre-stage for the treatment of these or related problems.

In fact the actual problem of interest is so entangled, that initially it seems appropriate to be content with an immense simplification: We only consider sentences of a certain type, sentences that we can write symbolically: $(a_1, \ldots, a_n) \rightarrow b$ and that can be expressed linguistically by formulations such as: If $(a_1, \ldots, a_n)$ altogether holds, so does b. In addition, a second simplification will be introduced in the present first part, by only considering sentences of type $a \rightarrow b$; however, we will liberate ourselves from this limitation in a following part. Further we assume rules according to which from certain sentences other ones follow: So, e.g., the validity of the sentences $a \rightarrow b$, $b \rightarrow c$ should result in the holding of the sentence $a \rightarrow c$.

However, what is actually meant by such a sentence, what the symbol $\rightarrow$ means in the combination of characters $a \rightarrow b$ or the word 'if' in the corresponding linguistic formulation, does not have to be indicated here. At this point, it cannot be our task to lay out where we take the inference rules from, in what sense such sentences appear in ordinary life and which relation our question might have with the configuration of a scientific discipline. Also, the reason cannot be given here, why our considerations do not even begin to reach that level of generality which would be necessary if we wanted to reach by them a full understanding of the connection between mathematical or physical sentences and

Translated by J. Legris, with kind permission, from: Hertz, P., "Über Axiomensysteme für beliebige Satzsysteme, Teil 1: Sätze ersten Grades", Mathematische Annalen Vol. 87(3–4), pp. 246–269.
© Springer 1922

not only being prepared for such an understanding. All those questions will be addressed on another occasion and at a different place[1]. It is completely sufficient if in this paper we stick to those inference rules according to which from certain sentences other ones follow. From the formal point of view assumed here, our sentences are just character strings, of which we examine certain sets; namely sets with the property that if they contain certain sentences, others necessarily occur in them, which are formed from those according to a certain rule.

As far as possible, the appeal to the direct geometrical intuition has been avoided in this paper. But, because of this, the easy understanding of the sentences to be dealt with might be jeopardized. That is why in this introduction the content of what is to be proved in the first part shall be summarized and explained by a geometrical illustration; although this illustration is only applicable in this way to laws of type $a \to b$ which are the ones we limit ourselves to in the first part[2].

The elements a, b, c... are presented as dots, and the sentences $a \to b$ as a solid arrow drawn from a to b; furthermore, it needs to be taken care that if such an arrow leads from a to b and another one from b to c, there also must lead an arrow from a to c. If now according to this rule all other sentences arise from certain sentences, then those sentences are called axioms and the corresponding arrows shall be drawn out, whereas the other arrows are drawn as dotted lines only.

Now Fig. 1 on the one hand and Figs. 2a and 2b on the other hand provide examples for the case in which there is only one axiomatic system and for the case in which there are several axiomatic systems. Figures 3a and 3b reveal that if the choice of the axiomatic system is not unique, a varying number of axioms may be employed to present the system of sentences. Particularly, they present the case in which every element is connected with every other element in both directions. If that is the case, one obtains the least number of axioms by connecting the elements with each other cyclically by axioms. Furthermore it emerges that the choice of the axiomatic system is unique if, and only if, there are no two elements that are mutually connected with each other by sentences. The proof for that given in this paper can be easily translated into the realm of geometry.

Fig. 1

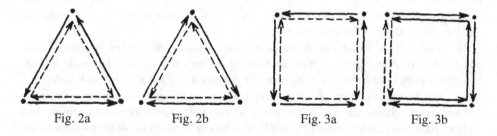

| Fig. 2a | Fig. 2b | Fig. 3a | Fig. 3b |

---

[1]It might be added though, that our sentences $a \to b$ are nothing other than formal "implications" in the sense of Russell [Whitehead–Russell **1** (1900), p. 15], and that the scheme of inference used as a base in the first part is the Theorem listed by Russell as No. 10, 3, p. 150, or put differently: Our sentences are judgements of subsumptions, our inferences are syllogisms of modus Barbara.

[2]Similar geometrical illustrations in the papers by Zaremba, Enseignement mathématique (1916), p. 5; G. Pólya, Schweizer Pädagog. Zeitschr. 1919, Hft. 2.

Additionally it will be possible to trace back the most general case to that one where there is only *one* independent axiomatic system, namely by at first substituting certain groups of interconnected elements with *one* element each, and then turning to the axiomatic system for the reduced system and subsequently reintroducing the omitted elements. Therefore, we only have to consider the case of uniqueness.

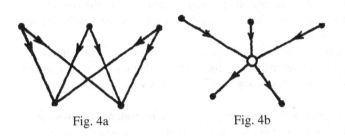

Fig. 4a   Fig. 4b

Now, in the case of uniqueness, however, one can reduce the number of axioms even further by introducing *ideal* elements. If, e.g., each of three elements is connected to each of two elements with one sentence, as it is visualized in Fig. 4a, then six sentences are needed; by introducing an ideal element—as indicated with a small circle in Fig. 4b—one can lower that number to five. Such an ideal element thus has the only purpose to avail us of a convenient presentation of the sentences between real elements; the sentences between ideal elements or between real and ideal elements have no meaning per se, but only those are followed from them obtaining between real elements.

Ideal elements are used constantly in all of the sciences, especially within physics and mathematics. Counted among them is, e.g., the concept of force, by which we can better describe the connection between the real elements: position of the acting bodies and movement of those bodies that are affected. From a philosophical point of view the ideal element has been made an object of thorough investigation by Vaihinger in his philosophy of the "As-If"[3].

One can ask now: How does one get to the least number of axioms if ideal elements are admitted? This problem will not be entirely solved here; it will only be shown for a very special case, how to get to that axiomatic system with the least number of sentences among those axiomatic systems in which no ideal elements are related to each other.

2 Problem Statement

1. (Definition.) We consider finitely many elements. By a sentence we understand the embodiment [*Inbegriff*] of a complex that may only consist of a single element, called *antecedens*, and an element, called *succedens*. We represent sentences by formulas like $(a_1, \ldots, a_n) \to b$ respectively $a \to b$. If the antecedens consists of only one element, then the sentence is called a *sentence of the first degree*.

In this part only sentences of the first degree are considered and therefore "sentence" is always to be understood as one of the first degree. Furthermore, we only want to consider here sentences of the first degree whose antecedens and succedens differ.

---

[3]2nd edition, Berlin 1913.

2. (Definition.) A system of sentences

$$\text{I. } a \to b$$
$$\text{II. } b \to c$$
$$\overline{\text{III. } a \to c}$$

is called an *inference*, I. the *minor sentence*, II. the *major sentence*, III. the *conclusion*. Minor sentence and major sentence together are called *premises*. Of the conclusion we also say that it follows from the premises.

3. (Definition.) Regarding a system $\mathfrak{T}$ of sentences, we denote by chain inference a simply ordered finite series of inferences with the property that each sentence in the inferences of this series which does not belong to $\mathfrak{T}$ coincides with a conclusion of an earlier sentence.

4. (Definition.) A given sentence is called properly *provable* from a given system of sentences $\mathfrak{T}$ if there is a chain inference belonging to $\mathfrak{T}$ in which the sentence occurs as conclusion. That chain inference itself is called proof of the sentence. A sentence is called provable from $\mathfrak{T}$, if it belongs to $\mathfrak{T}$ or if it is actually provable from $\mathfrak{T}$.

5. (Definition.) A system of sentences is called *closed* if each provable sentence regarding the system occurs within this system.

6. (Definition.) For every system of sentences $\mathfrak{S}$, a system of sentences $\mathfrak{A}$ is called *an axiomatic system* if each sentence of $\mathfrak{S}$ is provable from $\mathfrak{A}$.

7. (Definition.) An axiomatic system $\mathfrak{A}$ that belongs to a system of sentences $\mathfrak{S}$ is called *independent* if no sentence within $\mathfrak{A}$ is provable using other sentences within $\mathfrak{A}$.

8. (Proposition.) If every sentence within a system of sentences $\mathfrak{C}$ is provable from a system of sentences $\mathfrak{B}$, and if a sentence a is provable from $\mathfrak{C}$, then a is provable from $\mathfrak{B}$.

9. (Proposition.) For every closed system $\mathfrak{S}$ at least one independent axiomatic system exists.

PROOF. $\mathfrak{S}$ is an axiomatic system itself. If it is not independent, there has to be at least one sentence which follows from the others; we leave out that one; if the residual system is not independent, we leave out another sentence, and so on, until the method comes to an end. The remaining system is an axiomatic system as outlined by Proposition 8 and it is independent.

Now it will be our task to get an overview of the number of possibilities for independent axiomatic systems. We especially want to examine, when there is only one axiomatic system and how to systematically arrive at axiomatic systems that contain as few sentences as possible.

3 Possibility of a Unique Choice of the Axiomatic System

10. (Definition.) A system of elements and their connecting sentences is called a *net* if there are two sentences $x \to y$, $y \to x$ for each pair of elements x, y.

11. (Proposition.) All sentences within a net form a closed system of sentences.

12. (Proposition.) If a number of elements is ordered in a series, such that every following succedens is in a sentence in which the former is antecedens, and the first succedens is in a sentence in which the last is antecedens, then those sentences are axioms of the

net and inversely there exists an axiomatic system for every net whose sentences can be ordered in such a way.

13. (Proposition.) For all sentences of a net which consists of n elements, there exists an independent axiomatic system consisting of n sentences, and, if $n \geq 3$, there exist independent systems of axioms which consist of more than n sentences, but no independent axiomatic system which consists of less than n sentences.

PROOF. It follows from Proposition 12 that there are independent axiomatic systems which consist of n sentences, therefore an axiomatic system for a net of n elements is the following:

$$a_1 \to a_2, \; a_2 \to a_3, \ldots, a_n \to a_1.$$

An independent axiomatic system of more than n elements ($n \geq 3$) is the following:

$$a_1 \to a_2, \; a_2 \to a_3, \ldots, a_{n-1} \to a_n,$$
$$a_2 \to a_1, \; a_3 \to a_2, \ldots, a_n \to a_{n-1}.$$

One realizes that there is no independent axiomatic system with less than n elements by noticing that every element needs to be antecedens and succedens of an axiom. Therefore, one has at least $2n$ axioms that need to be found within the system and of those, at most 2 can be identical.

14. (Proposition.) If two nets share an element, they are parts of one and the same net.

15. (Definition.) Regarding a closed system of sentences $\mathfrak{S}$, a net which is not part of another one that can be found within $\mathfrak{S}$, is called *maximal*.

16. (Proposition.) Maximal nets are disjoint.

17. (Definition.) Two elements, x and y, of a system of sentences, are called *disconnected*, if it is possible to divide all elements into two classes such that x and y belong to different classes and if there is no sentence which connects two elements of different classes.

18. (Definition.) Sentences that belong to a system of sentences, are called *disconnected*, if it is possible to divide all sentences of the system into two disjoint classes, such that the given sentences belong to different classes.

19. (Definition.) Elements are said to be *connected*, if they are not disconnected.

20. (Definition.) Sentences are said to be *connected*, if they are not disconnected.

21. (Proposition.) If two elements are connected with a third element, they are connected with each other.

PROOF. Let a be connected with c and b be connected with c. If a and b were disconnected, then we could divide all elements into two classes A and B, such that a would belong to A and b would belong to B, and there would be no sentence that would contain both one element belonging to A and one belonging to B. Now then, if c belonged to A, c would be disconnected with b, that conflicts with our condition. In the same manner, it would also lead to a contradiction if c belonged to B.

22. (Proposition.) If two sentences are connected with a third sentence, they are connected with each other.

PROOF. Let the sentence $\mathfrak{a}$ be connected with sentence $\mathfrak{c}$, and $\mathfrak{b}$ be connected with sentence $\mathfrak{c}$. If $\mathfrak{a}$ and $\mathfrak{b}$ were disconnected, all sentences of the system could be divided into two disjoint classes $\mathfrak{A}$ and $\mathfrak{B}$. Class $\mathfrak{A}$ would contain $\mathfrak{a}$, and class $\mathfrak{B}$ would contain $\mathfrak{b}$. Now then, if $\mathfrak{c}$ belonged to $\mathfrak{A}$, $\mathfrak{c}$ would be disconnected with $\mathfrak{b}$, that conflicts with our condition. The same contradiction follows if we assume that $\mathfrak{c}$ belongs to $\mathfrak{B}$.

23. (Proposition.) The necessary and sufficient condition for two sentences to be connected, is that the four elements determinated by them are connected.

PROOF. 1. Let the sentences $e \equiv (a \rightarrow b)$; $f \equiv (c \rightarrow d)$ be connected. It is immediately clear that a, b and c, d are connected. If a, c were disconnected, there would be two disjoint classes A and C, to which a and c respectively would belong, such that there would be no sentence which connects an element of class A with an element of class C. All sentences could be divided in two disjoint classes $\mathfrak{E}$ and $\mathfrak{F}$ where the elements of these sentences would either belong only to A or to C. Hence, e belongs to $\mathfrak{E}$, and f to $\mathfrak{F}$, e and f were disconnected which would be contrary to the condition.

2. Let a and c be connected. If e and f were disconnected, there would be two disjoint classes of sentences $\mathfrak{E}$ and $\mathfrak{F}$ to which e and f would belong. The elements that belong to $\mathfrak{E}$ form a class A, while elements that belong to $\mathfrak{F}$ form class C. So every sentence either belongs to $\mathfrak{E}$, meaning it only contains elements of A, or $\mathfrak{F}$, meaning it only contains elements of C, then according to our assumption, there is no sentence that connects an element of A and C. Then the elements a and c which belong to A and C respectively, would be disconnected, which would conflict with our condition. That is why the assumption, that e and f are disconnected, is wrong. One can reason in a similar way if assuming a connected with d, or b with c, or b with d.

Remark. If two elements, say a and c are identical, then of course e and f are connected as well. That is because if e and f were disconnected, there would have to be two disjoint systems of sentences $\mathfrak{E}$ and $\mathfrak{F}$ which e and f belonged to. However, $\mathfrak{E}$ and $\mathfrak{F}$ would share elements because $e \equiv a \rightarrow b$; $f \equiv c \rightarrow d$ and $a \equiv c$.

24. (Definition.) Regarding a system of sentences $\mathfrak{S}$ we call a system of connected sentences a *chain*, if that system does not contain a sentence that belongs to a net of $\mathfrak{S}$.

Remark. However, a chain may contain elements that belong to a net.

25. (Proposition.) Two chains that share an element form a new chain together.

PROOF. Let $\mathfrak{K}'$ and $\mathfrak{K}''$ share an element x. Let t' and t'' be sentences that belong to $\mathfrak{K}'$ and $\mathfrak{K}''$ which contain the element x. Then t' and t'' are connected by the remark according to Proposition 23. Also by Proposition 22, each sentence of $\mathfrak{K}'$ is connected with each sentence of $\mathfrak{K}''$ or all sentences of $(\mathfrak{K}', \mathfrak{K}'')$ amongst each other. Now if within $(\mathfrak{K}', \mathfrak{K}'')$, there was a sentence that belonged to a net, this sentence would either have to be found in $\mathfrak{K}'$ or $\mathfrak{K}''$. Therefore these two would not both be chains.

26. (Definition.) Regarding a system of sentences $\mathfrak{S}$ we call a chain that is not contained in any other chain, a *maximal* chain.

27. (Proposition.) A maximal chain and a chain that is not contained within it have no element in common.

PROOF. Follows from Proposition 25 and Definition 26.

28. (Proposition.) In a closed system $\mathfrak{S}$ the conclusion of two sentences of a maximal chain belongs to the maximal chain as well.

PROOF. If $a \rightarrow b$, $b \rightarrow c$ belong to the maximal chain $\mathfrak{M}$, then $a \rightarrow c$ must belong to $\mathfrak{M}$ as well. If that were not the case, $a \rightarrow c$ could not be a chain by Proposition 27, $a \rightarrow c$, therefore, would belong to a net by Definition 24. That is why $c \rightarrow a$ would be a sentence of $\mathfrak{S}$, and because $b \rightarrow c$ is a sentence of $\mathfrak{S}$, $b \rightarrow a$ would be as well. Then $a \rightarrow b$ would belong to a net, and because this sentence is contained in $\mathfrak{M}$, so $\mathfrak{M}$ would be no chain according to Definition 24 because that would conflict with the condition.

29. (Proposition.) A maximal chain in a closed system is a closed system.

PROOF. Follows from Definition 5 and Proposition 28.

30. (Proposition.) Two maximal chains have no sentence in common.

PROOF. Follows from Proposition 27.

31. (Proposition.) A sentence in a closed system does not belong to a maximal net and a maximal chain at the same time.

PROOF. Follows from Definitions 24 and 26.

32. (Proposition.) Every sentence in a closed system belongs either to a maximal net or a maximal chain.

PROOF. If a sentence a does not belong to any net, it represents a chain by Definition 24. Let $\mathfrak{M}$ be the entirety of all sentences that are connected with a, and do not belong to any net. Then those sentences are connected amongst each other. Therefore, $\mathfrak{M}$ is a chain, and, as is easily recognizable, a maximal chain as well. But, if a belongs to a net, it belongs to a maximal net as well.

33. (Proposition.) The entirety of all sentences in a closed system together form maximal nets and maximal chains which have no sentence in common.

34. (Proposition.) If a sentence $a \equiv a \to b$ is provable from a system of sentences $\mathfrak{T}$, there is a series of sentences $a \to \alpha_1$, $\alpha_1 \to \alpha_2, \ldots, \alpha_{\rho-1} \to \alpha_\rho$, $\alpha_\rho \to b$ that all belong to $\mathfrak{T}$.

PROOF. If $a \equiv a \to b$ belongs to $\mathfrak{T}$, this sentence is already proved. In the other case, according to Definition 4, $a \to b$ has to be found as a conclusion in an inference that belongs to a chain inference whose premises $a \to x$, $x \to b$ would be provable from $\mathfrak{T}$. These two sentences again belong to $\mathfrak{T}$, or are conclusions of earlier sentences. In the last case, $a \to x'$, $x' \to x$, $x \to x''$, $x'' \to b$ would be provable from $\mathfrak{T}$. One can see that this claim is true by continuing this procedure, which has to be finite.

35. (Theorem.) There is only one independent axiomatic system for a closed system of sentences $\mathfrak{S}$ without nets.

Proof. According to Proposition 9 there is at least *one* independent axiomatic system. Assuming that for $\mathfrak{S}$ there were two independent axiomatic systems $\mathfrak{A}$ and $\mathfrak{A}'$, then each axiomatic system would have to contain an axiom which is not contained in the other. Let $a \equiv a \to b$ be an axiom of $\mathfrak{A}$ which is not contained in $\mathfrak{A}'$. Then this sentence would have to be provable from $\mathfrak{A}'$, and by Proposition 34 there must exist a series of at least two sentences that belong to $\mathfrak{A}'$:

$$a \to \alpha_1, \ \alpha_1 \to \alpha_2, \ldots, \alpha_{\rho-1} \to \alpha_\rho, \ \alpha_\rho \to b.$$

Each of those sentences, however, has to be provable from $\mathfrak{A}$ again. That is why we again obtain by Proposition 34, a series of at least two sentences:

$$a \to \beta_1, \ \beta_1 \to \beta_2, \ldots, \beta_{\sigma-1} \to \beta_\sigma, \ \beta_\sigma \to b,$$

which all belong to $\mathfrak{A}$. Amongst these sentences, however, a must be found, because if not, the axiomatic system $\mathfrak{A}$ would not be independent. From this it follows that the series of above sentences has to contain at least three sentences. Hence, three cases are possible:

1. It is $a \equiv a \to \beta_1$. Then $\beta_1 \equiv b$ and $\mathfrak{A}$ would contain the net $b \to \beta_2, \ldots, \beta_\sigma \to b$, therefore, $\mathfrak{S}$ would contain a net contrary to the condition.

2. It is $a \equiv a \to b_\sigma$. Then $b_\sigma \equiv a$ and $\mathfrak{A}$ would contain the net $a \to \beta_1, \ldots, \beta_{\sigma-1} \to a$, therefore, $\mathfrak{S}$ would contain a net contrary to the condition.

3. The sentence a is identical with one middle sentence $\beta_{\mu-1} \to \beta_\mu$, then $\beta_{\mu-1} \equiv a$, $b_\mu \equiv b$ and $\mathfrak{A}$ would contain the nets $a \to \beta_1, \ldots, \beta_{\mu-2} \to a$ and $b \to b_{\mu+1}, \ldots, \beta_\sigma \to b$, therefore $\mathfrak{S}$ would contain a net contrary to the condition.

36. (Definition.) To a given closed system of sentences $\mathfrak{S}$, we call $\mathfrak{S}'$ a *reduced system* of sentences with the following properties:

1. For every element a in $\mathfrak{S}$ there is an element a' in $\mathfrak{S}'$ that does not belong to a net.

2. To all elements $c_1, c_2, \ldots$ of a maximal net of $\mathfrak{S}$ together corresponds a single element c' in $\mathfrak{S}'$.

3. Every sentence $a \to b$ in $\mathfrak{S}$, where a and b both do not belong to a net, correspond to a sentence $a' \to b'$ in $\mathfrak{S}'$ where a' and b' are the corresponding elements to a and b.

4. All sentences together, $a \to \bar{c}_i$ respectively $\bar{\bar{c}}_j \to b$, where a and b do not belong to a maximal net, but where all $\bar{c}_i$ and $\bar{\bar{c}}_j$ belong to one and the same maximal net, correspond to a sentence $a' \to \bar{c}'$ respectively $\bar{\bar{c}}' \to b$ in $\mathfrak{S}'$.

All sentences $\bar{c}_k \to \bar{\bar{c}}_l$ together, where all $\bar{c}_k$ belong to one and the same maximal net, and where all $\bar{\bar{c}}_l$ belong to one and the same but another maximal net, correspond to one sentence $\bar{c}' \to \bar{\bar{c}}'$ in $\mathfrak{S}'$.

5. $\mathfrak{S}'$ contains no elements other than those corresponding to $\mathfrak{S}$, according to 1 and 2.

6. $\mathfrak{S}'$ contains no sentences other than those corresponding to $\mathfrak{S}$, according to 3 and 4.

37. (Assumption.) There exists a reduced system of sentences for every closed one.

Remark. Of course, Theorem 35 can be derived from simpler assumptions. One can point out, that if it is known how many elements the system contains and how these elements are connected, then it can be said that it exists. The meaning of this claim, however, will only be clear after discussing the concept of existence in detail. One could assume, for instance, that the elements a of $\mathfrak{S}$ and the elements $c_1, \ldots, c_n$ together are assigned to the symbols a' and c'. But these symbols exist in our minds. Or one could say that the totality of the things $c_1, \ldots, c_n$ is another thing again. Finally, one could also regard the reduced system as a means to portray the rules of a non-reduced system.

In Definition 36 we introduced the reduced system in hypothetical form and stated that we call a reduced system one that has certain relations to the given system. We also could have done that differently and might have said: instead of each element a we put an element a', instead of the elements $c_1, c_2, \ldots, c_n$ of a net we put an element c', and so on (the requirements of 5 and 6 would then be dispensable). We will call the resulting system a reduced system. This sentence, then, would not have the form of a definition but of a statement.

We had explained in the introduction that we only want to consider sentences of the form $a \to b$. This seems to be in contradiction with the expression we get now. But the reduced system on its own underlies the same rules as the initial one and is only used to be able to infer certain sentences about the initial one more easily.

38. (Proposition.) A reduced system $\mathfrak{S}'$ that belongs to a closed system $\mathfrak{S}$ contains no net.

PROOF. If $\mathfrak{S}'$ contained a net, there would have to be two elements a', b' such that $a' \to b'$, $b' \to a'$ would exist in $\mathfrak{S}'$. According to Definition 36, there would be two sentences $a_1 \to b_1$, $b_2 \to a_2$ in $\mathfrak{S}$ that would correspond to these sentences. But because b_1 and b_2 would correspond to the same element b' they then would also have to belong to the same maximal net (according to Definition 36), so $a_2 \to a_1$ would occur in $\mathfrak{S}$. Because it is closed, $\mathfrak{S}$ would then contain $b_1 \to a_1$, i.e. a_1 and b_1 would belong to a maximal net. Then different elements in $\mathfrak{S}'$ could not correspond to them, which would be contrary to the condition.

39. (Proposition.) The reduced system $\mathfrak{S}'$ which belongs to a closed system is a maximal chain regarding to $\mathfrak{S}$ or can be broken down into several maximal chains that do not have any element or sentence in common.

PROOF. Follows from Definition 26 and Proposition 38.

40. (Proposition.) A reduced system $\mathfrak{S}'$ which belongs to a closed system $\mathfrak{S}$ is itself closed.

PROOF. Let $\mathfrak{e}' \equiv a' \to b'$ and $\mathfrak{f}' \equiv b' \to c'$ be two sentences that are contained in $\mathfrak{S}'$, then $a \not\equiv c$, according to Proposition 38. Then there have to be two corresponding sentences by Definition 36 that correspond to $\mathfrak{e}'$ and $\mathfrak{f}'$ in $\mathfrak{S}$: $a_1 \to b_1$, $b_2 \to c_2$. Now either $b_1 \equiv b_2$, then $a_1 \to c_2$ would be a sentence in $\mathfrak{S}$. Or $b_1 \not\equiv b_2$, then b_1 and b_2 would belong to a maximal net by Definition 36. Therefore $b_1 \to b_2$ would be contained in $\mathfrak{S}$, so $a_1 \to c_2$ would be a sentence in $\mathfrak{S}$ again. But because $a' \not\equiv c'$, a_1 and c_2 do not belong to the same maximal net and, again by Definition 36, $a' \to c'$ is a sentence in $\mathfrak{S}'$.

41. (Proposition.) There is one and only one independent axiomatic system for every reduced system.

PROOF. Follows from Theorem 35 and Proposition 38.

42. (Proposition.) The following procedure may be used to obtain an axiomatic system $\mathfrak{A}$ from a closed system $\mathfrak{S}$: one constructs an axiomatic system $\mathfrak{A}'$ from $\mathfrak{S}'$, furthermore, for every sentence in $\mathfrak{A}'$, one looks for a corresponding sentence in $\mathfrak{S}'$ (Definition 36) and adds an axiomatic system for each maximal net that is contained in $\mathfrak{S}$.

PROOF. Let $\mathfrak{e} \equiv a \to b$ a sentence of $\mathfrak{S}$, it needs to be shown that it is provable from $\mathfrak{A}$. If $\mathfrak{e}$ is a sentence of a maximal net, then the proof is already accomplished. So let us assume that $\mathfrak{e}$ does not belong to a maximal net. Then, according to Definition 36, there has to be a corresponding sentence $\mathfrak{e}' \equiv a' \to b'$ in $\mathfrak{S}'$. This one has to be provable by axioms in $\mathfrak{A}'$. Thus, according to Proposition 34, a series of sentences must exist that belong to $\mathfrak{A}'$:

$$a' \to \alpha'_1,\ \alpha'_1 \to \alpha'_2, \ldots, \alpha'_{\rho-1} \to \alpha'_\rho,\ \alpha'_\rho \to b'.$$

Let the corresponding sentences in $\mathfrak{A}$ be the following:

$$\overline{a} \to \overline{\alpha}_1,\ \overline{\overline{\alpha}}_1 \to \overline{\alpha}_2, \ldots, \overline{\overline{\alpha}}_{\rho-1} \to \overline{\alpha}_\rho,\ \overline{\overline{\alpha}}_\rho \to \overline{b}.$$

Now either two consecutive elements $\overline{\alpha}_i$, $\overline{\overline{\alpha}}_i$ are identical to each other, or they belong to the same maximal net, because they correspond to the same α'_i, such that $\overline{\alpha}_i \to \overline{\overline{\alpha}}_i$ is provable from $\mathfrak{A}$. Additionally a and $\overline{a}$, as well as b and $\overline{b}$, are identical or belong to the same maximal net.

Consequently, there is a system of sentences

$$a \to \alpha_1,\ \alpha_1 \to \alpha_2, \ldots, \alpha_\rho \to b$$

that belong to $\mathfrak{A}$, so we have provided the proof.

43. (Proposition.) The following procedure may be used to obtain an independent axiomatic system $\mathfrak{A}$ from a closed system $\mathfrak{S}$: one constructs the independent axiomatic system (Proposition 41) of $\mathfrak{S}'$, furthermore one looks for a corresponding sentence in $\mathfrak{S}$ for every sentence in $\mathfrak{A}'$, and adds an independent axiomatic system for every maximal net that is contained in $\mathfrak{S}$.

PROOF. According to Proposition 42, one obtains an axiomatic system by using this procedure. Suppose that this system is not independent, and that $\mathfrak{e} \equiv a \to b$ is a sentence of $\mathfrak{A}$ which is provable from the other axioms. Then $\mathfrak{e}$ can either belong to a maximal net of $\mathfrak{S}$ or not.

a) If $\mathfrak{e}$ belonged to a maximal net of $\mathfrak{S}$, a series of two or more sentences would exist, according to Proposition 34:

$$a \to \alpha_1, \ \alpha_1 \to \alpha_2, \ldots, \alpha_\rho \to b,$$

which all would belong to $\mathfrak{A}$ and each would be different from $\mathfrak{e}$. Of these, according to the condition, not all could belong to the same maximal net as $\mathfrak{e}$. Therefore, not all could belong to the same maximal net amongst each other, which is why in the series there has to be a sentence which belongs to no maximal net. Let this sentence be $\alpha_i \to \alpha_{i+1}$.

On the other hand, $b \to a$, $a \to \alpha_1$ are sentences of $\mathfrak{S}$, because $a \to b$ would belong to a net, so $\alpha_{i+1} \to \alpha_i$ are sentences of $\mathfrak{S}$ as well, and therefore, $\alpha_i \to \alpha_{i+1}$ would belong to a net nevertheless.

b) If $\mathfrak{e}$ did not belong to a maximal net, again according to Proposition 34, there would be a series $\mathfrak{R}$ of two or more sentences $a \to \alpha_1, \ldots, \alpha_\rho \to b$, which would belong to $\mathfrak{A}$, which each would be different from $e \equiv a \to b$, and which could not all belong to a maximal net. Let $\alpha_k \to \alpha_{k+1}$ be such a sentence which does not belong to a maximal net and would be different from $a \to b$. However, such sentences of $\mathfrak{R}$ which belong to no maximal net, correspond to sentences in $\mathfrak{S}'$ that belong to $\mathfrak{A}'$, and which form a series $\mathfrak{R}'$ $a' \to \beta_1'$, $\beta_1' \to \beta_2', \ldots, \beta_\rho' \to b'$ which has the property that the antecedens of each sentence (except the first) is identical with the succedens of the preceding sentence. Furthermore the series $\mathfrak{R}'$ contains the sentence which is corresponding to $\alpha_k \to \alpha_{k+1}$ and which is different from $a' \to b'$ (because every sentence in $\mathfrak{A}'$ only corresponds to one sentence in $\mathfrak{A}$), thus it contains at least two sentences. If one of the sentences was $\mathfrak{R}' \equiv a' \to b'$, the series would contain a net, in contradiction with Proposition 38, but if $a' \to b'$ were not to be found in $\mathfrak{R}'$, then the system $\mathfrak{A}'$ would not be independent, contrary to the condition.

44. (Proposition.) For a given closed system of sentences there are no other independent axiomatic systems than those which can be found using the method described in Proposition 43.

PROOF. Let $\mathfrak{A}$ be an independent axiomatic system. It needs to be shown:

1. Those sentences of $\mathfrak{A}$ which belong to a maximal net, form an independent axiomatic system for this.

2. If $\mathfrak{A}^*$ is the embodiment of those sentences of $\mathfrak{A}$ which do not belong to a maximal net and $\mathfrak{A}^{*'}$ is the corresponding system to $\mathfrak{A}^*$ in $\mathfrak{S}'$, then $\mathfrak{A}^{*'}$ is the independent axiomatic system of $\mathfrak{S}'$.

3. $\mathfrak{A}$ contains only one of several sentences which correspond to the same sentence in $\mathfrak{S}'$.

ad 1. It is sufficient to show that the system $\overline{\mathfrak{A}}$ of those sentences in $\mathfrak{A}$, which do belong to a maximal net $\mathfrak{M}$, is an axiomatic system for $\mathfrak{M}$.

If that were not be the case, and if $a \to b$ were a sentence belonging to $\mathfrak{M}$ that was not provable from $\overline{\mathfrak{A}}$, then according to Proposition 34, a series of sentences $a \to \alpha_1, \ldots, \alpha_\rho \to b$ would exist that would belong to $\mathfrak{A}$ and of which at least one $\alpha_i \to \alpha_{i+1}$ would not belong to $\overline{\mathfrak{A}}$ and therefore also would not belong to $\mathfrak{M}$.

Then $\alpha_{i+1} \to b$, $b \to a$ would have to belong to $\mathfrak{S}$, which means that $\alpha_{i+1} \to \alpha_i$ would belong to $\mathfrak{S}$ as well, i.e. α_i and α_{i+1} would belong to a maximal net, and therefore, also to $\mathfrak{M}$ which is contrary to our assumption.

ad 2. It needs to be shown: a) that each sentence of $\mathfrak{S}'$ is provable from $\mathfrak{A}^{*'}$, b) that the sentences are independent of $\mathfrak{A}^{*'}$.

a) Suppose that the sentence $e' \equiv a' \to b'$ is not provable by $\mathfrak{A}^{*}$. But now the corresponding sentence $e \equiv a \to b$ is provable from $\mathfrak{A}$. Thus, there are sentences $a \to \alpha_1, \ldots, \alpha_\rho \to b$ which belong to $\mathfrak{A}$. By finding the corresponding sentences in $\mathfrak{S}'$ of those sentences among them that do not belong to a maximal net, one obtains a series of sentences that belong to $\mathfrak{A}^{*'}$ contrary to our assumption:

$$a' \to \beta'_1, \ldots, \beta'_\sigma \to b'.$$

b) If there was a sentence $a' \to b'$ in $\mathfrak{A}^{*'}$ which was provable from other sentences in $\mathfrak{A}^{*'}$, one would obtain a series of at least two sentences $a' \to \beta'_1, \ldots, \beta'_\sigma \to b'$ which would belong to $\mathfrak{A}^{*'}$. The corresponding series of at least two sentences in $\mathfrak{A}$ would be

$$\bar{a} \to \bar{\alpha}_1, \bar{\bar{\alpha}}_2 \to \bar{\alpha}_2, \ldots, \bar{\bar{\alpha}}_\sigma \to \bar{b}.$$

But if two neighbouring elements like $\bar{\alpha}_1$, $\bar{\bar{\alpha}}_2$ are different, they belong to a net, according to Definition 1, there exists a series of sentences that belong to $\mathfrak{A}$ which connects a and b. If $a \to b$ were found within this series, then a, b would belong to a net, but this is impossible because a sentence $a' \to b'$ corresponds to $a \to b$ in $\mathfrak{S}$. Therefore, $a \to b$ which belongs to $\mathfrak{A}$, would be provable from other sentences in $\mathfrak{A}$, and $\mathfrak{A}$ would not be independent.

ad 3. $\mathfrak{A}$ cannot contain two sentences that correspond to the same sentence in $\mathfrak{S}'$. Let $a_1 \to b_1$ and $a_2 \to b_2$ be two such sentences, then either a_1 and a_2 have to be identical and b_1 and b_2 would belong to the same net, or vice versa, or a_1 and a_2 on the one hand, and b_1 and b_2 on the other hand, would belong to the same net.

Let us examine the last case for instance. According to Definition 1, a series of sentences would exist, which would belong to $\mathfrak{A}$ and the same maximal net, which would therefore be different from $a_1 \to b_1$ and from which $a_1 \to a_2$ would be provable. We will call this series $a_1 \to \alpha_1, \ldots, \alpha_\rho \to a_2$. Likewise, there would be a series $b_2 \to \beta_1, \ldots, \beta_\sigma \to b_1$ which does not contain $a_1 \to b_1$. Thus, one could prove $a_1 \to b_1$ by the series:

$$a_1 \to \alpha_1, \ldots, \alpha_\rho \to a_2, a_2 \to b_2, b_2 \to \beta_1, \ldots, \beta_\sigma \to b_1,$$

which cannot contain $a_1 \to b_1$, i.e. $\mathfrak{A}$ would not be independent.

45. (Definition.) By a *sequence* we understand a system of elements and a closed system of sentences that exists between these elements, with the property that each pair of elements is connected by one and only one sentence.

46. (Proposition.) There is no net within a sequence.

PROOF. Follows from Definitions 10 and 45.

47. (Proposition and Definition.) There is one and only one element in a sequence which is succedens of no other element—we call this element the *highest element*—and there is one and only one element in a sequence which is antecedens of no other element—we call this element the *lowest element*.

PROOF. We consider an arbitrary element, if there exists an antecedens of it, so we locate that one, then the one which is the antecedens of that last one, and so on. Because the system of sentences is closed, one can never meet an element already encountered, thus the series has to terminate, therefore there has to be a highest element. In the same way one can recognize the existence of the lowest element. From the Definition 45 it is also clear that there can only be one highest and only one lowest element.

48. (Proposition and Definition.) The elements of a sequence can be arranged in one and only one way, such that each element, with the exception of the lowest, is antecedens to the following, and each, with the exception of the highest, is succedens to the preceding and that there is no element for two consecutive elements which can be antecedens to one and succedens to the other. We call this order the *natural order*.

PROOF. As first element we pick the highest, as the next one we pick the highest of the rest, and so on. If now x for example is one element and y is the following, so $x \to y$ has to be an element of the sequence, because if $y \to x$ was an element of the sequence, we would have committed a mistake when choosing x, because y was still available. Furthermore one can see, that if $x \to z$ is an element of the sequence where $z \not\equiv y$, $z \to y$ does not belong to the sequence. One can also easily show that this order is the only one that has the required property.

49. (Proposition.) One obtains an independent axiomatic system for the sequence, which is also the only one, by ordering the elements in natural order and using as axioms all those sentences which connect two consecutive elements.

PROOF. That these sentences form an axiomatic system is immediately apparent. That they are independent follows by Proposition 34 from the fact, that for two consecutive elements, there is none which is succedens of the higher one and antecedens of the lower one. That there is no other independent axiomatic system follows from Theorem 35 and Proposition 46.

50. (Definition.) By a *maximal sequence* in a closed system of sentences we understand a sequence that is contained in no other sequence.

51. (Proposition.) Every sentence in a closed system belongs to at least one maximal sequence.

52. (Proposition.) We can obtain the independent axiomatic system of a closed system of sentences without net, by forming axiomatic systems for all maximal sequences (thereby sentences might possibly occur as axioms of several maximal sequences).

PROOF. Let $\mathfrak{A}$ be the system of sentences which consists of all axiomatic systems of all maximal sequences.

1. Let $a \to b$ be a sentence of $\mathfrak{S}$. According to Proposition 51, it belongs to a maximal sequence, and is therefore provable from $\mathfrak{A}$.

2. Let $a \to b$ be a sentence of $\mathfrak{A}$. It needs to be shown that it cannot be proven from other sentences of $\mathfrak{A}$. Let us assume just that, however. Because $a \to b$ belongs to $\mathfrak{A}$, there must be a maximal sequence Z, in which b is the following element of a. But if $a \to b$ was provable from other sentences in $\mathfrak{A}$, a series of sentences would exist that would be different from $a \to b$ and would belong to $\mathfrak{A}$:

$$a \to \alpha_1, \ldots, \alpha_\rho \to b.$$

These sentences cannot belong to Z, because there should not be an element x of Z, for which $a \to x$, $x \to b$ holds. On the other hand we again get a sequence if we add these sentences to Z. Because for element x that belongs to Z, either $x \to a$, as well as $x \to \alpha_i$ holds, or $b \to x$, as well as $\alpha_i \to x$. Then Z would be no maximal sequence, which would be in conflict with our condition.

4 Ideal Elements

53. (Definition.) If $\mathfrak{S}$ is the totality of elements and of a closed system of sentences that exist between these elements, then we call the *extended system* $\hat{\mathfrak{S}}$ a totality of elements and of a closed system of sentences that exist between these elements if the following holds:

1. Each element a in $\mathfrak{S}$ corresponds to an element $\hat{a}$ in $\hat{\mathfrak{S}}$. We call such elements *real* elements of $\mathfrak{S}$, and we call the non-real elements of $\hat{\mathfrak{S}}$ *ideal* elements.

2. Each sentence $a \rightarrow b$ of $\mathfrak{S}$ corresponds to a sentence $\hat{a} \rightarrow \hat{b}$ of $\hat{\mathfrak{S}}$ between the corresponding real elements in $\hat{\mathfrak{S}}$.

3. Each sentence of $\hat{\mathfrak{S}}$ between real elements corresponds to a sentence in $\mathfrak{S}$ between corresponding elements.

Remark 1. Sentences in $\hat{\mathfrak{S}}$ between ideal elements or between a real and an ideal element correspond to no sentences in $\mathfrak{S}$.

Remark 2. Because of property 3, by examining the extended system, we gain an overview of the sentences of the original system at the same time.

Remark 3. In the future, for the sake of convenience, we shall consider the extended system as having been formed from elements of the original system on the one hand, which will become real elements of the extended system, and from added ideal elements on the other hand.

Remark 4. The extended system can have an independent axiomatic system which consists of fewer sentences than the original system. If, for instance, the original system consists of the elements a', a'', b', b'', b''', and of the six sentences between them:

$$a' \rightarrow b',$$
$$a' \rightarrow b'',$$
$$a' \rightarrow b''',$$
$$a'' \rightarrow b',$$
$$a'' \rightarrow b'',$$
$$a'' \rightarrow b''',$$

then there is an extended system of these five sentences:

$$a' \rightarrow \mathfrak{a},$$
$$a'' \rightarrow \mathfrak{a},$$
$$\mathfrak{a} \rightarrow b',$$
$$\mathfrak{a} \rightarrow b'',$$
$$\mathfrak{a} \rightarrow b''',$$

where $\mathfrak{a}$ is an ideal element.

Remark 5. By introducing ideal elements we circumvent the usage of the word 'or' in the antecedens, as well as the usage of the word 'and' in the succedens.

54. (Definition.) In a closed system of sentences $\mathfrak{S}$ without net, we call a *pair of groups* a pair of two disjoint groups A and B of elements with the property that there is a sentence between each element of A as antecedens and each element of B as succedens

which belongs to the independent axiomatic system (Theorem 35). A is called *group of antecedens*, B is called *group of succedens*.

Remark. For A and B to be a pair of groups, it is therefore not sufficient that there is a sentence between every element of A as antecedens and B as succedens.

55. (Definition.) A *maximal* pair of groups is a pair of groups A, B, if there is no element, except the elements of A, where there is an axiom between such an element, as antecedens, and each element of B. In the same way there must not be an element, except the elements of B, where there is an axiom between such an element, as succedens, and each element of A.

56. (Proposition.) Each axiom can be found in at least one maximal pair of groups.

57. (Definition.) A *small maximal* pair of groups is such a pair which contains only one element in either the antecedens group or the succedens group. A *medium maximal* pair of groups is such a pair, which contains two elements in each of the antecedens group and the succedens group. A *large maximal* pair of groups is such a pair, which is neither a small, nor a medium one.

58. (Proposition.) There are no sentences between two elements of an antecedens group and two elements of the succedens group of a closed system of sentences $\mathfrak{S}$ without net.

PROOF. Let e.g. $a_i \rightarrow a_k$ be a sentence that exists between elements of A, then there will be a series which belongs to the axiomatic system:

$$a_i \rightarrow \alpha_1, \ldots, \alpha_\rho \rightarrow a_k.$$

On the other hand there are axioms for $\mathfrak{S}$ $a_i \rightarrow b_l$, $a_k \rightarrow b_l$, where b_l is an element of B. We can also substitute $a_i \rightarrow b_l$ by a series of axioms

$$a_i \rightarrow \alpha_1, \ldots, \alpha_\rho \rightarrow a_k, \ a_k \rightarrow b_l.$$

Because these axioms should be independent from each other, $a_i \rightarrow b_l$ would have to occur in this series, i.e. it would contain a net contrary to the condition.

59. (Definition.) A pair of groups A, B is called *closed* if there are no elements, except those in A, where there is an axiom between such an element, as antecedens, and an element of B as succedens. In the same way there must not be an element, except the elements of B, where there is an axiom between such an element, as succedens, and an element of A as antecedens.

60. (Proposition.) Each closed pair of groups is a maximal pair of groups.

61. (Proposition.) The antecedens groups and succedens groups of two closed pair of groups are disjoint.

62. (Definition.) A closed system of sentences is called *simple* if it does not contain a net and each of its maximal pairs of groups is closed.

63. (Proposition.) In an arbitrary simple system of sentences each axiom belongs to one and only one maximal pair of groups.

64. (Proposition.) In a simple system of sentences, if there is a sentence $x \rightarrow b_i$, where b_i belongs to the succedens group of a maximal pair of groups, but x does not belong to the corresponding antecedens group of A, then there is a sentence $x \rightarrow a_l$ in $\mathfrak{S}$, where a_l belongs to A. Likewise: if there is a sentence $a_k \rightarrow y$ in $\mathfrak{S}$, where a_k belongs to A, but y does not belong to B, then there is a sentence $b_m \rightarrow y$ in $\mathfrak{S}$, where b_m belongs to B.

PROOF. According to Proposition 34, in the first case, there is a series of at least two sentences that consists of axioms

$$x \rightarrow \alpha_1, \ldots, \alpha_\rho \rightarrow b_i.$$

Because A, B are a closed pair of groups now, α_ρ has to belong to A. From that follows the first part of the proposition, and the second part as well.

65. (Definition.) If $\mathfrak{S}$ is a simple system of sentences and $\hat{\mathfrak{S}}$ an extended system to it, then an independent axiomatic system $\hat{\mathfrak{A}}$ of $\mathfrak{S}$, in which no two ideal elements are connected via axioms, is called *an axiomatic system of the first degree to* $\mathfrak{S}$.

66. (Proposition.) If $\mathfrak{S}$ is a simple system and $\hat{\mathfrak{A}}$ is an axiomatic system of the first degree of $\mathfrak{S}$, and if a, b are elements of an antecedens group and of a succedens group of a maximal pair of groups of $\mathfrak{S}$ (there is an axiom $a \to b$ of $\mathfrak{S}$), then in $\hat{\mathfrak{A}}$, there is either an axiom $\hat{a} \to \hat{b}$ or a pair of axioms $\hat{a} \to \mathfrak{a}$, $\mathfrak{a} \to b$, where $\mathfrak{a}$ is an ideal element.

PROOF. Let $\hat{\mathfrak{S}}$ be the extended system that $\hat{\mathfrak{A}}$ belongs to. According to Definition 53, there is a sentence $\hat{a} \to \hat{b}$ in $\hat{\mathfrak{S}}$. If this sentence does not belong to $\hat{\mathfrak{A}}$, there would be a series $\mathfrak{R}$ in $\hat{\mathfrak{A}}$,

$$\hat{a} \to \hat{\alpha}_1, \ \hat{\alpha}_1 \to \hat{\alpha}_2, \dots, \hat{\alpha}_\rho \to b,$$

of at least two axioms where $\hat{\alpha}_1 \not\equiv b$. If now $\hat{\alpha}_1$ is real, then by Definition 53.3 the sentence $\hat{a} \to \hat{\alpha}_1$ would correspond to $a \to \alpha_1$ in $\mathfrak{S}$, which would mean that by Proposition 64 there would have to be a sentence $b_m \to \alpha_1$ in $\mathfrak{S}$, where b_m would belong to B. $\mathfrak{S}$ therefore would contain both sentences $b_m \to \alpha_1$, $\alpha_1 \to b$. If now $b_m \equiv b$, then $\mathfrak{S}$ would contain a net, and if $b_m \not\equiv b$, there would be a conflict with Proposition 58. Both are impossible, α_1 must be ideal.

Now if $\mathfrak{R}$ contained more than two sentences, $\hat{\alpha}_2$ would have to be real, because $\hat{\mathfrak{A}}$ has to be of the first degree. Thus, there would have to be a sentence in $\mathfrak{S}$, $a \to \alpha_2$, and therefore, by Proposition 64, also a sentence $b_n \to a_2$, where b_n would belong to B. This would again be in conflict with $\alpha_2 \to b$. That is why the series $\mathfrak{R}$ does not contain another element.

67. (Definition.) If A, B are two disjoint complexes of elements we call a *connecting system* a system of these elements and other elements $\mathfrak{v}$ (which we call connecting elements), and of sentences between the elements $\mathfrak{v}$, the elements a of A, and the elements b of B with the following properties:

1. An element $\mathfrak{v}$ has only to be antecedens to one element of B and succedens to one element of A.

2. For each pair a, b there is a sentence $a_i \to b_k$ or a pair of sentences $a_i \to \mathfrak{v}_{ik}$, $\mathfrak{v}_{ik} \to b_k$.

68. (Definition.) A connecting system without connecting elements we call a *disparate* connecting system.

69. (Definition.) If a connecting system has only one connecting element and if this element is antecedens to all elements of B and succedens to all elements of A and if the connecting system does not contain any more sentences, then we call it a *centralized* connecting system.

70. (Proposition.) If $\mathfrak{S}$ is a simple system of sentences, $\hat{\mathfrak{A}}$ a corresponding axiomatic system of the first degree, furthermore A, B a maximal pair of groups of $\mathfrak{S}$, then regarding A, B, there is always a connecting system $\mathfrak{v}$ that belongs to $\hat{\mathfrak{A}}$. The connecting elements are ideal elements of $\hat{\mathfrak{S}}$.

PROOF. Follows from Proposition 66.

71. (Proposition.) In a simple system of sentences the connecting systems of two different maximal pairs of groups do not share an ideal element.

PROOF. From Proposition 61.

72. (Assumption.) If A, B are complexes of elements in a system $\mathfrak{T}$ of elements and sentences, then we can assume the existence of a different system $\check{\mathfrak{T}}$, such that each element and each sentence of $\mathfrak{T}$ corresponds to an element and a sentence in $\check{\mathfrak{T}}$, however, $\check{\mathfrak{T}}$ does additionally contain a connecting system to the complexes A, B.

About the character of this Assumption, the same can be said as about Assumption 37. For the following it will be more convenient if we express ourselves in such a way that we say: We can add a connecting system to the pair of complexes A, B.

73. (Proposition.) If $\mathfrak{S}$ is a simple system of sentences, A, B is a maximal pair of groups, $\hat{\mathfrak{A}}$ is an axiomatic system of the first degree of $\mathfrak{S}$, $\hat{\mathfrak{B}}$ is a connecting system to $\hat{\mathfrak{A}}$ for the maximal pair of groups A, B (Proposition 70), $\hat{\mathfrak{B}}^*$ some other added connecting system for A, B (see Assumption 72), then each sentence between real elements of $\hat{\mathfrak{S}}$, which is provable from $\hat{\mathfrak{A}}$ and $\hat{\mathfrak{B}}^*$, corresponds to a sentence of $\mathfrak{S}$.

(It is assumed that the added elements of $\hat{\mathfrak{B}}^*$ are different from those ideal elements that were originally contained in $\hat{\mathfrak{S}}$.)

PROOF. Let $\hat{c} \to \hat{d}$ be such a sentence which is provable by the series of sentences that belongs to $\hat{\mathfrak{A}}$ and $\hat{\mathfrak{B}}^*$:

$$\hat{c} \to \alpha_1, \ldots, \alpha_\rho \to \hat{d}.$$

If now $\alpha_\lambda \to \alpha_{\lambda+1}$ did not belong to $\hat{\mathfrak{A}}$ here, then this sentence would have to belong to $\hat{\mathfrak{B}}^*$, thus, it would have to have the form $a \to b$ or $a \to \mathfrak{a}$, $\mathfrak{a} \to b$.

(Now and in the following a shall always be an element of A and b an element of B, $\mathfrak{a}$ shall be a non-real element of $\hat{\mathfrak{S}}$, meaning an ideal or connecting element.)

In the first case $\alpha_\lambda \to \alpha_{\lambda+1}$ is provable by $\hat{\mathfrak{A}}$. In the second case $\alpha_{\lambda+1}$ is a connecting element, therefore different from $\hat{d}$, and there still is another sentence $\alpha_{\lambda+1} \to \alpha_{\lambda+2}$ that would have to have the form $\mathfrak{a} \to b$ necessarily. The pair of sentences $\alpha_\lambda \to \alpha_{\lambda+1}$, $\alpha_{\lambda+1} \to \alpha_{\lambda+2}$ can be substituted by axioms $\hat{\mathfrak{A}}$. The same is true in the third case and for all sentences of $\mathfrak{R}$. Now follows the claim from Definition 53.3.

74. (Proposition.) (Denomination as in Proposition 73.) If $\hat{\mathfrak{A}}^*$ is a system of sentences that is generated from $\hat{\mathfrak{A}}$, by removing the sentences of $\hat{\mathfrak{B}}$ and adding the sentences of $\hat{\mathfrak{B}}^*$, and if the sentence $\hat{c} \to \hat{d}$, which exists between real elements, is provable from $\hat{\mathfrak{A}}^*$ and $\hat{\mathfrak{A}}$, then the corresponding sentence $c \to d$ belongs to $\mathfrak{S}$.

PROOF. Follows from Proposition 73.

75. (Proposition.) (Denomination as in Propositions 73 and 74.) If $\hat{c} \to \hat{d}$ is provable from $\hat{\mathfrak{A}}^*$, then the sentence $c \to d$ holds in $\mathfrak{S}$.

PROOF. Follows from Proposition 74.

76. (Proposition.) (Denomination as in Propositions 73 and 74.) The embodiment $\hat{\mathfrak{S}}^*$ of all elements of $\hat{\mathfrak{A}}^*$ and all sentences provable from $\hat{\mathfrak{A}}^*$ is an extended system of $\mathfrak{S}$.

PROOF. It follows from the definition that $\hat{\mathfrak{S}}^*$ is a closed system. Furthermore it follows from Proposition 75 that each sentence between real elements $\hat{c} \to \hat{d}$ in $\hat{\mathfrak{S}}^*$ corresponds to a sentence $c \to d$ in $\mathfrak{S}$. So we still need to show that each sentence $c \to d$ corresponds to a sentence $\hat{c} \to \hat{d}$ in $\hat{\mathfrak{S}}^*$ or that each sentence $\hat{c} \to \hat{d}$ is provable from $\hat{\mathfrak{A}}^*$.

Because $\hat{c} \to \hat{d}$ is provable from $\hat{\mathfrak{A}}$, there is a series that belongs to $\hat{\mathfrak{A}}$

$$\hat{c} \to \alpha_1, \ldots, \alpha_\rho \to \hat{d}.$$

Now if a contained sentence $\alpha_\lambda \to \alpha_{\lambda+1}$ did not belong to $\hat{\mathfrak{A}}^*$, then it would have to belong to $\hat{\mathfrak{B}}$, it would therefore have to have the form $\hat{a} \to \hat{b}$ or $\hat{a} \to \mathfrak{a}$ or $\mathfrak{a} \to \hat{b}$.

In the first case we substitute the sentence with sentences that belong to $\hat{\mathfrak{B}}^*$, and thus to $\hat{\mathfrak{A}}^*$. In the second case $\alpha_{\lambda+1}$ cannot be the last element because it is ideal. Therefore, it has to follow a sentence $\alpha_{\lambda+1} \to \alpha_{\lambda+2}$. In this connection $\alpha_{\lambda+2}$ has to be real, because $\hat{\mathfrak{A}}$ is of the first degree. So the sentence $\alpha_{\lambda+1} \to \alpha_{\lambda+2}$ has the form $\mathfrak{a} \to \hat{y}$. If y belonged to B, one could substitute the pair of sentences $\alpha_\lambda \to \alpha_{\lambda+1}$, $\alpha_{\lambda+1} \to \alpha_{\lambda+2}$ with a sentence of the form $\hat{a} \to \hat{b}$ or with a sentences of $\hat{\mathfrak{A}}^*$. But if y did not belong to B, according to Proposition 64, there would have to be a sentence $b_m \to y$ in $\mathfrak{S}$, thus, there would have to be a series

$$\hat{b}_m \to \beta_1, \ \beta_1 \to \beta_2, \dots, \beta_\rho \to \hat{y}$$

which belongs to $\hat{\mathfrak{A}}$. In this case we substitute $\alpha_\lambda \to \alpha_{\lambda+1}$, $\alpha_{\lambda+1} \to \alpha_{\lambda+2}$ with the series

$$\hat{a} \to \overline{\mathfrak{a}}, \ \overline{\mathfrak{a}} \to \hat{b}_m, \ \hat{b}_m \to \beta_1, \ \beta_1 \to \beta_2, \dots, \beta_\rho \to \hat{y},$$

or with the series $\hat{a} \to \hat{b}_m$, $\hat{b}_m \to \beta_1$, $\beta_1 \to \beta_2, \dots, \beta_\rho \to \hat{y}$, where the first two sentences and the first sentence respectively belong to $\hat{\mathfrak{B}}^*$, thus also to $\hat{\mathfrak{A}}^*$. The other sentences, however, belong to $\hat{\mathfrak{A}}$. The procedure works the same way, if the sentence $\alpha_\lambda \to \alpha_{\lambda+1}$ has the form $\mathfrak{a} \to \beta$.

It is therefore shown: for each sentence or each pair of sentences in our series that does not belong to $\hat{\mathfrak{A}}^*$, we can introduce solely a sentence that belongs to $\hat{\mathfrak{A}}^*$ or a pair of sentences $\hat{a} \to \mathfrak{a}$, $\mathfrak{a} \to \hat{b}$ that belongs to $\hat{\mathfrak{A}}^*$, in connection with further sentences in $\hat{\mathfrak{A}}$. We want to repeat this procedure. It can only terminate if the series contains only sentences of $\hat{\mathfrak{A}}^*$. Now we want to assume that this procedure never terminated. From what was formerly said one realizes that this is only the case if we continuously introduce new sentences of the form $\hat{a} \to \hat{b}$, meaning pairs of sentences of the form $\hat{a} \to \overline{\mathfrak{a}}$, $\overline{\mathfrak{a}} \to \hat{b}$, into our series $\mathfrak{R}$. It has to happen then that our series contains two identical sentences or pairs of sentences, for instance $\hat{a}_i \to \mathfrak{a}$, $\mathfrak{a} \to \hat{b}_k$. Then by Proposition 74 the sentences $a_i \to b_k$, $b_k \to a_i$ would exist in $\mathfrak{S}$, thus $\mathfrak{S}$ would contain a net.

From this we can realize that our assumption was wrong. The series has to terminate, i.e. $\hat{c} \to \hat{d}$ has to be provable from $\hat{\mathfrak{A}}^*$.

Second proof.[4] For each sentence $c \to d$ there is a series of sentences $c \to \alpha_1, \dots, \alpha_\rho \to d$ that belongs to $\mathfrak{A}$. Each series belongs to one and only one maximal net, according to Proposition 63. If $\alpha_i \to \alpha_{i+1}$ belongs to a maximal pair of groups that is different from (A, B), then by Proposition 66, $\hat{\alpha}_i \to \hat{\alpha}_{i+1}$ belongs either to $\hat{\mathfrak{A}}$ and by Proposition 61 not to $\hat{\mathfrak{B}}$, so it does belong to $\hat{\mathfrak{A}}^*$, or it can be substituted by two sentences of $\mathfrak{A}$ that, according to Proposition 61 or 71, do not belong to $\hat{\mathfrak{B}}$, such that they also belong to $\hat{\mathfrak{A}}^*$. If $\alpha_\kappa \to \alpha_{\kappa+1}$ belongs to (A, B), then $\hat{\alpha}_\kappa \to \hat{\alpha}_{\kappa+1}$ belongs to $\hat{\mathfrak{B}}^*$ and such also to $\hat{\mathfrak{A}}^*$, or it is substitutable by two such sentences.

77. (Definition.) (Cf. Definition 57.) We call a pair of element complexes A, B a *small complex* pair, if each complex contains only one element, a *medium complex* pair, if A and B both contain two elements, and a *large complex* pair, if it is neither small, nor medium.

78. (Proposition.) For each complex pair with a number of sentences m and n there is always a connecting system with $m + n$ sentences.

PROOF. In fact it is a centralized connecting system (see Definition 69).

---

[4] Additional remark after proof-reading.

79. (Proposition.) A non-centralized connecting system $\mathfrak{B}$ for a large complex pair with m, n as number of elements always contains more than $m + n$ sentences.

PROOF. 1. $\mathfrak{B}$ is disparate (see Definition 68).

2. $\mathfrak{B}$ is not disparate and not centralized.

ad 1. Then the number of sentences is $m \cdot n$. However, if $m > n$, it is $m + n < 2m \leq m \cdot n$ and $m + n = 2m < n \cdot m$, if $m = n$.

ad 2. It is sufficient to show that the number of sentences can be reduced. Let $\mathfrak{v}'$, $\mathfrak{v}''$ be two connecting elements and let $a' \to \mathfrak{v}'$, $\mathfrak{v}' \to b'$ and $a'' \to \mathfrak{v}''$, $\mathfrak{v}'' \to b''$ be two corresponding pairs of sentences of $\mathfrak{B}$, where $a' \not\equiv a''$ and $b' \not\equiv b''$. Now we let $\mathfrak{v}'$ and $\mathfrak{v}''$ coincide and we can either drop a sentence that existed formerly between a' and b'', or, in case we already had the sentences $a' \to \mathfrak{v}'''$, $\mathfrak{v}''' \to \mathfrak{v}''$, we can have $\mathfrak{v}'''$ coincide with $\mathfrak{v}'$ and $\mathfrak{v}''$, which also leads to a reduction. One can reason similarly, if $a' \not\equiv a''$, $b' \equiv b''$ or $a' \equiv a''$, $b' \not\equiv b''$ or $a' \equiv a''$, $b' \equiv b''$. If there exists only one connecting element, and either only one antecedens element or only one succedens element corresponding to it, then the connecting element can be left out. If there is only one connecting element, and more than one antecedens element and more than one succedens element corresponding to it, and if x is an element of A or B respectively, that is not connected to this element, then adding the sentence $x \to \mathfrak{v}$ or $\mathfrak{v} \to x$ respectively, allows us a reduction.

80. (Proposition.) The number of sentences in a connecting system for a medium complex pair is 4, and 4 only then, if the connecting system is disparate or centralized.

81. (Proposition.) The number of sentences for a complex pair with $(m, 1)$ or $(1, m)$ number of elements is at least m, and m only then, if the connecting system is disparate.

82. (Definition.) For a closed system of sentences $\mathfrak{S}$ without a net, an independent axiomatic system of the first degree is called an independent minimal axiomatic system of the first degree, if there is no other axiomatic system for $\mathfrak{S}$ of the first degree which contains fewer sentences.

83. (Definition.) For a simple system of sentences $\mathfrak{S}$ without a net we define *canonical system* to be a system of sentences that consists of connecting systems of the maximal pairs of groups of $\mathfrak{S}$, which in fact contains for each large maximal pair of groups of $\mathfrak{S}$ a centralized connecting system, for each small maximal pair of groups of $\mathfrak{S}$ a disparate connecting system and for each medium maximal pair of groups of $\mathfrak{S}$ either a centralized or a disparate connecting system.

84. (Proposition.) For each closed system of sentences $\mathfrak{S}$ there is a minimal system of the first degree.

PROOF. There is an axiomatic system of the first degree, in fact the axiomatic system $\mathfrak{A}$ of $\mathfrak{S}$ itself. Therefore there also must be an independent axiomatic system of the first degree with the lowest number of sentences.

85. (Proposition.) Each minimal system of the first degree of a simple system of sentences $\mathfrak{S}$ is a canonical system.

PROOF. Let $\hat{\mathfrak{A}}$ be a minimal system of the first degree to $\mathfrak{S}$. By Proposition 70, each maximal pair of groups of $\mathfrak{S}$ is connected by a connecting system. Now if a connecting system of a maximal pair of groups would not satisfy the requirements of Proposition 83, then it could be substituted by a different one. In that way, by Propositions 79 to 81, the number of sentences would decrease, but according to Proposition 76, the system could not cease to be an axiomatic system. Thus, $\hat{\mathfrak{A}}$ could not be a minimal system. Therefore $\hat{\mathfrak{A}}$ contains a connecting system for each maximal pair of groups which fulfills the requirements of Definition 83. According to Proposition 56, however, all axioms of $\mathfrak{S}$ are already derivable from the sentences of these connecting systems. The axiomatic system $\hat{\mathfrak{A}}$ of the first degree can therefore not contain any more sentences, because it is supposed to be a minimal system.

86. (Proposition.) Every canonical system of sentences is a minimal axiomatic system of the first degree.

PROOF. We assume an arbitrary minimal axiomatic system of the first degree. According to Proposition 85, the canonical system at hand can differ from it only by the connecting system of the medium maximal pairs of groups. If one substitutes these connecting systems in the minimal system of the first degree in this sense, the number of sentences does not change and by Proposition 76 we get an axiomatic system of the first degree which has the lowest number of sentences. It is therefore independent and a minimal axiomatic system of the first degree.

Göttingen, September 15 1921

Acknowledgements   Javier Legris thanks Peter Arndt and Arnold Koslow for their support.

Translator
J. Legris
University of Buenos Aires and CEF/CONICET, Buenos Aires, Argentina
e-mail: jlegris@retina.ar

Part 2
Paul Bernays (1926)

Paul Bernays and the Eve of Non-standard Models in Logic

Walter Carnielli

Abstract The role of Paul Bernays' "Axiomatische Untersuchung des Aussagenkalküls des Principia Mathematica" (cf. [Bernays, 1926]), abridged version of his *Habilitationsschrift* of 1918 is here discussed, emphasizing its share in the foundations of contemporary logic and in universal logic. The intention is to complement, rather than to controvert, previous work on Bernays' trust as e.g. [Zach, 1999] and [Sieg, 1999].

Keywords Non-standard models · Universal logic · Independence proofs · Completeness

Mathematics Subject Classification (2000) 03A05 · 00A30 · 03-02

1 The Dawn of Completeness, and Its Role in Universal Logic

Paul Bernays was born from a Jewish family on October 17th, 1888 in London as a Swiss citizen, and died in Zürich on September 17th, 1977. His almost ninety yeas of life coincided with a conspicuous period in the history of logic. Bernays studied mathematics during four semesters in Berlin, having met mathematicians of the stature of Edmund Landau, Issai Schur and Ferdinand Georg Frobenius, neo-kantian philosophers of science as Alois Riehl, Carl Stumpf (who had an important influence on Edmund Husserl, the founder of modern phenomenology), Ernst Cassirer, as well as Max Planck in physics. As Landau moved to Göttingen, Bernays decided to go to the University of Göttingen as well, where after six semesters he got a degree in mathematics, with philosophy and theoretical physics as subsidiaries. In the Spring of 1912 he received a doctor degree on analytic number theory (binary quadratic forms) supervised by Landau. This training on formal number theory surely influenced his approach to logic, as I intend to show. In Göttingen he also studied under Felix Klein, Herman Weyl and David Hilbert.

At the end of the same year he moved to Zürich, where Zermelo was a professor, obtaining his *Habilitationsschrift* on function theory and becoming a *Privatdozent*. In 1917 Hilbert came to Zürich to deliver his lecture on "Axiomatisches Denken" and invited Bernays to work with him as his assistant on his research on the foundations of arithmetic. This was how Bernays was brought back to Göttingen, where he was involved in helping Hilbert on crafting his lectures, in the preparation of typewritten notes and in fruitful (for both sides) discussions with Hilbert. Besides this work, Bernays was also involved in giving lectures on various areas of mathematics at the University of Göttingen, which helped him to obtain his *venia legendi* in 1919, after a second *Habilitationsschrift* in 1918 on the completeness of the propositional calculus of "Principia Mathematica" and the investigation on the independence of its axioms, "Beiträge zur axiomatischen Behandlung des Logik-Kalküls". This work remained unpublished for 8 years, being published only in abridged form as [3].

Thanks to Hilbert's efforts, Bernays was promoted (without tenure) to the position of Professor Extraordinarius in Göttingen in 1922, a status held until 1933, when as a "non-Aryan" he was deprived of his position. Hilbert even employed him privately as his assistant (at his own expenses) for the next six months, but then Bernays and his family moved to Switzerland, whose nationality they had kept from Bernays' father, and he held a position at the ETH in Zürich from then on. It is revealing of his personality, and it helps to explain his concern with the philosophy of mathematics, the fact that Bernays was an important link between Gödel and Hilbert, during more than four decades (from 1930 to 1975) of an intense correspondence (see [7]).

The material of Bernays' *Habilitationsschrift* of 1918 is what is summarized in [3]. Bernays begins by remarking that the propositions of the *Principia Mathematica* by A. Whitehead and B. Russell can be all written in terms of negations and disjunctions. The five axioms (*Grundformeln*) are the following, replacing the usage of dots, as in $p \vee .q \vee r \supset q \vee .p \vee r$, by the more familiar parenthesis when necessary, and by taking into account that $\supset$ separates more than $\vee$ and $\sim$, and $\vee$ more than $\sim$:

Taut: $p \vee p \supset p$
Add: $q \supset p \vee q$
Perm: $p \vee q \supset q \vee p$
Assoc: $p \vee (q \vee r) \supset q \vee (p \vee r)$
Sum: $(q \supset r) \supset (p \vee q \supset p \vee r)$

The only rules are Insertion (as he calls it, *Einsetzung*) or Replacement, which permits to replace any variable by any arbitrary sentence, and Modus Ponens (MP): from A and $A \supset B$ one obtains[1] B. Bernays remarks that the axioms are valid (*algemeingültige*), and that applications of rules preserve validity. Having defined a system of axioms as *complete* when it permits to obtain, with the help of the rules, all the valid sentences, the problem is how to prove completeness. Actually, this notion of semantical completeness (and a corresponding proof) had been already given in [2], at least three years before the work of Emil Post (cf. [13]); besides, the completeness proof is notoriously distinct from Post's. The proof is done by noting that any sentence can be rewritten in conjunctive normal form (as conjunctions of disjunctions of variables and their negations), using the axioms and the properties of conjunction.

Conjunctive normal forms give a direct analysis (apart of computational complexity issues) of whether a sentence is a tautology. If all clauses contain complementary literals, then the formula is a tautology. Otherwise, the formula is falsifiable. This is the spirit, much in contemporary terms, of Bernays proof, whose argument runs as follows:

Theorem 1 (Semantic Completeness) *Let A be any sentence; if A is a tautology, then A is derivable from the axioms by means of the rules.*

---

[1] I avoid here the hard-reading Gothic letters.

Proof Firstly, convert A to B in conjunctive normal form. Clearly, A is derivable if and only if B is derivable. Since A (and B) are tautologies, each of its clauses C are also tautologies. Now, either C, being a disjunction of literals, has as elements an atomic sentence and its negation (as p and $\sim p$), or not. If not, just replace each occurrence or a positive atomic sentence by p, and each occurrence or an atomic sentence under negation by $\sim p$. Therefore, C turns out to be replaced by a disjunction whose elements are p or $\sim (\sim p)$, and thus C is nothing else than p; however, from p (by replacement) any sentence whatsoever is derivable.

In the case C has among its elements an atomic sentence and its negation (as p and $\sim p$), then C is indeed derivable; with no loss of generality, suppose C is $D \vee (\sim p \vee p)$ (if not, just apply *Perm* and *Assoc*):

1. $\sim p \vee (p \vee p)$ (*Add*);
2. $p \vee (\sim p \vee p)$ from 1, (*Assoc*);
3. $\sim p \vee (p \vee (\sim p \vee p))$ from 2, (*Add*), MP;
4. $(\sim p \vee p) \vee (\sim p \vee p)$ from 3, (*Ass*) and (*Perm*), MP;
5. $\sim p \vee p$ from 4, (*Taut*), MP;
6. $D \vee (\sim p \vee p)$ from 5, (*Add*) and Replacement.

Thus each clause C is derivable, and B in CNF is derivable from the properties of conjunction. Therefore the axioms and rules can derive any tautology. $\qquad\square$

Bernays did not find it necessary to explicit the above derivations; we can feel, however, in his (informal) argumentation a sort of interplay between semantical and syntactical completeness, a notion that Bernays himself helped to carve, now called Post completeness. The meaning of the distinction between semantical and syntactical completeness for Bernays and Hilbert is far from being yet elucidated (see e.g. [12] and [19]), but what is relevant for our purposes is that Bernays' proof of completeness also provides an immediate decision procedure for propositional logic, which is surprisingly close to the modern resolution algorithm, introduced by John Alan Robinson in [15]. Moreover, Bernays' proof is constructive[2], anteceding L. Kalmár's completeness proof for propositional calculus of 1935.

Since 1917 Bernays continued to improve on the axiomatization and decidability of propositional and first-order logic which eventually led to his paper with Moses Schönfinkel (see [4]).

Universal logic, regarded as a general theory of logical structures, concerns general methods and tools, as well as the fostering of concepts that can be applied to all logics. In particular, investigating domains of validity of metatheoretical concepts (such as of completeness or consistency) are topics of higher significance for universal logic, but of course the notion of model itself will be touchy for universal logic; in this way, the intuitions of Paul Bernays on non-standard models occupies a pivotal position. Bernays pioneered the perspective that logical concepts are not necessarily one-sided, and that patronizing a logical doctrine requires a philosophical price. Bernays can be legitimately considered, in his way, also a founder of universal logic.

---

[2]I owe this observation to J. Bueno-Soler.

2 Non-standard Models: Room for Independence

What concerned Bernays very much, and what constituted a major impetus for non-standard models and consequently for non-standard logics, were the questions of dependence and independence of the assumed logic principles. Four systems of axioms are considered:

System 1: *Taut*, *Add*, *Perm*, *Sum*;
System 2: *Taut*, *Add**, *Assoc*, *Sum*, where *Add** is $p \supset (p \vee q)$;
System 3: *Taut*, *Add*, *Add**, *Assoc**, *Sum*, where *Assoc** is $p \vee (q \vee r) \supset (p \vee q) \vee r$;
System 4: *Taut*, *Simp*, *Perm*, *Sum*, where *Simp* is $q \supset (p \supset q)$.

The main dependence and independence results concerning such system can be summarized as follows:

1. From System 1 *Assoc* is derivable, and moreover the four properties are all indispensable, even if *Assoc* were assumed. So System 1 is complete.
2. From System 2 *Perm* is derivable, and *Add** is essential, although *Add* and *Add** are equivalent in the presence of *Perm* and *Sum*. In this way, System 2 is complete (i.e, equivalent to System 1).
3. *Perm* is however not derivable if *Assoc* is replaced by the closely related *Assoc**; on the other hand, *Perm* is indeed derivable from System 3, where the complementary character of *Add** and *Assoc** makes System 3 equivalent to System 1 (and thus complete).
4. From System 4 *Add* is derivable, thus System 4 is also complete.
5. Considering two other basic forms: *Comm*: $p \supset (q \supset r) \supset (q \supset (p \supset r)$ and *Syll*: $(q \supset r) \supset (p \supset q) \supset (p \supset r)$, a natural question is: can *Assoc* be derivable from *Comm*, or *Sum* be derivable from *Syll*? The answer is that neither *Assoc* in System 2 nor *Assoc** in System 3 are replaceable by *Comm*. Equally in the negative, *Sum* cannot be replaced by *Syll* in the Systems 1, 2, 3 or 4.
6. Further, in System 3, *Add* cannot be replaced by *Simp*.
7. Finally, consider *Id*: $p \supset p$, or $\sim p \vee p$, the *einfachste allgemein richtige Aussagen-verknüpfung*, or the simplest all non-atomic formulas. What is the dependence status of *Id*? Actually, *Id* is derivable from *Add* and *Assoc*, or from *Simp* and *Comm*, or from *Taut*, *Add** and *Assoc**, or from *Add**, *Perm* and *Assoc**. But *Id* is underivable without resource to *Add*, *Simp* or *Add**.
8. Evaluating *Id* a bit further, one finds that *Id* cannot replace either *Add* or *Add**, since without them *Id* together with *Taut*, *Perm*, *Assoc*, *Sum* would not derive *Id**: $p \supset (p \vee p)$; moreover, except when *Id* or *Id** are assumed, *Add*, *Add** and *Simp* remain underivable.

The meaning of those (nowadays perhaps regarded as garden-variety) observations is deeper than it seems to be: they show that there may be many ways to warrant a logical concept, and that endorsing a certain way is not a logically or philosophically indifferent choice.

Section 2 of the paper is dedicated to the (nowadays almost obvious, after 8 decades of elementary logic) proofs of such *Abhängigkeit* (dependence, or mutual derivations). Although simple, the proofs are elucidative: Bernays remarks, for instance, that in the proofs of Assoc from *Taut, Add, Perm, Sum* (item 1 above), of *Perm* from *Taut, Add**, *Assoc, Sum* (item 2 above), and *Assoc* from *Taut, Add, Assoc**, *Sum* no negation is involved; the purely positive proofs make it possible to take implication as a system primitive.

Section 3 is dedicated to the topic of *Unabhängigkeit,* or independence proofs; this is the innovative contribution of Bernays, not only to many-valued logic but to the methods of logic as we conceive them today. Bernays explicitly refers to it as the *Methode der Aufweisung,* or "method of displaying": in each case, set up a finite collection of arbitrary elements, and binary and unary operations on this collection interpreting disjunction and negation (which, he carefully adverts, need not be commutative, neither cyclic, nor associative). A subset of the collection makes the distinguished values, and a sentence is "correct" (*richtige*) when any assignment outputs distinguished values, under the following condition (called *Bedingung B*) concerning disjunction and negation simultaneously: the value of the disjunction $\sim p \vee q$ made up of the negation of a distinguished value p and an arbitrary element q is distinguished only if q is distinguished. This is precisely what we know today as "logic matrices", and more generally "algebraic semantics".

The explicit independence proofs are the following (where symbols between parentheses denote the elements of the matrix, while symbols between curly brackets denote its distinguished elements); in each case the underivable sentence gets a non-distinguished value, while other sentences are correct:

I. *Taut* cannot be derived from *Add, Perm, Assoc, Sum.*

Use $(0, 1, 2)$, $\{0\}$, and the matrices:

$\vee$	0	1	2
0	0	0	0
1	0	1	2
2	0	2	0

	0	1	2
$\sim$	1	0	2

II. *Sum* cannot be derived from *Taut, Add, Add*, Perm, Assoc, Assoc*, Syll.*

Use $(0, 1, 2, 3)$, $\{0\}$, and the matrices:

$\vee$	0	1	2	3
0	0	0	0	0
1	0	1	2	3
2	0	2	2	0
3	0	3	0	3

	0	1	2	3
$\sim$	1	0	3	0

III. *Simp* cannot be derived from *Taut, Perm, Assoc, Sum, Id, Id*.*

Use $(0, 1, 2, 3)$, $\{0, 2\}$, and the matrices:

$\vee$	0	1	2	3
0	0	0	0	0
1	0	1	1	1
2	0	1	2	2
3	0	1	2	3

	0	1	2	3
$\sim$	1	0	3	2

IV. *Id** cannot be derived from *Taut, Perm, Assoc, Sum, Id*.
Use (0, 1, 2, 3), {0, 2}, and the matrices:

∨	0	1	2	3
0	0	0	0	0
1	0	1	1	1
2	0	1	1	2
3	0	1	2	3

	0	1	2	3
~	1	0	3	2

V. *Id* cannot be derived from *Taut, Perm, Assoc, Sum*.
Use (0, 1, 2), {0}, and the matrices:

∨	0	1	2
0	0	0	0
1	0	1	1
2	0	1	1

	0	1	2
~	1	0	2

VI. *Add* (and consequently also *Assoc*) cannot be derived from *Id, Taut, Add*, Simp, Assoc*, Comm, Sum*.
Use (0, 1, 2), {0}, and the matrices:

∨	0	1	2
0	0	0	0
1	0	1	2
2	2	2	2

	0	1	2
~	1	0	0

VII. *Add** cannot be derived from *Id, Taut, Add, Assoc*, Sum*.
Use (0, 1, 2), {0, 2}, and the matrices:

∨	0	1	2
0	0	0	0
1	0	1	2
2	0	1	2

	0	1	2
~	1	0	1

VIII. *Perm* (and consequently *Assoc* and *Assoc**) cannot be derived from *Id, Taut, Add, Add*, Comm, Sum*.
Use (0, 1, 2, 3), {0}, and the matrices:

∨	0	1	2	3
0	0	0	0	0
1	0	1	2	3
2	0	2	2	0
3	0	3	3	3

	0	1	2	3
~	1	0	0	2

IX. *Assoc** (and consequently *Add** and *Perm*) cannot be derived from *Taut, Add, Assoc, Sum*.
Use (0, 1, 2, 3), {0, 2}, and the matrices:

∨	0	1	2	3
0	0	0	0	0
1	0	1	2	3
2	0	1	2	2
3	0	1	2	3

	0	1	2	3
~	1	0	3	2

We see here Bernays' eyes trained on analytic number theory: he points that, in the case of item I, the table for disjunction coincides with multiplication of 0, 1, 2 modulo 4, while in the case of item II the table for disjunction coincides with multiplication of 0, 1, 3, 4 modulo 6. He does not show any interpretation in terms of finite arithmetic for negations in such terms, nor he could: $\mathbf{Z}_4$ and $\mathbf{Z}_6$ are not fields, and the lack of inverting operations impeaches interpreting negations in such case as functions. However, it is known in contemporary combinatorics that, for q a power of a prime p, any function $f : A^n \mapsto A$ (for A a collection with q or fewer elements) can be equated to a polynomial $p(x_1, x_2, \cdots, x_n)$ with coefficients in the Galois field $\mathbf{GF}(q)$ (see e.g., [10], section 5.7 for the mathematical relevance of this property, and [5] for its significance and applications in logic). The tables of item I, for instance, are just the functions $\sim (x) = 2x + 1$ from $\mathbf{GF}(3)$ (i.e., $\mathbf{Z}_3$) on $\mathbf{GF}(3)$ for the case of negation, and $\vee(x, y) = 2x^2y^2 + x^2y + xy^2$ from $\mathbf{GF}(3)^2$ on $\mathbf{GF}(3)$ for the case of disjunction. All other tables can be rephrased in analogous terms as polynomials over finite fields, an observation that not only is akin with the spirit of the work of Bernays, but also with the idea of many-valued algebraic structures.

It is an widespread (though not consensual) opinion that the roots of many-valued logic, in the strict sense of considering more than two truth-values, inherits from Aristotle's discussion of future contingents and of "tomorrow's sea battle". I do not wish to engage on any historical dispute, nor to contend with e.g. G. Malinowski (cf. [11]) when he says that medieval philosophers as Duns Scotus, William of Ockham and Peter de Rivo would be prepared to accept a third truth-value.

However, some care should be taken when claiming that logic as many-valued was born at the turn of the twentieth century with Hugh MacColl, or with the scrawls on "trychotomic mathematics" or "triadic logic" of Charles S. Peirce of 1902, or with the *Imaginary logic* of Nicolai A. Vasil'ev. This view, in any case, is far from unanimous: [6] for instance, argues against the role of Vasil'ev as a precursor to many-valued (and to paraconsistent) logics. Simons in [17] defends that Alexius Meinong, whose work was influential on Lukasiewicz can be considered a founding father of many-valued logics, with greater title than Hugh MacColl, while [18] plainly discards MacColl's logic as legitimately many-valued.

As I. Anellis, referring to the pages containing Peirce's truth-table device found in a manuscript discovered by Shea Zellweger at the Harvard library (cf. [1]) and while discussing, among Wittgenstein and Russell, who learned about the truth-table device from whom, peremptorily puts it (p. 66 his italics):

> Peirce's manuscript of 1902 *does* permit us to unequivocally declare with certitude that *the earliest, the first recorded, verifiable, cogent, attributable and complete truth-table device in modern logic attaches to Peirce*, rather than to Wittgenstein's 1912 jottings and Eliot's notes on Russell's 1914 Harvard lectures.

But to be the inventor of the truth-table device is not necessarily the same as to be the creator of many-valued logic. Firstly, the mention to a 'complete truth-table device' by Anellis is no more than a *façon de parler*, as Peirce did not have any axioms or any syntactical device in the triadic case (as he had with his graphs for dyadic logic). Secondly, Peirce seems to be the first perpetrator of the famous mistake of conflating modal and many-valued logics, as his triadic reasoning was an attempt to insert indeterminacy between his "potentiality" and "real potentiality" (according to [8]).

The parallel between many-valued logics (as Lukasiewicz, MacColl, Peirce, Vasil'ev, etc.) and non-euclidean geometries is coercive, to say the least. All such authors force an extra state into logic, while Lobachevskian geometries, for instance, were not born with the intention to provide any "non-straight line", but around the dispute about the independence of an axiom: Euclidean geometry assumes Euclid's Fifth Postulate, while Lobachevskian geometry assumes its negation, and a most important fact is that both sides are substantiated by expressive models. What Lobachevsky found was essentially a *non-standard model*, or at least a *non-standard interpretation* of the Euclidean plane. It was Hilbert's *Grundlagen der Geometrie* of 1899 the responsible for clarifying the notion of models through his "classes of isomorphic construction", and for showing how models can separate what is essential from what is accidental in a mathematical notion. This was the kernel of the idea of non-standard models, in the sense of models that subtly deviate from the standard ones. Undoubtedly, this is what is behind the work of Bernays and his many-valued models: it is remarkable that Bernays never talks about truth-values, only about function values.

The notion of non-standard models is a hallmark of contemporary logic and the foundations of mathematics: Thoraf Skolem's non-standard models of set theory and arithmetic, and Abraham Robinson's models of non-standard are good examples, and even forcing is a sophisticated tool for providing non-standard models. In this sense, the *soi-disant* precursors of many-valued logics, including MacColl, Peirce, Vasil'ev, Post and Lukasiewicz, are just looking for "non-straight lines", what the geometers did not do, while Bernays, in the pure geometrical spirit, is making room for broader mathematical concepts.

References

1. Anellis, I.: The genesis of the truth-table device. Russell: the Journal of Bertrand Russell Studies **24**(1), 55–70 (2004). Available at: http://digitalcommons.mcmaster.ca/russelljournal/vol24/iss1/5
2. Bernays, P.: Beiträge zur axiomatischen Behandlung des Logik-Kalküls. Habilitationsschrift, Universität Göttingen. Unpublished typescript, 1918
3. Bernays, P.: Axiomatische Untersuchungen des Aussagen-Kalküls der 'Principia Mathematica'. Math. Z. **25**, 305–320 (1926). Abridged version of [2]
4. Bernays, P., Schönfinkel, M.: Zum Entscheidungsproblem der mathematischen Logik. Math. Ann. **99**, 342–372 (1928)
5. Carnielli, W.: Polynomizing: Logic inference in polynomial format and the legacy of boole. In: Model-Based Reasoning in Science, Technology and Medicine. Studies in Computational Intelligence, vol. 64, pp. 349–364 (2007)
6. Cavaliere, T.: Review-essay of N.A. Vasil'ev, Imaginary logic (Russian), edited by V.A. Smirnov, Mod. Log. **2**(1), 52–76 (1991)
7. Feferman, S.: Lieber Herr Bernays!, Lieber Herr Gödel! Gödel on finitism, constructivity and Hilbert's program. Dialectica **62**(2), 179–203 (2008)
8. Fisch, M., Turquette, A.: Peirce's triadic logic. Trans. Charles S. Peirce Soc. **11**, 71–85 (1966)
9. Hilbert, D.: Grundlagen der Geometrie, 1st edn. Teubner, Leipzig (1899)

10. Kung, J.P.S., Rota, G.-C., Yan, C.H.: Combinatorics: The Rota Way. Cambridge University Press (2009)
11. Malinowski, G.: A philosophy of many-valued logic. The third logical value and beyond. In: Lapointe, S., Wolénski, J., Marion, M., Miskiewicz, W. (eds.) The Golden Age of Polish Philosophy. Logic, Epistemology, and the Unity of Science, vol. 16, pp. 81–92. Springer, Netherlands (2009)
12. Moore, G.H.: Hilbert and the emergence of modern mathematical logic. Theoria (Segunda época) **12**, 65–90 (1997)
13. Post, E.L.: Introduction to a general theory of elementary propositions. Am. J. Math. **43**, 163–185 (1921)
14. Rescher, N.: Many-Valued Logic. McGraw-Hill, New York (1969)
15. Robinson, J.A.: Machine-Oriented Logic Based on the Resolution Principle. J. ACM **12**(1), 23–41 (1965)
16. Sieg, W.: Hilbert's programs: 1917–1922. Bull. Symb. Log. **5**(1), 1–44 (1999)
17. Simons, P.M.: Lukasiewicz, Meinong, and many-valued logic. In: Szaniawski, K. (ed.) The Vienna Circle and the Lvov–Warsaw School, pp. 249–292. Kluwer (1989)
18. Simons, P.M.: MacColl and many-valued logic: An exclusive conjunction. Nord. J. Philos. Log. **3**(1), 85–90 (1999)
19. Zach, R.: Completeness before Post: Bernays, Hilbert, and the development of propositional logic. Bull. Symb. Log. **5**(3), 331–366 (1999)

W. Carnielli (✉)
Centre for Logic, Epistemology and the History of Science – CLE and Department of Philosophy – UNICAMP, P.O. Box 6133, 13083-970, Campinas SP, Brazil
e-mail: walter.carnielli@cle.unicamp.br

Axiomatic Investigations of the Propositional Calculus of the "Principia Mathematica"[1]

Paul Bernays

§1

In Whitehead and Russell's "Principia Mathematica," the systematic development of the logical propositional calculus is carried out by starting with five propositional formulas as basic formulas ("primitive propositions"), from which every generally valid combination of propositions [*Aussagenverknüpfung*], i.e., one which is correct for arbitrary values of the propositional variables occurring in it, can be obtained by *substitutions* and the application of a single formal *rule of inference*.

This reduction of the theorems to the basic formulas is combined with a reduction of the logical connectives to the two operations:

of negation

$$\sim p \quad (\text{not } p)$$

and of disjunction

$$p \lor q \quad (p \text{ or } q; \text{ "or" in the sense of the latin "vel"}),$$

from which are formed by combination:[2]

the implication

$$p \supset q \quad (\text{if } p, \text{ then } q), \text{ as abbreviation for} \quad \sim p \lor q$$

Translated by R. Zach, with kind permission, from: Bernays, P., "Axiomatische Untersuchung des Aussagenkalküls des Principia Mathematica". Mathematische Zeitschrift Vol. 25(1), pp. 305–320. © Springer 1926

[1] The content of this article is taken to a large extent from an unpublished *Habilitationsschrift*, which was submitted by the author to the Faculty of Mathematics and Natural Sciences at Göttingen in 1918. The questions on the possibility of replacing formulas by inference rules investigated there has been left out of the present investigation.

[2] In order to avoid a proliferation of parentheses, dots shall be used as separating symbols, just as in the Princip. Math., where the rule is that more dots separate more strongly than fewer dots. This rule, however, applies only to the separation of symbols of the *same kind*. Whenever two different symbols from among $\sim$, $\lor$, $\supset$ compete, in general the convention shall hold that for the separation of parts of the formula, $\supset$ takes precedence over $\lor$ and $\sim$, as well as $\lor$ over $\sim$, as long as no other separation is indicated by *parentheses*. Accordingly, the formula

$$p \supset \sim q \lor : p \lor \cdot q \lor r \cdot \supset \cdot p \supset p \lor : \sim q \lor \cdot q \lor r$$

J.-Y. Béziau (ed.), *Universal Logic: An Anthology*, 43–56
Studies in Universal Logic, DOI 10.1007/978-3-0346-0145-0_4, © Springer Basel AG 2012

and the conjunction[3]

$$p \,\&\, q \quad (p \text{ and } q), \text{ as abbreviation for } \sim(\sim p \vee \sim q).$$

The five basic formulas are

(*Taut*)	$p \vee p \supset p,$
(*Add*)	$q \supset p \vee q,$
(*Perm*)	$p \vee q \supset q \vee p,$
(*Assoc*)	$p \vee \cdot q \vee r \supset q \vee \cdot p \vee r,$
(*Sum*)	$q \supset r \cdot \supset \cdot p \vee q \supset p \vee r.$

When replacing the implication by the defining expression, they read:

(*Taut*)	$\sim(p \vee p) \cdot \vee p,$
(*Add*)	$\sim q \vee \cdot p \vee q,$
(*Perm*)	$\sim(p \vee q) \cdot \vee \cdot q \vee p,$
(*Assoc*)	$\sim(p \vee \cdot q \vee r) : \vee : q \vee \cdot p \vee r,$
(*Sum*)	$\sim(\sim q \vee r) \cdot \vee : \sim(p \vee q) \cdot \vee \cdot p \vee r.$

The rule of inference can be formulated as follows: From two formulas[4]

$$\mathfrak{A}, \quad \mathfrak{A} \supset \mathfrak{B} \quad (\text{resp. } \sim\mathfrak{A} \vee \mathfrak{B})$$

the formula $\mathfrak{B}$ is to be obtained.

The substitution rule says that any arbitrary formula can be substituted for a propositional variable.

One may now easily convince oneself that each of the five basic formulas represents a generally valid combination of propositions when interpreted contentually. And that the application of the rules only yields formulas *of this kind*.

At the same time, the system of basic formulas is *complete* in the sense that from it *all* generally valid combinations of propositions can be obtained using the rules. This claim can even be strengthened as follows:

If $\mathfrak{A}$ is any formula formed from the symbols introduced, then either it is derivable from the five basic formulas according to the rules, or *any arbitrary formula* $\mathfrak{B}$ becomes derivable if $\mathfrak{A}$ is added to the basic formulas.

The justification of this claim shall be indicated briefly.[5] It results from considering the *conjunctive normal form*. A conjunctive normal form is a formula which is formed from one or more "simple disjunctions" combined using & (conjunction), i.e., such disjunctions in which every disjunct is either a variable or a variable to which a negation is applied.

---

has to be read in the same way as would be given without any further conventions by the following way of putting parentheses:

$$\Big(p \supset \big((\sim q) \vee (p \vee (q \vee r))\big)\Big) \supset \Big(p \supset \big(p \vee ((\sim q) \vee (q \vee r))\big)\Big).$$

[3] In the Princip. Math. the symbol for conjunction is simply a dot.

[4] We shall use uppercase Fraktur letters as symbols used to indicate formulas of an indefinite form.

[5] A slightly different proof can be found in the treatise by E.L. Post, "Introduction to a general theory of elementary propositions" (American Journal of Math. **43** (1921)).

For instance,

$$p \mathbin{\&} (p \vee \sim q \vee r)$$

is a conjunctive normal form.

Now the familiar theorem holds, that every formula can be brought into a conjunctive normal form; i.e., we have a procedure to find, for any given formula $\mathfrak{A}$, a normal form $\mathfrak{N}$ such that

$$\mathfrak{A} \supset \mathfrak{N} \quad \text{as well as} \quad \mathfrak{N} \supset \mathfrak{A}$$

is derivable from the basic formulas. (As an aside, the determination of $\mathfrak{N}$ for $\mathfrak{A}$ is not unique.)

One now sees immediately: If $\mathfrak{A}$ is derivable, so is $\mathfrak{N}$, and conversely; furthermore, if $\mathfrak{N}$ is derivable, then so is each one of the simple disjunctions occurring in $\mathfrak{N}$ (as conjuncts).

Our claim is thus reduced to the corresponding claim for simple disjunctions. For those, however, it follows immediately; for a simple disjunction either contains two disjuncts one of which is the negation of the other (such as p and $\sim p$); then it is derivable; or the disjunction does not contain two such disjuncts; then from it one can obtain, by substitution, a disjunction every disjunct of which is either the variable p or the double negation of p, i.e., $\sim(\sim p)$. From such a formula, however, the formula consisting only of the variable p is derivable, and from this one can obtain any arbitrary formula by substitution.

This consideration at the same time yields a simple procedure to determine whether a given formula $\mathfrak{A}$ represents a generally valid combination of propositions, or—and this amounts to the same thing—is derivable from the basic formulas: find a conjunctive normal form $\mathfrak{N}$ of $\mathfrak{A}$. If in each simple disjunction occurring in $\mathfrak{N}$ two disjuncts occur one of which is the negation of the other, then the given formula $\mathfrak{A}$ is generally valid, otherwise it is not.

This decision procedure solves the main problem of the propositional calculus completely, and if one only wanted to characterize the generally valid logical combinations of propositions, one would be able to obtain this result more directly than by the method of Principia Mathematica.

The insight, that the listed five basic formulas suffice for the derivation of all generally valid combinations of propositions (using the rules), is nevertheless significant in itself. In this light, the further question now arises whether the five basic formulas are mutually independent in the sense of formal derivability.

This is indeed *not* the case, rather, the formula *Assoc* can be derived from the other four. Therefore, the system of basic formulas can be replaced by that of the four formulas

$$\textit{Taut, Add, Perm, Sum} \qquad \text{(System 1)}$$

Of these four formulas, none are redundant, not even if one retains *Assoc*.

These claims shall be proved below. Our investigation of dependencies is not exhausted by it, however. A number of further questions shall also be considered, to which this result naturally leads and which arise from two remarks.

The first remark is that, in the event that *Assoc* is retained, the formula *Perm* becomes provable, if instead of

$$\textit{Add:} \quad q \supset p \vee q$$

the formula

$$Add^*: \quad p \supset p \vee q$$

is taken as a basic formula.

These two formulas are known to be mutually equivalent using *Perm* and *Sum*, i.e., one can, using *Perm* and *Sum*, pass from the formula *Add* to *Add**, and also back.

It will thus become apparent that the following four formulas suffice as a system of basic formulas:

$$\text{Taut, Add}^*, \text{ Assoc, Sum} \qquad\qquad \text{(System 2)}$$

This fact is not so surprising, since an exchange occurs in the formula *Assoc*. One will therefore ask if the formula *Perm* is still dependent if instead of *Assoc* the formula

$$Assoc^*: \quad p \vee \cdot q \vee r \supset p \vee q \cdot \vee r,$$

which expresses the associative character of disjunction more purely, is chosen as a basic formula.

It can be shown that this is not the case; i.e., in system 2 the formula *Assoc* cannot be replaced by *Assoc**. It can, however, be replaced by *Add* and *Assoc** together, so that the five formulas

$$\text{Taut, Add, Add}^*, \text{ Assoc}^*, \text{ Sum} \qquad\qquad \text{(System 3)}$$

suffice as a system of basic formulas. (That the formulas *Assoc* and *Assoc** do not have the same inferential power when *Perm* is removed does of course not contradict the known fact that these two formulas are equivalent in the presence of *Perm* and *Sum*.)

The second remark is that in system 1, the formula *Add* can be replaced by the specialized formula

$$Simp: \quad q \supset \cdot p \supset q$$

resp.

$$\sim q \vee \cdot \sim p \vee q,$$

so that the four formulas

$$\text{Taut, Simp, Perm, Sum} \qquad\qquad \text{(System 4)}$$

suffice as basic formulas.

By an analogous specialization as that of *Add* to *Simp*, the formula

$$Comm: \quad p \supset \cdot q \supset r : \supset : q \supset \cdot p \supset r$$

resp.

$$\sim(\sim p \vee \cdot \sim q \vee r) \vee : \sim q \vee \cdot \sim p \vee r$$

results from *Assoc*, and the formula

$$Syll: \quad q \supset r \cdot \supset : p \supset q \cdot \supset \cdot p \supset r$$

resp.

$$\sim(\sim q \vee r) \vee : \sim(\sim p \vee q) \vee \cdot \sim p \vee r$$

from *Sum*. Now the question arises whether *Assoc* might not also be replaced by *Comm*, or *Sum* by *Syll*. This question is decided in the negative. Neither *Assoc* in system 2, nor *Assoc** in system 3 can be replaced by *Comm*, and *Sum* cannot be replaced by *Syll* in any of the systems 1, 2, 3, or 4. Furthermore, it is shown that the formula *Add* cannot be replaced by *Simp* in system 3.

Finally the relation of the formula

$$Id: \quad p \supset p$$

resp.

$$\sim p \lor p$$

to the formulas of systems 1, 2, 3, 4 shall be considered. *Id* is distinguished by being the simplest generally valid combination of propositions. One would therefore hope to use it as a basic formula. However, it cannot replace any of the formulas in any of the systems 1, 2, 3, 4.

The simplest derivation of *Id* is that from

$$Taut, \; Add, \; Syll$$

where one first obtains from *Add* by substitution the formula

$$Id^*: \quad p \supset p \lor p$$

resp.

$$\sim p \lor \cdot p \lor p.$$

Without applying *Sum*, one can derive *Id*

from   *Add* and *Assoc*,

even already from   *Simp* and *Comm*;

moreover from   *Taut, Add*, Assoc**,

and also from   *Add*, Perm, Assoc**.

Carrying out these derivations, which are found easily, shall be left to the reader. In every one of these derivations either *Add* or *Simp* or *Add** is used. We will show that without applying one of these formulas, the formula *Id* is *no longer* provable from the remaining formulas in systems 1, 2, 3, 4.

Of course the formula *Id* is not a sufficient replacement for either of the formulas *Add*, *Add**. It will be shown that by adding *Id* to the formulas

$$Taut, \; Perm, \; Assoc, \; Sum$$

not even the formula *Id** is provable, and moreover, that even if *Id** is taken in addition to *Id*, the formulas *Add* and *Add** as well as *Simp* remain unprovable.

The claims put forward shall now be established, specifically, the claimed *dependencies* on the basis of which the formula systems 1, 2, 3, 4 are recognized to be sufficient systems of basic formulas will be proved in §2; in §3, proofs for the *independence* claims will be given, and it shall be shown that *none of the systems 1, 2, 3, 4 contains a redundant formula*.

For the sake of convenience, let us collect the systems here again:

System 1:   *Taut, Add, Perm, Sum,*

„   2:   *Taut, Add*, Assoc, Sum,*

„   3:   *Taut, Add, Add*, Assoc*, Sum,*

„   4:   *Taut, Simp, Perm, Sum.*

§2

By way of explanation of the derivations below, we start with a few remarks.

If a formula is obtained by literal repetition or by substitution from a basic formula or a formula already derived, then the reference label of said formula is given on the *left*. A *new reference label* for a formula obtained is given on its *right*.

The application of the inference rule follows the schema

$$\frac{\mathfrak{A}}{\mathfrak{A} \supset \mathfrak{B}}{\mathfrak{B}}$$

Furthermore, we will use—in order to give an *abbreviated description* of proofs—the schema

$$\frac{\mathfrak{A} \supset \mathfrak{B}}{\mathfrak{B} \supset \mathfrak{C}}{\mathfrak{A} \supset \mathfrak{C}}$$

This is explained as follows: From the basic formula *Sum* one obtains, as mentioned already, the formula

Syll: $q \supset r \cdot \supset : p \supset q \cdot \supset \cdot p \supset r$

by substitution. By using this formula twice in conjunction with the inference rule, one can derive the formula

$$\mathfrak{A} \supset \mathfrak{C}$$

from the formulas

$$\mathfrak{A} \supset \mathfrak{B} \quad \text{and} \quad \mathfrak{B} \supset \mathfrak{C}.$$

Therefore, we may proceed, wherever *it is permitted to use the basic formula Sum*, as if we had an inference rule according to which the formula

$$\mathfrak{A} \supset \mathfrak{C}$$

can be obtained from

$$\mathfrak{A} \supset \mathfrak{B}, \quad \mathfrak{B} \supset \mathfrak{C}.$$

And this we shall do, in order to avoid unnecessary complexities.

Reference labels introduced in a proof for derived formulas are given by numerals in brackets; these labels need to be fixed only *within a proof*.

Furthermore let us remark that two formulas which result from one another by applying the abbreviation

$$p \supset q \quad \text{for} \quad \sim p \vee q$$

shall count as equivalent in our considerations. Passing from one of the ways of writing the formula to the other is indicated by "resp."

We will now proceed to the proofs of the claimed dependencies. The theorems will be numbered in such a way that *the number of each theorem coincides with that of the system of formulas* which is established as complete by it.[6]

1. Derivation of *Assoc* from *Taut, Add, Perm, Sum.*

(Add)	$r \supset p \vee r$
(Sum)	$\dfrac{r \supset p \vee r \cdot \supset \cdot q \vee r \supset q \vee \cdot p \vee r}{q \vee r \supset q \vee \cdot p \vee r}$
(Sum)	$\dfrac{q \vee r \supset q \vee \cdot p \vee r \cdot \supset \cdot p \vee \cdot q \vee r \supset p \vee : q \vee \cdot p \vee r}{p \vee \cdot q \vee r \supset p \vee : q \vee \cdot p \vee r}$
(Perm)	$\dfrac{p \vee : q \vee \cdot p \vee r \supset q \vee \cdot p \vee r : \vee p}{p \vee \cdot q \vee r \supset q \vee \cdot p \vee r : \vee p} \qquad (1)$

(Add)	$p \supset r \vee p$
(Perm)	$\dfrac{r \vee p \supset p \vee r}{p \supset p \vee r}$
(Add)	$\dfrac{p \vee r \supset q \vee \cdot p \vee r}{p \supset q \vee \cdot p \vee r}$
(Sum)	$\dfrac{p \supset q \vee \cdot p \vee r \cdot \supset \cdot q \vee \cdot p \vee r : \vee p \supset q \vee \cdot p \vee r : \vee : q \vee \cdot p \vee r}{q \vee \cdot p \vee r : \vee p \supset q \vee \cdot p \vee r : \vee : q \vee \cdot p \vee r}$
(Taut)	$\dfrac{q \vee \cdot p \vee r : \vee : q \vee \cdot p \vee r \supset q \vee \cdot p \vee r}{q \vee \cdot p \vee r : \vee p \supset q \vee \cdot p \vee r} \qquad (2)$

(1)	$p \vee \cdot q \vee r \supset q \vee \cdot p \vee r : \vee p$
(2)	$\dfrac{q \vee \cdot p \vee r : \vee p \supset q \vee \cdot p \vee r}{p \vee \cdot q \vee r \supset q \vee \cdot p \vee r}$

2. Derivation of *Perm* from *Taut, Add*, Assoc, Sum.*

(Taut)	$p \vee p \supset p$
(Sum)	$\dfrac{p \vee p \supset p \cdot \supset \cdot q \vee \cdot p \vee p \supset q \vee p}{q \vee \cdot p \vee p \supset q \vee p} \qquad (1)$

---

[6]Note that negation does *not occur explicitly anywhere* in the derivations 1, 2, 3. The dependencies 1, 2, 3 therefore still obtain if implication is considered as a *primitive connective* instead of as a combination of negation and disjunction.

(Add*)
$$q \supset q \vee p$$

(Sum)
$$\frac{q \supset q \vee p \cdot \supset \cdot p \vee q \supset p \vee \cdot q \vee p}{p \vee q \supset p \vee \cdot q \vee p}$$

(Assoc)
$$\frac{p \vee \cdot q \vee p \supset q \vee \cdot p \vee p}{p \vee q \supset q \vee \cdot p \vee p}$$

(1)
$$\frac{q \vee \cdot p \vee p \vee p \supset q \vee p}{p \vee q \supset q \vee p}$$

3. Derivation of *Assoc* from *Taut, Add, Assoc*, Sum*.

(Add)
$$r \supset p \vee r$$

(Sum)
$$\frac{r \supset p \vee r \cdot \supset \cdot q \vee r \supset q \vee \cdot p \vee r}{q \vee r \supset q \vee \cdot p \vee r}$$

(Add)
$$\frac{q \vee \cdot p \vee r \supset r \vee : q \vee \cdot p \vee r}{q \vee r \supset r \vee : q \vee \cdot p \vee r}$$

(Sum)
$$\frac{q \vee r \supset r \vee : q \vee \cdot p \vee r \cdot \supset \cdot p \vee \cdot q \vee r \supset p \vee \therefore r \vee : q \vee \cdot p \vee r}{p \vee \cdot q \vee r \supset p \vee \therefore r \vee : q \vee \cdot p \vee r}$$

(Assoc*)
$$\frac{p \vee \therefore r \vee : q \vee \cdot p \vee r \supset p \vee r \cdot \vee : q \vee \cdot p \vee r}{p \vee \cdot q \vee r \supset p \vee r \cdot \vee : q \vee \cdot p \vee r}$$

(Add)
$$\frac{p \vee r \cdot \vee : q \vee \cdot p \vee r \supset q \vee \therefore p \vee r \cdot \vee : q \vee \cdot p \vee r}{p \vee \cdot q \vee r \supset q \vee \therefore p \vee r \cdot \vee : q \vee \cdot p \vee r}$$

(Assoc*)
$$\frac{q \vee \therefore p \vee r \cdot \vee : q \vee \cdot p \vee r \supset q \vee \cdot p \vee r : \vee : q \vee \cdot p \vee r}{p \vee \cdot q \vee r \supset q \vee \cdot p \vee r : \vee : q \vee \cdot p \vee r}$$

(Taut)
$$\frac{q \vee \cdot p \vee r : \vee : q \vee \cdot p \vee r \supset q \vee \cdot p \vee r}{p \vee \cdot q \vee r \supset q \vee \cdot p \vee r}$$

4. Derivation of *Add* from *Taut, Simp, Perm, Sum*.

(Taut)
$$\sim p \vee \sim p \supset \sim p$$

(Sum)
$$\frac{\sim p \vee \sim p \supset \sim p \cdot \supset \cdot \sim(\sim p) \vee \cdot \sim p \vee \sim p \vee \sim p \supset \sim(\sim p) \vee \sim p}{\sim(\sim p) \vee \cdot \sim p \vee \sim p \supset \sim(\sim p) \vee \sim p} \qquad (1)$$

(Simp)
$$\sim p \supset \cdot p \supset \sim p$$

resp.

$$\sim(\sim p) \vee \cdot \sim p \vee \sim p$$

(1)
$$\frac{\sim(\sim p) \vee \cdot \sim p \vee \sim p \supset \sim(\sim p) \vee \sim p}{\sim(\sim p) \vee \sim p}$$

(Perm)
$$\sim(\sim p) \vee \sim p \supset \sim p \vee \sim(\sim p)$$

$$\sim p \lor \sim(\sim p)$$

resp.

$$p \supset \sim(\sim p)$$

(Sum)
$$\frac{p \supset \sim(\sim p) \cdot \supset \cdot q \lor p \supset q \lor \sim(\sim p)}{q \lor p \supset q \lor \sim(\sim p)}$$

(Perm)
$$\frac{q \lor \sim(\sim p) \supset \sim(\sim p) \lor q}{q \lor p \supset \sim(\sim p) \lor q}$$

resp.

$$q \lor p \supset \cdot \sim p \supset q \tag{2}$$

(Taut)
$$p \lor p \supset p$$

resp.

$$\sim(p \lor p) \lor p$$

(Perm)
$$\frac{\sim(p \lor p) \lor p \supset p \lor \sim(p \lor p)}{p \lor \sim(p \lor p)}$$

(2)
$$\frac{p \lor \sim(p \lor p) \supset \cdot \sim(\sim(p \lor p)) \supset p}{\sim(\sim(p \lor p)) \supset p}$$

(Sum)
$$\frac{\sim(\sim(p \lor p)) \supset p \cdot \supset \cdot q \lor \sim(\sim(p \lor p)) \supset q \lor p}{q \lor \sim(\sim(p \lor p)) \supset q \lor p} \tag{3}$$

(Perm)
(3)
$$\frac{\sim(\sim(p \lor p)) \lor q \supset q \lor \sim(\sim(p \lor p))}{q \lor \sim(\sim(p \lor p)) \supset q \lor p}$$
$$\frac{}{\sim(\sim(p \lor p)) \lor q \supset q \lor p}$$

(Sum)
$$\frac{\sim(\sim(p \lor p)) \lor q \supset q \lor p \cdot \supset \cdot \sim q \lor \cdot \sim(\sim(p \lor p)) \lor q \supset \sim q \lor \cdot q \lor p}{\sim q \lor \cdot \sim(\sim(p \lor p)) \lor q \supset \sim q \lor \cdot q \lor p} \tag{4}$$

(Simp)
$$q \supset \cdot \sim(p \lor p) \supset q$$

resp.

$$\sim q \lor \cdot \sim(\sim(p \lor p)) \lor q$$

(4)
$$\frac{\sim q \lor \cdot \sim(\sim(p \lor p)) \lor q \supset \sim q \lor \cdot q \lor p}{\sim q \lor \cdot q \lor p}$$

resp.

$$q \supset q \lor p$$

(Perm)
$$\frac{q \lor p \supset p \lor q}{q \supset p \lor q}$$

§3

The task is now to prove the theorems claimed in §1 about *independence*—to which also the claims about non-replaceability belong. That a formula $\mathfrak{A}$ is not dispensable within a system of formulas, resp., that it cannot be replaced by another formula $\mathfrak{B}$, is established if we show that some generally valid formula $\mathfrak{C}$ cannot be derived from the formula system which remains after deleting the formula $\mathfrak{A}$, resp., which results by replacing the formula $\mathfrak{A}$ by the formula $\mathfrak{B}$.

Taking the established *dependencies* into account, one sees that the claimed independence claims are justified, provided the following independencies are shown:

I.	*Taut*	cannot be derived from				*Add, Perm, Assoc, Sum*;
II.	*Sum*	„	„	„	„	*Taut, Add, Add*, Perm, Assoc,*
						Assoc*, Syll;
III.	*Simp*	„	„	„	„	*Taut, Perm, Assoc, Sum, Id, Id**;
IV.	*Id**	„	„	„	„	*Taut, Perm, Assoc, Sum, Id*;
V.	*Id*	„	„	„	„	*Taut, Perm, Assoc, Sum*;
VI.	*Add*	„	„	„	„	*Taut, Add*, Simp, Assoc*, Sum*;
VII.	*Add**	„	„	„	„	*Taut, Add, Assoc*, Sum*;
VIII.	*Assoc**	„	„	„	„	*Taut, Add, Add*, Comm, Sum*;
IX.	*Assoc**	„	„	„	„	*Taut, Add, Assoc, Sum.*

The proofs of these theorems will be given according to the usual method used for such investigations, viz., the *method of exhibition:* in each case, a finite group (in the extended sense of the word) is given, i.e., a finite totality of elements, for which the "disjunction" is defined as a two-place operation, and "negation" as a one-place operation, by giving the course-of-values (the operation in general need not be associative, uniquely invertible, nor commutative).[7]

Furthermore, a subtotality of "designated values" is singled out from the group.

A formula is then called a "correct formula" with respect to the group under consideration, if it always yields a designated value for any substitution of elements of the group for the propositional variables.

From this it is first of all clear that every propositional formula which results from a correct formula by substitution is also correct. Furthermore, the groups will be set up in such a way that applying the inference rule to two correct formulas yields another correct formula. For this it is sufficient that the following condition B is satisfied:

The disjunction $\sim p \vee q$, which is formed from the negation of a *designated value p* and an arbitrary element q, has a designated value only if q is a designated value. If this condition is satisfied, it follows that every formula which is derivable from correct formulas is also correct.

In order to prove that a formula $\mathfrak{F}$ is independent of certain other formulas $\mathfrak{A}, \ldots,$ $\mathfrak{K}$, one only has to find a group which satisfies condition B and for which $\mathfrak{A}, \ldots, \mathfrak{K}$ are correct formulas while $\mathfrak{F}$ is not a correct formula.

For each of the independence theorems I through IX we will now give a corresponding group. They are specified as follows: First the elements are enumerated within parentheses, and the designated values between braces. Then negation and disjunction are defined.[8] (If a defining equation contains a propositional variable, this means that the equation shall hold for *every* value of the variable.) Subsequently the formulas that have to be verified as correct are listed.

---

[7] The elements will be designated by lowercase Greek letters.

[8] Here the equality sign is used in the sense of definitional equality.

This verification is left to the reader; for the more complex formulas it can be abbreviated significantly by suitable case distinctions.

Finally, a specific substitution of values is given for the formula to be shown independent, under which the formula yields a *non-designated value*, which shows that it is not a correct formula.

In order to facilitate the verification, the following should be remarked: If the formula *Id* is a correct formula for a group, then

if disjunction is commutative, *Perm* is also a correct formula;

if disjunction is associative, *Assoc** is also a correct formula;

if disjunction is commutative as well as associative, *Assoc* is also a correct formula;

if $p \vee p = p$ (for every value of p), then *Taut* is a correct formula.

For *each of the groups, condition B is satisfied.* This has to be established for each group separately, but we will not mention it every time.

Group I

$$(\alpha, \beta, \gamma); \quad \{\alpha\}.$$
$$\sim\alpha = \beta, \quad \sim\beta = \alpha, \quad \sim\gamma = \gamma,$$
$$\alpha \vee p = p \vee \alpha = \alpha,$$
$$\beta \vee p = p \vee \beta = p,$$
$$\gamma \vee \gamma = \alpha.$$

The disjunction so defined can be represented arithmetically as multiplication of the three congruence classes 0, 1, 2 mod 4.

Id, Add, Perm, Assoc, Sum are correct formulas; by contrast, *Taut* is not a correct formula, since

$$\sim(\gamma \vee \gamma) \vee \gamma = \sim\alpha \vee \gamma = \beta \vee \gamma = \gamma.$$

Group II

$$(\alpha, \beta, \gamma, \delta); \quad \{\alpha\}.$$
$$\sim\alpha = \beta, \quad \sim\beta = \alpha, \quad \sim\gamma = \delta, \quad \sim\delta = \alpha$$
$$\alpha \vee p = p \vee \alpha = \alpha,$$
$$\beta \vee p = p \vee \beta = p,$$
$$p \vee p = p,$$
$$\gamma \vee \delta = \delta \vee \gamma = \alpha.$$

The disjunction can be represented arithmetically as multiplication of the congruence classes 0, 1, 3, 4 mod 6.

Id, Taut, Add, Add*, Perm, Assoc, Assoc*, Syll are correct formulas; but not *Sum*, since

$$\sim(\sim\delta \vee \beta) \vee \cdot \sim(\gamma \vee \delta) \vee (\gamma \vee \beta)$$
$$= \sim(\alpha \vee \beta) \vee \cdot \sim\alpha \vee \gamma = \sim\alpha \vee \cdot \beta \vee \gamma$$
$$= \beta \vee \gamma = \gamma.$$

Group III

$$(\alpha, \beta, \gamma, \delta); \quad \{\alpha, \gamma\}.$$
$$\sim\alpha = \beta, \quad \sim\beta = \alpha, \quad \sim\gamma = \delta, \quad \sim\delta = \gamma$$
$$\alpha \vee p = p \vee \alpha = \alpha,$$
$$\beta \vee p = p \vee \beta = \beta \quad \text{for } p \neq \alpha,$$
$$p \vee p = p,$$
$$\gamma \vee \delta = \delta \vee \gamma = \gamma.$$

Id, *Id**, *Taut*, *Perm*, *Assoc*, *Sum* are correct formulas; but not *Simp* (and consequently neither is *Add*), since

$$\sim\gamma \vee \cdot \sim\alpha \vee \gamma = \delta \vee \cdot \beta \vee \gamma = \delta \vee \beta = \beta.$$

Group IV

This group differs from group III only in that

$$\gamma \vee \gamma = \beta$$

is specified:

$$(\alpha, \beta, \gamma, \delta); \quad \{\alpha, \gamma\}.$$
$$\sim\alpha = \beta, \quad \sim\beta = \alpha, \quad \sim\gamma = \delta, \quad \sim\delta = \gamma$$
$$\alpha \vee p = p \vee \alpha = \alpha,$$
$$\beta \vee p = p \vee \beta = \beta \quad \text{for } p \neq \alpha,$$
$$\gamma \vee \gamma = \beta,$$
$$\gamma \vee \delta = \delta \vee \gamma = \gamma$$
$$\delta \vee \delta = \delta.$$

Id, *Taut*, *Perm*, *Assoc*, *Sum* are correct formulas; but not *Id**, since

$$\sim\gamma \vee \cdot \gamma \vee \gamma = \delta \vee \beta = \beta.$$

Group V

$$(\alpha, \beta, \gamma); \quad \{\alpha\}.$$
$$\sim\alpha = \beta, \quad \sim\beta = \alpha, \quad \sim\gamma = \gamma$$
$$\alpha \vee p = p \vee \alpha = \alpha,$$
$$\beta \vee \beta = \beta \vee \gamma = \gamma \vee \beta = \gamma \vee \gamma = \beta.$$

Taut, *Perm*, *Assoc*, *Sum* are correct formulas; but not *Id*, since

$$\sim\gamma \vee \gamma = \gamma \vee \gamma = \beta.$$

Group VI

$$(\alpha, \beta, \gamma); \quad \{\alpha\}.$$
$$\sim\alpha = \beta, \quad \sim\beta = \alpha, \quad \sim\gamma = \gamma$$
$$\alpha \vee p = \alpha, \quad \beta \vee p = p, \quad \gamma \vee p = \gamma.$$

Id, Taut, Add*, Simp, Assoc*, Comm, Sum are correct formulas; but not *Add* (and consequently neither is *Assoc*), since

$$\sim\alpha \vee \gamma \vee \alpha = \beta \vee \gamma = \gamma.$$

Group VII

$$(\alpha, \beta, \gamma); \quad \{\alpha, \gamma\}.$$
$$\sim\alpha = \beta, \quad \sim\beta = \alpha, \quad \sim\gamma = \beta$$
$$\alpha \vee p = \alpha,$$
$$\beta \vee p = \gamma \vee p = p.$$

Id, Taut, Add, Assoc*, Sum are correct formulas; but not *Add**, since

$$\sim\gamma \vee \gamma \vee \beta = \beta \vee \beta = \beta.$$

Group VIII

$$(\alpha, \beta, \gamma, \delta); \quad \{\alpha\}.$$
$$\sim\alpha = \beta, \quad \sim\beta = \alpha, \quad \sim\gamma = \alpha, \quad \sim\delta = \gamma$$
$$\alpha \vee p = p \vee \alpha = \alpha,$$
$$\beta \vee p = p \vee \beta = p,$$
$$p \vee p = p,$$
$$\gamma \vee \delta = \alpha, \quad \delta \vee \gamma = \delta.$$

Id, Taut, Add, Add*, Comm, Sum are correct formulas; but not *Perm* (and consequently neither are *Assoc* and *Assoc**), since

$$\sim(\gamma \vee \delta) \vee \delta \vee \gamma = \sim\alpha \vee \delta = \beta \vee \delta = \delta.$$

Group IX

$$(\alpha, \beta, \gamma, \delta); \quad \{\alpha, \gamma\}.$$
$$\sim\alpha = \beta, \quad \sim\beta = \alpha, \quad \sim\gamma = \delta, \quad \sim\delta = \gamma$$
$$\alpha \vee p = p \vee \alpha = \alpha,$$
$$\beta \vee p = \delta \vee p = p,$$
$$\gamma \vee \beta = \beta,$$
$$\gamma \vee \gamma = \gamma \vee \delta = \gamma.$$

Id, Taut, Add, Assoc, Sum are correct formulas; but not *Assoc** (and consequently neither are *Add** and *Perm*), since

$$\sim(\gamma \vee \cdot \beta \vee \delta) \vee : \gamma \vee \beta \cdot \vee \delta$$
$$= \sim(\gamma \vee \delta) \vee \cdot \beta \vee \delta$$
$$= \sim\gamma \vee \delta = \delta \vee \delta = \delta.$$

(Received April 7, 1925)

Translator
R. Zach
University of Calgary, Calgary, Canada
e-mail: rzach@ucalgary.ca

Part 3
Alfred Tarski (1928)

Tarski's First Published Contribution to General Metamathematics

Jan Zygmunt
Translated and edited by Robert Purdy

Abstract The paper makes some historical comments on Tarski's first published contribution to the theory of consequence operations (1928).

Keywords Alfred Tarski · Methodology of deductive sciences · General metamathematics · Consequence operation · Lindenbaum's lemma

Mathematics Subject Classification (2000) Primary 01A60 · Secondary 03-03, 08-03—History of mathematics and mathematicians of the 20th century (Mathematical logic and foundations; General algebraic systems) · Primary 03B22—General logic: Abstract deductive systems

1

Remarks on Fundamental Concepts of the Methodology of Mathematics is a summary of Tarski's lecture to the Polish Mathematical Society, Warsaw Section, given on the 14th of December, 1928.[1] It is listed as item [28ᵃb] in Steven Givant's *Bibliography of Alfred Tarski*.[2] It is Tarski's first published contribution to general metamathematics. In it he uses the expresion 'methodology of mathematics' to stake out the field. In subsequent works he calls the field, variously, '(general) methodology of the deductive sciences'; '(general) metamathematics'; and 'metalogic'. The paper is also a harbinger of Tarski's whole body of works on the methodology of deductive sciences, a corpus which starts from [30c], [30d] (with J. Łukasiewicz), and [30e], and goes on to include [35a], [36d], and the books [36ᵐ] and [41ᵐ]. (See also [95].) Since there are several excellent expositions of Tarski's metamathematics available to today's readers,[3] we restrict ourselves here only to a few particular comments.

---

[1] Tarski delivered another talk under a similar title, 'Uwagi o kilku podstawowych pojęciach metamatematyki' ('Remarks on Some Fundamental Concepts of Metamathematics'), to the First Congress of Mathematicians of Slavic Countries in September 1929. (See *Comptes-rendus du I Congrès des Mathématiciens des Payes Slaves, Varsovie 1929*, edited by F. Leja, Książnica Atlas T. N. S. W., Warszawa 1930, where on p. 24 the title of this lecture is listed; the lecture itself was never published.)

[2] Published in *The Journal of Symbolic Logic*, vol. 51 (1986), pp. 913–941. In the present note we shall stick to Givant's two-digit bibliographical designators when we refer to Tarski's works.

[3] To name but a few... 1) W.J. Block and Don Pigozzi, *Alfred Tarski's work on general metamathematics*, *The Journal of Symbolic Logic*, vol. 53 (1988), pp. 36–50; 2) P. Suppes, *Philosophical implications of Tarski's work*, ibid., pp. 80–91; 3) R.L. Vaught, *Tarski's work in model theory*, ibid., vol. 51 (1986),

Tarski's published works up to 1928 were at first on Leśniewski's systems of protothetic and mereology (not surprisingly, as Tarski had been Leśniewski's doctoral student), and then mainly on set theory. These early works on set theory established a number of distinctive results, notably on finite sets, equivalents of the axiom of choice, cardinal and ordinal arithmetic (with Adolf Lindenbaum), and measure theory (with Stefan Banach). But Tarski's interest in the methodology of the deductive sciences had begun even earlier, in 1921, the year of his very first publication. This was a seminar paper titled *Przyczynek do aksjomatyki zbioru dobrze uporządkowanego* (*A Contribution to the Axiomatics of Well-Ordered Sets*);—see [21]. This paper has a clear methodological flavour, as it considers the problem of the independence of various sets of postulates for the notion of a well-ordered set. Moreover, at the beginning of 1921, Tarski, then a third year university student, acquainted himself with a newly published small book by Kazimierz Ajdukiewicz titled *Z metodologii nauk dedukcyjnych* (*From the Methodology of the Deductive Sciences*),[4] and in particular, with Part 1 of this book, an essay titled 'O pojęciu dowodu w znaczeniu logicznym' ('On the Notion of Proof in a Logical Sense').[5] Inspired by it, on April 14th of that same year Tarski delivered a lecture to the Logic Section of the Warsaw Philosophical Institute titled 'O pojęciu dowodu—z powodu rozprawy K. Ajdukiewicza' ('On the Notion of Proof—after an Essay by K. Ajdukiewicz'). In the lecture Tarski formulated the deduction theorem (with respect to the formalism of *Principia Mathematica*) which in its abstract form: *If $X \subseteq S$, $y \in S$ and $x \in Cn(X \cup \{y\})$, then $c(y, x) \in Cn(X)$* was used in [30c] as one of the axioms of "the second group" of a deductive theory.[6] Amost forty years later, in 1960, Ajdukiewicz himself remarked on Part 1 of his own book, that it

> "was the first Polish work in the field of methodology of deductive sciences to have been undertaken from a mathematical-logic perspective. It inaugurated—in Poland at least—the structural method of defining methodological notions (such as, for instance, the notion of proving something, or of one thing's being a con-

---

pp. 869–882; 4) A. Mostowski, *Tarski, Alfred*, **The Encyclopedia of Philosophy**, ed. by P. Edwards, McMillan, London 1967, vol. 8, pp. 77–81; 5) W.A. Pogorzelski and S.J. Surma, review of [56[m]], **The Journal of Symbolic Logic**, vol. 34 (1969), pp. 99–106; and 6) J. Corcoran, *Editor's introduction to the revised edition*, in [56[m]](1), pp. xv–xxvii.

See also J. Czelakowski and G. Malinowski, *Key notions of Tarski's methodology of deductive systems*, **Studia Logica**, vol. 44 (1985), pp. 321–351. [Cf. MR832393 (87f:03083), reviewed by J. Dawson.]

[4]K. Ajdukiewicz, **Z metodologii nauk dedukcyjnych**, Wydawnictwo Polskiego Towarzystwa we Lwowie, vol. 10, Nakładem Polskiego Towarzystwa Filozoficznego, Lwów 1921, 63 pp. [Reviewed by T. Kotarbiński in **Ruch Filozoficzny**, vol. 7 (1922–1923), pp. 11a–13b. Translated into English by J. Giedymin, and edited and preceded by Editor's note by L. Borkowski as *From the methodology of the deductive sciences*, **Studia Logica**, vol. 19 (1966), pp. 9–45. For reviews of this translation see Zbl.0301.02040 and MR0197276 (33#5447); *nota bene* the latter review misinforms in stating that only the first essay of Ajdukiewicz's book is translated.]

[5]Cf. also K. Ajdukiewicz, *Definicja dowodu w znaczeniu logicznym* (A Definition of Proof in a Logical Sense), **Ruch Filozoficzny**, vol. 5, No. 3 (1919), pp. 59b–60a. [Translated into English by J. Giedymin as *A definition of the logical concept of proof*, **Studia Logica**, vol. 19 (1966), p. 46.] This short item is a summary of a lecture Ajdukiewicz gave on the 29th of November 1919 in Lwów. Tarski almost certainly didn't attend this lecture, but he might have read the summary.

[6]For historical comments by Tarski himself, see [30c](1), p. 32, footnote †.

sequence of another), which subsequently played a pivotal role in the marvelous flowering of the science of deductive systems, known eventually as metamathematics".[7]

2

At the centre of Tarski's work in the methodology of deductive sciences is a definition of a theory of consequence operation. The definition is developed in several stages in the papers cited above. This is an axiomatic theory expressed in the language of set theory, and presupposing general set-theoretical principles. Its basic part, usually called a general theory of finitary consequence operations or—in modern terms—of algebraic closure spaces, is only sketched in *Remarks on Fundamental Concepts of the Methodology of Mathematics*. Tarski assumes that the two concepts—of sentence and of consequence—are the only primitive concepts of the theory. Let us denote by the symbol 'S' the set of all sentences, and the set of consequences of the set X of sentences by the symbol '$Cn(X)$'.[8] Then in [28^{a}b] the following axioms (or better, 'elementary properties of Cn', as Tarski calls them) are explicitly stated:

Axiom A. *If $X \subseteq S$, then $X \subseteq Cn(X) \subseteq S$;*

Axiom B. *If $X \subseteq Y \subseteq S$, then $Cn(X) \subseteq Cn(Y)$;*

Axiom C. *If $X \subseteq S$, then $Cn(Cn(X)) \subseteq Cn(X)$;*

Axiom D. *If, for every natural number n, $X_n \subseteq X_{n+1} \subseteq S$, then*

$$Cn\left(\sum_{n=1}^{\infty} X_n\right) = \sum_{n=1}^{\infty} Cn(X_n).$$

In modern terms, Axioms A–D say, respectively, that Cn is an expanding, monotonic, idempotent, and countably continuous function on the power set of S ordered by inclusion.

Using the notion of consequence operation Tarski defines among other things the following terms: 'deductive system', 'consistent system', 'complete system', 'independent system' and 'axiomatizable system'. In his subsequent publications he will replace the expression 'deductive system' by the expression 'closed system' or simply 'system'. At the same time he will use the words 'consistent', 'complete', 'independent', and 'axiomatizable' as pure adjectives.

By narrowing down the use of the word 'system' in this way, he nips a nascent ambiguity in the bud. In the formative terminology of [28^{a}b], a set of sentences may

---

[7]K. Ajdukiewicz, *Język i poznanie*; *Tom I, Wybór pism z lat 1920–1939* (Language and Knowledge; Volume I, Selected Papers from the Years 1920–1939), PWN, Warszawa 1960, see p. v.

[8]In [28^{a}b] Tarski uses the symbol 'X_κ', and in [30c, 30e] and [35a] the symbol '$Fl(X)$' to denote the set of consequences of the set X; in [56^m] the symbol '$Cn(X)$' appears, and it has been commonly used in the logical literature since then.

qualify as a 'consistent system' without being closed under the consequence operation. In later publications, for instance in [30c], if he calls something a 'consistent system', he means to be understood as saying not only that it's consistent but also that it's closed.

In our comments here we employ the modern terms 'Cn-system', 'Cn-consistent', 'Cn-complete', etc., in order to make clear that these notions are always to be understood as relative to some given consequence operation Cn.

Next Tarski lists certain theorems concerning the concepts introduced. Since he gives no proofs, the reader may naturally ask whether the theorems really follow from Axioms A–D. In a moment we will briefly consider this question in connection with Lindenbaum's lemma.

3

As [28ᵃb] clearly contains the germ of both [30c] and [30e]—see for example the bibliographical notes in [56ᵐ] on pages 30 and 60—then let us consider it from the points of view of those two papers.

In [30c] Tarski explicitly sets out the following five axioms as the basis of his theory of consequence operation:

Axiom 1. $\overline{\overline{S}} \leq \aleph_0$;

Axiom 2. *If $X \subseteq S$, then $X \subseteq Cn(X) \subseteq S$;*

Axiom 3. *If $X \subseteq S$, then $Cn(Cn(X)) = Cn(X)$;*

Axiom 4. *If $X \subseteq S$, then $Cn(X) = \sum \{Cn(Y) : Y \subseteq X \text{ and } \overline{\overline{Y}} < \aleph_0\}$;*

Axiom 5. *There exists a sentence $x \in S$ such that $Cn(\{x\}) = S$.*

According to Axiom 4, Cn is a finitary (or compact, or algebraic) operation, while Axiom 5 states that the set S is finitely Cn-axiomatizable by means of one sentence. The finitary character of consequence is mentioned in [28ᵃb] where Tarski explains the intended meaning of the operation Cn—$Cn(X)$ is the smallest set containing X and closed under certain operations, "which are performed on a finite number of elements of the set $S \ldots$".

In [30e] Tarski changes his mind about Axiom 5, demotes it to an *ad-hoc* posit, and weakens it, to the effect that *the set S of all sentences is finitely Cn-axiomatizable*, rather than Cn-axiomatizable by a single sentence. In [56ᵐ], on page 92, he explains that although the erstwhile axiom was satisfied by all then-known formalized disciplines "it does not seem desirable to include this formula among the axioms, on account of its special and, in a certain sense, accidental character". Instead he posits it only as needed for establishing specific results (e.g., Theorems 48, 49, 51, 56 (Lindenbaum's lemma), and 57), and in these cases he states the dependency explicitly.

1. It is well known that Axioms 2–5 suffice to prove[9] the following formulation of Lindenbaum's lemma (see also [30c], Theorem 12): *If X is a Cn-consistent set of sen-*

---

[9] In ZFC set theory. Having Axiom 1, Tarski can avoid the Axiom of Choice.

tences, then there is a set of sentences Y *such that* $X \subseteq Y$ *and* Y *is a Cn-consistent and Cn-complete Cn-system* (or simply, *every consistent set of sentences can be extended to a consistent and complete system*). In [30e] Tarski stresses in his very formulation of the lemma that its proof requires positing the erstwhile Axiom 5, albeit in its weakened form (see Theorem I. 56, p. 394, or [30e](1), p. 98). Interestingly, only in the second edition of [56^m], which appears in the year of his death, 1983, does Tarski finally add essentially the same comment on Lindenbaum's lemma retrospectively to [30c] (see [56^m](1), p. 34, footnote):

> "This theorem was originally established by Lindenbaum for incomplete systems of sentential calculus. When his proof was subsequently reconstructed within a more general framework, in fact on the basis of the first group of our axioms [i.e., Axioms 1–4], we have observed that, to assure the validity of the proof, it is necessary to insert in the axiom set a statement of a more special character than all the remaining axioms, viz. Ax. 5. Actually, a weaker statement to the effect that there is a finite set $X \subseteq S$ for which $Cn(X) = S$ would suffice for this purpose. It may be noticed that, on the basis of our full axiom set including both groups of axioms, Ax. 5 can be derived from other axioms and can therefore be omitted."[10]

2. That Axioms 1–4, and hence Axioms A–D, do not suffice to prove the Lindenbaum's lemma was first demonstrated in print by Tadeusz Kubiński.[11] He exhibited a counterexample consisting of a countably infinite set S and a finitary consequence operation Cn on S such that: (i) no Cn-consistent Cn-system may be extended to a Cn-consistent Cn-complete Cn-system; and (ii) there exists a non-empty monotonic family of Cn-consistent Cn-systems whose set-theoretic union is not Cn-consistent.

3. Let us consider the following two propositions closely related to Axiom D:

D_1 *The family of Cn-systems is closed under unions of denumerable chains.*

D_2 *The family of Cn-systems is closed under unions of upward directed families of sets.*

In [28^{a}b] proposition D_1 is presented as a theorem. In fact one can go further, and state that in the presence of Axioms A–C, Axiom D is equivalent to proposition D_1. But not to proposition D_2. On the strength of Axioms A–C alone, D_1 fails to imply D_2. To see this, let S be an uncountable (sic!) set, and let $Cn(X) = X$ if X is a countable subset of S, and $Cn(X) = S$ otherwise. However, if we assume S to be countable (Axiom 1), then on the basis of Axioms A–C, proposition D_2 and Axiom D are equivalent. In [30e] (see Theorem I. 12) it is proved that D_2 follows from Axioms 1–4.

In the light of these remarks a historian of logic may wonder how far was Tarski in the 1930s from the following theorem: *The system of Axioms A–C plus D_2 is equivalent to*

---

[10]Tarski's axioms of the second group, mentioned in the last sentence, are Axioms 6*–10* in [56^m](1), pp. 31–32. They describe mutual connections between the consequence operation and the propositional connectives of negation and implication.

[11]See T. Kubiński, *O zasięgu twierdzenia Lindenbauma o nadsystemach zupełnych* (On the scope of Lindenbaum's theorem on complete supersystems), **Studia Logica**, vol. 12 (1961), pp. 83–96. [MR0153588 (27#3551), reviewed by S. Jaśkowski.]

the system of Axioms 2–4, or in a different wording, *if S is an arbitrary set and Cn is a closure operator on S* (*i.e. Cn satisfies Axioms* A, B, C), *then Cn is finitary* (*i.e. satisfies Axiom* 4) *if and only if Cn satisfies* D_2.[12]

4

In conclusion, we think the following comment by Jerzy Łoś,[13] heretofore not widely known, may interest the reader...

"It is fair to say that logicians of the time[14] knew most of the theorems in these papers (with the possible exception of *Grundzüge*),[15] or at least were becoming aware of them. So here I draw attention to one, in my opinion very important, aspect of these works. In logic we have syntax and semantics. Implication[16] and everything pertaining to it belongs to syntax, that was the fundamental rule of modern logic. Drawing conclusions does not depend on the content but only on the form of a proposition, only structural rules are permissible. Materialists attacked this view, but their aversion to formalizing propositions lay in a failure to see that the form of propositions and statements is precisely one of their material attributes, that in addressing the form of expressions we are dealing precisely with that material substrate of our thought processes which takes only the simplest of sense acts to apprehend. So one way or another, implication is syntax. And thus in these papers Tarski, as it were, bumped the theory of consequence up a notch. He showed that one can use it independently of a concrete language, that it has its own autonomous laws and properties. It can be said that Tarski algebraized that part of syntax pertaining to implication. He algebraized it to such an extent that syntactic properties of sentences[17] became irrelevant, so far-reaching was his abstraction, yet as is often the way with abstractions, by so doing made matters extremely straightforward and understandable. On the other hand, algebraization doesn't work at all when turning from syntax to semantics. The latter has to be grounded in the detailed syntax of a language. This is already evident in the first chapters of Tarski's great work, *Pojęcie prawdy w językach nauk dedukcyjnych* (The Concept of Truth in the Languages of the Deductive Sciences)."

---

[12]This is known as Jürgen Schmidt's theorem; see J. Schmidt, *Über die Rolle der transfiniten Schlußweisen in einer allgemeinen Idealtheorie*, **Mathematische Nachrichten**, vol. 7 (1952), pp. 165–182. [MR0047628 (13, 904b), reviewed by H.B. Curry; Zbl. 049.166, reviewed by J. Riguet.]

[13]J. Łoś, *O Alfredzie Tarskim* (On Alfred Tarski), **Ruch Filozoficzny**, vol. 43 (1986), pp. 3–10.

[14]By "logicians of the time" he means logicians of the 1920s and the 1930s.

[15]By "these papers" he means [30c], [30e], [35a] and [36d] in the bibliography below.

[16]The English word "implication" comes closest to capturing Łoś's meaning here. The Polish original uses the word "konsekwencja".

[17]By "sentences" he means sentences in concrete languages.

Cited works of Tarski, in chronological order

[21]   *Przyczynek do aksjomatyki zbioru dobrze uporządkowanego* (A contribution to the axiomatics of well-ordered sets), **Przegląd Filozoficzny**, vol. 24 (1921), pp. 85–94.

[28ᵃb]   *Remarques sur les notions fondamentales de la Méthodologie*[18] *des Mathématiques*, **Rocznik Polskiego Towarzystwa Matematycznego** (= *Annales de la Société Polonaise de Mathématique*), vol. 7 (1928—published 1929), pp. 270–272.

[30c]   *Über einige fundamentalen Begriffe der Metamathematik*, **Sprawozdania z posiedzeń Towarzystwa Naukowego Warszawskiego, Wydział III nauk matematyczno-fizycznych** (= *Comptes Rendus des séances de la Société des Sciences et des Lettres de Varsovie, Classe III*), vol. 23 (1930), pp. 22–29. (See [30e] for a more detailed exposition.) [Abstract [28ᵃb]; JFM 57.1318.03, reviewed by W. Ackermann.]

   (1)   *On some fundamental concepts of metamathematics*, in [56ᵐ], pp. 30–37. (Revised English translation of [30c].)

[30d]   *Untersuchungen über den Aussagenkalkül* (co-author: J. Łukasiewicz), **Sprawozdania z posiedzeń Towarzystwa Naukowego Warszawskiego, Wydział III nauk matematyczno-fizycznych** (= *Comptes Rendus des séances de la Société des Sciences et des Lettres de Varsovie, Classe III*), vol. 23 (1930), pp. 30–50. [JFM 57.1319.01, reviewed by W. Ackermann.]

   (1)   *Investigations into the sentential calculus*, in: [56ᵐ], pp. 38–59. (Revised English translation of [30d].)

[30e]   *Fundamentale Begriffe der Methodologie der deduktiven Wissenschaften. I*, **Monatshefte für Mathematik und Physik**, vol. 37 (1930), pp. 361–404. (See [30c].) [Abstrakt [28ᵃb]; JFM 56.0046.02, reviewed by R.(sic!) Skolem).]

   (1)   *Fundamental concepts of the methodology of the deductive sciences*, in [56ᵐ], pp. 60–109. (Revised English translation of [30e].)

[33ᵐ]   **Pojęcie prawdy w językach nauk dedukcyjnych** (The Concept of Truth in the Languages of the Deductive Sciences), Prace Towarzystwa Naukowego Warszawskiego, Wydział III Nauk Matematyczno-fizycznych (= Travaux de la Société des Sciences et des Lettres de Varsovie, Classe III Sciences Mathématiques et Physiques), no. 34, Warsaw, 1933, vii + 116 pp. (See [56ᵐ] for English translation.)

[35a]   *Grundzüge des Systemenkalküls. Erster Teil*, **Fundamenta Mathematicae**, vol. 25 (1935), pp. 503–526. [JFM 62.0038.01, reviewed by T. Skolem; Zbl. 0014.38701, reviewed by A. Schmidt.]

   (1)   *Foundations of the calculus of systems*, in [56ᵐ], pp. 342–383. (Revised English translation of [35a] and [36d].)

---

[18]"Méthologie" in the original, which is obviously a misprint.

[36d] *Grundzüge des Systemenkalküls. Zweiter Teil*, **Fundamenta Mathematicae**, vol.
       26 (1936), pp. 283–301. [JFM 62.0038.02, reviewed by T. Skolem; JSL 1, pp.
       71–72, reviewed by W.V. Quine.]

[36ᵐ] *O logice matematycznaej i metodzie dedukcyjnej* (On mathematical logic and the
       deductive method), Bibljoteczka matematyczna 3–5, Książnica-Atlas, Lwów and
       Warszawa, 1936, 167 pp.

    (1)   *Einführung in die mathematische Logik und in die Methodologie der*
             Mathematik, Julius Springer Verlag, Vienna, 1937, x + 166 pp. (Ger-
             man translation of [36ᵐ].) [JFM 63.0022.01, reviewed by W. Ackermann;
             Zbl. 0018.00101, reviewed by H.B. Curry; JSL 3, pp. 51–52, reviewed by
             S. MacLane.]

[41ᵐ] **Introduction to Logic and to the Methodology of Deductive Sciences**, Oxford
       University Press, Oxford and New York, 1941, xviii + 239 pp. (Enlarged and re-
       vised English translation of [36m] (1) by O. Helmer.) [JFM 67.0033.03; MR 2, p.
       209; JSL 6, pp. 30–32.]

[56ᵐ] **Logic, Semantics, Metamathematics. Papers from 1923 to 1938**, Clarendon
       Press, Oxford 1956, xiv + 471 pp. (English translation of [30c, 30e, 35a, 36d] and
       several other of Tarski's papers by J.H. Woodger.) [MR 17, p. 1171 (registered);
       JSL 34, 99–106, reviewed by W.A. Pogorzelski and S.J. Surma.]

    (1)   Second, revised edition, with editor's introduction and an analytic index
             (J. Corcoran, editor), Hackett Publishing Company, Indianapolis, Indiana
             1983, xxx + 506 pp. [MR736686 (85e: 01065), reviewed by E. Mendelson;
             JSL 54, 281–282, reviewed by I. Grattan-Guiness.]

[95]   *Some current problems in metamathematics*, edited by J. Tarski and J. Woleński,
       History and Philosophy of Logic, vol. 16 (1995), pp. 159–168. [An edited ver-
       sion of the text of a lecture given by Tarski at Harvard University in the 1939/40
       academic year. MR1468343 (98g:01055), reviewed by I. Angelelli.]

J. Zygmunt (✉)
Katedra Logiki i Metodologii Nauk, Uniwersytet Wrocławski, ul. Koszarowa 3/20,
51-149 Wrocław, Poland
e-mail: Logika@uni.wroc.pl

Remarks on Fundamental Concepts of the Methodology of Mathematics

Alfred Tarski

Translators' Note: This article forms part of the record of proceedings of a meeting of the Polish Mathematical Society which convened in Warsaw on December 14th, 1928, the other part being an abstract of a talk given by Kazimierz Kuratowski at the same meeting. Though the speakers write these abstracts themselves, the conventional format of such records requires that speakers be referred to in the third person. Thus Tarski writes, of himself, that: "Monsieur T. les appelle"; "Monsieur T. signale"; "Monsieur T. illustre". The translators have chosen to keep the narrator's voice in the third person.

R. Purdy and J. Zygmunt

The subject-matter of the Methodology of Mathematics (also called "Metamathematics", following the Hilbert school) are mathematical theories in much the same sense in which numbers are the subject-matter of arithmetic and figures of geometry. From the standpoint of this methodology mathematical theories are sets of expressions or inscriptions constructed according to certain rules of procedure. Tarski calls these expressions *meaningful sentences* (a term introduced by Leśniewski) and denotes their set by S; it is not possible to make precise the notion "meaningful sentence" unless one chooses as one's object of study a well-determined mathematical theory.

To each set X of such sentences there corresponds another set X_K called *the set of consequences of the set X*. It is defined as the smallest set containing X and closed under certain operations; again, how to make precise these operations depends on specific properties of the envisaged theory. In any case, these are operations (not necessary univocal) which are performed on a finite number of elements of the set S and giving as a result also elements of S.

Tarski indicates several elementary properties of the concept X_K:

$X \subset S$ implies $X \subset X_K \subset S$;

$X \subset Y$ implies $X_K \subset Y_K$;

for all X, $X_{KK} = X_K$;

Translated by R. Purdy & J. Zygmunt, with kind permission of his son, Jan Tarski, from: Tarski, A., "Remarques sur les notions fondamentales de la Méthodologie des Mathématique". Annales de la Société Polonaise de Mathématique, Vol. VII, pp. 270–272, 1929

if, for every natural number n, $X_n \subset X_{n+1}$, then $\left(\sum\limits_{n=1}^{\infty} X_n\right)_{\kappa} = \sum\limits_{n=1}^{\infty} (X_n)_{\kappa}$.

On the basis of these two primitive concepts, namely, (1) meaningful sentence, and (2) the consequence-set of a set of meaningful sentences, a series of other concepts important for the Methodology of Mathematics can be defined. Thus one calls a *deductive system* every set X of meaningful sentences which satisfies the condition: $X = X_{\kappa}$. The set X of meaningful sentences is said to be a *consistent system* if $X_{\kappa} \neq S$; it is called a *complete system* if $X \subset Y \subset S$ implies $X_{\kappa} = Y_{\kappa}$ or $Y_{\kappa} = S$; it is said to be an *independent system* (or system of independent sentences) if the conditions $Y \subset X$ and $X_{\kappa} = Y_{\kappa}$ imply the equality $X = Y$. One calls a *basis* of a set X every independent system of sentences Y such that $X_{\kappa} = Y_{\kappa}$; every set X which admits a finite basis is known as an *axiomatizable system*.

Next, Tarski establishes a series of theorems concerning the concepts introduced above. Here are a few examples of these theorems:

If for every natural number n, X_n is a deductive system (or a consistent system) and $X_n \subset X_{n+1}$, then $\sum_{n=1}^{\infty} X_n$ is also a deductive system (or a consistent system).

For every consistent system X there exists a consistent and complete system Y such that $X \subset Y$ (Lindenbaum's lemma).

If X admits an infinite basis then it admits no finite basis and consequently it is not an axiomatizable system.

Tarski illustrates his considerations by a concrete example of the simplest mathematical theory, namely by the so called *Theory of deduction*. He shows in particular that the class of all consistent and complete deductive systems, formed of meaningful sentences in the sense of the Theory of deduction, is of power $2^{\aleph_0}$ (while Lindenbaum had demonstrated that the set of all consistent deductive systems, complete or incomplete, is of that power).

Translators
R. Purdy
Toronto, Canada

J. Zygmunt
University of Wrocław, Wrocław, Poland
e-mail: Logika@uni.wroc.pl

Part 4
Kurt Gödel (1933)

On Gödel's Modal Interpretation
of the Intuitionistic Logic

Itala M. Loffredo D'Ottaviano and Hércules de Araújo Feitosa

Abstract The aim of this paper is to analyse Gödel's paper "Eine Interpretation des intuitionistischen Aussagenkalküls", presented in 1932 at the Mathematical Colloquium in Vienna and published in 1933. This paper presents an interpretation from the intuitionistic propositional logic into a certain modal expansion $\mathcal{G}$ of the classical propositional logic introduced by Gödel. We discuss Gödel's results and several known important extensions of his conjectures. We also analyse a second paper presented at the Vienna Mathematical Colloquium in 1932 and published in 1933, in which Gödel introduces an interpretation from classical propositional logic into intuitionistic propositional logic, that he extends to the corresponding arithmetics in order to prove the relative consistency of one relative to the other. We emphasize the originality and relevance of Gödel's results and their meaningful extensions, and analyse them under the scope of the study of interrelations between logical systems through translations between them.

Keywords Gödel's interpretations between logics · Intuitionistic logic · Modal logic · Classical logic · Relative consistency of arithmetic · Extended results · Translations between logics · Conservative translations · Contextual translations

Mathematics Subject Classification (2000) 03A05 · 03B22

1 Introduction

Among several papers published by Kurt Gödel in 1933, two are of special interest for the study of interrelations between logical systems.

Gödel participated actively in the Mathematical Colloquium in Vienna chaired by Karl Menger, his former teacher. In 1932, Gödel presented at the Colloquium two relevant results, that would become important references for several posterior studies and results in the logic literature.

At one session, Gödel described an interpretation of the intuitionistic propositional logic **IPC** into a certain modal expansion $\mathcal{G}$ of the classical propositional logic, showing that the notions and formulas of **IPC** could be "translated" into $\mathcal{G}$, such that if a formula φ is provable in **IPC** then its "translation" φ' is provable in $\mathcal{G}$. Gödel also conjectured the converse, that an intuitionistic formula φ should be provable in the intuitionistic propositional calculus **IPC** if and only if its "translation" were provable in the modal expanded classical calculus $\mathcal{G}$. Gödel's conjecture was later proved to be true.

J.-Y. Béziau (ed.), *Universal Logic: An Anthology*, 71–88
Studies in Universal Logic, DOI 10.1007/978-3-0346-0145-0_7, © Springer Basel AG 2012

On June 28th, during another session of the Colloquium, Gödel showed that if φ is a classically valid connective formula built with the connectives $\neg$ and $\wedge$, then the direct "translation" (that is, $\neg$ by – and $\wedge$ by Δ) of φ is an intuitionistically valid formula; as the other classical connectives $\vee$, $\rightarrow$, $\leftrightarrow$ are definable from $\neg$ and $\wedge$, it results that the set of all valid formulas of the propositional classical logic according to such translation is a subset of the set of the valid formulas of the propositional intuitionistic logic. In addition, Gödel extended his translation from the Herbrand [36] formal classical arithmetic into the Heyting [37] formal intuitionistic arithmetic and proved that the formal classical arithmetic can be included into the intuitionistic arithmetic, presenting a proof of the relative consistency of the classical arithmetic (and classical logic) relative to intuitionistic arithmetic.

Gödel's talks were published in 1933 in *Ergebnisse eines mathematischen Kolloquiums*, number 4 (1931–32), edited by K. Menger [25, 26].

The first talk was published as a short note, under the title "Eine Interpretation des intuitionistischen Aussagenkalküls (An interpretation of the intuitionistic propositional calculus)", pp. 39–40. The original text was translated into English as 'An interpretation of the intuitionistic sentencial logic' in *The philosophy of mathematics*, pp. 128–129, edited by Hintikka in 1969 [40]; and was reprinted in 1971, by Berka and Kreiser [2], in *Logic-Texte. Kommentierte Auswahl zur Geschichte der modern Logic*, pp. 187–188.

The second one was published in the same volume, under the title "Zur intuitionistischen Arithmetik und Zahlentheorie (On intuitionistic arithmetic and number theory)", pp. 34–38. There is a translation into English by Martin Davis in [10, pp. 75–81].

The first edition in any language of Gödel's complete works is the Spanish translation, *Kurt Gödel: Obras Completas* (*Kurt Gödel: Complete Works*), edited by Jesús Mosterín in 1981 and published by Alianza Editorial, Spain [50]. Aside from preparing the edition, Mosterín wrote a special introductory note before each one of the papers and personally translated from German into Spanish the two papers mentioned above. A review by Rolando Chuaqui of *Kurt Gödel: Obras Completas* was published in 1983 in *The Journal of Symbolic Logic*, number 48, pp. 1199–1201 [7]. The second edition of the book was published in 1989, and in the third edition of 2006 the two papers appear on pp. 136–137 and 126–135, respectively.

In 1986, the book *Kurt Gödel: Collected Works. Volume I: Publications 1929–1936* [29] was edited by S. Feferman (Editor-in-chief), J.W. Dawson Jr., S. Kleene, G. Moore, R. Solovay and J. van Heijenoort, under the auspices of the Association for Symbolic Logic. This volume brings together Gödel's original texts and presents a bilingual version of "Eine Interpretation des intuitionistischen Aussagenkalküls" (the original German paper and a translation into English [28, pp. 300–303]), with an "Introductory note" by A.S. Troelstra, pp. 296–299; and also a bilingual version of "Zur intuitionistischen Arithmetik und Zahlentheorie" [27, pp. 286–295], with an "Introductory note" by A.S. Troelstra, pp. 282–285.

In spite of the originality and relevance of Gödel's results, his interpretations involving intuitionistic logic, modal logic and classical logic are not the first known results in the literature concerning the study of inter-relations between logical systems through the study of translations between them. The first known "translations" involving classical, intuitionistic and modal logic were presented by Kolmogorov [44], Glivenko [24] and Lewis and Langford [47]. Gentzen also published a paper on the subject in 1933 [20], the same year Gödel's results were published.

Kolmogorov's aim "is to explain why" the illegitimate use of the principle of excluded middle in the domain of transfinite arguments "has not yet led to contradictions". He introduces the intuitionistic formal logic **B** and the classical propositional calculus **H**, based on Hilbert system of 1923 [39], and defines inductively a function K that associates to every formula α of **H** a formula α^k of **B** by adding a double negation in front of every subformula of α. It is then proven that, given a set of axioms $A = \{\alpha_1, \ldots, \alpha_n\}$,

$$A, \mathbf{H} \vdash \alpha \quad \text{implies} \quad A^k, \mathbf{B} \vdash \alpha^k,$$

where $A^k = \{\alpha_1^k, \ldots, \alpha_n^k\}$. Kolmogorov suggests that a similar result can be extended to quantificational systems and, in general, to all known mathematics, anticipating Gödel's and Gentzen's results on the relative consistency of classical arithmetic with respect to intuitionistic arithmetic (see [18, 19] and [6]).

Also related to intuitionism is [24], where Glivenko proves that, if α is a theorem of classical propositional logic **CPL**, then the double negation of α is a theorem of intuitionistic propositional logic **IPL**.

In "Zur intuitionistischen Arithmetik und Zahlentheorie", Gödel uses Glivenko [24] results. However, apparently not aware of Kolmogorov [44], he does not mention Kolmogorov's paper.

In "Eine Interpretation des intuitionistischen Aussagenkalküls" [26], Gödel only mentions Kolmogorov's paper in a footnote, and interprets it incorrectly.

Kolmogorov [44] was published in Russian and his results were not published in any other languages, until its English translation, "On the principle of excluded middle", appeared in *From Frege to Gödel*, edited by van Heijenoort in 1967 (corrected third printing, 1977) [45, 55]. By the time Kolmogorov's paper finally became widely known, his results had been superseded by Gödel's and Gentzen's results.

Our aim in this paper is to analyse and discuss Gödel's "Eine Interpretation des intuitionistischen Aussagenkalküls" [26], which is among the selected collection of papers of this book.

We will analyse the "interpretation" presented by Gödel and will discuss some posterior results and extensions by several other authors concerning Gödel's conjectures and results.

We will also briefly present a general discussion of Gödel's "Zur intuitionistischen Arithmetik und Zahlentheorie" [25], aiming at analysing both of Gödel's papers as forerunners of the study of interrelations between logical systems through translations between them, that is, from a universal logic approach.

2 "Eine Interpretation des intuitionistischen Aussagenkalküls": Gödel's "Interpretation" from the Intuitionistic Propositional Calculus into the Modal System $\mathcal{G}(S_4)$

In this short paper of 1933 [26], Gödel interprets the intuitionistic propositional calculus **IPC** (as introduced by Heyting [37]) into the modal system $\mathcal{G}$.

The classical propositional logic **CPL** is enriched with an additional unary operator B, that may be interpreted either epistemically or as a modal operator, and with three axioms and an inference rule for B. The letter B stands for "beweisbar", that is, "Bp" can be read as "p is provable", "provable by any correct means", an intuitive interpretation; the operator B must not be interpreted as "provable in a given formal system", and if Bp is understood as "p is necessary" the expanded system results as the Lewis modal system S_4, with B written for the necessity operator $\Box$ (or N) [47]).

Hence, Gödel's result shows that there is an immersion of the intuitionistic propositional logic into the modal logic S_4, or into the corresponding epistemic logic. It thus can be expressed by:

$$\text{If } \mathbf{IPC} \vdash \varphi, \quad \text{then } \mathbf{S}_4 \vdash \varphi',$$

where φ' is Gödel's "translation" of φ.

Following Gödel's conjecture in the paper, we can claim that

$$\mathbf{IPC} \vdash \varphi \quad \text{if, and only if, } \mathbf{S}_4 \vdash \varphi'.$$

2.1 The System $\mathcal{G}$

The primitive "notions" of the system **IPC** are $\neg, \supset, \vee$ and $\wedge$.

Gödel introduces the system $\mathcal{G}$ by adding to the primitive notions $\sim, \rightarrow, \vee$ and $\bullet$, and to the axioms and rules of inference of the classical propositional calculus, the notion "p is provable", denoted by Bp, and the following three new axioms and a special inference rule:

$$\text{Axiom} \quad \mathcal{G}_1 : Bp \rightarrow P$$

$$\text{Axiom} \quad \mathcal{G}_2 : Bp \rightarrow B((p \rightarrow q) \rightarrow Bq)$$

$$\text{Axiom} \quad \mathcal{G}_3 : Bp \rightarrow BBp$$

$$\text{Rule} \quad \mathcal{G} : \varphi/B\varphi.$$

Gödel then defines his "translation" Gd_1 from **IPC** into $\mathcal{G}$:

$$Gd_1 : \mathbf{IPC} \rightarrow \mathcal{G}^1$$

$$(p)^{Gd_1} =_{df} p$$

$$(\neg p)^{Gd_1} =_{df} \sim Bp$$

$$(p \supset q)^{Gd_1} =_{df} Bp \rightarrow Bq$$

$$(p \vee q)^{Gd_1} =_{df} Bp \vee Bq$$

$$(p \wedge q)^{Gd_1} =_{df} p \bullet q.$$

---

[1]Gödel does not use any special denotation for his interpretation function.

We can naturally obtain the following "translation" from the set of formulas of the language of **IPC** into the formulas of $\mathcal{G}$:

$$(p)^{Gd_1} = p$$

$$(\neg\varphi)^{Gd_1} =\sim B(\varphi^{Gd_1})$$

$$(\varphi \supset \psi)^{Gd_1} = B(\varphi^{Gd_1}) \to B(\psi^{Gd_1})$$

$$(\varphi \vee \psi)^{Gd_1} = B(\varphi^{Gd_1}) \vee B(\psi^{Gd_1})$$

$$(\varphi \wedge \psi)^{Gd_1} = \varphi^{Gd_1} \bullet \psi^{Gd_1}.$$

Gödel observes, without proving, that:

(i) We could obtain another "translation" by defining

$$(\neg p)^{Gd} =_{df} B(\sim B(p))$$

$$(p \wedge q)^{Gd} =_{df} Bp \wedge Bq.$$

(ii) The "translation" of $(p \vee \neg p)$, $(p \vee \neg p)^{Gd_1}$, is not derivable in $\mathcal{G}$ and, in general, for any formulas φ and ψ, if $B\varphi$ and $B\psi$ are not provable in $\mathcal{G}$ then $B\varphi \vee B\psi$ is not provable in $\mathcal{G}$.

Gödel then claims that if a formula is provable in the intuitionistic logic, then its "translation" is derivable in $\mathcal{G}$, that is:

$$\text{If } \textbf{IPC} \vdash \varphi, \quad \text{then } \mathcal{G} \vdash \varphi^{Gd_1}.$$

He conjectures that the converse also holds, and thus we should have:

$$\textbf{IPC} \vdash \varphi \quad \text{if, and only if, } \mathcal{G} \vdash \varphi^{Gd_1}.$$

Gödel's justification for why the operator B can not be interpreted as "is provable in a formal system", is that this would collapse with the second incompleteness theorem if φ is the sentence that formalizes the consistency of first-order arithmetics.

Gödel observes, mentioning Parry [51], that his system $\mathcal{G}$ is equivalent to Lewis's system of strict implication if Bp is "translated" by "necessary p" ($N(p)$ or, equivalently, $\Box p$) and we add to the Lewis axioms the additional axiom of Becker [1, p. 497]:

$$\textit{Becker Axiom:} \quad N\varphi \to NN\varphi.$$

Hence, by interpreting B as the modal operator $\Box$ (for necessity), Gödel's system $\mathcal{G}$ results in the Lewis modal system $\textbf{S}_4$ (see [47]).

As noted above, Gödel mentions Kolmogorov's paper in a footnote: "Kolmogorov (1932) has given a some what different interpretation of the intuitionistic propositional calculus, without, to be sure, specifying a precise formalism". In fact, Kolmogorov's paper [44] published in 1925, presents for the first time in the literature an axiomatization of Brouwer's intuitionistic logic, anticipating Heyting's axiomatization published in 1930. Kolmogorov's system **B** is the same system posteriorly introduced in 1936 by Johansson [42], and known as *Johansson's minimal logic*.

2.2 On the Converse of Gödel's Result

Gödel's conjecture concerning the converse of his result was proved by J.C.C. McKinsey and A. Tarski [49], in a rigorous and elucidative text. These authors present a strict implication propositional calculus and prove its equivalence to Lewis's system S_4 (with the modal operators $\square$ and $\Diamond$). It is verified that the closure algebra is an adequate (sound and complete) model for S_4.

In addition, the Heyting propositional calculus **IPC** is studied, and the authors prove that the Heyting algebra (called Brouwer algebra in the paper) is an adequate semantics for the system **IPC**.

Finally, based on a certain duality between the closure algebras and the Heyting algebras, McKinsey and Tarski define three functions from **IPC** into S_4, the first one being:

$$TM : \textbf{IPC} \to S_4$$

$$(p_i)^{TM} =_{df} \square p_i$$

$$(\varphi \vee \psi)^{TM} =_{df} \varphi^{TM} \vee \psi^{TM}$$

$$(\varphi \wedge \psi)^{TM} =_{df} \varphi^{TM} \bullet \psi^{TM}$$

$$(\varphi \supset \psi)^{TM} =_{df} \square(\varphi^{TM} \to \psi^{TM})$$

$$(\neg\varphi)^{TM} =_{df} \sim \Diamond\varphi^{TM} (\square \sim \varphi^{TM}).$$

Gödel's conjecture is then proved.

Theorem 1 **IPC** $\vdash \varphi$ *if, and only if,* $S_4 \vdash \varphi^{TM}$.

Observe that there is a relationship between the interpretations Gd_1 and TM:

$$S_4 \vdash \varphi^{TM} \leftrightarrow \square\varphi^{Gd_1}.$$

According to Hacking [33], one can also weaken S_4, namely to S_3, in order to prove Gödel's conjecture. Hacking's proof is based on cut-elimination.

Several further results are proved in McKinsey and Tarski [49], in particular the remarkable following theorem.

Theorem 2 *If* $S_4 \vdash \square\varphi \vee \square\psi$, *then* $S_4 \vdash \square\varphi$ *or* $S_4 \vdash \square\psi$.

By means of the proved Gödel's conjecture, this property of S_4 implies the disjunction property of **IPC**, proved by Gentzen [21] using cut-elimination.

Theorem 3 *If* **IPC** $\vdash \varphi \vee \psi$, *then* **IPC** $\vdash \varphi$ *or* **IPC** $\vdash \psi$.

In the same paper, McKinsey and Tarski analyse some extensions of Lewis's system, with special attention to the system S_5, classified as a normal extension of S_4.

Rasiowa and Sikorski [53], by using algebraic semantics, extended Tarski and McKinsey's result to the first-order predicate calculus. After presenting algebraic semantics, in the Lindenbaum style, for various logical systems, the authors obtain the extended result by using the following function from Heyting predicate calculus **IQC** into Lewis predicate calculus **QS$_4$**:

$$RS : \mathbf{IQC} \rightarrow \mathbf{QS_4}$$

$$(F_m^k(x_{i_1}, \ldots, x_{i_k}))^{RS} =_{df} \square(F_m^k(x_{i_1}, \ldots, x_{i_k}))$$

$$(\varphi \vee \psi)^{RS} =_{df} \varphi^{RS} \vee \psi^{RS}$$

$$(\varphi \wedge \psi)^{RS} =_{df} \varphi^{RS} \bullet \psi^{RS}$$

$$(\varphi \supset \psi)^{RS} =_{df} \square(\sim \varphi^{RS} \vee \psi^{RS})$$

$$(\neg\varphi)^{RS} =_{df} \square(\sim \varphi^{RS})$$

$$(\exists x\varphi)^{RS} =_{df} \exists x\varphi^{RS}$$

$$(\forall x\varphi)^{RS} =_{df} \square\forall x\varphi^{RS}.$$

The same result can be obtained by extending Gödel's function

$$Gd_1 : (\exists x\varphi)^{RS} =_{df} \exists x\square\varphi^{RS}$$

$$(\forall x\varphi)^{RS} =_{df} \forall x\varphi^{RS}.$$

The result proved by Rasiowa and Sikorski was independently obtained by Maehara [48], using cut-elimination.

Prawitz and Malmnäs [52] proved the same result using normalization for adequate natural deduction systems for **IQC** and **QS$_4$**.

Due to the above mentioned symmetry between the closure algebras and the Heyting (Brouwer) algebras, a question that naturally appears is whether there is an interpretation from S$_4$ into any other modal system, or into the intuitionistic system. At present the answer to such a question is still only a partial one, for there are known functions from a modal system into classical arithmetic and into intuitionistic arithmetic, and not only into the underlying logics.

2.3 Other Extensions of Gödel's Result

The first result in the latter sense, extending Rasiowa and Sikorski's result, is given by Solovay [54], using proof-theoretical methods. Solovay defines the following function from S$_4$ into Peano classical arithmetic **PA**:

$$S : \mathbf{S_4} \rightarrow \mathbf{PA}$$

$$(p_i)^S =_{df} p_i$$

$$(\bot)^S =_{df} \bot$$

S commutes with $\sim, \vee, \bullet$ and $\rightarrow$

$$(\square\varphi)^S =_{df} Bew(\ulcorner\varphi^S\urcorner),$$

where $\ulcorner \varphi \urcorner$ denotes the numeral of the Gödel number of φ, Bew is the canonically defined predicate expressing arithmetized provability (Prov) in **PA** and $\ulcorner k \urcorner$ is the Gödel number of the sentence φ^S; so $Bew(\ulcorner k \urcorner)$ is the formula expressing that k is the Gödel number of a theorem of **PA**. The following theorem in then proved.

Theorem 4 $\mathbf{S_4} \vdash \varphi$ *if, and only if,* $\mathbf{PA} \vdash \varphi^S$.

Goldblatt [30] indicates a problem in Solovay's result. According to Solovay's interpretation, it is not the case that every translation $(\Box\varphi \rightarrow \varphi)^S$ of the formula $(\Box\varphi \rightarrow \varphi)$ of $\mathbf{S_4}$ is a theorem of **PA**, for it is not the case that $\mathbf{PA} \vdash Bew(\ulcorner\varphi\urcorner)^S \urcorner \rightarrow \varphi^S$ (this occurs only when $\mathbf{PA} \vdash \varphi^S$, but by the incompleteness of **PA** it is known that there are true sentences of arithmetic that are not theorems). Hence the scheme $\Box\varphi \rightarrow \varphi$ would not be valid in $\mathbf{S_4}$.

Goldblatt suggests a slight modification in Solovay's interpretation, in order to maintain valid the above scheme by defining

$$(\Box\varphi)^{GS} =_{df} \varphi \bullet Bew(\ulcorner\varphi\urcorner),$$

with the meaning "φ is true and provable in **PA**".

Let us consider Solovay's original interpretations, but defining

$$(p_i)^S =_{df} \emptyset_i, \quad \text{with } \emptyset_i \text{ a sentence of } \mathbf{PA},$$

$$\text{for } p_i \text{ a proposition letter of the language of } \mathbf{S_4}.$$

Write $\vDash^* \varphi$ if we have $\mathbf{PA} \vdash (\varphi)^S$ for all possible interpretations S. Let $\mathbf{G}$ be the modal system containing all classical tautologies, modus ponens, the necessitation rule (from φ infer $\Box\varphi$), the schema

$$\Box(\varphi \rightarrow \psi) \rightarrow (\Box\varphi \rightarrow \Box\psi)$$

and the following *Löb's Axiom*:

$$\Box(\Box\varphi \rightarrow \psi) \rightarrow \Box\varphi.$$

Solovay's result (see Troelstra, in [29]) can be stated as

$$\mathbf{G} \vdash \varphi \quad \text{if, and only if,} \quad \vDash^* \varphi.$$

Goldblatt [30] presents other interesting translations. The first one involves the modal systems $\mathbf{K_4W}$ and $\mathbf{S_4Grz}$, with $\mathbf{K_4W}$ extended from **CPC** by the following axioms and rule:

$$(RN)\Box(\varphi \rightarrow \psi) \rightarrow (\Box\varphi \rightarrow \Box\psi)$$

$$(K)\Box\varphi \rightarrow \Box\Box\varphi$$

$$(W)\Box(\Box\varphi \rightarrow \varphi) \rightarrow \Box\varphi \quad (\textit{Löb's Axiom})$$

$$\textit{Necessitation Rule:} \quad \varphi/\Box\varphi.$$

The system $\mathbf{S_4Grz}$ is obtained, from $\mathbf{S_4}$, by adding the following axiom due to Grzegorczyk [32]

$$(Grz)\Box(\Box(\varphi \rightarrow \Box\varphi) \rightarrow \varphi) \rightarrow \varphi.$$

The function Gl_1 is defined by:

$$Gl_1 : \mathbf{S_4Grz} \to \mathbf{K_4W}$$

$$(p_i)^{Gl_1} =_{def} p_i$$

Gl_1 commutes with $\sim, \vee, \bullet, \to$

$$(\Box\varphi)^{Gl_1} =_{def} \varphi \bullet \Box\varphi^{Gl_1}.$$

The following result is proved.

Theorem 5 $\mathbf{S_4Grz} \vdash \varphi$ *if, and only if,* $\mathbf{K_4W} \vdash \varphi^{Gl_1}$.

From another interpretation function,

$$SG : \mathbf{K_4W} \to \mathbf{PA}$$

$$(p_i)^{SG} =_{df} p_i$$

SG commutes with $\sim, \vee, \bullet, \to$

$$(\Box\varphi)^{SG} =_{df} \varphi \wedge Bew(\ulcorner\varphi^{SG}\urcorner),$$

the following theorem is proved.

Theorem 6 $\mathbf{K_4W} \vdash \varphi$ *if, and only if,* $\mathbf{PA} \vdash \varphi^{SG}$.

The same result is also proved relative to Solovay's interpretation.

Theorem 7 $\mathbf{K_4W} \vdash \varphi$ *if, and only if,* $\mathbf{PA} \vdash \varphi^S$.

Hence, we have the following theorem, for every formula φ in a modal language.

Theorem 8 $\mathbf{S_4Grz} \vdash \varphi$ *if, and only if,* $\mathbf{PA} \vdash (\varphi^{Gl_1})^{SG}$.

Finally, Goldblatt observes that the interpretation TM of McKinsey and Tarski can be extended to $\mathbf{S_4Grz}$ and, by the composition of the interpretations, he defines the following function Gl_2 from $\mathbf{IPC}$ into $\mathbf{PA}$:

$$Gl_2 : \mathbf{IPC} \to \mathbf{PA}$$

$$Gl_2(\varphi) =_{df} SG \circ Gl_1 \circ TM(\varphi)$$

$$\underbrace{\mathbf{IPC} \xrightarrow{TM} \mathbf{S_4Grz} \xrightarrow{Gl_1} \mathbf{K_4W} \xrightarrow{SG} \mathbf{PA}.}_{Gl_2}$$

The following result is then proved.

Theorem 9 $\mathbf{IPC} \vdash \varphi$ *if, and only if,* $\mathbf{PA} \vdash \varphi^{Gl_2}$.

Boolos [4, 5] presents a synthesis of the results by Solovay and Goldblatt. The function B from $\mathbf{S_4Grz}$ into Gödel's $\mathcal{G}$ is defined as follows:

$$B : \mathbf{S_4Grz} \to \mathcal{G}$$

$$(p_i)^B =_{df} p_i$$

$$(\perp)^B =_{df} \perp$$

B commutes with $\sim, \vee, \bullet, \to$

$$(\Box\varphi)^B =_{df} (\Box\varphi^B \bullet \varphi^B).$$

Theorem 10 *The following assertions are equivalent:*

(i) $\mathbf{S_4Grz} \vdash \varphi$;
(ii) φ *is valid in every weak finite partial order*;
(iii) $\mathcal{G} \vdash \varphi^B$;
(iv) *Every Goldblatt translation of φ is a theorem of* **PA**.

According to Boolos [4, Chapter 13], we have the following result.

Theorem 11 $\mathbf{IPC} \vdash \varphi$ *if, and only if,* $\mathbf{S_4Grz} \vdash \varphi^{G_1}$.

Goodman [31] presents a new extension of the results involving intuitionistic and modal systems. He considers a first-order arithmetic based on $\mathbf{S_4}$ and named *epistemic arithmetic* (**EA**), and the Heyting arithmetic **HA** [38]. **EA** is to be understood as a conservative extension of **HA**.

The following interpretation is defined thus:

$$Gm : \mathbf{HA} \to \mathbf{EA}$$

$$(\varphi)^{Gm} =_{df} \varphi, \quad \text{if } \varphi \text{ is atomic}$$

Gm commutes with $\neg$ and $\wedge$

$$(\varphi \vee \psi)^{Gm} =_{df} \Box(\varphi)^{GM} \vee \Box\varphi^{Gm}$$

$$(\varphi \supset \psi)^{Gm} =_{df} \Box(\varphi^{GM} \to \psi^{Gm})$$

$$(\forall x\varphi)^{Gm} =_{df} \Box\forall x\varphi^{Gm}$$

$$(\exists x\varphi)^{Gm} =_{df} \exists x\Box\varphi^{Gm}.$$

As **EA** is an extension of **HA**, we have that if **HA** $\vdash \varphi$, then **EA** $\vdash \varphi^{Gm}$. The main results of Goodman's paper is the proof, via cut-elimination, of the converse of the previous result.

Theorem 12 **HA** $\vdash \varphi$ *if, and only if,* **EA** $\vdash \varphi^{Gm}$.

3 "Zur intuitionistischen Arithmetik und Zahlentheorie": Gödel's Intuitionistic Interpretation of the Classical Propositional Calculus and Classical Arithmetic

In this also very well known paper Gödel defines a translation from the classical propositional calculus **CPC** into the intuitionistic propositional calculus **IPC**, which is extended to a translation between classical and intuitionistic arithmetic.

Gödel distinguishes the classical and intuitionistic connective symbols. He considers the negation $\neg$ and the conjunction $\wedge$ as the classical primitive connectives, $-$ and Δ being the corresponding intuitionistic ones. As the other classical connectives ($\vee, \to, \leftrightarrow$) are definable from the $\neg$ and $\wedge$, Gödel introduces the following "translation" from **CPC** into **IPC**:

$$Gd_2 : \mathbf{CPC} \to \mathbf{CPI}$$

$$(p)^{Gd_2} =_{df} p$$

$$(\neg\varphi)^{Gd_2} =_{df} -\varphi^{Gd_2}$$

$$(\varphi \wedge \psi)^{Gd_2} =_{df} \varphi^{Gd_2} \Delta \psi^{Gd_2}$$

$$(\varphi \vee \psi)^{Gd_2} =_{df} -(-\varphi^{Gd_2} \Delta - \psi^{Gd_2})$$

$$(\varphi \to \psi)^{Gd_2} =_{df} -(\varphi^{Gd_2} \Delta - \psi^{Gd_2}).$$

Glivenko [24] had proved the following results.

Theorem 13 *If φ is a theorem of* **CPC**, *then* $--\varphi$ *is a theorem of* **IPC**.

Corollary 14 *A formula* $\neg\varphi$ *is a theorem of* **CPC** *if, and only if,* $-\varphi$ *is a theorem of* **IPC**.

Gödel explicitly mentions Glivenko's Corollary, observing that such a result obtained for the connective calculus could not be extended to number theory. He uses it in the following version.

Theorem 15 *If φ is a theorem of* **CPC**, *built only with the connectives* $\neg$ *and* $\wedge$, *then the corresponding formula* $(Gd_2(\varphi))$ *in* **IPC** *is a theorem of* **IPC**.

Based on this result, Gödel shows that the function Gd_2 is a "translation", in the sense that it preserves theoremhood.

Theorem 16 *If φ is a theorem of* **CPC**, *then* $Gd_2(\varphi)$ *is a theorem of* **IPC**.

Aiming at extending these results to the classical arithmetic, built on the first-order classical predicate logic, Gödel takes the semi-formal Herbrand arithmetic system [36], with some additional conditions in order to make it the formal system **HA**. As the corresponding intuitionistic arithmetic, Gödel assumes the Heyting [37] system and introduces new variables $x'_1, x'_2, \ldots$ for the natural numbers, introducing the system **H'**. The interpretation of these new variables of **H'** is given by: a formula of type $\forall x \varphi(x')$ is equivalent to a formula $\forall x(x \in \mathbb{N} \to_{\mathbf{H'}} \varphi(x))$ and, if in $\varphi(x_1, \ldots, x_k)$ the variables $x_1, \ldots, x_k$ occur free then $\varphi(x'_1, \ldots, x'_k)$ is equivalent to $x_1, \ldots, x_k \in \mathbb{N} \to_{\mathbf{H'}} \varphi(x_1, \ldots, x_k)$. Thus, every formula that contains new variables is equivalent to a usual formula.

Next, Gödel extends the function Gd_2.

$$Gd_2 : \mathbf{HA} \to \mathbf{H'}$$

$$(x_i)^{Gd_2} =_{df} \mathbf{x}'_i$$

$$(f_i)^{Gd_2} =_{df} \mathbf{f}_i \quad \text{(functional symbols)}$$

$$(=)^{Gd_2} =_{df} = \quad \text{(identity symbol of } \mathbf{H'})$$

$$(0)^{Gd_2} =_{df} \mathbf{1}$$

$$(+1)^{Gd_2} =_{df} \mathbf{S} \quad \text{(successor)}$$

$$(\forall x \varphi)^{Gd_2} =_{df} \forall \mathbf{x}' \varphi^{Gd_2}.$$

The following results are then proved.

Lemma 17 *For every $\mathbf{H'}$-numerical formula φ^{Gd_2}, we have that*

$$\mathbf{H'} \vdash --\varphi^{Gd_2} \to_{\mathbf{H'}} \varphi^{Gd_2}.$$

Lemma 18 *If φ^{Gd_2} and ψ^{Gd_2} are $\mathbf{H'}$-numerical formulas, then*

$$\mathbf{H'} \vdash (\varphi^{Gd_2} \to_{\mathbf{H'}} \psi^{Gd_2}) \leftrightarrow_{\mathbf{H'}} -(\varphi^{Gd_2} \Delta - \psi^{Gd_2}).$$

Theorem 19 *If the formula φ is deductible in the Herbrand extended system $\mathbf{HA}$, then its translation φ^{Gd_2} is deductible in $\mathbf{H'}$.*

It is interesting to observe that in Gödel's text the expressions "φ is valid", "φ is deductible" and "φ is a theorem" are used indistinctly.

Gödel comments on his results: "Intuitionism appears to introduce genuine restrictions only for analysis and set theory; these restrictions, however, are due to the rejection, not of the principle of the excluded middle, but of notions introduced by impredicative definitions". He explicitly states that Theorem 19 gives a proof of the relative consistency of classical logic and arithmetic relative to the respective intuitionistic theories. If the classical theories were contradictory, then the intuitionistic ones would also be contradictory.

Hence, if we are sure that the intuitionistic arithmetic is consistent, then we can be sure that the classical arithmetic is also consistent.

But Gödel explicitly calls our attention that its proof is not "finitary", in the sense given to this word by Herbrand [34, 35], following Hilbert.

Let us finally emphasize, as mentioned in the Introduction, that Gödel was apparently not aware of Kolmogorov's paper [44] and does not mention it in his paper.

In March 1933, Gentzen published a rigorous and complete paper, with a simpler translation from **CPC** into **IPC**. The aim of [20] (see [22, 23]) is to show that "the applications of the law of double negation in proofs of classical arithmetic can in many instances be eliminated". He introduces a "transformation" Ge from the language of **CPL** into **IPC** and proves that φ is derivable in the intuitionistic predicate calculus if, and only if, $Ge(\varphi)$ is derivable in the classical predicate calculus; as a consequence, a proof of the consistency of classical elementary arithmetic with respect to intuitionistic arithmetic is obtained. It must be observed that Gentzen's proof, differently of Gödel's, is constructive, and thus acceptable from the intuitionistic point of view.

4 Final Remarks: Gödel's Interpretations from the Point of View of Translations Between Logics

In spite of dealing with interrelations among intuitionistic, modal and classical logic, Kolmogorov [44], Glivenko [24], Gödel [25, 26] and Gentzen [20] are not interested in the meaning of the concept of translation between logics, several distinct terms having been used by them such as interpretation and transformation. Since their papers were written, interpretations between logics have been used for different purposes (see [16, 19] and [6]).

Prawitz and Malmnäs [52] survey these historical papers and introduce the first general definition for the concept of translation between logic systems. For these two authors, a *translation* from a logic system S_1 into a logic system S_2 is a function t that maps the set of formulas of S_1 into the set of formulas of S_2, such that, for every formula φ of S_1,

$$S_1 \vdash \alpha \quad \text{if, and only if, } S_2 \vdash t(\alpha).$$

The system S_1 is then said to be *interpretable* in S_2 by t. Moreover, S_1 is said to be *interpretable in S_2 by t with respect to derivability* if, for every set $\Gamma \cup \{\alpha\}$ of formulas in S_1,

$$S_1 \cup \Gamma \vdash \alpha \quad \text{if, and only if, } S_2 \cup t(\Gamma) \vdash t(\alpha),$$

where $t(\Gamma) = \{t(\beta) : \beta \in \Gamma\}$.

Two books, by Wójcicki [56] and Epstein [17], respectively, can be considered as the first works envisioning a general systematic study on translations between logics. For Wójcicki, logics are defined as algebras with consequence operators: a logic is a pair (A, C) such that A is a formal language and C is a Tarskian consequence operator in the free algebra of formulas of A. Given two propositional languages S_1 and S_2, with the same sets of variables, a mapping t from S_1 into S_2 is a *translation* if, and only if:

(i) There is a formula $\varphi(p_0)$ in S_2 in one variable p_0 such that, for each variable $p, t(p) = \varphi(p)$;

(ii) For each connective μ_i in S_1 of arity k there is a formula φ_i in S_2 in the variables $p_1, \ldots, p_k$, such that $t(\mu_i(\alpha_1, \ldots, \alpha_k)) = \varphi_i(t(\alpha_1), \ldots, t(\alpha_k))$, for every $\alpha_1, \ldots, \alpha_k$ in S_1.

A *propositional calculus* is defined as a pair $C = \langle S, C \rangle$, where C is a consequence operator over the language S. Finally, $C_1 = \langle S_1, C_1 \rangle$ is said to be *translatable* into $C_2 = \langle S_2, C_2 \rangle$ if there is a mapping t from S_1 into S_2, such that for all $X \subseteq S_1$ and all $\alpha \in S_1$,

$$\alpha \in C_1(X) \quad \text{if, and only if, } t(\alpha) \in C_2(t(X)).$$

Epstein [17] defines a *translation* of a propositional logic **L** into a propositional logic **M**, in semantic terms, as a map t from the language of **L** into the language of **M** such that

$$\mathbf{L}, \Gamma \vDash \alpha \quad \text{if, and only if, } \mathbf{M}, t(\Gamma) \vDash t(\alpha),$$

for every set $\Gamma \cup \{\alpha\}$ of formulas.

Da Silva, D'Ottaviano and Sette [9], motivated by D'Ottaviano [11][2] and Hoppmann [41][3] and explicitly interested in the study of inter-relations between logic systems in general, propose a general definition for the concept of translation between logics, in order "to single out what seems to be in fact the essential feature of a logical translation": logics are characterized as pairs constituted by an arbitrary set (without the usual requirement of dealing with formulas of a formal language), and a Tarskian consequence operator; translations between logics are then defined as maps preserving consequence relations.

Definition 20 A *logic* **A** is a pair $\langle A, C \rangle$, where the set A is the *domain* of **A** and C is a Tarskian *consequence operator* in A[4].

The usual concepts and known results on closure spaces are here assumed. The general definition of translation between logics is then proposed.

Definition 21 A *translation* from a logic **A** into a logic **B** is a mapping $t : A \to B$ such that, for any $X \subseteq A$,

$$t(C_A(X)) \subseteq C_B(t(X)).$$

Of course, it is possible to consider logics defined over formal languages.

Definition 22 A *logical system* or *deductive system* defined over L is a pair $\mathbf{L} = \langle L, C \rangle$, where L is a formal language and C is a structural consequence operator in the free algebra **Form**(L) of the formulas of L.

Naturally, Definition 22 can be presented in terms of consequence relations. If **A** and **B** are logics with associated consequence relations $\vdash_{C_A}$ and $\vdash_{C_B}$, respectively, then a function $t : A \to B$ is a translation if, and only if, for every $\Gamma \cup \{\alpha\} \subseteq$ **Form**(A):

$$\Gamma \vdash_{C_A} \alpha \quad \text{implies } t(\Gamma) \vdash_{C_B} t(\alpha).$$

An initial treatment of a theory of translations between logics is presented by da Silva, D'Ottaviano and Sette in [9], and by Feitosa [18]. An important subclass of translations, the conservative translations, is investigated by Feitosa and D'Ottaviano [12–16, 18, 19].

---

[2]In this paper variants of Tarskian closure operators characterized by interpretations are studied.

[3]This is apparently the first paper in the literature where the term "translation between general logic systems" is used to mean a function preserving derivability.

[4]It is not hard to see that a logic could be defined along the lines of *universal logic*, as for instance by Béziau in [3]; that is, a logic is basically a pair formed by a set of entities called formulas and a consequence relation, without assuming any properties.

Definition 23 Let **A** and **B** be logics. A *conservative translation* from **A** into **B** is a function $t : A \to B$ such that, for every set $X \cup \{x\} \subseteq A$,

$$x \in C_A(X) \quad \text{if, and only if, } t(x) \in C_B(t(X)).$$

Note that, in terms of consequence relations, $t : \mathbf{Form}(L_1) \to \mathbf{Form}(L_2)$ is a conservative translation when, for every $\Gamma \cup \{\alpha\} \subseteq \mathbf{Form}(L_1)$:

$$\Gamma \vdash_{C_1} \alpha \quad \text{if, and only if, } t(\Gamma) \vdash_{C_2} t(\alpha).$$

Coniglio [8] proposes a different approach to translations by means of *meta-translations*, called *contextual translations* by Carnielli, Coniglio and D'Ottaviano [6]. *Contextual translations* are mappings between languages preserving certain meta-properties of the source logics, that are defined in a formal first-order meta-language, acting as a kind of sequent calculus whose rules govern the consequence relation of the logic. Carnielli, Coniglio and D'Ottaviano present a simplified version of the definitions given in [8], obtain some related results and analyse several examples.

A contextual translation is a translation in the sense of Definition 21. The authors present examples showing that contextual translations and conservative translations are essentially independent concepts, and that neither of them entails the other. But paradigmatic examples of both kinds of translations, conservative and contextual, are also introduced and discussed.

The general notion of translation (Definition 21) introduced by da Silva, D'Ottaviano and Sette [9], that encompasses the notions of conservative translation and contextual translation, accommodates certain maps that seem to be intuitive examples of translations, such as the identity map from intuitionistic into classical logic and the forgetful map from modal logic into classical logic: such cases would be ruled out if the stricter notion of conservative translation (or the general notions of Wójcicki and Epstein) or of contextual translation were imposed.

In particular, the identity function $i : \mathbf{IPL} \to \mathbf{CPL}$, both logics considered in the connectives $\neg, \wedge, \vee, \to$, is a translation, is a contextual translation, but is not a conservative translation: it suffices to observe that

$$p \vee \neg p \notin C_{\mathbf{IPC}}(\phi), \quad \text{while } i(p \vee \neg p) = (p \vee \neg p) \in C_{\mathbf{CPL}}(\phi).$$

Translations in the sense of Prawitz and Malmnäs [52] do not coincide with conservative translations, nor with translations in the sense of Definition 21, for they preserve only theoremhood, and not derivability. Translations in Wójcicki's sense [56] are particular cases of conservative translations. Epstein's translations [17] are instances of conservative translations.

It can be seen that the interpretations of Kolmogorov [44], Glivenko [24] and Gentzen [20] are translations in the sense of Prawitz and Malmnäs, Wójcicki and Epstein, and are conservative translations according to Definition 23 (see [19]). Such translations are also examples of contextual translations.

Although Gödel's papers of 1933 and their meaningful extensions by others, discussed earlier in this paper, are important and relevant, Gödel's interpretations $Gd_1 : \mathbf{IPL} \to \mathbf{S}_4$ and $Gd_2 : \mathbf{CPL} \to \mathbf{IPL}$ are translations only in Prawitz's sense. They do not preserve derivability, even on the propositional level, and hence are not translations in the sense of our Definition 21.

In the case of the Gödel function $Gd_2 : \mathbf{CPC} \to \mathbf{IPC}$, even at the propositional level, it is not a translation. If Gd_2 were a translation, as

$$\neg\neg p \vdash_{\mathbf{CPC}} p, \quad \text{then} \quad --p \vdash_{\mathbf{IPC}} p.$$

Hence, by the Deduction Theorem of **IPC**, we should have

$$\vdash_{\mathbf{IPC}} \neg\neg p \to p,$$

what is very well known not to be a theorem of $\mathbf{IPC}^5$.

Gödel's interpretation of "Eine Interpretation des intuitionistischen Aussagenkalküls", $Gd_1: \mathbf{IPL} \to \mathbf{S}_4$, also is not a translation, in the sense of da Silva, D'Ottaviano and Sette [9]: it suffices to consider that

$$p, p \supset q \vdash_{\mathbf{IPC}} q \quad \text{and} \quad p, \Box p \to \Box q \not\vdash_{\mathcal{G}/\mathbf{S}_4} \Box q.$$

Although being not a translation, in the sense of Wójcicki, Epstein and da Silva, D'Ottaviano and Sette, Gödel's papers [25, 26], following Kolmogorov [44] and Glivenko [24], play a very important role in the study of inter-relations between logical systems through the analysis of interpretations (translations) between them, and hence in the comparative strength and extensibility of a logic with respect to another.

For Gödel, the interest of his result in "Eine Interpretation des intuitionistischen Aussagenkalküls" is that it gave for the intuitionistic propositional calculus an interpretation that is also meaningful from a non-intuitionistic point of view. Since $\mathbf{S}_4$ is a system based on classical logic, Gödel's interpretation was not apparently expectable from Brouwer's conceptions and from Heyting's interpretation of intuitionistic logic.

According to Troelstra [29, p. 297], the discussion of Gödel's system $\mathcal{G}$ is also "really a topic in the history of Gödel's second incompleteness theorem".

Aside from this, Gödel's axiomatization of $\mathbf{S}_4$ was new and led to a much simpler axiomatization of systems of modal logics (see [46, pp. 6–7]). Furthermore, Gödel's result was "instrumental" in the development of Kripke's semantics for intuitionistic logic. According to the Introduction to Kripke [43], once the semantics for $\mathbf{QS}_4$ had been formulated, Gödel's interpretation showed how to obtain a semantics for $\mathbf{IQC}$.

The two papers here discussed may be considered among the Gödel's many relevant contributions to the history of logic in the 20th Century.

References

1. Becker, O.: Zur Logik der Modelitäten. Jahrb. für Philos. Phänomenolog. Forsch. **11**, 497–548 (1930)
2. Berka, K., Kreiser, L.: Logik-Texte. Kommentierte Auswahl zur Geschichte der modernen Logik. Akademie-Verlag, Berlin (1971)
3. Béziau, J.-Y.: Recherches sur la logique universelle (Excessivité, Négation, Séquents). Ph.D. Thesis, Paris 7 (1994)

---

^{5}We could present another simple proof using the algebraic structures associated to the corresponding logics.

4. Boolos, G.: Solovay's completeness theorems. In: The Unprovability of Consistency: An Essay in Modal Logic, pp. 151–158. Cambridge University Press, Cambridge (1979a)
5. Boolos, G.: An S₄-preserving proof-theoretical treatment of modality. In: The Unprovability of Consistency: An Essay in Modal Logic, pp. 159–167. Cambridge University Press, Cambridge (1979b)
6. Carnielli, W.A., Coniglio, M.E., D'Ottaviano, I.M.L.: New dimensions on translations between logics. Logica Universalis 3, 1–18 (2009)
7. Chuaqui, R.: Review of Kurt Gödel: Obras Completas. J. Symb. Log. 48, 1199–1201 (1983)
8. Coniglio, M.E.: Recovering a logic from its fragments by meta-fibring. Logica Universalis 1(2), 377–416 (2007)
9. da Silva, J.J., D'Ottaviano, I.M.L., Sette, A.M.: Translations between logics. In: Caicedo, X., Montenegro, C.H. (eds.) Models, Algebras and Proofs, Lectures Notes in Pure and Applied Mathematics, vol. 203, pp. 435–448. Marcel Dekker, New York (1999)
10. Davis, M.: The Undecidable. Basic Papers on Undecidable Propositions, Unsolvable Problems and Computable Functions. Raven Press, Hewlett (1965)
11. D'Ottaviano, I.M.L.: Fechos caracterizados por interpretações (Closures characterized by interpretations), in Portuguese. Master Dissertation, IMECC, State University of Campinas (1973)
12. D'Ottaviano, I.M.L., Feitosa, H.A.: Conservative translations and model-theoretic translations. Manuscrito – Rev. Int. Filos. XXII(2), 117–132 (1999a)
13. D'Ottaviano, I.M.L., Feitosa, H.A.: Many-valued logics and translations. J. Appl. Non-Class. Log. 9(1), 121–140 (1999b)
14. D'Ottaviano, I.M.L., Feitosa, H.A.: Paraconsistent logics and translations. Synthèse 125, 77–95 (2000)
15. D'Ottaviano, I.M.L., Feitosa, H.A.: Translations from Lukasiewicz logics into classical logic: Is it possible. In: Malinowski, J., Pietrusczak, A. (eds.) Essays in Logic and Ontology, Poznan Studies in the Philosophy of the Sciences and the Humanities, vol. 91, pp. 157–168 (2006)
16. D'Ottaviano, I.M.L., Feitosa, H.A.: Deductive systems and translations. In: Béziau, J.-Y., Costa-Leite (Org.), A. (eds.) Perspectives on universal logic, pp. 125–157. Polimetrica International Scientific Publisher, Monza (2007)
17. Epstein, R.L.: Propositional logics. The Semantic Foundations of Logic, vol. 1. Kluwer Academic Publishers, Dordrecht (1990)
18. Feitosa, H.A.: Traduções conservativas (Conservative translations), in Portuguese. Ph.D. Thesis, IFCH, State University of Campinas (1997)
19. Feitosa, H.A., D'Ottaviano, I.M.L.: Conservative translations. Ann. Pure Appl. Log. 108(1–3), 205–227 (2001)
20. Gentzen, G.: Über das Verhältnis zwischen intuitionistischen und klassischen Arithmetik. Manuscript set in type by Mathematische Annalen, but not published (eigegangen am 15.03.1933) (1933)
21. Gentzen, G.: Untersuchungen über das logische Schliessen. II. Mat. Z. 39(3), 405–431 (1935)
22. Gentzen, G.: Die Widerspruchsfreiheit der reinem Zahlentheorie. Math. Ann. 112, 493–565 (1936). Translation into English in [23]
23. Gentzen, G.: On the relation between intuitionist and classical arithmetic (1933). In: Szabo, M.E. (ed.) The Collected Papers of Gerhard Gentzen, pp. 53–67. North-Holland, Amsterdam (1969)
24. Glivenko, V.: Sur quelques points de la logique de M. Brouwer. Bull. Cl. Sci., Acad. R. Belg. 15, 183–188 (1929)
25. Gödel, K.: Zur intuitionistischen Arithmetik und Zahlentheorie. Ergebnisse eines mathematischen Kolloquiums 4, 34–38 (1933a)
26. Gödel, K.: Eine Interpretation des intuitionistischen Aussagenkalküls. Ergebnisse eines mathematischen Kolloquiums 4, 39–40 (1933b)
27. Gödel, K.: On intuitionistic arithmetic and number theory (1933e). In: Feferman, S., et al. (eds.) Collected Works, vol. 1, pp. 282–285. Oxford University Press, Oxford (1986)
28. Gödel, K.: An interpretation of the intuitionistic propositional calculus (1933f). In: Feferman, S., et al. (eds.) Collected Works, vol. 1, pp. 300–303. Oxford University Press, Oxford (1986)
29. Gödel, K.: Publications 1929-1936 vol. I: In: Feferman, S., Dawson, J.W. Jr., Kleene, S., Moore, G., Solovay, R., van Heijenoort, J. (eds.) Collected Works. Oxford University Press, Oxford (1986)
30. Goldblatt, R.: Arithmetical necessity, provability and intuitionistic logic. Theoria 44, 36–38 (1978)
31. Goodman, N.D.: Epistemic arithmetic is a conservative extension of intuitionistic arithmetic. J. Symb. Log. 49, 192–203 (1984)

32. Grzegorczyk, A.: Some relational systems and the associated topological spaces. Fundam. Math. **60**, 223–231 (1967)
33. Hacking, I.: What is strict implication? J. Symb. Log. **28**, 51–71 (1963)
34. Herbrand, J.: (1930a) Recherches sur la théorie de la démonstration. Ph.D. Thesis, presented to the Faculty of Sciences of the University of Paris (Partial translation into English in [55, pp. 525–581])
35. Herbrand, J.: Les bases de la logique hilbertienne. Rev. Métaphys. Morale **37**, 243–255 (1930b)
36. Herbrand, J.: Sur la non-contradiction de l'arithmétique. J. Reine Angew. Math. **166**, 1–8 (1931)
37. Heyting, A.: Die formalen Regeln der intuitionistischen Logik. In: Sitzungsberichte der Preussischen Akademie der Wissenschaften. Physikalisch-Mathematische Klasse, II, pp. 42–56 (1930a)
38. Heyting, A.: Die formalen Regeln der intuitionistischen Mathematik. In: Sitzungsberichte der Preussischen Akademie der Wissenschaften. Physikalisch-Mathematische Klasse, II, pp. 57–71 (1930b)
39. Hilbert, D.: Die logischen Grundlagen der Mathematik. Math. Ann. **88**, 151–165 (1923)
40. Hintikka, J. (ed.): The Philosophy of Mathematics. Oxford University Press, Oxford (1969)
41. Hoppmann, A.G.: Fecho e imersão (Closure and embedding), in Portuguese. Ph.D. Thesis, FFCL, São Paulo State University, Rio Claro, SP (1973)
42. Johansson, I.: Der Minimalkalkül, ein reduzierter intuitionistischer Formalismus. Compos. Math. **4**, 119–136 (1936)
43. Kripke, S.A.: Semantical analysis of intuitionistic logic, I. In: Crossley, J.N., Dummett, M.A.E. (eds.) Formal Systems and Recursive Functions, pp. 92–1130 (1965)
44. Kolmogorov, A.N.: O printsipe tertium non datur, Mam. cb. **32**, 646–667 (1925) In Russian (Translation into English in [55, pp. 416–437])
45. Kolmogorov, A.N.: On the principle of excluded middle (1925). In: Heijenoort, J. (ed.) From Frege to Gödel: A Source Book in Mathematical Logic 1879–1931, pp. 414–437. Harvard University Press, Cambridge (1977)
46. Lemmon, E.J., Scott, D.: An Introduction to Modal Logic. Basil Blackwell, Oxford (1977)
47. Lewis, C.I., Langford, C.H.: Symbolic Logic. New York (1932) (Reprinted in 1959)
48. Maehara, S.: Eine Darstellung der intuitionistischen Logik in der klassischen. Nagoya Math. J. **3**(4), 105–110 (1954)
49. McKinsey, J.C.C., Tarski, A.: Some theorems about the sentential calculi of Lewis and Heyting. J. Symb. Log. **13**(1), 1–15 (1948)
50. Mosterín, J. (ed.): Kurt Gödel: Obras Completas. Alianza Editorial, Madrid (2006) (1st edn. in (1981), 2nd edn. in (1989))
51. Parry, W.T.: Zum Lewisschen Aussagenkalkül. Ergebnisse eines mathematischen Kolloquiums **4**, 15–16 (1933)
52. Prawitz, D., Malmnäs, P.E.: A survey of some connections between classical, intuitionistic and minimal logic. In: Schmidt, H., et al. (eds.) Contributions to Mathematical Logic, pp. 215–229. North-Holland, Amsterdam (1968)
53. Rasiowa, H., Sikorski, R.: Algebraic treatment of the notion of satisfiability. Fundam. Math. **40**, 62–95 (1953)
54. Solovay, R.M.: Provability interpretations of modal logic. Israel J. Math. **25**, 287–304 (1976)
55. van Heijenoort, J. (ed.): From Frege to Gödel. A Source Book in Mathematical Logic. 1879–1931. Harvard University Press, Cambridge (1967) (corrected 3rd edn. in 1977)
56. Wójcicki, R.: Theory of Logical Calculi: Basic Theory of Consequence Operations. Kluwer, Dordrecht (1988) (Synthese Library, v. 199)

I.M.L. D'Ottaviano (✉)
Centre for Logic, Epistemology and the History of Science – CLE, Department of Philosophy –
IFCH, State University of Campinas – UNICAMP, São Paulo, Brazil
e-mail: itala@cle.unicamp.br

H.A. Feitosa
Department of Mathematics – FC, Bauru, São Paulo State University – UNESP, São Paulo, Brazil
e-mail: haf@fc.unesp.br

An interpretation
of the intuitionistic propositional calculus
(1933f)

One can interpret[1] Heyting's propositional calculus by means of the notions of the ordinary propositional calculus and the notion 'p is provable' (written Bp) if one adopts for that notion the following system $\mathfrak{S}$ of axioms:

1. $Bp \to p$,

2. $Bp \to . B(p \to q) \to Bq$,

3. $Bp \to BBp$.

In addition, for the notions $\to$, $\sim$, $.$, $\vee$ the axioms and rules of inference of the ordinary propositional calculus are to be adopted, as well as the new rule of inference: From A, BA may be inferred.

Heyting's primitive notions are to be translated as follows:

$\neg p$	$\sim Bp$
$p \supset q$	$Bp \to Bq$
$p \vee q$	$Bp \vee Bq$
$p \wedge q$	$p . q.$

One could also translate $\neg p$ by $B\sim Bp$ and $p \wedge q$ by $Bp . Bq$ with equal success. The translation of an arbitrary formula that holds in Heyting's system is derivable in $\mathfrak{S}$; on the other hand, the translation of $p \vee \neg p$ is not derivable in $\mathfrak{S}$, nor in general is any formula of the form $BP \vee BQ$ for which neither BP nor BQ is already provable in $\mathfrak{S}$. Presumably a formula holds in Heyting's calculus if and only if its translation is provable in $\mathfrak{S}$.

The system $\mathfrak{S}$ is equivalent to Lewis' system of strict implication if Bp is translated by Np (see *Parry 1933a*) and one supplements Lewis' system by the following additional axiom[2] of Becker: $Np < NNp$.

It is to be noted that for the notion "provable in a certain formal system S" not all of the formulas provable in $\mathfrak{S}$ hold. For example, $B(Bp \to p)$

---

[1]Kolmogorov (*1932*) has given a somewhat different interpretation of the intuitionistic propositional calculus, without, to be sure, specifying a precise formalism.

[2]*Becker 1930*, p. 497.

J.-Y. Béziau (ed.), *Universal Logic: An Anthology*, 89–90
Studies in Universal Logic, DOI 10.1007/978-3-0346-0145-0_8, © Springer Basel AG 2012

never holds for that notion, that is, it holds for no system S that contains arithmetic. For otherwise, for example, $B(0 \neq 0) \rightarrow 0 \neq 0$ and therefore also $\sim B(0 \neq 0)$ would be provable in S, that is, the consistency of S would be provable in S.

Kurt Gödel

Reprinted from: Gödel, K., "An interpretation of the intuitionistic propositional calculus (1933f)". In: Kurt Gödel – Collected Works. Volume I Publications 1929–1936, pp. 301+303, Oxford University Press, 1986

Part 5
Louis Rougier (1941)

An Early Defence of Logical Pluralism: Louis Rougier on the Relativity of Logic

Mathieu Marion

One of the central planks of universal logic is the thesis of 'logical pluralism', i.e., the thesis that there is more than one genuine deductive consequence relation, e.g., classical, intuitionist, relevant.[1] Today, the variety of logics is certainly a fact of science, but one might argue on *philosophical* grounds for 'logical monism', i.e., the thesis that there is only 'one true logic'.[2] In 'The Relativity of Logic',[3] Louis Rougier (1889–1982) provided arguments in favour of an earlier version of logical pluralism, which is related to both to Lewis and Langford's claim that "there is no such thing as 'logic'" [16, p. 256] and Carnap's well-known 'principle of tolerance':[4]

> *It is not our business to set up prohibitions, but to arrive at conventions.* [...] *In logic, there are no morals.* Everyone is at liberty to build up his own logic, i.e., his own form of language, as he wishes. All that is required of him is that, if he wishes to discuss it, he must state his methods clearly, and give syntactical rules instead of philosophical arguments. [6, §17]

Lewis and Langford's book was published in 1932, the same year as Lewis' discussion in 'Alternative Systems of Logic' [14], while Carnap's *Logical Syntax of Language* appeared in 1934 (with an English translation in 1937). On the other hand, Rougier's first defence of pluralism, 'La relativité de la logique' appeared in the journal of the Vienna Circle, *Erkenntnis*, only in 1939 [32]. Although he owes a lot to both Lewis and Carnap, whom, strangely enough, he does not cite, Rougier's arguments are unmistakably his own. His paper was reprinted in la *Revue de métaphysique et de morale* a year later [33], and an English translation appeared in *Philosophy and Phenomenological Research* in 1941, which is reprinted here. Since the beginning of that year, Rougier was in New York, having obtained a fellowship of the Rockefeller Foundation at the New School of Social Research. As the only French member of the Vienna Circle, as well as one of a key actor in the origins of the 'neo-liberal' doctrine in political economy, Rougier was a

---

[1] For this version of the thesis and a recent defence, see [1] and [2].

[2] The expression is taken from [27].

[3] In what follows, references to page numbers in parenthesis are to [34].

[4] The claim here is not that Lewis and Carnap are the inventors of 'logical pluralism' that title should go to Hugh MacColl, see [26]—rather that they form the immediate background to Rougier.

J.-Y. Béziau (ed.), *Universal Logic: An Anthology*, 93–99
Studies in Universal Logic, DOI 10.1007/978-3-0346-0145-0_9, © Springer Basel AG 2012

marginal figure in French 20th-century philosophy of science. After returning to France in 1947, Rougier was further marginalized when dismissed from his university position.[5] He was re-instated in 1954 and he published in following year a *Traité de la connaissance*, dedicated to the memory of Moritz Schlick, which included the content of these papers, revised and augmented [35, pp. 137–170]. Rougier's early defence of pluralism would have gone unnoticed, however, but for Arthur Pap's essay review of the *Traité*, 'Logical Empiricism and Rationalism' [20]. Alas, Rougier was by then 67, he appears to have had lost interest in theoretical debates and he never wrote a rejoinder.

As opposed to other pioneers of pluralism such as MacColl, Lewis or Carnap, Rougier made no contribution to logic but he wrote extensively about it, being one of the earliest defenders of the new 'formal' logic in his own country. Not being a logician by training, his writings are occasionally marred by blunders, e.g., when he confuses here propositional functions with universal sentences (pp. 104–105), and they do not always display the highest standards of clarity, e.g., when he presents the nature of truth preservation in both modal (p. 101) and formal (p. 105) terms. Having written *inter alia* a short book on *La structure des théories déductives* [31] and a fine, undeservedly ignored study of *La philosophie géométrique de Poincaré* [30], Rougier was by the early 1930s already well versed in Hilbert's formalism and Poincaré's conventionalism, which were to remain fundamental influences on his philosophy, so that when he came into contact with members the Vienna Circle in 1931, he had already developed a particular form of conventionalism at odds with the central tenets of the early logical positivists.[6] This much comes out clearly in the deeply pragmatist nature of Rougier's arguments, e.g., when he writes that "[d]epending on the order of research, depending on the practical aims that we set ourselves, the choice of such and such a logic [...] is more or less convenient" (p. 109).[7] Thus, according to Rougier, whose views have here a surprisingly modern ring, although we are at liberty to construct a variety of logics, our choice is not totally free:

> We find ourselves, in fact, to face with a great number of logics among which to choose. Our choice, although free, is not arbitrary. It will in each case be adapted to the domain of facts to which the research is relevant, or to the aim that we set for ourselves in our action. (pp. 114–115)

One should also note that this pragmatism does not sit well with Carnap's view that empirical reality cannot constrain logical rules since it is only when we adopt them that the idea that our statements are responsible to empirical evidence acquires a definite sense.[8]

Rougier is the author of *Les paralogismes du rationalisme* [29], which was appreciated by members of the Vienna Circle for its critique of '*a priori* rationalism'. A critique of

---

[5]Rougier's life and work were largely overshadowed by his links with the Vichy Regime in the Autumn of 1940, and by his post-war political involvement at the right wing fringes of the political spectrum, from which his reputation as a philosopher greatly suffered. For a recent set of studies on Rougier, see [21] and [22]. The former includes an intellectual biography, [3], but see also his own [36]. For his connections with the Vienna Circle, see [3, pp. 29–35] and [19].

[6]See, e.g., [19].

[7]Rougier got his pragmatism directly from Poincaré, but one should note that this was also Lewis' view [14, p. 507]. The only reference to Poincaré in the paper is to his idea of *définitions déguisées*, i.e., the idea that experience suggests the laws that we raise "to the dignity of principles, that is, by transforming them into definitions in disguise" (p. 117).

[8]For a critical discussion of conventionalism that takes into account [30], see [4].

the rationalist tradition was indeed Rougier's hobby horse, and, although he does not state this explicitly in 'The Relativity of Logic', his critique of logical monism was only part and parcel of this overall critical project: as he would put in his *Traité*, "the concept of an a priori logic, unique, universal and normative, imposing itself to all minds and in all domains, is not acceptable" [35, pp. 169–170]; it is the "last idol" of *a priori* rationalism [35, p. 143]. So Rougier's reasons for rejecting pluralism were fundamentally related to his own agenda against '*a priori* rationalism'. His nemesis, Arthur Pap was instead a 'rationalist'.[9]

Rougier's paper is divided in eight sections: after a succinct, standard review in section I of the Aristotelian theory of deduction and the changes caused by the discovery of non-Euclidean geometries, Rougier launches in section II–III into a critical discussion of the notion of tautology in Wittgenstein's *Tractatus Logico-Philosophicus* and the 'radical empiricism' that it purportedly supports. The purpose of section I is better captured by a quotation from his *Traité*:

> [the classical theory of knowledge] consisted essentially in the distinction of knowledge: rational knowledge, which is *a priori*, universal and necessary, founded on self-evident principles, and empirical truths, *a posteriori*, singular, revisable and contingent, founded on observation and experiment. The new theory of knowledge retains this distinction but interprets it differently. The rational truths of the scholastics and the rationalists become formal truths, founded not on truths 'self-evident', *a priori* and universal but on consistent, freely chosen conventions, although more often than not suggested by experience and based on an intuitive interpretation. [35, p. 309]

There are, of course obvious problems with the analogy between alternative geometries and alternative logics that are glossed over by Rougier.[10] In sections II–III, Rougier is not in the business of providing a scholarly commentary of Wittgenstein's notoriously hermetic book; he takes instead for granted a reading reminiscent of the early interpretation within the Vienna Circle, as voiced, e.g., by Hans Hahn, when he argued that modern logic vindicates 'pure empiricism' [12, pp. 20–42]. In Rougier's argument against this view, the 'linguistic theory of logical truth' plays an important role; it is nicely stated by Carnap in *The Logical Structure of the World*:

> Logic (including mathematics) consists solely of conventions concerning the use of symbols, and of tautologies on the basis of these conventions. Thus, the symbols of logic (and mathematics) do not designate objects, but merely serve as symbolic fixations of these conventions. [7, §107]

According to this view, any language that makes empirical claims about the world, such as A or $\neg A$ contains devices, the logical constants, whose (semantic) function, properly understood, allow us to see that this language must allow for sentences, such as $A \vee \neg A$ that are always true but say nothing about the world, i.e., tautologies. These logical truths are thus a *by-product*, to use Ian Hacking's expression [11]. It suffices now to point out that the language is set up by the adoption of a set of conventions in order to argue that the set of tautologies will vary as the conventions vary, a point about which Rougier makes

---

[9]Pap was deeply influenced by his teacher Ernst Cassirer. For a broad outline of his philosophy, see [37].

[10]The point is argued in [20, p. 158].

heavy weather. While some, like Hahn, saw this account of logic (and, provided logicism is vindicated, of mathematics) as lifting one major hurdle towards 'pure empiricism', Rougier saw this instead as the proof that the latter cannot be sustained:

> [Radical empiricism] supposes that the properties of *tautological, synthetic, contradictory,* and *meaningless,* applied to propositions, have an *absolute meaning.* Now this is not the case at all, because there exists an infinity of possible logics; and, according to the logic adopted, the same proposition is *tautological* or *contradictory, analytic* or *synthetic, meaningful* or *meaningless.* These properties, just like those of *indemonstrable* or *demonstrable* as applied to propositions, and of *indefinable* and *definable* as applied to concepts, are *relative to the language we adopt.* (p. 108)[10]

However, the major defect with Rougier's clearly dated approach to the topic is that it presupposes (p. 119) what John Etchemendy has called the 'template theory of logic' [8, p. 322]. Roughly put, under this view a 'language' or 'language system' consists of a vocabulary along with (1) grammatical rules that determine what constitutes a sentence, (2) the rules of logical syntax, which determines what logical properties and relations holds, and, since these rules are meant to be syntactical, i.e., specified without reference to the meaning of terms, (3) semantic rules, that determine the meanings of our sentences, i.e., the circumstances in which they are true or false. But these last are somehow disregarded, while valid arguments display a particular syntactic structure, they fit a 'template', which is proclaimed by the syntactic rules as 'valid'. However, as Pap clearly saw [20, pp. 158–159], if one moves from 'uninterpreted' to 'interpreted' languages, i.e., if one adopts instead a semantic approach (paradoxically pioneered by Carnap himself in his *Introduction to Semantics* [5]), the whole view threatens to collapse and we are led to entertain the now dominant semantic account of logical consequence which rejects the claim that validity is based on logical form.[11] Even as late as his *Traité* of 1955, Rougier seems to have been unaware of this problem.

Rougier then proceeds in section IV with a cursory classification of logics, which covers 'rivals' such as intuitionist logic, minimal logic, many-valued logics, Reichenbach's and Destouches–Février's logics for quantum mechanics and, as 'supplement',[12] modal logics; this very brief presentation concludes, oddly, on a mention of Curry's combinatory logic (p. 113).[13] Rougier's claim that "there exists an infinity of possible logics" (p. 108)

---

[10]One should note here Rougier's early critique of the analytic-synthetic distinction. It is repeated a little bit further: "The designations analytic or synthetic are not absolute; they are relative to the previous choice of such and such a logic" (p. 109). This argument can be seen as supported by the claims that the division between logical and non-logical vocabulary plays a significant role in the determination of the concept 'analytical' and that, since Tarski's 'On the Concept of Logical Consequence' [38, pp. 418–420], it is generally agreed that there are no objective grounds for such a distinction.

[11]One should note that Pap also criticized the semantic account in 'Logic and the Concept of Entailment', [13, pp. 197–212]. As noted in [37, p. 8], he prefigured the views expressed by Etchemendy, e.g., in [8] and [9].

[12]The terms 'rivals' and 'supplements' in this sentence are taken from [10], they are not in Rougier although he makes the same distinction.

[13]There is also a brief discussion of a paper by Paul Weiss for which no reference is given but which is most probably [39]. One should note that Weiss was, however, a monist. For his arguments, see [40], and for a rejoinder, see [15]. This is one rare discussion of this topic prior to Quine's notorious arguments.

surely comes from Lewis' 'Alternative Systems of Logic', and, beyond, from a result by Łukasiewicz and Tarski concerning systems of many-valued logic,[14] cited by Lewis [14, p. 482]. At all events, the implicit argument here seems to be that the existence of alternative logics must be taken as a fact of science, against which the philosopher cannot legitimately argue, this being a *reductio* of the rationalist stance against pluralism. Here Rougier is, of course, on strong grounds, inasmuch as philosophy has no basis of its own from which to criticize and reject scientific achievements.

The remaining sections, V to VIII, concern philosophical issues raised by the existence of these alternative logics, some of which being of lesser interest today. I would like to conclude with some remarks on one central philosophical issue, which was raised not by Rougier but, instead, by Pap in his review. Contrary to modern versions of logical pluralism, where one defines different consequence relations over the same language, Carnap's principle as stated in the above quotation clearly links the choice of logic with a choice of language: "Everyone is at liberty to build up his own logic, i.e., *his own form of language*" (my underlining).[15] Under this view, to admit of another logical consequence relation is literally to speak a different language. To see this, let us take 'explosion', i.e., $A, \neg_c A \vdash B$, which does not occur in systems of paraconsistent logics: for Carnap, one would have to write something like this: $A, \neg_c A \vdash B$ and $A, \neg_p A \nvdash B$, i.e., A together with its classical negation entail B, while it does not when taken together with its paraconsistent negation; the subscripts indicate that one has changed languages, so that there are now two concepts of negation.[16] Quine had a well-known argument against this view, most famously put in *Philosophy of Logic*:

> My view of this dialogue is that neither party knows what he is talking about. They think they are talking about negation, '$\neg$', 'not'; but surely the notation ceased to be recognizable as negation when they took to regarding some conjunction of the form '$p \& \neg p$' as true, and stopped regarding such sentences as implying all others. Here, evidently, is the deviant logician's predicament: when he tries to deny the doctrine he only changes the subject. [25, p. 81]

According to this argument, the so-called 'deviant' logics result from a change of meaning of the logical constants, e.g., negation or '$\neg$', and where there is a change of meaning, there is no conflict.[17] Of course, this is a late statement, but the argument occurs in one of Quine's earliest classics, 'Truth by Convention', published in 1936 [24, p. 97] and it is served to Rougier by Arthur Pap, when discussing three-valued logic in his review of the *Traité*:

> Has the rationalist who maintains that the law of the excluded middle is *absolutely* necessary, not just *relatively* to a system based on arbitrary initial conventions, been refuted? By no means. For the same symbols have *different* meanings in the three-valued logic. [20, p. 152][18]

---

[14] See [18], as translated in [38, p. 47f.] and, for the philosophical discussion referred to by Lewis [17].

[15] For a discussion of the differences between modern logical pluralism in references given in footnote 1 and Carnap's 'tolerance', see [28].

[16] This is taken from [28, 432].

[17] See [10, p. 8].

[18] Part of Pap's point here is that three-valued logic is compatible with the Law of Excluded Middle, and he is right in claiming this against Rougier's claim that "the principle of excluded third is replaced

Rougier, who appears not to have known Quine's paper, left this argument unanswered. The modern avenue (consonant with universal logic), which consists in a shift to the consequence relation, i.e., saying instead that $A, \neg A \vdash_c B$ and $A, \neg A \nvdash_p B$, was not available to him. But Quine's 'change of meaning' argument is not, in and of itself, sufficient close the door to pluralism, as it only purports to immunize classical logic against criticism launched from the standpoint of rival logics. In order fully to argue for monism, Quine had further to appeal to his indeterminacy of translation argument.[19] Therefore, although Pap was right, as against Rougier, that the sheer existence of alternative logics does not refute the rationalist *qua* monist, the 'change of meaning' argument does not suffice, on the other hand, to undermine the possibility of alternative logics.

References

1. Beall, J.C., Restall, G.: Logical pluralism. Aust. J. Philos. **778**, 475–493 (2000)
2. Beall, J.C., Restall, G.: Logical Pluralism. Oxford University Press, Oxford (2006)
3. Berndt, C., Marion, M.: Vie et œuvre d'un rationaliste engagé : Louis Rougier (1889–1982). Philosophia Scientiae **10**(2), 11–90 (2006)
4. Black, M.: Conventionalism in geometry and the interpretation of necessary statements. Philosophy of Science **9**, 335–349 (1942)
5. Carnap, R.: Introduction to Semantics. Harvard University Press, Cambridge (1942)
6. Carnap, R.: The Logical Syntax of Language. Littlefield Adams, Paterson (1959)
7. Carnap, R.: The Logical Structure of the World. University of California Press, Los Angeles (1967)
8. Etchemendy, J.: The doctrine of logic as form. Linguist. Philos. **6**, 319–334 (1983)
9. Etchemendy, J.: The Concept of Logical Consequence. CSLI Publications, Stanford (1990)
10. Haack, S.: Deviant Logic Fuzzy Logic. Beyond the Formalism. University of Chicago Press, Chicago (1996)
11. Hacking, I.: What is logic? J. Philos. **66**, 285–319 (1979)
12. Hahn, H.: Empiricism, Logic and Mathematics. Philosophical Papers. Kluwer, Dordrecht (1980)
13. Keupink, A., Shieh, S. (eds.): The Limits of Logical Empiricism. Selected Papers of Arthur Pap. Springer, Dordrecht (2006)
14. Lewis, C.I.: Alternative systems of logic. Monist **42**, 481–507 (1932)
15. Lewis, C.I.: Paul Weiss on alternative logics. Philos. Rev. **43**, 70–74 (1934)
16. Lewis, C.I., Langford, C.H.: Symbolic Logic, 2nd edn. Dover, New York (1959)
17. Lukasiewicz, J.: Philosophische Bemerkungen zu mehrwertigen Systemen des Aussagenkalküls, Comptes Rendus des séances de la Société des Sciences et des Lettres de Varsovie, cl. III, **23**, 51–77 (1930)
18. Lukasiewicz, J., Tarski, A.: Untersuchungen über den Aussagenkalkül, Comptes Rendus des séances de la Société des Sciences et des Lettres de Varsovie cl. III, **23**, 30–50 (1930); English translation: Investigations into the Sentential Calculus, in Tarski (1983), 38–59
19. Marion, M.: Louis Rougier, the Vienna circle, and the unity of science. In: Nemeth, E., Roudet, N. (eds.) Paris-Wien. Enzyklopädien im Vergleich, pp. 151–174. Veröffentlichungen des Institut Wiener Kreis, Vienna (2005)
20. Pap, A.: Logical empiricism and rationalism. Dialectica **10**, 148–166 (1956)
21. Pont, J.-C., Padovani, F. (eds.) Louis Rougier (1889–1982) : Vie et œuvre d'un philosophe engagé. Philosophia Scientiae **10**(2) (2006)

---

by the principle of excluded fourth" (p. 109). He also could have mentioned C.I. Lewis, who confused intuitionist logic with three-valued logic in [14, p. 505], and probably influenced Rougier on that point.

[19] As Quine says of a deviant logician in *Philosophy of Logic*: "We impute our orthodox logic to him, or impose it upon him, by translating his deviant dialect" [25, p. 81]. The allusion here is to chapter 2 of *Word and Object* [23, §§12–13].

22. Pont, J.-C., Padovani, F. (eds.) Louis Rougier (1889–1982) : Vie et œuvre d'un philosophe engagé. Témoignages - écrits politiques. Philosophia Scientiae. cahier spécial no. 7 (2007)
23. Quine, W.v.: Word and Object. Harvard University Press, Cambridge (1960)
24. Quine, W.v.: The Ways of Paradox, 2nd edn. Harvard University Press, Cambridge (1976)
25. Quine, W.v.: Philosophy of Logic, 2nd edn. Harvard University Press, Cambridge (1986)
26. Rahman, S., Redmond, J.: Hugh MacColl and the birth of logical pluralism. In: Woods, J., Gabbay, D. (eds.) Handbook of History of Logic, vol. 4, pp. 535–606. Elsevier, Amsterdam (2008)
27. Read, S.: Monism: The one true logic. In: Devidi, D., Kenyon, T. (eds.) A Logical Approach to Philosophy, pp. 193–209. Springer, Dordrecht (2006)
28. Restall, G.: Carnap's tolerance, meaning and logical pluralism. J. Philos. 99, 426–443 (2002)
29. Rougier, L.: La philosophie géométrique de Henri Poincaré. Alcan, Paris (1920)
30. Rougier, L.: Les paralogismes du rationalisme. Essai sur la théorie de la connaissance. Alcan, Paris (1920)
31. Rougier, L.: La structure des théories déductives. Théorie nouvelle de la déduction. Alcan, Paris (1921)
32. Rougier, L.: La relativité de la logique. Erkenntnis 8, 193–217 (1939)
33. Rougier, L.: La relativité de la logique. Revue de métaphysique et de morale 47, 305–330 (1940)
34. Rougier, L.: The relativity of logic. Philos. Phenomenol. Res. 2, 137–158 (1941)
35. Rougier, L.: Traité de la connaissance. Gauthier-Villars, Paris (1955)
36. Rougier, L.: Itinéraire philosophique. In: Revue libérale, pp. 6–79 (October 1961)
37. Shieh, S.: Introduction. In: Keupink & Shieh (2006), pp. 3–43 (2006)
38. Tarski, A.: Logic, Semantics, Metamathematics, 2nd edn. Hackett, Indianapolis (1983)
39. Weiss, P.: Relativity in logic. Monist 38, 536–548 (1928)
40. Weiss, P.: On alternative logics. Philos. Rev. 42, 520–525 (1933)

M. Marion (✉)
Chaire de recherche du Canada en philosophie de la logique et des mathématiques, Département de philosophie, Université du Québec à Montréal, C.P. 8888, Succursale Centre-Ville, Montréal, Québec, Canada H3C 3P8
e-mail: marion.mathieu@uqam.ca

PHILOSOPHY AND PHENOMENOLOGICAL RESEARCH
A Quarterly Journal
VOLUME II, No. 2   DECEMBER, 1941

THE RELATIVITY OF LOGIC

Logic is defined as the art of thinking well, the art of reasoning justly. Reasoning consists in showing that certain propositions are necessarily true on the assumption that other propositions, called premises, are true. It was in the sciences of reasoning, that is in mathematics, that logic came to be employed for the first time. If we are to believe Proclus, it was in the sixth century before our era that the intuitive and empirical mathematics of the Orientals was transformed into an abstract, deductive discipline, and this transformation was due to Pythagoras: "Pythagoras came and transformed geometry into a liberal study, for he went back to the first principles and sought out theorems abstractly and by pure intellect." This transformation consists essentially in substituting for sensible evidence, which bears only on the determination of particular cases, intelligible evidence which rests on the reasoning power and attains the universal.

I. THE APODICTIC OF ARISTOTLE

Aristotle, in the *Posterior Analytics*, analyzed the logical procedure of deductive science, called by him *Apodictic,* as the Greek mathematicians of his time conceived it. His analysis found favor with logicians until the end of the nineteenth century. According to him, a deductive science, like the Pythagorean geometry, rests on principles evident in themselves which he calls its *proper principles,* and demonstration has for its aim to transfer the evidence step by step from the proper principles to their most remote consequences. *It is the self-evidence of the proper principles that underlies the evidence of the theorems. The proper principles are more evident than the theorems deduced from them,* by virtue of a principle that can be called *the principle of the eminence of the cause,* according to which a quality manifested by an effect ought to be found in a more eminent degree in the cause of that eff ct: "For example, the reason that makes one love an object is even more loved that the object. Since, then, we know and believe by virtue of principles, we ought to know and believe them better than the conclusions drawn from them" (I *Post. An.* 1, 2, 71b, 19-25). Further, "If we have correctly established what knowing is, it necessarily follows that demonstrative science proceds from propositions which are true, primary, and immediate, and, relative to the conclusion, better known and anterior. Such are the *proper principles* of demonstrations; for without them there is no syllogism, therefore no demonstration, therefore no science" (*Meta.* $\triangle$ 6, 1015b, 17). The necessity of theor-

J.-Y. Béziau (ed.), *Universal Logic: An Anthology*, 101–122
Studies in Universal Logic, DOI 10.1007/978-3-0346-0145-0_10, © Springer Basel AG 2012

ems derives from that of the *proper principles:* "We call that demonstration necessary which draws its necessity from that of the premises."

The Aristotelian theory of demonstration can be analyzed into four propositions:

1. There exist principles which are indemonstrable by nature, being by nature primary and immediate. These are the *proper principles* of the different demonstrative sciences.

2. The necessity of the *proper principles* proceeds from their self-evidence.

3. The evidence of the proper principles is passed on to the theorems by means of demonstration.

To these three statements we can add a fourth which Aristotle does not explicitly formulate but which results from the preceding ones:

4. When demonstration has passed on to a theorem the evidence of the proper principles, the statement of this theorem can be detached from its demonstration. It constitutes a proposition true in itself, provided that one remembers having correctly deduced it from the *proper principles.*

The theory of Aristotle, professed by all the Scholastics, survived the Cartesian revolution. In the *Regulae ad directionem ingenii* (Reg. XII), René Descartes states in effect: "There are no other ways open to man to arrive at certain knowledge of truth except self-evident intuition and necessary deduction." Intuition gives us possession of the *evident principles,* that is, the prime truths which impose themselves on every attentive mind as soon as their terms are understood. Deduction, which calls memory into play, is only a *deferred intuition,* for "all the propositions which we deduce the ones from the others, provided that the deduction be evident, are thereby brought back to a veritable intuition." Deduction is only a detour, a discursive procedure due to the weakness of our minds which cannot embrace in a single glance the sum of all truths, but have to pass from one to another in a regular order, to pass on the self-evidence of the principles to their remotest consequences.

This conception of the deductive sciences ruled until the last quarter of the nineteenth century. It is still to be found at the foundation of the classic work of Duhamel, *Methods in the Sciences of Reason,* whose second edition is dated 1875. More recently, such well-informed logicians as Bertrand Russell and Brouwer still call on self-evidence to justify the choice of principles on which they found mathematics, the one invoking their rational, the other their intuitive self-evidence.

Aristotle's theory of deduction implies a theological conception of the world. If the primary propositions of demonstrative science are *true in themselves* and if the same is so for the theorems once they have been demonstrated, it must be that these truths in themselves subsist somewhere eternally, independent of their momentary grasp by our minds. This leads to the affirmation of a separate world of truths in themselves, the world of the Platonic ideas, which the theologians have referred back to the divine intellect by virtue of the following reasoning: eternal truths presuppose an eternal mind where they are perpetually understood. Such is the famous argument for the existence of God by the eternal truths, of which Bossuet has given the best exposition: "All of these truths and those which I deduce from them by infallible reasoning subsist independently of all time; in whatever time I place a human understanding it will discover them; but in coming to know them, it will not make them what they are: for our knowledge does not make its objects, but it presupposes them. Thus, these truths subsist before all ages and before there is an human understanding. . . . If I now ask in what place they subsist eternal and immutable as they are, I am obliged to confess a being where the truth is eternally subsistent and where it is always known: This being must be truth itself and must be all truth."

It was the discovery of the non-Euclidian geometries that ruined the *Aristotelian Apodictic* and substituted for it the *Axiomatic of David Hilbert.*

1. First of all, the attempts to demonstrate Euclid's postulate have shown that it makes no difference whether you accept as one of the primary propositions Euclid's postulate or any other proposition logically equivalent, such as that the sum of the angles of a triangle equals two right angles or that there are similar figures. This means that no absolute meaning attaches to the properties *indefinable* or *definable* as applied to concepts, or of *indemonstrable* or *demonstrable* as applied to propositions. A concept is indefinable or a proposition indemonstrable only with reference to a certain system of definitions and a certain order of demonstrations; with another system or according to another order, the same concepts could be defined or the same propositions demonstrated anew. Thus, for the deductive exposition of the metric geometry of Euclid, an infinity of systems of primary concepts and of primary propositions, all equivalent to each other, are possible. Peano takes as primary the concepts of *point* and *segment;* Pieri, those of *point* and *movement;* Veblen, *point* and *order;* Padoa, *point* and *distance of two points;* Hilbert, *point, straight, plane, situated on, situated between, congruent, parallel.*

All of these systems are equivalent; that is, the same body of propositions can be deduced from them. To ask if a concept is definable or if a proposition is indemonstrable, without specifying according to what system of concepts and propositions, is a question as meaningless as to ask if a body is at rest or in motion without indicating the system of reference. *There are no concepts which are most simple; hence, there are no primary concepts; and, since there are no propositions which are most evident, there are no primary propositions.* Aristotle's first epistemological proposition is to be rejected.

2. So must the second be rejected: that *proper principles impose themselves on us by virtue of their self-evidence.* If this were so, the non-Euclidian geometries and the infinite variety of geometries which correspond to the metric spaces of Riemann, and to functional and abstract spaces, would not be possible. *For the criterion of self-evidence, applied to each primary proposition taken in isolation, is substituted the criterion of coherence applied to a set of propositions freely chosen as primary.* All that is required of a system of primary concepts and propositions is that it be sufficient and coherent.

3. But then the third Aristotelian proposition falls for the same reason. *What makes a theorem necessary is not the self-evidence of the proper principles of a theory; it is the necessity of the ordinary rules of deduction.* The deductive sciences are not apodictic sciences as Aristotle understood the term; they are hypothetico-deductive systems.

4. At once Aristotle's fourth proposition is eliminated: *a demonstrated theorem is a proposition true in itself.* All that demonstration establishes is the logical dependence between a theorem and the system of axioms from which one has started out. The aim of geometry is not to prove that the Pythagorean theorem is true; its aim is to demonstrate that *if one accepts the axioms of Euclid, one is forced to admit, by virtue of the rules of deduction, the Pythagorean theorem.* The same proposition, by reference to one or another system of axioms, is true, false, or meaningless. For example, the theorem "There are similar figures" is true with reference to the axioms of Euclid, false with reference to the axioms of Lobachevski or Riemann. The theorems of the deductive sciences are *relative truths;* they are formulated in the form of *hypothetical propositions.* All this appears even more clearly if we consider the axiomatic method, founded by Pasch in 1882 and formalized by David Hilbert in 1899. This method consists in treating the primary concepts of a deductive theory as symbols devoid of intuitive meaning, as simple variables empty of content, characterized equivocally by the obligation to satisfy the logical relations stated in the axioms. The axioms take on the aspect of propositional functions

which are neither true nor false, but which become true or false depending upon the different values of their variables, that is, depending upon the different interpretations of the undefined symbols which occur in them. The statement that the logical product of the axioms implies such and such a proposition is not a propositional function, but a formal implication true for any values attributed to the arguments; it is a universal proposition. It is always true, *because it imposes itself on us by virtue of its form, independent of the truth or falsity of the axioms from which one has started out.* It is a law of logic.

II. TAUTOLOGICAL CHARACTER OF THE LAWS OF LOGIC

If we expel the necessity of proper principles from a deductive theory, it takes refuge in the logical rules of the deduction itself, and a problem then rises: Whence comes the necessity of the logical rules? The solution of this problem was given only in 1921, in the *Tractatus Logico-Philosophicus* of Ludwig Wittgenstein. The solution is as follows: *The rules of logic are unconditionally necessary, because their truth is independent not only of the meaning but even of the truth-value, the truth or falsity, of the elementary propositions which compose them.*

Let us agree to call a *proposition* any statement able to be true or false. Let us designate by p and q any two propositions whatsoever, treated as propositional variables. The affirmation: "p and q express similar ideas" can only be verified if we make use of the *meaning* of p and q. On the contrary, the affirmation: "Of p and q, one is true, the other false" does not make use of the meaning of p and q, but only of their *truth-value:* it is true if the propositions p and q are neither simultaneously true or false. We shall call such an affirmation, which depends only on the truth-value of the propositions composing it, a *truth-function.* Let us consider finally the following affirmation: "The proposition p is true or false." Such an affirmation is true whatever be the *meaning* and also the *value* of p, for p can by convention take on only two values, true and false, and this affirmation exhausts all the possible alternatives. We shall say that such an affirmation is necessarily true, true by virtue of its proper form: it is a *tautology* and, further, conforming to the definition of Wittgenstein, it is a law of logic, called the *principle of the excluded middle.*

The logical operations: negation, conjunction, disjunction, implication, and equivalence, are *truth-functions.* In effect, the truth or the falsity of the negation of p depends only on the truth or the falsity of p; the affirmation that p and q are simultaneously true depends only on the truth or the falsity of p and q taken in isolation. We can draw up the exhaustive table of all the *truth-functions* of two propositions. Since each proposition can take on two truth-values, true and false, there will be four possible distributions of truth-value between the two propositions, which we can represent by the following table:

I

P	Q
T	T
T	F
F	T
F	F

But since each one of these distributions can itself be true or false, we have for two propositions, taken as arguments, sixteen truth-functions, represented by Table II:

II

P	Q	1	2	3	4	5	6	7	8	9	10	11	12	13	14	15	16
T	T	T	T	T	T	T	T	T	T	F	F	F	F	F	F	F	F
T	F	T	T	T	T	F	F	F	F	T	T	T	T	F	F	F	F
F	T	T	T	F	F	T	T	F	F	T	T	F	F	T	T	F	F
F	F	T	F	T	F	T	F	T	F	T	F	T	F	T	F	T	F

This table of the truth-functions of the two propositions p and q exhausts all the possible combinations of truth-values. It is therefore exhaustive.

We can give great precision to Wittgenstein's proposition. When a truth-function is true, whatever may be the truth or falsity of the propositions which compose it, it is called a *tautology*; when a truth-function is false, whatever may be the truth or the falsity of the propositions which compose it, it is called a *contradiction*. Wittgenstein shows that the laws of logic are nothing but *tautologies*. *A law of logic is a proposition true in all the cases conceived as possible by definition.* In the case of any proposition whatsoever, we can verify immediately, without intermediate deductions, if it is a law of logic. It suffices to construct the table of values of the propositions which compose it and to see if the resulting proposition is true in all possible cases.

Consider the proposition: $p \lor (-p)$, which is read: "A proposition is true or false:

III

P	—P	P V (—P)
T	F	T
F	T	T

Such a proposition is called *the principle of the excluded middle*.

Let us take the more complex proposition:

(—p—q) v (p v q), which is read: "Two propositions are either both false or one of them is true."

IV

P	Q	—P	—Q	—P.—Q	P V Q	(—P.—Q) V (P V Q)
T	T	F	F	F	T	T
T	F	F	T	F	T	T
F	T	T	F	F	T	T
F	F	T	T	T	F	T

The laws of logic are nothing but the tautologies which regulate the implication of propositions.

If we agree to divide all living beings into mortal and immortal, we are certain not to err in affirming of a living cell that it is mortal or immortal, that it cannot be mortal and immortal at the same time, that if it is mortal, then it is not immortal. These propositions impose themselves necessarily on us as a result of the conventions of language which regulate the usage of the logical constants *or, not, if . . . then,* and of the initial convention to distribute all living beings into mortal or immortal. They are propositions relative to the syntax of the words *or, not, if . . . then.* Logical deduction is itself tautological; it does nothing but carry out tautological transformations. The conventions for the use of the conjunctions *not* and *or* are such that if I formulate the two propositions: "*x* is mortal or immortal" and "*a* is not mortal," it is equivalent to saying "*a* is immortal." In short, the rules of logic allow us to submit a system of propositions to all the tautological transformations into statements equivalent by virtue of the linguistic conventions adopted.

Let us consider the philosophical consequences of Wittgenstein's theory. The laws of logic are necessarily true, because they are true by virtue of their proper form, independent of the content and the truth-value of the elementary propositions that compose them. They give us no information about the world, about the events which constitute it, about the bodies which compose it, or about their proper behavior; *they propose no restriction on the universe.* They refer, not to things, but to the language with which we talk about things. They are true *a priori* because they do not depend on experience but on our linguistic conventions. They are *rules of syntax* which define the conditions of the use of *logical conjunctions* and allow us to submit a system of propositions to operations which transform them into other statements tautologically equivalent. To state the Pythagorean theorem is another way of stating the Euclidian system of axioms.

On the other hand, propositions endowed with content, whether they be particular propositions describing singular facts, or general propositions proposed as hypotheses, are all *synthetic*. *Thus the most radical empiricism is justified.* Synthetic propositions express our whole knowledge of the universe. Logic is a kind of auxiliary calculus for the manipulation of synthetic propositions. It allows us to carry out linguistic transformations by which we can deduce from certain synthetic statements other synthetic statements easier to confront with experience. Every *meaningful* proposition is either *synthetic*, or *tautological*, or *contradictory*. If a statement does not fall into one of these three categories, it is *meaningless*. In the last case, the result is what Wittgenstein calls a bad grammar, a false logic. Metaphysics is nothing but a disease of language.

Thus, Aristotle's Apodictic led us to a metaphysical conception of the world, since it forced us to admit a divine understanding, not part of the world of our immediate perceptions. On the other hand, the tautologism of Wittgenstein leads us to the most radical empiricism.

III. THE IMPOSSIBILITY OF RADICAL EMPIRICISM

I should like to show how such an empiricism cannot be maintained. It supposes that the properties of *tautological, synthetic, contradictory,* and *meaningless,* applied to propositions, have an *absolute meaning.* Now this is not the case at all, because there exists an infinity of possible logics; and, according to the logic adopted, the same proposition is *tautological* or *contradictory, analytic* or *synthetic, meaningful* or *meaningless.* These properties, just like those of *indemonstrable* or *demonstrable* as applied to propositions, and of *indefinable* and *definable* as applied to concepts, *are relative to the language we adopt.*

Let us imagine that we are making use of a barometer whose scale is divided by a line into two parts with the indices: *fair weather, foul weather.* The proposition "It is fair or foul weather" is tautological; there will be no need to consult the barometer to affirm it. Such a division of our barometer suffices, if, for instance, we only have to take a walk in the neighborhood; it will suffice to tell us whether we need to take an umbrella or not. But if we must take a dangerous trip into the mountains, it will be better to consult a barometer with three divisions: *fair weather, foul weather, doubtful weather,* and make a decision only if the needle is at fair weather. In this case the proposition "It is fair or foul weather" ceases to be tautological. It does not exhaust all the possibilities, since the weather might be doubtful; it is a synthetic proposition which experience may belie. Similarly, the proposition "It is neither fair nor foul weather" is a contradiction in our first system of divisions; it is a synthetic proposition in the second, since doubtful weather is weather which is by definition neither fair nor foul. If, now, we go up in an airplane, we shall have to push the division of our scale further, for we shall have to take account in our flight of quantitative variations of pressure. In this case, a very large number of values will have to be considered for the actual weather. To adopt the first bipartite division is equivalent to adopting a two-value logic, ruled by the principle of the excluded middle, where a proposition is necessarily true or false. To adopt the second case is equivalent to adopting a three-value logic, where the principle of the excluded third is replaced by the principle of the excluded fourth; where, consequently, a proposition can be neither true nor false, but indeterminate or possible. To adopt the third hypothesis is tantamount to taking a logic of as many values as there are divisions on the scale. We may conceive a logic involving a denumerable infinite set of values. Now, the choice of a bivalent, trivalent, or polyvalent logic is a question of pure convenience, as free as is the graduation of a barometer into two, three, or many divisions. Depending on the order of research, depending on the practical aims that we set ourselves, the choice of such and such a logic and of such and such a graduation is more or less convenient. The designations *analytic* or *synthetic* are not absolute; *they are relative to the previous choice of such and such a logic involving such and such a number of truth-values.*

Consider the possibility of a certain event occurring in the future. Three cases are presented: we can demonstrate, starting from the set-up of present circumstances and known physical laws, that this event will necessarily come to pass; or that it is impossible for it to come to pass; or we shall be able to demonstrate neither one nor the other, which case we express by saying that it is still simply possible that the different eventualities being considered are contingent. A single proposition, namely a prediction, can take on four different modalities: necessary, impossible, possible, and contingent, which, added to the two truth-values of classical logic, true and false, give six modalities. We then have a *modal logic of six values,* that is, a logic in which a proposition can be *true* or *false, necessary* or *contingent, possible* or *impossible.* If we now conceive, in the absence of certainty as to what will happen, among all the possible eventualities numbering n that there are those numbering m which will produce the event expected or feared, we can calculate the probability of the occurrence of this event, a probability equal to $\frac{m}{n}$. We can, besides, either be content to say that the probability of one outcome is greater or less than that of another, or we may quantitatively evaluate its degree of probability relative to all the cases possible. In the first case, we have to do with a *topological logic of probability*, in the second with a *metric logic of probability*. We can also say that an event will occur if there previously comes to pass a first series of circumstances, in turn subordinate to the occurrence of a series of circumstances, and so forth. We shall then have to do with logics of *irreducible complex modalities* or *subordinate probabilities*. Now, in everyday life we spontaneously make use of such *modal or probability logics*. Speaking of the likelihood of the end of the war in the spring of 1942, some demonstrate that it will be inevitable, others, impossible; there are those who more modestly are content to say that it is simply possible; others, more ambitious, tell us that it will come in nine chances out of ten. The prudent restrict themselves to declaring that it will happen if such and such circumstances come to pass, etc. Different minds, constrained by different psychological demands, will unconsciuosly make use, in speaking of the end of the war, of different modal logics.

Logicians of the twentieth century have studied these logics. They have even created a considerab!e number of them, which we may attempt to classify as follows.

IV. CLASSIFICATION OF LOGICS

Let us take our start from the classical bivalent logic in the form in which Russell and Whitehead expounded it in *Principia Mathematica*.

A first step in generalizing classical logic consists in freeing ourselves from the limitation of considering only *two values*, true and false. We can construct logics of two, three, four, five, or any finite number, or of a denumerable infinite number of exhaustive values. A logic of n exhaustive values will always involve a principle analogous to the excluded middle, but stated in the form of a principle of the excluded $n + 1$. We shall call these logics *polyvalent exhaustive logics*.

Lukasiewicz and his students particularly have studied the logic of three exhaustive values; then, further, the logic of a denumerable infinite of exhaustive values. Post has shown that a logic of $n + 1$ values can be interpreted as a synthetic representation of the re!ations between frequencies, truth-probabilities, or truth-levels of n two-valued propositions.

The preceding logics of exhaustive values are only particular cases of the probability-logic of Hans Reichenbach. This probability-logic is a logic with an infinite scale of values, presented as a generalization of ordinary logic, analogous to passing from the geometry of Euclid to that of Riemann. This generalization is arrived at by substituting for the concept of proposition the concept of propositional sequence, and for the concept of truth that of probability. Statements of probability play the same role as statements of truth in ordinary logics. They are presented not in the form of true propositions, but in the form of wagers possessing a *certain weight*.

A second group of logics is made up of the *modal logics*. These are also possible interpretations of certain polyvalent logics. We may arrive at the modal logics by two different methods: either we introduce the different modalities proposed all at once; or, following the method of Gödel, we start from the classical bivalent logic and complete it by addition of new postulates introducing new modalities. In this case we modify, for example, the proposition *p*, which is read "*p* is true" by modal operators, A *p*, E *p*, which are read "It is necessary that *p* is true," "It is possible that *p* is true." These operators allow us to introduce *modal functions*.

The most famous of the modal logics is that which calls into play the six modalities studied by Aristotle in his book *On Interpretation* and in the *Prior Analytics*, namely: *true, false, necessary, contingent, possible, impossible*. This logic of six modalities is reduced to five, if we identify possibility and contingency; it is reduced to four modalities, if we consider that every proposition must be necessary, impossible, possible, or contingent. The modal logic of four values is reduced to three values, if we again identify possibility and contingency. The logic of six modalities has been formalized by MacColl, Lewis, and O. Becker; the logic of five modalities, by O. Becker; the logic of three modalities, by Lukasiewicz.

We arrive at logics of more than six modalities, if we modify the modal propositions in their turn by a modality. We can, for example, consider as necessary, impossible, possible, or contingent, the modal affirmation, itself. We say, for example, "It is necessary that *p* is necessarily true," "It is possible that *p* is necessarily false." These *superposed modalities* can be interpreted as *levels of circumstances;* a first level on which would depend the truth of the circumstances on which *p* depends, and so forth. It is thus that Becker constructed a logic of ten values, interpreted as a logic involving only two levels of circumstances. One might propose logics involving an *infinity of irreducible superpositions of modalities*, distinguished according to whether the order of the modalities is or is not entirely determined. Although Lewis, Lewis and Langford, Becker, Gödel, and Smith have studied or made use of such logics, none of them has been the object of systematic formalization.

A third category of logics will be obtained not, now, by complicating the rules of classical logic, but by weakening these rules. They are arrived at by dropping some of these rules, by declining to use all the methods of reasoning allowed by classical logic.

The logic of Heyting, called "intuitionist logic," is obtained by accepting all the fundamental rules of classical logic except the principle of the excluded middle. This simple suppression leads, it is true, to considerable changes in the employment of the fundamental logical operations. Restricting ourselves to the rules of negation, we find that the principle of contradiction is kept, but affirmation is no longer equivalent to double negation; affirmation implies double negation, but not reciprocally. Contrariwise, negation is equivalent to triple negation; an odd number of superposed negations is equivalent to a simple negation.

Starting from the logic of Heyting, Johansson created a new logic called by him "minimal calculus." It lops off from the logic of Heyting the principle "ex falso sequitur quodlibet," the false implies the true.

The logic L_1 of J. L. Destouches, the logic L_F of Paulette Fevrier, the logic L_2 of Destouches, which contains L_1 and L_F, are all three obtained by weakening the classical logic.

We have not exhausted all the possible logics. Bertrand Russell in *Principia Mathematica* bases the principle of deduction, "If p is true and p implies q, then q is true," on a certain definition of implication:— p v q, characterized by the following table where p expresses the negation of p:

		v
P	Q	P Q defined as —P V Q
T	T	T
T	F	F
F	T	T
F	F	T

Now, this mode of implication, called *material implication,* is only one of the cases of the table of the sixteen truth-functions possible for two propositions, enumerated in the table of Wittgenstein. We can ask ourselves whether there is not something arbitrary in basing the principle of deduction on this definition alone. This is the opinion of Paul Weiss when he declares that, alongside the system of Russell, there is place for as many systems as truth functions, omitting tautology which allows all truths, and contradiction which allows no possibility of truth; this brings the total number of logical systems to fourteen. For example, instead of basing the principle of deduction on the material implication which is defined, starting with negation and addition, by p v q, we could base it on —p v —q, defined by the following table:

VI

P	Q	P Q defined as —P V —Q
T	T	F
T	F	T
F	T	T
F	F	T

We obtain fourteen systems of deduction among which *Principia Mathematica* is one possible case. A new generalization consists in considering the truth-functions of three, four, five . . . *n* propositional variables, instead of two. We then obtain an increasing number of truth-possibilities, starting from which an increasing number of truth-functions can be defined; and each one of these can be the basis of a principle of deduction, so that an unlimited number of systems of deduction is possible.

Paul Weiss has advanced the idea of a still greater generalization, the idea of a logic which would not be concerned solely with propositions and their properties, but with any objects whatever endowed with any properties whatever. Instead of saying: "If such and such a proposition is true, another proposition is tautologically true," one would say: "If this or that is the case, then this or that is tautologically the case."

A project of this kind is carried out in the *Combinatory Logic* of Curry, whose operations bear on absolutely any elements whatever, on *etwas*, as he says; for example, on the very signs of operations represented by capital letters.

Principia Mathematica, which we took as a point of departure for our successive generalizations, is only a very particular specification of such a logic, when we take as objects propositions to which we do not attribute other properties than the values true or false, of which we do not conceive other combinations than truth-functions, and which are connected by means of the Russellian principle of material implication.

V. USE OF THE LOGICS

The logics which we have just adumbrated seem at first like simple intellectual games. But an even superficial acquaintance with the history of ideas reveals that certain among them were spontaneously applied, and explains to us why people passed from one to another.

The bivalent logic of propositions was the first utilized. It suffices to explain the presence or absence in a place of a usual object, the coming to pass or the failure at a given instant of an expected event. If we deprive an object of all its qualities, there remain no other properties than being or not-being, these two properties being exclusive of each other. Thus, Gonseth was able to interpret the bivalent logic of propositions as the physics of any object whatsovere. This logic symbolizes also the observation that an expected event occurs or does not occur at a given instant; it appears thus as the natural canon of our perceptual judgments of existence. But such a logic does not apply to those kinds of weakened objects which we call *fields,* which at the same moment fill all space and can penetrate each other, whereas two ordinary objects cannot coexist in the same place, and the same object cannot act by its presence in different places. It does not apply, either, to the corpuscles of microphysics, of which only the probability of the presence in such or such a place can be determined, and not the presence or absence. It does not apply, further, to future events, concerning which one does not know whether they will or will not occur, and which Aristotle calls future contingents. With respect to them, judgments such as "It is possible that such an event will come to pass" or "It is possible that such an event will not come to pass" are true at the same time. Such judgments are relevant to a logic of three values, implying the principle of bilateral possibility like that of Lukasiewicz.

If we distinguish, further, empirical truth obtained by the sensible observation of a fact from the formal truth obtained by the demonstration of a proposition, we come to admit a logic of five or six modalities, sketched out by Aristotle and formalized by Oscar Becker. This is the logic customarily employed.

If we evaluate the chance of a favorable eventuality coming to pass among a series of possible eventualities, or the frequency of a determinate event in a series of events, we are brought to the use of probability logics, whose propositions are no longer affirmations but presumptions, that is, kinds of affirmations of a weaker sort. The logic of an infinite number of values comes into play everywhere, as in microphysics, where rigid laws of classical mechanics give way to statistical laws. The great merit of the probability-logic of Hans Reichenbach is that it takes into account the progress of our intellect in scientific research, interpreting the concept of probability as a relative frequency of events. The laws of nature, founded on the Baconian induction, appear to us as the most probable wagers that we can make as guides of our action, and these we ameliorate ceaselessly by secondary wagers, improving the primary odds. A scientific theory seems like a concatenation of approximating bets corresponding to systems of connected Baconian inductions.

We find ourselves, in fact, face to face with a great number of logics among which we can choose. Our choice, although free, is not arbitrary. It will in each case be adapted to the domain of facts to which the research is relevant, or to the aim that we set for ourselves in our action.

VI.   REASONS WHICH LEAD ONE TO PASS FROM ONE LOGIC TO ANOTHER

It is historically observable that we substitute for classical logic a stronger or a weaker logic in order to escape a contradiction inherent in a system of ideas which cannot be otherwise surmounted. If the bivalent logic is applied to the most general concept, that of *being,* and this concept is carried to the end, we fall inevitably into the monism of the Eleatics. Parmenides argued thus: "Only being exists, for if another thing that being exists, this thing, being different from being, is non-being, and as such it does not exist." Aristotle was able to escape the monism of Parmenides only by introducing, between absolute non-being and actual being, potential being, the place of the ambiguity of contraries; and the Doctors of the Church were able to escape pantheism only by introducing contingent being, which is created being, to distinguish it from necessary being; and this implies the use of logics of several modalities.

In the theory of sets we stumble on certain antinomies, when we are forced to assert the equivalent of two contradictory propositions. Lukasiewicz believed that he could get around them by means of his three-value logic, for in his logic the principle that a proposition is called false if it entails the equivalence of two contradictory propositions is rejected. The logic of Heyting, however, comes to the same result with much less expense by rejecting the absolute validity of the principle of the excluded middle, but not that of the principle of contradiction.

The theory of complementaries of Nils Bohr implies the validity of mutually exclusive propositions. This contradiction is intolerable in the field of classical logic. It ceases to be so in the field of three-value logic, where we can attribute to both theses, the wave and the corpuscular, the logical value of possibility. In more general fashion, J. L. Destouches showed in his doctor's thesis that the contradictions inherent in wave-mechanics can be overcome only on condition that we adopt a weaker logic which will forbid the steps of reasoning that lead just to the contradictions feared. According to him, two physical theories can always be unified: either by uniting them in a more general theory, if their axiom sets are compatible, the axioms of the new theory being then formed by the logical product of the two others; or, if the two axiom sets are contradictory, that is, if a proposition appears in one theory whose contradictory appears in the other, we may weaken the logic we adopt, so as to give up the logical operations which lead to the contradiction.

VII. philosophical consequences

What philosophical consequences must be drawn from all this?

First of all, the nature and the necessity of the laws of logic cease to be a mystery. The laws of logic are tautologies, which are true *a priori* because they exhaust all the alternatives which we have posed *by convention*, by attributing two, three, several, or an infinite number of values to propositions, or by considering a certain number of modalities. *Propositions have two values in the tradition logic, not because there exists a true principle a priori which imposes itself necessarily on our thought, and which we call the principle of the excluded middle, but simply because we have adopted the convention of calling propositions just those sentences which are capable of assuming only these two values.* We could have proposed other conventions. In general, the laws of logic result from the conventions by means of which we define the propositions, their values, their modalities, and the logical operations; that is, the syntactical structure or the formal syntax of the language which we employ.

The *tautological character* of logic condemns any attempt at panlogism, like that of Leibniz or Hegel, that is, any attempt to draw knowledge from the laws of the logic of our thought. In fact, the laws of logic are empty of all content; there is, in thought, no matter *a priori*.

The *relative character* of logic, on the other hand, condemns any radical empiricism. In fact, whether a proposition has the character of being *analytic* or *synthetic, a priori* or *empirical, tautological* or *contradictory,* depends on the logic adopted. The proposition "The weather is fair or foul" is tautological in a bivalent logic; it is synthetic in a trivalent logic. Its negation "The weather is neither fair nor foul" is contradictory in the first case, non-contradictory in the second. Now, classical logic is constructed with the figure 2 as conventionally as our metric system is constructed with the figure 10. The affirmation that every proposition is true or false, called *the principle of the excluded middle,* is no more necessary than this other: "Every number higher than 99 is written with the aid of at least 3 figures."

If we divide the statements which occur in a science into two groups, the empirical statements and the theoretical statements not directly verifiable by experience; and if we subdivide the latter again into physical principles resulting from a schematizing of experience, and logical laws which allow us to draw from the principles a series of theorems: *we shall always be able to transform a part of the empirical statements into theoretical statements, as Poincaré has well shown; and we shall likewise be able to transform part of the physical principles into logical rules.* These divisions depend on our choice, but this choice is not arbitrary. It is justified by reasons of intellectual temper, of theoretical or practical convenience.

Henri Poincaré showed how we withdraw certain empirical laws from being decided by experience by raising them to the dignity of principles, that is, by transforming them into definitions in disguise; thus the law of the conservation of energy has taken the rank of a principle by becoming the very definition of energy. Two examples, borrowed from Paulette Fevrier and Hans Reichenbach, show how we transform a physical principle into a logical law.

The uncertainty-relations of Heisenberg indicate in quantum physics not a purely psychological failing on the part of the scientist, but a fundamental physical indeterminacy, a law of nature. They express the fact that we cannot precisely measure at the same time both of a pair of complementary physical magnitudes. Take the two propositions: "The particle X has a coördinate-value that falls between q and q + q" and "The particle X has a moment whose value falls between p and p + $\triangle$ p." Either of these propostions taken by itself may be true, but they cannot be true at the same time if $\triangle$ p $\triangle$ q $<<$ h, ($\triangle$ p and $\triangle$ q are less than h), h being Planck's constant. Heisenberg's principle, by introducing the concept of "non-simultaneous observation," ends by rejecting the rule of the propositional calculus, that "the logical product of two true propositions is a true proposition."

Paulette Fevrier, in a note in the *Comptes Rendus de l'Academie des Sciences,* has undertaken to construct a new logic which allows the adequate formalization of the statements of quantum theory, especially those relative to the uncertainy-relations.

Her leading idea is that there are pairs of propositions of which some are *combinable* and some *incombinable;* that is, propositions whose logical product is in the first case possible, in the second case impossible. She is thus brought to construct a logic with three values, T, F, A, namely: *true, false* in the sense of what is possible but is not the case, and *absurd,* in the sense of what could not be the case. The operation corresponding to the product in classical logic is then defined by two matrices, according to whether we have to do with combinable or incombinable propositions:

Combinable Propositions					Incombinable Propositions			
&	T	F	A		&	T	F	A
T	T	F	A		T	A	A	A
F	F	F	A		F	A	A	A
A	A	A	A		A	A	A	A

(A preponderant over F, which   (A preponderant
is preponderant over T.)   absolutely.)

Miss Paulette Fevrier finds that the law that the logical product is distributive is not observed, but it seems to her insufficient to restrict oneself to its suppression. The essential thing seems to be the conception of pairs of incombinable propositions, a conception which is the formal description of physical complementarity in the sense of Bohr. Combinable propositions come into play only in the scale of macrophysical magnitudes; in this case the three-value logic becomes identified with the particular case of the classic bivalent logic used up to now in this field.

Just as the characteristic of a proposition of being *tautological* or *synthetic* varies with the logic adopted, so we can confer only a relative meaning on the qualities of being *empirical* and *formal* as applied to the statements of theoretical physics. The same statement will appear in turn as a generalization of experience, as a definition, or as a logical rule, depending on the conceptual construction adopted. But the preference accorded to one or another such procedure is not arbitrary. There are cases where it is possible to overcome contradictions inherent in a theory only by a change of logic. Carrying it to a conclusion the results of Miss Fevrier, J. L. Destouches has established that it is possible to overcome the contradictions inherent in wave mechanics only on the condition of adopting a weakened logic which forbids the steps of reasoning that lead to the observed contradictions. The modifications, to which the logical rules connecting the different statements are submitted in order to avoid contradictions, introduce indeterminism.

The principle of induction of Bacon and Hume seems the very type of a physical principle incapable of reduction to only logical rule. In fact, it is not guaranteed by logic, since it reaches a conclusion by going from the truth of a particular proposition to the truth of the corresponding universal, which is contrary to the rules of the subalternation of propositions. It is not even defended by experience, since to affirm that the uniformity and frequency observed in the past will be maintained in the future adds something new to the brute data of experience. The principle of induction seems to be either a working hypothesis employed so long as it succeeds, in the manner of Hume, or a synthetic principle *a priori,* in the manner of Kant. Now, if we substitute for the bivalent logic of propositions of Aristotle the probability logic of wagers of Reichenbach, the novelty stated by inductive inference is no longer given as a *true proposition,* but as a *more or less probable wager;* and it is then demonstrable, by means of simple tautologies, that the principle of induction corresponds to the most favorable wager that we can make in predicting the future.

Thus the conceptual and theoretical organization of our knowledge of the world is in no way univocal. Several systems of axioms can be elaborated, several logics adopted. The syntactical character of a statement, the fact that it is *tautological, contradictory, or synthetic,* that it is a *true or false proposition,* or *a more or less probable wager,* changes with the theoretical construction we adopt.

VIII. MEANINGLESS STATEMENTS

We can go further and ask ourselves if the fact that a statement is neither tautological, nor contradictory, nor synthetic, that is, is *meaningless,* possesses an *absolute* character. What has just been said allows us to guess the solution. A statement can be called meaningless *only within a given language and with reference to the logical structure of that language. The same statement can be meaningful in one language and meaningless in another.*

We completely determine the logical structure of a language when we have specified: its *basic vocabulary,* that is the elementary symbols designating the words, numbers, punctuation marks, of which the complex expressions of the language are composed; its *formation rules,* which stipulate which sets of elementary symbols are considered to be propositions; its *transformation rules* or *logical rules of deduction,* which determine which of its propositions, called logical consequences, are considered as tautologically equivalent to other propositions called premises. *The formation and transformation rules make up the syntactic rules of the language adopted.* A statement will be called *meaningless within a given language* in the two following cases: when it is formed of words which cannot be defined starting out from the vocabulary of the language; when these words, definable in the language, are combined contrary to the rules of its syntax. For example, in the language of formal logic, the implication-sign can unite only two propositions; if I write "Man is a rational animal implies Peter," the statement is meaningless. In the language of arithmetic, an exponent can be written only above a number; if I write "$+^2=4$," the set of signs is meaningless.

But it can happen that a statement meaningless in one language is meaningful in another. The language of physics is reducible to statements of coincidences of events, namely to readings of instruments which consist in noting the coincidence of a pointer with a line on a scale. At this degree of abstraction physics can pretend to a universal intersubjective meaning, for the coincidences of events and their causal order have an absolute meaning for all groups of observers. But this universality is paid for at the following price: the physicist drops out the qualitative content of his concepts. For example, he retains only so much of colors as a blind man could understand; for him a color will be a certain wave-length; the colored spectrum of a gas will reduce to the structure of its spectrum. Such sentences as "The union of these two shades is lyrical," "The dress of Pope Leo X in Raphael's picture is not so scarlet as the wine of Spain," will be meaningless in the language of pure physics. They will be meaningful in the less universal language of persons of good vision, since a painter and a paint salesman, a dressmaker and a dyer, do come to understand each other; that is, to equate their linguistic conventions in an intersubjective way, coherent relatively to the sensorial domain of colors. If I say now, "The dark night of the senses is the first of the states of prayer," the statement is meaningless in the language of colors, but it has meaning in the much more restricted language of the mystics, since they come to understand each other, to the point of writing treatises on a asceticism and mysticism. If I state, finally, a dogma such as the hypostatic union of the two natures in Jesus Christ, which is formulated, "In Jesus Christ human nature and divine nature are consubstantially united," the proposition has no meaning for an empiricist who pretends to construct every concept of his language by starting from external perceptions; but it has meaning for a Thomistic theologian, and for those who will agree to consider certain books as revealed, who will agree on the rules of interpretation of these books, and who will accept, furthermore, a certain first philosophy and a certain logic; on condition, however, that the sum of the conventions thus proposed would be coherent.

We are in fact free to take as the domain of our language either pointer readings alone, like the physicist; or the sum of sensory perceptions, like the empirical philosopher; or the sum of our perceptions of color, like the painter; or certain states of the soul, like the mystic; or the contents of certain books, like the theologian. To coördinate as conveniently as possible our pointer readings, our external sensations, our mystical states of the soul, the texts of certain books, we are quite free to construct the language which we consider most apt, in obedience to the syntax that we judge most favorable. We are bound only by three conditions: the syntactical rules that we have chosen must not lead to contradictions; when we wish to be understood by another, we must reveal to him in what manner we have constructed our language;

we are bound, finally, once our language has been chosen, to respect our own conventions.

Once we have delimited the domain of the facts that we wish to study and the proper method of observing them; once we have chosen a language containing a coherent logic; once we have posed in a univocal fashion the correspondence between the symbols of our language and the facts of the domain to be studied; we are bound by our conventions, and *the agreement between the sum of our propositions and the sum of the facts being studied* can be confirmed only with a higher or lower index of frequency. But our conventions rest on previous subjective choices, and objectivity is attained only via these subjective points of view. Science is not "the blue-print employed by the Great Architect of the Universe;" science remains a human adventure which in certain fields is successful.

LOUIS ROUGIER.

NEW SCHOOL FOR SOCIAL RESEARCH.

BIBLIOGRAPHY

1. O. Becker, *Zur Logik der Modalitäten,* Jahrb. f. Phil. u. Phänom. Forschung, 1930, vol. II, pp. 496-548.

2. Curry, *Grundlagen der Kombinatorischen Logik,* Am. J. of Math. 1930.

3. J. L. Destouches, *Essai sur la forme générale des théories physiques; Essai sur l'unité de la physique théorique;* Cluj, Romania, 1938.

4. P. Février, *Les Relations d'incertitude de Heisenberg et la logique,* C. R. Ac. Sc., Paris, 1937, vol. 204, pp. 481-483; *Sur une forme générale de la définition d'une logique,* Ibid., pp. 958-959.

5. R. Feys, *Les Logiques nouvelles des modalités,* Rev. néo-scol. de Phil., nov. 1937 et mai 1938.

6. K. Gödel, *Zur intuitionistischen Arithmetik und Zahlentheorie,* Ergebnisse eines math. Kolloquiums, 1933, Heft 4, pp. 34-38; — *Eine Interpretation des intuitionistischen Aussagenkalküls,* Ibid., pp. 39-40; — *Zum intuitionistischen Aussagenkalküls,* Ibid., pp. 40.

7. F. Gonseth, *Les Mathématiques et la Réalité,* Paris, 1936; — *Qu' est-ce que la Logique?* Paris, 1937.

8. A. Heyting, *Die formalen Regeln der intuitionistischen Logik,* Sitzungsber. Preuss. Akad. Wiss. (phys. math. Klasse), 1930, pp. 42-56; — *Sur la logique intuitionniste,* Acad. Belge Sc., 1930, vol. 16, pp. 157-963; — *Sur la logique intuitionniste,* Enseign. Math., 1932, vol. 31, pp. 271-272 et 274-275; — *Mathematische Grundlagenforschung, Intuitionismus, Beweistheorie,* Berlin, 1934; — *Bemerkungen zu dem Aufsatz von Herrn Freudenthal. . . ,* Compositio Mathematica, 1936, vol. 4, pp. 117-118.

9. I. Johansson, *Der Minimalkalkül, ein reduzierter intuitionistischer Formalismus,* Compositio Mathematica, 1936, vol. 4, pp. 119-136.

10. J. Lukasiewicz, *Philosophische Bemerkungen zum mehrwertigen System des Aussagenkalküls,* C. R. Soc. Sc. L., Varsovie, Classe III, 1930, vol. 23, pp. 51-77; — *Ein Vollständigkeitsbeweis des zweiwertigen Aussagenkalküls,* Ibid., 1923, vol. 24, pp. 153-183; — *Zur Geschichte der Aussagenlogik,* Erkenntnis, 1935-36, vol. 5, pp. 111-131; — *Zur vollen dreiwertigen Aussagenlogik,* Ibid., p. 176.

11. J. Lukasiewicz and A. Tarski, *Untersuchungen über das Aussagenkalkül,* C. R. Soc. Sc. L., Varsovie, Class III, 1930, vol. 23, pp. 30-50.

12. C. I. Lewis, *A Survey of symbolic logic*, Berkeley, 1918; — *Strict implication, An emendation*, Journ. Philos., 1920, vol. 17, pp. 300-302; — *Emch's calculus and strict implication*, Journ. Symb. Log., 1936, vol. I, pp. 77-86.

13. C. I. Lewis and C. H. Langford, *Symbolic Logic*, New York, 1932.

14. E. L. Post, *Introduction to a general theory of elementary proposition*, Amer. Journ. Math., 1921, vol. 43, pp. 163-185.

15. H. Reichenbach, *Wahrscheinlichkeitslehre*, Leiden, 1935.

16. H. B. Smith, The Algebra of propositions, Philosophy of Science, 1936, vol. 3, pp. 551-578.

17. A. Tarski, *Uber die Erweiterungen der unvollständigen Systeme des Aussagenkalküls*, Ergenbnisse eines math. Koll., 1936. Heft 7, pp. 51-57.

18. Paul Weiss, *Two Valued Logic, Another Approach*, Erkenntnis, Bd II, p. 248; The Monist, 1928, p. 542.

19. A. N. Whitehead and B. Russell, *Principia Mathematica*, vol. I, Cambridge, 1925.

20. L. Wittgenstein, *Tractatus logico-philosophicus*, New-York et Londres, 1922.

21. Z. Zawirski, *Ueber das Verhältnis der mehrwertigen Logik zur Wahrscheinlichkeitsrechnung*, Studia Philosophica, 1935, vol. I, pp. 407-442; — *Les rapports de la logique polyvalente avec le calcul des probabilités*, Actes Congrès internat. phil. scient., 1936, IV, pp. 40-45.

Part 6
Haskell B. Curry (1952)

Logical Algebras as Formal Systems: H.B. Curry's Approach to Algebraic Logic

Jonathan P. Seldin

Abstract Nowadays, the usual approach to algebras in mathematics, including algebras of logic, is to postulate a set of objects with operations and relations on them which satisfy certain postulates. With this approach, one uses the general principles of logic in writing proofs, and one assumes the general properties of sets from set theory. This was not the approach taken by H.B. Curry in (Leçons de Logique Algébrique, 1952 [25]) and (Foundations of Mathematical Logic, 1984, Chapter 4 [28]). He took algebras to be formal systems of a certain kind, and he did not assume either set theory or the "rules of logic". I have not seen this approach followed by anybody else.

The purpose of this paper is to explain Curry's approach. To do this, I will begin with a discussion of his approach to formal systems in Sect. 1. Then, in Sect. 2, I will take up his view of algebras.

1 Formal Systems

Today formal systems are usually regarded as formal systems of propositional or predicate logic. In systems of this kind, the objects under consideration are usually defined as certain strings of symbols called *well-formed formulas* (wffs). The systems do not depend on the "rules of logic" because all the axioms and rules are given explicitly. In fact, one of the conditions for being a truly formal system in this sense is that there be a mechanical procedure (which could be computed on a computer like the ones we all use except that there is to be no limitation on memory or the time it takes for the computation to finish) for deciding whether or not a sequence of wffs constitutes a valid proof.

Curry had an approach to formal systems which differs from this picture in some respects. One of those has to do with the way wffs are defined. Curry noticed that strings of symbols to not have unique constructions if the string is more than two symbols long. Notice the following two constructions for the string '*abc*':

$$\frac{\dfrac{a \quad b}{ab} \quad c}{abc} \qquad \frac{a \quad \dfrac{b \quad c}{bc}}{abc}$$

However, in all the usual definitions of wffs, each wff does have a unique construction. Consider, for example, the basic propositional calculus with atomic wffs $p_1, p_2, \ldots$. Computer scientists usually write this definition by saying that the wffs have the syntax

$$F \longrightarrow p_i \,|\, \neg F \,|\, F \wedge F \,|\, F \vee F \,|\, F \supset F.$$

J.-Y. Béziau (ed.), *Universal Logic: An Anthology*, 125–132
Studies in Universal Logic, DOI 10.1007/978-3-0346-0145-0_11, © Springer Basel AG 2012

This is a shorthand way of writing the following inductive definition:

Definition 1 *Well-formed formulas* (wffs) are defined by induction as follows:

1. Every atomic formula p_i is a wff;
2. If F is a wff, so is $\neg F$;
3. If F_1 and F_2 are wffs, then so are $F_1 \wedge F_2$, $F_1 \vee F_2$, and $F_1 \supset F_2$; and
4. Nothing is a wff unless its being so follows from the above three clauses.

It is clear from this definition that there is only one way to construct any given wff using these clauses.

Curry took this property that each wff has only one construction as basic to his idea of formal system.

But he also generalized formal systems beyond those that are usually taken as systems of logic. For Curry, a formal system was any structure with formal objects that are inductively generated so that each one has a unique construction and which also has predicates which apply to the formal objects and whose truth conditions are given explicitly by axioms and rules. Curry explains this in [25, Chapter I], of which a translation of the chapter involved appears elsewhere in this anthology, and also in other works, of which [20, 24, 28, 29] are probably the most important. An exposition can also be found in [31].

One simple example of a formal system is the system of *sams*. It has one primitive formal object, namely 0. It also has one unary operation that forms a new formal object from a given one, indicated by writing the prime symbol "′" after the argument. These are all the formal objects. So the syntax for these formal objects is given by

$$S \longrightarrow 0 \mid S'.$$

There is one predicate, indicated by writing "=" as an infix. This leads to the class of *elementary propositions*, which are all propositions of the form

$$x = y$$

where x and y are formal objects. There is one axiom, namely

$$0 = 0.$$

And there is one rule of inference:

$$\text{If } x = y, \quad \text{then } x' = y'.$$

It is easy to see that the theorems are all those elementary propositions of the form

$$x = x,$$

where x is one of the formal objects.[1] The system is also to be found in [28, p. 256].

In a system like the system of sams, it is not really appropriate to call the formal objects wffs or formulas. Curry felt a need for a general term to refer to the formal objects of a formal system. In his earliest work [1], he called them "entities". When he wrote his dissertation [2] in German, he used the German word *"Etwas"* as a noun, intending to use

---

[1]This is the system called $\mathcal{N}_1$ in [31], where it is incorrectly stated that Curry introduced the system in [28]. The system is given without the name sams in [24] (which was written in 1939) and [29, §1A3].

the word "entity" as an English translation for it. But he then had a conversation with a philosopher who told him that his use of the word "entity" implied some philosophical conclusions that Curry did not want to conclude.[2] He tried calling them *'terms'*[3], as we now refer to the formal objects of λ-calculus or combinatory logic, but this use of the word "term" would conflict with its use in quantifier logic, and in quantifier logic the "terms" of the formal system would be what are usually called "formulas", or wffs. So he invented a new term by taking the first syllable of the English word "object", and started calling the formal objects of a formal system *obs*.[4] Thus, an *ob* is a formal object of any formal system in which the formal objects are inductively defined so that each ob has a unique construction.[5] Since the obs are all inductively defined, they are generated from *atomic obs*, or simply *atoms*, by means of *operations*. In the system of sams, 0 is an atom and the operation is denoted by "*'*" as a suffix.

Later in his career, Curry decided to allow for the possibility that the formal objects of a formal system might be strings of symbols instead of obs, and so he defined what he called a *syntactic formal system* for such systems.[6] When he did this, he started calling the other kind of formal system an *ob system*.

In addition to the formal objects, every formal system has at least one predicate. In systems of propositional and predicate calculus, that predicate, which is unary, is expressed by the English verb "is provable" or by the symbol "⊢" used as a prefix. Curry called systems in which there is only one predicate is of this kind *logistic* systems. In the system of sams, that predicate, which is binary, is denoted by the symbol "=" used as an infix. There are also systems which have a binary predicate used as an infix which represents an ordering relation, which can be denoted by "<" or "≤". Curry called systems with such binary predicates *relational* systems. The *elementary propositions* are those which state that one of the predicates of the system applies to the formal objects. In the system of sams, the elementary propositions are those of the form $x = y$ where x and y are arbitrary obs.

In an ob system, it is not necessary that the obs be defined from strings of symbols; see [14, pp. 230–231], where he discusses obs as physical objects. Thus, it is only in a syntactical system that the obs are required to be linguistic objects. But since it is necessary to communicate information about a formal system, it is necessary to have a language to use for this purpose. Curry called this language the *U language*.[7] It consists of ordinary English (or whatever natural language is the main medium of communication) plus special symbols to name the formal objects and predicates of the formal system. These special symbols, which are part of the U language, is called the A language. The A language is,

---

[2] See Curry's explanation in [3, p. 157]. He continued to use the word "entity" in the papers [4–8], i.e., through the middle 1930s.

[3] See [9–21, 24], i.e., from the late 1930s through about 1950.

[4] Curry seems to have started using the word "ob" during his visit in 1950–1951 to the University of Louvain in Belgium. The word "ob" appears in French in [22, 23, 25, 27] and in English in [26] and later works.

[5] Some people have concluded that obs are the formal objects only of a system of combinatory logic, but this is false. Obs are the formal objects of any formal system defined as indicated above.

[6] See [28, §2C2].

[7] I.e., the language which is used.

in a sense, isomorphic to the formal system. In an ob system, the A language contains symbols to represent the atoms, the operations, and the predicates.

The study of formal systems is not limited to deriving theorems within the system. Instead, general results are proved about the system. An example of such a general result in the system of sams is the result that the theorems consist of the statements of the form $x = x$ for every ob x. Most logicians would call such results *metatheorems*, and so did Curry before about 1948. But Kleene [30] objected that the prefix "meta-" should be reserved for discussions of languages (as in the word "metalanguage"), and Curry, who always liked to avoid arguments over the use of words, decided to replace "meta-" with "epi-" when he was not discussing metalanguages.[8] This tendency he had to coin his own terminology makes many of his works, especially the later ones, hard to read. He did say to me once in his later life that he wished he had stuck with the prefix "meta-", but by then he had written most of what he was going to write in his lifetime. So Curry would have said that the general result about the system of sams mentioned earlier in this paragraph is an *epitheorem*. More complicated epitheorems would include such theorems as soundness results and completeness results, especially for formal systems of logic in the usual sense.

Some epipropositions are very complicated; complicated enough that large parts of mathematics, including calculus and analysis, can be taken as part of the epitheory of the system of sams (if classical logic is used). Usually, Curry limited himself to constructive logic in the epitheory.[9] Others are relatively simple. For example, some epipropositions can be obtained by starting with elementary propositions and building them up using some simple epitheoretic connectives and quantifiers. As elementary epitheoretic connectives, Curry used the following:

&: A & B means "A and B".

or: A or B means "A or B".

$\rightarrow$: $A \rightarrow B$ means "if A then B". This was later (see [29, §2B5c–d]) refined to mean that a proof of A could be converted into a proof of B by constructive means, and the connective $\Rightarrow$ was introduced so that $A \Rightarrow B$ meant that if A is added to the system as a new axiom, then B is provable. In this usage, A and B must be elementary propositions.

$\leftrightarrows$: $A \leftrightarrows B$ means "$A \rightarrow B$. & . $B \rightarrow A$".

Curry's work on Gentzen methods in [20, 28] was based on the idea of formalizing this elementary part of the epitheory.

Note that the connective "not" is not included in the above list. This is because Curry regarded negation as more complicated than the other connectives.[10] Saying that there is no proof of an elementary proposition is not saying anything of logical importance, since it only means that there is no proof *yet*. So Curry dealt with two kinds of negation:

---

[8]See the preface to [24]. See also [20].

[9]I think Curry's original motivation for this was to make his work acceptable to the greatest possible number of mathematicians and logicians. In this, he ignored the fact that to many mathematicians, constructive arguments tend to involve matters that seem obscure.

[10]See [28, Chapter 6].

1. An elementary proposition E is said to be *absurd* if, when it is adjoined as a new axiom, every elementary proposition becomes provable.
2. An elementary proposition E is said to be *refutable* if it implies one of a number of *counteraxioms*, which are assumed to be postulated.

There is one other symbol that Curry used regularly: the symbol "$\equiv$". He used $A \equiv B$ to mean that A and B are the same ob.

2 Algebras

Curry defined an algebra as a formal system with free variables but no bound variables. To fully understand this, we need to look at his view of variables.

In Curry's view, there are three kinds of variables:

1. An *indeterminate* is an ob about which the system postulates nothing except that it is an ob. This idea is now familiar from programming languages, in which a variable is just an identifier for a location in memory in which the program does not specify what is stored in that location, and what is stored there may change as the program is executed. Indeterminates are *free variables*.
2. A *substitutive variable* is one of a set of variables for which one has a rule of substitution. With such a rule, there must be a class of obs (perhaps the class of all obs) of which any arbitrary one may be substituted for the variable. Substitutive variables are also free variables.
3. A *bound variable* is one used with a *binding operator* to indicate a function. The simplest example is the x in $\lambda x \, . \, x^2$, which is the way the squaring function is written in λ-calculus. Bound variables are also used with the quantifiers $(\forall x)A$ and $(\exists x)A$ and with definite and indefinite integrals.

Curry explained that one difference between elementary algebra and calculus is that calculus involves bound variables as well as free variables, whereas elementary algebra involves only free variables.

The difference between indeterminates and substitutive variables can be illustrated by Curry's formal system for group theory. This is the formal system G, which is defined as follows:

1. *Primitive ideas*:
    An infinite sequence of atomic obs: $a_0, a_1, a_2, \ldots$.
    A binary operation, denoted by $\circ$ as an infix.
    A unary operation, denoted by $'$ as a suffix.
    A binary relation, denoted by "$=$" as an infix.
2. *Axioms*: If A, B, and C are arbitrary obs,
    a. $A \circ (B \circ C) = (A \circ B) \circ C$,
    b. $A \circ a_0 = A$,
    c. $A \circ A' = a_0$.

3. *Rules of deduction*: If A, B, and C are arbitrary obs,
   a. If $A = B$, then $B = A$,
   b. If $A = B$ and $B = C$, then $A = C$,
   c. If $A = B$, then $A \circ C = B \circ C$,
   d. If $A = B$, then $C \circ A = C \circ B$.

By 1 above, obs are defined by the syntax

$$A \longrightarrow a_i \,|\, A \circ A \,|\, A',$$

and elementary propositions are those of the form

$$A = B,$$

where A and B are arbitrary obs.

Now the ob a_0 appears in the axioms, and so there is something specified about it besides the fact that it is an ob. But for the obs a_i for $i \geq 1$, nothing is said in the axioms, so nothing is specified about them except that they are all obs. Hence, the obs a_i for $i \geq 1$ are all indeterminates.

The system G can be formulated with substitutive variables instead of indeterminates. This is done by reformulating the axioms as follows:

2. *Axioms*:
   1. $a_1 \circ (a_2 \circ a_3) = (a_1 \circ a_2) \circ a_3$,
   2. $a_1 \circ a_0 = a_1$,
   3. $a_1 \circ a_1' = a_0$.

Then a rule is added which allows the substitution of any ob for a_1, a_2, or a_3 in these axioms. But in no case may an ob be substituted for a_0.

Note that in this formulation, the symbol "=" does not mean identity as obs. An instance of axiom 2 is $a_1 \circ a_0 = a_1$. But $a_1 \circ a_0$ is *not* the same ob as a_1. In other words,

$$a_1 \circ a_0 \not\equiv a_1.$$

This means that what Curry is doing is similar to taking the free algebra generated from the atomic obs by the operations and defining $=$ to be an equivalence relation on the resulting objects. But more than that, by making a formal system out of the algebra, he is not just using the "rules of logic", but is specifying more precisely exactly what steps can be made in a valid proof.

This was Curry's approach to all algebras. For the logical algebras he used relational systems whose predicates were either $=$ or $\leq$. The operations always included $\wedge$ and $\vee$, both of which are idempotent, i.e., we have

$$a \vee a = a, \qquad a \wedge a = a.$$

The following properties are also usually provable:

$$a \leq a \vee b, \qquad a \wedge b \leq a$$

$$b \leq a \vee b, \qquad a \wedge b \leq b$$

$$a \leq x \,\&\, b \leq x \rightarrow a \vee b \leq x, \qquad x \leq a \,\&\, x \leq b \rightarrow x \leq a \wedge b.$$

Algebras with these operations and in which these results are provable are called *lattices*. Curry doubted any other algebras had any logical importance.

References

1. Curry, H.B.: An analysis of logical substitution. Am. J. Math. **51**, 363–384 (1929)
2. Curry, H.B.: Grundlagen der kombinatorischen Logik. Am. J. Math. **52**, 509–536, 789–834 (1930). Inauguraldissertation
3. Curry, H.B.: The universal quantifier in combinatory logic. Ann. Math. **32**, 154–180 (1931)
4. Curry, H.B.: Some additions to the theory of combinators. Am. J. Math. **54**, 551–558 (1932)
5. Curry, H.B.: Apparent variables from the standpoint of combinatory logic. Ann. Math. **34**, 381–404 (1933)
6. Curry, H.B.: Functionality in combinatory logic. Proc. Natl. Acad. Sci. USA **20**, 584–590 (1934)
7. Curry, H.B.: Some properties of equality and implication in combinatory logic. Ann. Math. **35**, 849–850 (1934)
8. Curry, H.B.: First properties of functionality in combinatory logic. Tôhoku Math. J. **41**, 371–401 (1936)
9. Curry, H.B.: Remarks on the definition and nature of mathematics. J. Unified Sci. **9**, 164–169 (1939). All copies of this issue were destroyed during World War II, but some copies were distributed at the International Congress for the Unity of Science in Cambridge, MA, in September, 1939. Reprinted with minor corrections in Dialectica **8**, pp. 228–333 (1954). Reprinted again in P. Benacerraf and H. Putnam (eds.) Philosophy of Mathematics: Selected Readings, pp. 152–156. Prentice Hall, Englewood-Cliffs (1964)
10. Curry, H.B.: The consistency and completeness of the theory of combinators. J. Symb. Log. **6**, 54–61 (1941)
11. Curry, H.B.: A formalization of recursive arithmetic. Am. J. Math. **63**, 263–282 (1941)
12. Curry, H.B.: The paradox of Kleene and Rosser. Trans. Am. Math. Soc. **50**, 454–516 (1941)
13. Curry, H.B.: A revision of the fundamental rules of combinatory logic. J. Symb. Log. **6**, 41–53 (1941)
14. Curry, H.B.: Some aspects of the problem of mathematical rigor. Bull. Am. Math. Soc. **47**, 221–241 (1941)
15. Curry, H.B.: The combinatory foundations of mathematical logic. J. Symb. Log. **7**, 49–64 (1942)
16. Curry, H.B.: The inconsistency of certain formal logics. J. Symb. Log. **7**, 115–117 (1942)
17. Curry, H.B.: Some advances in the combinatory theory of quantification. Proc. Natl. Acad. Sci. USA **28**, 564–569 (1942)
18. Curry, H.B.: Languages and formal systems. In: Proceedings of the Tenth International Congress of Philosophy, Amsterdam, August 11–18, 1948, pp. 770–772. North-Holland, Amsterdam (1949)
19. Curry, H.B.: A simplification of the theory of combinators. Synthese **7**, 391–399 (1949)
20. Curry, H.B.: A Theory of Formal Deducibility. Notre Dame Mathematical Lectures Number, vol. 6. University of Notre Dame, Notre Dame (1950). 2nd edn. (1957)
21. Curry, H.B.: L-semantics as a formal system. In: Congrès International de Philosophie des Sciences, Paris, 1949, II Logique, Actualités Scientifiques et Industrielles, vol. 1134, pp. 19–29. Hermann & Cie, Paris (1951)
22. Curry, H.B.: La logique combinatoire et les antinomies. Rend. Mat. Appl., Ser. V **10**, 360–370 (1951)
23. Curry, H.B.: La théorie des combinateurs. Rend. Mat. Appl., Ser. V **10**, 347–359 (1951)
24. Curry, H.B.: Outlines of a Formalist Philosophy of Mathematics. North-Holland, Amsterdam (1951)
25. Curry, H.B.: Leçons de Logique Algébrique. Gauthier-Villars and Nauwelaerts, Paris and Louvain (1952)
26. Curry, H.B.: On the definition of substitution, replacement and allied notions in an abstract formal system. Rev. Philos. Louvain **50**, 251–269 (1952)
27. Curry, H.B.: Les systémes formels et les langues. In: Les Méthodes Formelles en Axiomatique, Paris, Décembre 1950, Colloques Internationaux du Centre National de la Recherche Scientique, vol. 36, pp. 1–10. Paris (1953)
28. Curry, H.B.: Foundations of Mathematical Logic. McGraw-Hill, New York (1963). Reprinted by Dover (1977) and (1984)
29. Curry, H.B., Feys, R.: Combinatory Logic, vol. 1. North-Holland, Amsterdam (1958). Reprinted (1968) and (1974)

30. Kleene, S.C.: H. B. Curry, some aspects of the problem of mathematical rigor, Bull. Am. Math. Soc. **47**, 221–241 (1941) (review). J. Symb. Log. **6**, 100–102 (1941)
31. Seldin, J.P.: Arithmetic as a study of formal systems. Notre Dame J. Form. Log. **16**, 449–464 (1975)

J.P. Seldin (✉)
Department of Mathematics and Computer Science, University of Lethbridge, Lethbridge, Alberta,
Canada
e-mail: jonathan.seldin@uleth.ca

Lessons on Algebraic Logic.
Introduction and Chapters I and II

Haskell B. Curry

Translator's Note: In this translation, the footnotes are Curry's except for those which begin with the words "Translator's note:". The only footnote of Curry's that has been omitted is one in which he discusses which French word he is using for the English word "completeness".

I have retained Curry's original numbers for displayed formulae. Since there is no formula number (9) in Chapter II, I have left out that formula number here as well.

The sections "Remarques Complémentaires" at the end of the introduction and of each chapter have been omitted. This is partly because these sections are references to literature which is now 50 years out of date.

I would like to thank my wife, Goldie Morgentaler, for helping with the proof-reading of this translation.

Jonathan P. Seldin

Introduction

Definition of Mathematical Logic   Algebraic logic, which is the subject of this course, is conceived here as the most elementary part of mathematical logic. Later we will make precise what we intend to signify by the word "algebraic". But we must indicate immediately what constitutes mathematical logic, of which algebraic logic is the first part.

For this purpose, we remember that the word "logic" has three different meanings in almost all languages.

In the first case, logic is the science of reasoning; it has long been used to distinguish between arguments which are valid and those which are not valid, and to explain the relation between premise and conclusion. Logic, taken in this sense, has since ancient times been a part of philosophy. I use the term *"philosophical logic"* for logic taken in this sense.

Translated by J.P. Seldin, with kind permission of his grandchildren, p.p. Lisa Piper Warren, from: Curry, H.B., "Leçons de Logique Algébrique". Collection de Logique Mathématique, Série A, No. 2. © Gauthier-Villars 1952

In the study of philosophical logic, it has been useful to use mathematical methods; that is, to construct mathematical systems which have some connection to philosophical logic. What mathematical systems are and what their possible connections are to philosophical logic are questions that we will examine later. At this point, we observe that we may study these systems for themselves and that we sometimes call "logic" the study of these systems. Logic so conceived is a part of mathematics; I call it "*mathematical logic*".

In the two senses just discussed, the word "logic" is a proper noun. But we may use it also as a common noun. We frequently call a system of mathematical logic a logic; and, as a result, there are logics of different kinds. We can also use the word to designate philosophical theories, such as the logic of Aristotle, the logic of Hegel, etc.

We can clarify the three senses of the word "logic" by considering the analogous senses of the word "geometry". In the first sense, geometry is the study of space. The word "geometry" is derived from two Greek words meaning measure and earth, and the most ancient geometry must have been the rules of measurement of the ancient Egyptian surveyors. Geometry as the study of space is a branch of physics or the philosophy of nature. But alongside this there is geometry as a branch of mathematics. In this sense, one considers systems of mathematics which are suggested by the study of natural space or which have some analogy with systems so suggested. Finally, there are diverse geometries that one studies in mathematical geometry—Euclidean, non-Euclidean, projective, differential, finite, non-Archimedian etc. geometries.[1]

This analogy with geometry suggests a number of points for discussion. The term "geometry" as the name for a part of the science of nature is almost obsolete; nowadays the term almost always means mathematical geometry. In the latter sense, one has systems which have hardly any connection with the theory of space. Mathematical geometry has developed to such an extent that it has really become a new subject; and for the theory of natural or real space there is a need for another term. Some thinking such as this will hold for logic. It is certain that in mathematical logic one must consider, for reasons of analogy or structure, systems that have little in common with philosophical logic. But no analogy can be pushed too far. There is currently a unity between the two aspects of logic. What will happen about this in the future we do not need to settle here.

We study here mathematical logic so conceived. We also study it from the mathematical point of view. The matters which we treat are fundamental from this point of view. I suppose we have so far shed little light on logic or on mathematics, but we do not yet have the technical knowledge to do more.

We will intentionally not define more precisely the word "logic". In fact, one cannot define any science by exactly describing its boundaries; rather, one defines the central ideas and leaves the boundaries indeterminate. Thus, we speak below of "logical systems", "algebraic logics", etc., without giving a criterion for deciding whether a given system is or is not logic. We will say only that these systems etc. have some sort of connection to philosophical logic. The significance of the word will be made clear in due course.

---

[1] Some of these terms sometimes designate parts of geometry and are thus proper nouns. But one also speaks for example of a finite projective geometry. It is this use of these nouns as common nouns which we take account of in the text.

Chapter I
Formal Systems

As the fundamental idea of mathematical logic we take that of formal system. We must explain what this idea consists of. Before getting to the general definition, it will be instructive to sketch the evolution of this concept from more naive ideas. We will consider first a very simple example. After these preliminaries the general definition should be easy to understand. Finally we will discuss the significance of formal systems and certain of their properties.

1 Evolution of the Concept of Formal System

We begin our discussion with what we might call the naive axiomatic method. You already know that one begins the study of elementary geometry by enunciating certain propositions called axioms, which are accepted without proof. All the propositions which are accepted after that must be proved. A theory is called axiomatic when the theorems, i.e. the accepted propositions, are arranged in a deductive order so that each one is either an axiom or else is a consequence of some preceding theorems. The propositions of the theory treat certain concepts which are also arranged in a quasi-deductive order; some—such as point, line, and geometry—are primitive concepts, the others—such as square, circle—are derived by definition. At the level of naive thought, the primitive concepts are regarded as ideas given intuitively, and the axioms, and hence the theorems, as intuitively true.

One may regard the notion of formal system as the result of an evolution of this first idea. First, one notices that the theorems are formally deduced from the axioms. Less important than this are the primitive concepts, provided that they satisfy the axioms. One may regard these concepts as variable entities and, in a certain sense, the theory as a function for which these variables are the arguments. If one gives definite values to these arguments, one has an *interpretation* of the theory, and for every interpretation, if the interpreted axioms are true, the theorems are also true. One may have many interpretations of this kind. In plane geometry, for example, one may regard a point as a pair of numbers, a line as a linear equation, and the incidence of a point on a line as the fact that the equation is satisfied. But the interpretation is in essence of no importance. One arrives thus at the conception of an abstract axiomatic theory, which is defined, like Euclidean or plane geometry, by its axioms and its primitive concepts. The theory consists essentially— leaving aside the concepts and definitions—in the axioms and the theorems which one deduces following the rules of logic.

One arrives at a second step of the evolution when one notes that the expression "rules of logic" contains something vague. Just what are these rules? Certainly not those of traditional logic, because it has long been known that those rules are not sufficient. Furthermore, the antinomies which have been discovered show that one cannot accept without criticism the reasoning that is usual in mathematics. Finally, if one wishes to use the axiomatic method for logic itself, the conception indicated below creates a vicious circle; at the same time we study logic, we use logic. We must thus replace the "rules of logic" with rules given explicitly. When we admit that these rules, like the axioms, may be arbitrary, we have arrived at the notion of formal system.

2 An Example

Before going any further, we discuss a very simple example. Consider a class of objects which we will call *sams* (the term is a neologism).[2] We suppose that there is a particular sam which we call 0; that there is also an operation which forms from a given sam a new sam, and that every sam except 0 is formed from 0 by applying this operation one or more times. We denote this operation by a sign "'" (prime) written after the argument.[3] Then the sams are $0, 0', 0'', \ldots$, and there are no others than those in this sequence. We postulate also a relation between two sams, and that relation we designate by the usual symbol "=", equality. We thus have a class of propositions of the form

$$x = y, \tag{1}$$

where x and y are sams. We call these propositions elementary propositions. We choose as the only axiom the proposition

$$0 = 0, \tag{2}$$

and we postulate the rule of deduction

$$\text{If } x = y, \quad \text{then } x' = y'.$$

In this system, the proposition

$$0'' = 0''$$

is a theorem; because we have the proof

$$0 = 0,$$
$$0' = 0',$$
$$0'' = 0'',$$

where the first proposition is an axiom, the second is deduced from the first by the rule, and the third similarly from the second. We may now convince ourselves that all elementary propositions of the form

$$x = x, \tag{3}$$

and only those, are theorems.

There is clearly an analogy between sams and numbers. This is why we employ the word "sam", which is a modification of the Hungarian word "szám". (I chose a word in a little known language expressly to underline the abstract and arbitrary character of what we are doing.)

3 General Definition of a Formal System

We may now proceed in a more general fashion. We say that we have a *formal system* just when we enunciate conventions which determine a class of propositions, the *elementary*

---

[2]Translator's footnote: This means that it is a new term coined by the author.

[3]For an explanation of the quotation marks, see Sect. 4 below.

propositions, and which engender within it a subclass, the *elementary theorems*. It is necessary that the conventions be objective. For the elementary propositions it is necessary that the conventions determine, for each given proposition, whether it is or is not an elementary proposition; we express this situation by saying that the elementary propositions form a *definite class*.[4] For the theorems, it is necessary that we begin with a definite class, the *axioms*, and also that we have the definite *rules of deduction*; so that the elementary theorems are the axioms and the propositions which one deduces from them according to the rules. In the above example the elementary theorems also form a definite class, but this is not the case in general. In the most interesting systems of logic and of mathematics there is no definite procedure for deciding whether any given proposition is or is not an elementary theorem. But the notion of proof is definite in the sense that one can always decide whether a sequence of elementary propositions is or is not a proof of its last proposition. Employing a more technical language, the elementary propositions forms a *recursive class*,[5] but the elementary theorems are *recursively generated* from the axioms by the rules.

For enunciating the definitions of this sort, one needs some auxiliary notions. We need a class of objects, like the sams in the above example; in the general case, I call these things *obs*.[6] One may have more than one class of obs in more complicated systems. We also need *operations* for forming new obs from old ones, and we need *predicates*[7] for forming elementary propositions from obs; in our example there is only one operation, designated by the prime, and one predicate, known as equality. With regard to the obs, they must form a recursive class which is, in the end, generated by the operations from the primitive obs; but it is not necessary to specify particular objects for the obs, it is enough to say that they exist. These notions—primitive obs, operations, and predicates—constitute the collection of primitive ideas; by contrast the axioms and the rules form the collection of *postulates*. The collection of conventions which define a formal system is called its *primitive frame*; these frames thus specify the primitive ideas and give the postulates.

4 Semiotic Aspect: The Languages U and A

One may make more precise the notion of formal system in regard to the point of view of language. For this it is necessary to explain some technical terms which have proven useful in the discussion of symbols.

---

[4]A class is definite if one can always decide, by a finite procedure, if a given object is or is not a member of the class; a rule is definite if, given a sequence of propositions, one can always decide, by a finite procedure, whether the last of the sequence of propositions is or is not a consequence of the others in virtue of the rule.

[5]The word "recursive" is not applied rigorously except to sets of natural numbers. But it has been shown that one can "arithmetize" formal systems, such that the ideas are applicable to the numerical images of propositions etc., and thus indirectly to these notions themselves.

[6]Translator's footnote: Curry coined this word "ob" by taking the first syllable of the English word "object".

[7]Here I use the word "predicate" in the sense of modern logic, where a predicate may have more than one argument.

The word "symbol" is not defined: we suppose only that the symbols are objects which can be produced and repeated such as letters and the usual signs. The word "language" is used here in a generalized sense, as the name for an entirely arbitrary system of symbols. Less important is whether or not one uses a language to communicate; in the first case, the language is called a *communicative language*. An *expression* of a language is any finite sequence of symbols. In speaking of expressions the current usage is to employ an example of an expression between quotations marks as the name for this expression, for example: Paris is a city in France, but "Paris" is a word with five letters. This usage is convenient and avoids certain confusions; therefore we conform to it here.

The construction of a formal system, like every intellectual activity, must be explained in a communicative language mutually understood by the author and the listener. I call this language the *U language*. It is a well determined language, but it is not a rigidly fixed one. One may introduce new expressions by definitions, make precise old expressions, etc.

One may identify the establishment of a formal system with the introduction of new symbols into U. These symbols form by themselves a language which I call the A language. This A language contains the names for the primitive obs and the symbols for the operations and the predicates; its grammar is such that it can name all the obs and enunciate all the elementary propositions. Thus, in our example, the primitive symbols of the A language are three in number, namely "0", "'" (i.e., the prime symbol), and "=". Here "0" is a noun, "'" is a suffix which forms from a noun another noun, and "=" is a verb which, placed between two nouns, forms a sentence. But the A language cannot enunciate the rules of the formal system; in particular the symbols 'x" and "y" of Sect. 2 are not part of A. The primitive frame of a formal system may be regarded as a collection of rules which specify the structure of the A language and its usage as an addition to U. The essence of the formal method is that this A language must be employed in U following precise rules, without the significance of certain words (the nouns) being completely determined.

Our last sentence contradicts a common opinion according to which one cannot reason formally without speaking of the symbols. It is doubtless useful to regard the obs as symbols or expressions of a language, called the *object language*, completely separate from U. Thus, one may agree that 0 is an asterisk, "*", and that "'" designates the operation of adding on the right an exclamation point, "!", such that the three first sams will be the expressions on the right of the following table (on the left are the names of these expressions):

$$
\begin{array}{ll}
0 & * \\
0' & *! \\
0'' & *!!
\end{array}
$$

When one assigns particular objects to obs, as we have done for sams in this table, one says that the system is *represented* in the collection of given objects. One may, in effect, represent an arbitrary system in any given object language provided that it contains at least two distinct symbols. With such a representation one may mitigate the above conditions of exactitude and profit from the intuitive properties of the sequences of symbols. But if one maintains, as we do here, the rigor of the conception of a formal system, such a representation has no advantage.

5 Meta-languages

The preceding considerations lead naturally to the notion of a meta-language. Although we will have no need for this concept in the following chapters, one encounters it almost all the time in the literature, and its discussion will explain several points on the nature of a formal system. We thus devote a section to its discussion; but this discussion is not necessary for what follows.

We begin by illustrating this concept for our system of sams. We consider the A language for this system, a language which for the moment we will call the language L. We wish to isolate L completely from U and to regard it in turn as an object language. We also wish to describe, in U naturally, and without any use of L, which expressions of L express elementary propositions and which of those express elementary theorems. For this we introduce the following table in U, the symbols on the left being names for the symbols on the right.

$$
\begin{array}{cc}
z & 0 \\
p & {}' \\
e & =
\end{array}
$$

For the grammar of L we avoid the terms "noun", "sentence", and "sentence designating a theorem", of which the usual sense may imply that we have constructed a language that we use. In place of these, we use respectively the terms "well-formed expression" or "numeral",[8] "formula", and "provable formula". We also use the symbol "$\wedge$" as an infix[9] symbol for the operation of concatenation, i.e., the operation of joining two expressions (which are written on the two sides of the symbol) immediately one after the other. Thus, if X and Y are expressions, $X \wedge Y$ is the expression which is formed by writing Y immediately to the right of X; for example, if X is "aba" and Y is "cacb", then $X \wedge Y$ is "abacacb". The description that we have sought is the following:

1° *Expressions.* They are recursively generated by the rules:
    a) the symbols z, p, and e are expressions,
    b) if X and Y are expressions, so is $X \wedge Y$.
2° *Numerals.* They are recursively generated by the rules:
    a) the symbol z is a numeral,
    b) if X is a numeral, so is $X \wedge p$.
3° *Formulas.* They are expressions of the form $(X \wedge e) \wedge Y$, where X and Y are numerals.
4° *Provable formulas.* They are generated recursively by the two rules:
    a) the formula $(z \wedge e) \wedge z$ is provable.
    b) if $(X \wedge e) \wedge Y$ is a provable formula, then so is the formula $((X \wedge p) \wedge e) \wedge (Y \wedge p)$.

We now have a formal system which we call a *syntactic system*. The language M, which plays here the role of an A language, is a *meta-language* for L. We will soon make this preliminary definition more precise.

We observe next that the above syntactic system is not completely satisfying as a formulation of the grammar of L. As a matter of fact, the expressions $(z \wedge e) \wedge z$ and

---

[8]The first term is employed for systems in general, the second only for the present one.

[9]"Infix" is a term of grammar; it is distinguished from prefix and suffix; it designates a particle introduced in the middle of a word.

$z \wedge (e \wedge z)$ are both identical to the expression "$0 = 0$". Thus, the expressions of the syntactic system are not the expressions of L, but the basis for constructing these expressions by concatenation. To avoid all confusion, it is necessary to replace the word "expression" by another; we will choose "compound". To create a satisfactory grammar we must have a relation of equality, with the postulates which give its ordinary properties and the associativity of concatenation.[10] With this addition we have a formal system which we may call the syntax of L; its A language, i.e., the language M, is a syntactic meta-language of L. It contains the nouns "z", "p", "e", and the operation "$\wedge$". But with regards to predicates there is still something not determined in it.

The resolution of this lack of determination offers us a new clarification of the concept of a formal system. It is often convenient to call the part of a system which deals with the elementary propositions its *morphology*, by contrast with the *theory proper* which deals with the theorems. One may take in more than one way the line which separates these two parts. If we take the compounds for the obs, our language M will have four predicates— i.e., to be a numeral, to be a formula, to be a provable formula, and equality—and the elementary propositions will be formed by applying these predicates to obs. On the other hand, we can take for the elementary propositions only those of the form "x is a provable formula" and relegate the other predicates to the morphology.

Each of these two methods has its advantages. The former is more abstract and explicit; the latter is more practical and convenient. The morphology of a system must be intuitively comprehensible. Thus, for the strictest formalization one tries to reduce the morphology to a minimum. This is the case when one has only one kind of obs, and one can apply the operations and predicates to these obs without distinction. Then the morphology consists only of the enumeration of the operations and the predicates, with the specification of the number of arguments in each case, and the primitive obs. A system of this kind is called *completely formal*. This is the ideal conception; we adopt as a basis of the most abstract and rigorous studies. But our definition of a formal system admits the possibility of a more complicated morphology. The only essential restriction is that the morphology be definite. In practice one takes no account of details one considers trivial. It will be convenient to consider systems in which some of the conditions of the definition of a formal system are relaxed; we call these systems *partially formal*. We can see that there is no technical difference between the two conceptions for the case of a syntactic system, and that the set of predicates of the language M depends on the manner of separating the morphology and the theory; but this makes no essential difference in the result.

In a general way, if L is a given language, a *meta-language* of L is the A language of a (partially) formal system called the *meta-system* which has L for its subject. If this meta-language M and the system which it expresses treats L as a system of objects without meaning (as is the case here), one speaks of a syntactic language or meta-language; if it contains L (or expressions synonymous with those of L) and the predicates express relations between L and its objects, one speaks of a *semantic* system. In both cases, one must avoid a confusion between the conceptions of meta-language and the U language, although the two conceptions are sometimes confused in the current literature.

---

[10]For the system which we are considering here, this is not quite necessary, because we can specify a unique method for constructing each expression. (We have done this in the above statement.) But then the theory will appear incomplete, and besides, a more complicated addition may be necessary in more complicated cases.

We can see now that the construction of a meta-language is a special case of the construction of an A language of a formal system. It has the peculiarity that the obs are concrete and well determined objects, where it follows that one has the possibility of satisfying only approximately the demands of formal methods. But the construction of a meta-language is only the most formal of those of an A language of an *abstract formal system*, i.e., a system in which we say nothing about the obs. For example, for our language L, we show that the meta-system, if it is formal, has essentially the same content as the original system. If there are advantages in the meta-linguistic method, they are of a psychological character. In the two cases, the construction of the A language is the important point. The conceptual language which we have used to describe an abstract formal system is not essential, and if one prefers, one can replace it by a more nominalistic language.[11]

6 Epitheorems

We have spoken of elementary theorems. Are there other theorems? Certainly. In so far as the propositions are expressed in the U language, there is the possibility of forming other propositions by the regular procedures of U. We distinguish these new propositions, theorems, etc. by the prefix "epi".[12] For example in the system of sams, the assertion that all propositions of the form (3)[13] are elementary theorems is an epitheorem. The most important results of modern mathematical logic are epitheorems.

The criteria for the truth of epitheorems is naturally more complicated than that of elementary theorems. To tell the truth, there are many different types of epitheorems, and each type has it own criterion. The simplest epitheorems are probably the derived rules. For these, a proof must be constructive, i.e., such that given a definite procedure permitting the construction and following the general instructions, the proof covers every special case. In these proofs one often uses a principal of mathematical induction. For example, one proves as follows the epitheorem that we cited above: the proposition (2)[14] is true (i.e., is a theorem) because it is an axiom; if (3) is true, then by virtue of the rule, the proposition

$$x' = x'$$

is also true. This proves the epitheorem because the proof of every special case of (3) can be constructed by repeating the second step above. This application of mathematical induction is justified, by definition, because the obs of the system are recursively generated. An induction of this kind is called a *structural induction*.

---

[11]One may also emphasize that the replacement of the symbols of the A language by other convenient symbols changes nothing essential. One may say in this case that we have two different A languages, or two presentations.

[12]Up to 1948 I used the prefix "meta". But Kleene (see his note [6]) has indicated that this manner of speaking may be a source of confusion, because the notion indicated by "epi" is not the same as that that indicated by "meta" in Sect. 5. Although the two notions are connected, the criticism is justified. Thus, I began in [2] to use the prefix "epi". Its meaning is more extensive than that of "meta", which is henceforth reserved for considerations which have to do explicitly with a meta-language.

[13]Translator's note: $x = x$.

[14]Translator's note: $0 = 0$.

Among the epitheorems, those that express consistency, completeness, and decidability or undecidability are especially interesting. A system is said to be *consistent* if there is an elementary proposition which is not a theorem; it is called *complete* if the addition as a new axiom of any elementary proposition whatever, provided that it is not a theorem, makes the extended system inconsistent; and it is *decidable* if there is a constructive procedure for deciding for every elementary proposition whether it is or is not a theorem; i.e., if the class of elementary theorems is definite. To prove consistency, for example, one must find a property of elementary propositions such that one can prove constructively that it applies to every elementary theorem, and then produce an elementary proposition which does not have this property. The first of these requirements requires an induction on the steps of a proof; one proves that the axioms have the property and then one proves that the conclusion of a rule has the property, provided that the premises also have it. We call an induction of this kind a *deductive induction*.

These procedures of definition may equally be regarded as epitheoretic. A definition is an introduction of a new symbolism in U, and it is controlled by the general conventions of U. We note only that it is possible to have definitions by recursion (normally structural).

It is not useful to consider here epimethods in more detail. We consider in this work only very simple epimethods, and we shall discuss these methods when we get to them. But I must emphasize that epitheorems are the heart and the spirit of formal methods.

7 Significance of a Formal System

We may define formal systems in a completely arbitrary manner. But we are interested in only a very small subclass of all imaginable systems. These are the systems which lead in one fashion or another to an objective which is worthy of being taken seriously. I call *acceptability* the property of a system to be adopted to its objective. One may consider it as a kind of truth which is applicable to a formal system as a whole; but it is a concept relative to its objective, and there is no way to say that a system is absolutely acceptable. The simplest case of acceptability is that in which one has an interpretation of some sort. We say that we have a *direct interpretation*[15] when there is a correspondence not only between the obs and real objects, but also between the predicates of the system and intuitive predicates. To each elementary proposition corresponds then an intuitive proposition about reality. The system is then acceptable when all the intuitive propositions which correspond to elementary theorems are true. But a system may be acceptable in more general circumstances than these, such as wave mechanics, for example.

Because of the fact that the rules are given explicitly one can conclude that the truth of an elementary proposition does not depend on knowledge of the principles of logic, except, of course, the principles which are necessary for knowledge and human communication in general. For the control of the validity of a demonstration is a definite procedure; one has only to see whether the rules are used correctly. This is the importance of

---

[15]In a more general sense, one has an *interpretation* when one has a correspondence between all the elementary propositions and some intuitive propositions without the necessity of having a correspondence between the obs and real things. For example the elementary propositions of the system G of Chapter II Sect. 2 may be interpreted as the general laws of groups.

the formal method for logic. One may study logical systems without presupposing logic. One may imagine different systems of logic; one may imagine that different systems of logic are acceptable for different purposes and then examine the consequences of this hypothesis. But we do not have to decide these questions of acceptability here; these are questions of philosophical logic.

One may adopt the notion of formal system as the central idea of mathematics. Mathematics may be defined as the science of formal systems. Mathematical logic is thus the study of formal systems which have, vis-a-vis logic, the same connection that geometry has vis-a-vis space.

8 Reduction of a Formal System

We consider now how one may reduce a formal system to special forms.

The reduction that is most important for us is that to a system with only one predicate, one which has only one argument. Let S be a given formal system. We form now a new formal system T as follows: The obs of T will be the same as those of S, and each operation of S will be adopted as an operation of T. For each predicate f of S, we adjoin to T a new operation F with the same arguments of f. The only predicate of T will be a predicate of one argument, which we will designate by Frege's sign "⊢". Then we identify each proposition of S of the form

$$f(a_1, a_2, \ldots, a_n),$$

where $a_1, a_2, \ldots a_n$ are obs, with the proposition

$$\vdash F(a_1, a_2, \ldots a_n)$$

of T. This creates a bijective correspondence between the theorems of the two systems, and one sees that the system T is, in a certain sense, essentially the same as S.

If S is completely formalized, T will be also; but T contains obs and elementary propositions which are not obs or elementary propositions of S. (For example, $F(a_1, a_2, \ldots, a_n)$ is not an ob of S, and $\vdash a_1$ is not an elementary proposition.) To avoid this, one may formulate in T a special class of obs, the *formulas*, which are the images of the elementary propositions of S. But this complication is not always necessary, because in many cases the obs and the adventitious propositions cause no harm.

This reduction is usually made in logic. We call a system in which there is only one predicate and a predicate of this kind a *logistic system*. We call this predicate affirmation, because the proposition $\vdash X$ indicates that the formula X is proved or affirmed. The predicate is usually expressed by the words "is provable" or "is affirmed" or often by the sign "⊢".

The reduction is natural for syntactic meta-systems. Actually, these systems are in general, by definition, logistic. But it is often artificial for mathematical systems. For example, if we reduce in this way our systems of sams, we must replace the relation of equality by a symbol "□" used as infix, and the rules become the following:[16]

---

[16]We have here a system which can be transformed into a meta-system of our system of sams by simply changing the symbols. In fact, if we replace "□" by "e" and the other symbols in Sect. 5, and if we

0 is an ob.

If x is an ob, x' is also an ob.

If x and y are obs, $x \square y$ is a formula.

$0 \square 0$ is a provable formula.

If $x \square y$ is a provable formula, then $x' \square y'$ is also a provable formula.

Here the formulas are ordered pairs of numbers, the provable formulas are a subclass of these pairs. A similar reduction of systems of ordinary algebra will be truly artificial.

If one restricts oneself to logistic systems, one may give a definition of formal systems a bit different from that of Sect. 3. We will then speak of formulas instead of elementary propositions, provable formulas in place of elementary theorems. The axioms will be a class of formulas, and the rules will transform formulas into other formulas. This a bit closer to the conception of Hilbert. But the conception adopted here is more natural for certain mathematical theories.

Another reduction is that to one sole operation, which is a binary application called "application". It can be done by a similar method. It has great importance for combinatory logic, where one tries to simplify as much as possible the ultimate foundations. But it does not interest us here. We content ourselves with the reduction to a logistic form, and we do not always use it.

Chapter II
Logical Algebras

We will study here a special class of systems which we call algebras. They are systems with free variables (but without bound variables) and a transitive fundamental relation.

1 Variables

We consider first the meanings of the word "variable". I say "the meanings" because there is more than one. Moreover, their discussion is difficult, and certain misconceptions in this subject are perhaps responsible for the ridiculous idea that in mathematics we speak about symbols.

Originally, it was indeed true that a variable was a symbol. In the U language we distinguish between constant symbols, which have a fixed meaning, and variable symbols, whose meaning can change. Thus the symbols "x" and "y" in the definition of the primitive frame of our system of sams are variables. We use variables of this type each time we give a general rule. I call these symbols *intuitive variables*. An intuitive variable is thus necessarily a symbol—a symbol of the U language which we use in this way.

---

regard "p" and "e" as operational symbols, we have a meta-system much simpler than that of Sect. 5. This method may be generalized, and it reformulates the basis of one of the methods used for formalizing syntax. We have followed a more complicated method in Sect. 5 because it is more naive, and it illustrates certain points of our discussion.

But this does not take account of all uses. Consider, for example, the sentences:

$$x^2 - 1 = (x - 1)(x + 1), \tag{1}$$

$$x^2 \text{ is a function of } x, \tag{2}$$

$$\int_0^3 x^2 \, dx = 9, \tag{3}$$

$$\frac{dx^2}{dx} = 2x. \tag{4}$$

In each of these cases we usually say that x is a variable (or the variable). What does this mean? One often says that this affirmation is false, we must say that "x" is a variable. If one means by that that "x" is an intuitive variable in the same sense as in I Sect. 2, the phrase can only be true for (1), and in the other cases it is certainly false. Similarly, for (1), the sentence can only be true if (1) is an intuitive equation. But there is another explanation. In all cases we employ "x" in a particular way. We may regard the phrase "x is a variable" as a paraphrase for the description of the way we use it. We have had the same situation for the symbols of an A language. Thus a possible explanation of the notion of a variable is that a variable is a special kind of ob of a (partially) formal system. We call a variable so conceived a *formal variable*.

If we examine (1)–(4) more attentively, we will soon convince ourselves that these equations say nothing about x or "x". The symbol "x" is a symbol which we use to say something about other objects. These objects are functions, i.e., the laws which cause to correspond to each given object another object which is uniquely determined when the first is given. All the equations (1)–(4) say something about certain functions. For example, (1) says that the functions $x^2 - 1$ and $(x - 1) \cdot (x + 1)$ are identical, i.e., that the law which causes to correspond to each number X ("X" is here an intuitive variable) the number $X^2 - 1$ is the same law which causes $(X - 1) \cdot (X + 1)$ to correspond to X. The usual notation is obscure and confusing because we do not have a clear and systematic notation for functions. In introducing a notation of this kind we will see that all the equations (1)–(4) really say something about functions.

The notation which we will use was suggested by my colleague Professor Church of Princeton University. Let "M" be an abbreviation for an expression $\mathcal{M}$ (of a certain A language) which contains the symbol "x". Church designates by

$$\lambda x \cdot M$$

the function which gives M as a function of x, i.e., the law which associates to each ob X the ob which is designated by the expression obtained in substituting X for x in $\mathcal{M}$. For example, $\lambda x \cdot x^2$ is the squaring function. If we designate the operations of differentiating and integrating from 0 to 3 respectively by "D" and "J", the sentences (1)–(4) become

$$\lambda x \cdot x^2 - 1 = \lambda x \cdot (x - 1)(x + 1), \tag{5}$$

$$\lambda x \cdot x^2 \text{ is a function}, \tag{6}$$

$$J(\lambda x \cdot x^2) = 9, \tag{7}$$

$$D(\lambda x \cdot x^2) = \lambda x \cdot 2x. \tag{8}$$

The sentences (5)–(8) are clearer than (1)–(4), but they are not a definite response to our question. They contain the symbol "x" although they do treat of x, and we still have a need to explain the notion of formal variable. It must be possible to explain these propositions directly without using artifices such as formal variables. This is indeed possible, and constitutes the fundamental principle of combinatory logic. But we will not attempt here to achieve such a deep analysis. We content ourselves with formalizing precisely the systems with formal variables.

Formal variables are of three kinds, which are: 1^0 indeterminates—2^0 substitutive variables—3^0 bound variables.

An *indeterminate* is an ob in the primitive frame about which we say nothing except that it is an ob. It is not necessary that this ob be labelled as a variable; one may have indeterminates in a completely formal system.

Substitutive variables are those for which one has a rule of substitution. With this rule there must be a class of obs for which arbitrary obs may be substituted. The obs are substitutive variables. This class must be specified as a special class of primitive obs, and thus the system cannot be completely formal.

Bound variables are those used to designate functions as in the examples (5)–(8). Intuitively, one has bound variables when one has operations or predicates which take functions as arguments. These variables require a rule of substitution and involve some additional complications.

By contrast with bound variables, we call the other variables *free variables*. We will see that the two kinds of free variables have much in common, whereas free variables and bound variables are essentially different. A system with free variables but without bound variables may be called an *algebra*, a system with bound variables a *calculus*. Although this usage is not current in logic, it is highly worthy of consideration, and it is adopted here.[17]

In this work, we leave bound variables and calculi aside and consider especially algebras.

2 An Example from the Theory of Groups

Before going any further, it will be useful to consider an example. The system of sams has no kind of variables, and thus we will need to consider another. A convenient example is available from the theory of groups. We seek to formalize this theory by constructing a system whose elementary theorems are, in interpretation, the equations satisfied by the unspecified elements of an arbitrary group.[18]

---

[17]It corresponds also to mathematical usage; in fact, it is the presence of bound variables that differentiates infinitesimal calculus from algebra.

[18]This is an example of the more general kind of interpretation which was mentioned in I Sect. 7 footnote 15. If we use intuitively the quantifiers of higher symbolic logic, we may clarify this as follows: the interpretation for example of the elementary proposition

$$a_1 \circ a_2 = a_2 \circ a_1,$$

is the intuitive proposition

$$(x)(y) \, . \, x \circ y = y \circ x$$

of the theory of groups.

We call this system the system G. To formulate this as a formal system, we formalize the postulates of Dickson[19] as follows:

Primitive ideas:

An infinite sequence of primitive obs: $a_0, a_1, a_2, \ldots$
A binary operation, designated by "$\circ$" as an infix.
A unary operation, designated by a prime as a suffix.
A binary relation, designated by "$=$" as an infix.

Obs:

The primitive obs $a_0, a_1, a_2, \ldots$ are obs.
If A and B are obs, then $A \circ B$ is an ob.
If A is an ob, then A' is an ob.

Elementary propositions:

They are those of the form
$A = B$ where A and B are arbitrary obs.

Axioms:

They are all propositions of the following forms, where A, B, and C are arbitrary obs:
1/ $A \circ (B \circ C) = (A \circ B) \circ C$,
2/ $A \circ a_0 = A$,
3/ $A \circ A' = a_0$.

Rules of deduction:

Let A, B, and C be arbitrary obs
1/ If $A = B$, then $B = A$,
2/ If $A = B$ and $B = C$, then $A = C$,
3/ If $A = B$, then $A \circ C = B \circ C$,
4/ If $A = B$, then $C \circ A = C \circ B$.

The system G may be regarded as a formalization of certain properties of ordinary addition. In this case the operation "$\circ$" is the same as "$+$", a_0 is the number 0, and A' is $-A$. A mathematician will know many other interpretations; but these will not concern us here.

The elementary theorems of the system G are the equations of the theory of groups which are valid for every group and for arbitrary values of $a_1, a_2, \ldots$. Here are some examples:

$$a_1 = a_1,$$
$$a_1' \circ a_1 = a_0,$$
$$a_0' = a_0,$$
$$(a_1 \circ a_2)' = a_2' \circ a_1'.$$

One may observe that already the proof of the second of these examples is difficult enough. As simple epitheorems we have for example, if A and B are arbitrary obs:

---

[19]Translator's note: See [4].

$$A = A,$$
$$A' \circ A = a_0,$$
$$\text{If } A = B, \quad \text{then } A' = B'.$$

The theorems of the part of higher algebra which is called the theory of groups may be regarded as complicated epitheorems of the system G.

The axioms of the system G are infinite in number. They are arranged in three schemes, which are called, after von Neumann [9], axiom schemes. The primitive obs are also infinite in number. The obs $a_1, a_2, \ldots$ are indeterminates; but a_0 is not, since it appears in the axiom schemes 2/ and 3/.

Although there are no substitutive variables, we have an epitheorem of substitution. To understand this situation, we must define substitution more precisely then we have so far. Let us denote the result of the substitution of an ob A for a primitive ob a_i in the ob B by the expression $(A/a_i)B$.[20] Let us abbreviate this expression for the moment by C. Then C is to be defined for every ob B. Intuitively, B is constructed from the primitive obs following a certain procedure; if we use the same procedure with A in place of a_i, we have $(A/a_i)B$. This is equivalent to a definition by recursion:

a/ If B is a_i, then C is A,
b/ If B is a primitive ob distinct from a_i, then C is B,
c/ If B is $B_1 \circ B_2$, then C is $((A/a_i)B_1) \circ (A/a_i)B_2)$,
c/ If B is B_1', then C is $((A/a_i)B_1)'$.

It follows by structural induction that C is defined for all B. Then the epitheorem of substitution is the following: Let a be one of $a_1, a_2, \ldots$, and let A, B, and C be arbitrary obs; then if $B = C$ it follows that $(A/a_i)B = (A/a_i)C$. We may establish this theorem by a deductive induction. It is clear that a similar epitheorem is true for every system which has indeterminates.

We will consider now the modification of the system G in which we adopt the theorem of substitution as a rule of deduction. Naturally this will oblige us to treat the primitive obs other than a_0 as a special category. In this case, we do not need an infinite number of axioms. It suffices to have three particular axioms:

$$a_1 \circ (a_2 \circ a_3) = (a_1 \circ a_2) \circ a_3,$$
$$a_1 \circ a_0 = a_1,$$
$$a_1 \circ a_1' = a_0.$$

We call the new system G'. Thus G' has substitutive variables. But G and G' have the same elementary theorems.

This situation is present generally. We may formulate the same system (in the sense in which a system is a set of theorems) either with indeterminates and axiom schemes or with substitutive variables and a rule of substitution. This equivalence was first noticed by von Neumann [9].

We will now make some auxiliary observations.

---

[20]Translator's note: Curry's usual notation is $[A/a_i]B$. I suspect that the square brackets were not available on the typewriter on which the text of the book was typed for the original printing.

Although $a_1, a_2, a_3, \ldots$ are formal variables, the symbols "a_1", "a_2", "a_3", $\ldots$ are constants of the U language. On the other hand, the symbols "A", "B", "C", etc., which we have used to state the rules and the epitheorems, are intuitive variables. (This may serve as an example of the fundamental difference between the two notions.)

The interpretation of a system with formal variables is a slightly complicated matter. One may not easily find a direct interpretation in a system of symbols. The interpretations important for mathematics are of another kind. In the interpreted system the symbols "a_1", "a_2", $\ldots$ may become intuitive variables, although they are considered as constants in the formal system.[21]

We have remarked that there are an infinite number of primitive obs. To speak truly, the intuitive interpretation which we have taken demands an infinite number. For if we have only a finite number, we may find general equations of the theory of which groups which are not elementary theorems of the system. This is the general situation for systems with formal variables.

Indeterminates may be important for epitheoretic studies. To study the structure of a system it is frequently convenient to extend it by adding new obs which will be indeterminates in the extended system. We speak here of adjoined indeterminates. The elementary theorems of the augmented system, which really contain the *adjoined indeterminates*, are epitheorems of the original system. The system G may be regarded as so formed from trivial systems which contain only one primitive ob a_0. More specialized systems of groups may be formed by adjoining additional axioms.

As we have formalized the theory of groups by the system G, we may also formalize other mathematical systems, especially all of elementary algebra. These systems will be, like the system G, completely formalized. Consequently, they do not presuppose the systems of logic. Elementary mathematics may be expressed in the form of a formal system, and it will thus have the independence from logic that we have seen in I Sect. 7.

To present a system with an infinite number of obs we have need of numerical indices (or something equivalent). One can say that the notion of a positive integer is presupposed. Poincaré has, we believe, made a criticism of this kind. To formulate the system in completely finite terms one has need of our systems of sams, which is thus the most simple system that one can conceive. But if one regards the indeterminates as adjoined, this difficulty disappears, and one can completely eliminate formal variables by combinatory logic. We will not enter into these questions here; we are content to accept an infinite sequence intuitively.

3 Some Notational Conventions

Before going any further, it is useful to explain here some conventions of notation.

First, we give some conventions on nominative symbols, i.e., those which we use as nouns (or pronouns). We reserve the symbols

$$e_1, e_2, e_3, \ldots$$

as proper nouns for the formal variables. The other lower case Latin letters will be, unless something is said to the contrary, intuitive variables; they designate obs which are

---

[21] See footnote 18.

unspecified or specified only by the context; one may regard them, if one wishes, as ad-joined indeterminates. We prefer this form of presentation for either formal variables or indeterminates and we do not have a rule of substitution. We give the axioms by schemes which contain intuitive variables. Thus we will have almost no occasion to mention for-mal variables; almost all of our theorems are epitheorems where the intuitive variables designate arbitrary obs.

We now give some conventions for the symbols we use to designate functions, i.e., operations, predicates, etc. We call these symbols *functors*. Here there are three kinds;

1) *connective functors* which form sentences from other sentences (they designate the connectives which form propositions from other propositions);
2) *predicate functors* which form sentences from nouns (they designate predicates for forming propositions from obs);
3) *operators* which form nouns from other nouns (they designate the operations which form obs from other obs).

Each of these kinds may be either unary, i.e., of one argument, or binary, i.e., of two arguments. We usually write a unary functor in front of its argument, so that the argument is to the right of the symbol, and we always write binary functors between the arguments. The functors we will use are those in the following table:

	binary	unary
Connective functors	$\rightarrow, \leftrightarrows$	
	&	
Predicative functors	$R, \geq, \leq, =$	$\vdash$
Operators	$\supset, \frown, \square$	
	$\wedge, \vee, +, -$	$\neg, -, \#, \Diamond, \sim, \div$

If one forms complex expressions with these symbols, special conventions will be needed to avoid a mass of parentheses creating confusion. We sometimes use the dot notation. A group of dots at the side of a functor indicates that on this side the functor has precedence over all functors with a smaller number of dots, as equality takes precedence over addition in elementary algebra. For functors with the same number of dots, we agree that a functor always has precedence over one which appears in a lower line in the table above, and that, in the same line, a binary functor always takes precedence over a unary functor. For functors which have the same precedence we adopt the rule of association to the left. These conventions reduce the number of dots which are necessary. But to obtain maximum clarity, we sometimes use dots or ordinary parentheses which are not strictly necessary.

We sometimes use these symbols as names for the functions themselves. Thus, we speak of the operation $\wedge$, the relation $\leq$, the connective $\rightarrow$. I believe that this usage is clear enough. It reduces the need to use special artifices for functions.[22]

As in elementary algebra we indicate a certain binary operation by simple juxtaposition without a special operator. In this case a group of dots between the two arguments is to be thought of as an operator for this operation with the number of dots indicating the number on both sides. As a functor this juxtaposition has lower precedence than all the others.

---

[22]These more refined notations were used in Curry [2] and [3]; but I have found that they are not really necessary.

We also use the symbol "≡" for definitions. This sign indicates that the expression on its left is synonymous with that to the right.[23] As a functor this sign takes precedence over all the symbols in the table. Also, one may use "≡" to define a connective functor, a predicative functor, an operator, or also a noun.

We now give some definitions of special functors.

The functor "→" indicates the connection between propositions which one usually expresses by the words "if ..., then ...". More exactly, if A and B are propositions, $A \rightarrow B$ is the proposition which says that B results from A in virtue of the rules. We use this functor to give rules (primitive or derived). For example the rules 1/ and 3/ of the system G may be given thus:

1/ $A = B \rightarrow B = A$
3/ $A = B \rightarrow A \circ C = B \circ C$

The functor "&" indicates logical conjunction.[24] Thus, if A and B are propositions, $A \& B$ is the proposition which says that A and B are both true. For example, rule 2/ of G says that

$$A = B \ \& \ B = C \ \rightarrow \ A = C.$$

The functor "⇆" is defined formally thus:

$$A \leftrightarrows B \ . \ \equiv \ . \ A \rightarrow B \ . \ \& \ . \ B \rightarrow A.$$

These three functors are, of course, connective functors in the U language, i.e. "A" and "B" may be replaced by sentences of the U language.

As regards predicative signs, we use "R" as an intuitive variable for an unspecified relation. The relation R is called *reflexive* if one always has

$$x \, R \, x;$$

it is called *transitive* if one has

$$x \, R \, y \ \& \ y \, R \, z \ \rightarrow \ x \, R \, z;$$

it is called *symmetric* if one has

$$x \, R \, y \ \rightarrow \ y \, R \, x.$$

4 Relational Algebras

The algebras which we will study here are not the most general which one can conceive. An important class of these algebras has a special character, which we will consider now.

---

[23] We also use this sign to indicate substitutions for intuitive variables. This is in principle a kind of definition. For example see the proof of III Theorem 1. (Translator's note: This chapter is not translated here.)

[24] This is the current usage of "conjunction" in logic. In grammar one calls "conjunctions" words that correspond approximately to our "connective functors". I use this last expression to avoid a confusion between the two senses of "conjunction".

These algebras have a binary relation that is reflexive and transitive. We call this relation the *fundamental relation* and indicate it by the signs "$\leq$" or "$=$". The elementary propositions are thus of one of the forms

$$x \leq y, \qquad x = y, \tag{10}$$

where x and y are obs or obs of a certain class. It may happen that there are other predicates, but they will have an auxiliary character; the fundamental relation is always the principal predicate. One may also have a somewhat complex morphology. These complications, though imaginable, will not be presented in this work. All the systems which we will introduce in the present work are completely formal and the fundamental relation here is the only predicate. But our conclusions will be applicable to more general systems, because one understands by "ob" an ob of the class associated with elementary propositions of the form (10).

We call an algebra of this kind a *relational algebra*. If the fundamental relation is symmetric, we denote it by the usual sign of equality ("$=$"), and we call the algebra an *equational algebra*. If symmetry is not postulated, we denote the fundamental relation by the sign "$\leq$", we speak of a *quasi-ordered algebra*. We also use the sign "$\geq$" for the converse of the relation "$\leq$", and define it by

$$x \geq y . \equiv . y \leq x. \tag{11}$$

It should be noted that everything that we can say about a quasi-ordered algebra may be said also about a general relational algebra, because nothing prevents us from adjoining the supplementary supposition of symmetry, and thus our theorems are equally valid for an equational algebra.

In ordinary mathematics, one sometimes distinguishes between quasi-ordered systems and partially ordered systems. The latter have the supplementary property

$$x \leq y \ \& \ y \leq x \ \rightarrow \ x = y,$$

whereas for the former this property is not postulated. But this distinction presupposes that one has a previously defined relation of equality. In the formal systems themselves we do not have equality until we define it. We adopt the definition

$$x = y . \equiv . x \leq y \ \& \ y \leq x. \tag{12}$$

One proves easily (as an epitheorem) that this relation is reflexive, symmetric, and transitive, and that it coincides with the fundamental relation if that relation is symmetric. Our systems are thus, in a trivial manner, partially ordered. But this is not necessarily the case with regard to an identity in an interpretation. Otherwise the relation of equality that we introduce may be interpreted as a relation other than identity. If this circumstance causes a confusion in a later application, the symbol "$=$" must be replaced by another symbol for equivalence. We do not use the expression "partially ordered" when speaking of applications, and thus we always suppose that the equality defined by (12) is the same as the previously defined identity.

5 The Replacement Theorem

We prove in this section a general theorem of replacement which holds for all quasi-ordered systems.

Let ϕ be an operation of one argument. The sentence "ϕ is *directly monotone*" means that

$$x \leq y \;\to\; \phi x \leq \phi y$$

is true, and "ϕ is *inversely monotone*" means that

$$x \leq y \;\to\; \phi y \leq \phi x$$

is true. Suppose that B is constructed from an ob A by a sequence of operations which is either directly monotone or inversely monotone; and that B' is constructed from A' by the same operations.

If the number of inversely monotone operations is even, one has

$$A \leq A' \;\to\; B \leq B';$$

if the number is odd one has

$$A \leq A' \leq B' \leq B.$$

This theorem is easily proved by an induction on the number of operators. When the fundamental relation is symmetric, there is no distinction between direct monotony and inverse monotony.[25]

In applications of this theorem there almost always intervene some binary operations, such as, for example, addition. These operations have the following monotonic properties:

$$x \leq y \to a + x \leq a + y,$$
$$x \leq y \to x + a \leq y + a.$$

Thus if one considers addition as an operation of argument x, which forms either $a + x$ or $x + a$, it is in either case monotone in the sense we are adopting. One says that it is monotone with respect to all its arguments, or, if there is no possibility of confusion, simply monotone. All the operations that are treated in algebraic logic are monotone in one or the other sense with respect to all their arguments. Thus, the theorem says that if one replaces a component A in an ob B by an ob A' such that $A \leq A'$ or $A \geq A'$, one may deduce immediately a relation between B and the B' formed by the replacement. All theorems on the replacement of equivalents are special cases of this theorem.

6 Two Kinds of Logical Algebras

Almost all the algebras which are considered in mathematical logic are relational algebras or are systems derived from relational algebras by the reduction of I Sect. 8. We call the latter *logistic algebras*. In a logistic algebra we use the symbol "$\supset$" for the operation which replaces "$\leq$" and we use the signs "$\curlywedge$" and "$\Box$" (depending on the application and convenience, etc.), for the operation which replaces "$=$". We thus have two kinds of logical algebras: relational algebras and logistic algebras.

---

[25] If an operation is monotone with respect to $\leq$ it is monotone in the same sense with respect to $\geq$ and, as a consequence, also with respect to $=$.

These two kinds of algebras are in principle equivalent. In fact we have seen in I Sect. 8 that we may reduce a very general system to a logistic system which is essentially equivalent to it. Conversely, if one has a logistic algebra such that

$$\vdash x \supset x, \tag{13}$$

$$\vdash x \supset y . \& \vdash y \supset z . \to . \vdash x \supset z, \tag{14}$$

and if one defines

$$x \leq y . \equiv \vdash x \supset y, \tag{15}$$

every theorem or every proposition of the form

$$\vdash x \supset y \tag{16}$$

is transformed into a relational form. For propositions which are not of the form (16), we introduce[26] an ob such that

$$\vdash 1. \tag{17}$$

If the principles

$$\vdash x \supset . y \supset x \tag{18}$$

$$\vdash x \supset y . \& . \vdash x . \to . \vdash y . \tag{19}$$

are valid, one always has

$$\vdash x \leftrightarrows \vdash 1 \supset x. \tag{20}$$

Thus every proposition of the system can be expressed in a relation form which is equivalent to it. If the postulates are transformed according to the same method our system becomes a relational algebra.[27]

It is not immediately obvious that these two transformations are inverses of each other. But if the logistic algebra is obtained by the reduction of I Sect. 8 from a relational algebra T, the inverse transformation does not need (13) and (14) for us to return to T. Similarly, if T is obtained from S without using (13), (14), the transformation of I Sect. 8 leads to a system S' which differs from S only in that the principle $\supset$ of each proposition is replaced by a new operation which may be designated by "$\supset'$". There is a bijective correspondence between the elementary theorems of S' and those of S;[28] S' is thus essentially the same as S. Finally, suppose that T is obtained from S using also (17)–(19). One has thus the same situation for theorems of the form (16). A theorem $\vdash x$ which is not of this form is replaced by $\vdash 1 \supset x$ which is equivalent to it.[29] The theorems remain essentially the same.

---

[26]One can introduce it by a definition.

[27]Cf. IV Sect. 5 (below). (Translator's note: this section is not translated here.)

[28]More meticulously, let x' be defined so that if x is $y \supset z$, x' is $y \supset' z$, and if x is not of the form $y \supset z$, x' is x. Thus the correspondence $\vdash x \sim \vdash x'$ is bijective.

[29]The correspondence such that $\vdash x$ corresponds to $\vdash x'$ if x is of the form $y \supset z$ and to $\vdash 1 \supset x'$ otherwise is not bijective, because $\vdash x$ and $\vdash 1 \supset x$ may correspond to the same $\vdash 1 \supset' x$. But $\vdash x$ and $\vdash 1 \supset x$ are equivalent in S.

We will now see that there is no essential difference between the two kinds of logical algebras, because the principles (13), (14), (18), (19) are valid in logical algebras in a general manner.[30] One may speak of the relational and logistic forms of the same algebra. Besides, neither of these forms is strictly more fundamental than the other. Of course there may be significant differences from the psychological point of view. For certain applications one form may appear more natural than the other, and one form may sometimes suggest a set of postulates which is not so simple with the other. Sometimes the same algebra has been presented in the two forms for different purposes. For example, the classical algebra of Boole is used in the relational form for classes and in the logistic form for propositions; another algebra has been used by Heyting in the logistic form for intuitionistic propositional logic and by Tarski in the other form.[31]

But these two forms are only different aspects of that which is, at bottom, the same thing.

For propositional logic, the logistic form is naturally preferable. One might think that a propositional logic must precede every other study of logic, because one uses propositions in the other studies. But this seems to me to result from a confusion between two senses of the word "proposition" in the U language. This word is for us a term which we use in its usual sense. In propositional algebra one uses the same word to designate the obs. To avoid this confusion one must have another word, perhaps "statement" for the first meaning. Thus we use statements in the study of an arbitrary formal system, but we have no need of propositions. Thus, the argument for the priority of propositional algebra fails and, at last, the argument for the logistic form also fails.[32] The interpretation of a formal system has no connection with the priority of thoughts; every truly formal system may be taken on its own standing.

From the mathematical point of view, on the other hand, there are reasons to favor the opposite order, which has been, we may say in passing, the historical order. The relational form is the form of usual mathematics, whereas the logistic form appears to be a technical artifice. Moreover, because of the confusion over the word "proposition", the propositional interpretation is more subtle and difficult. These are reasons which have value from the didactic point of view. But in other respects, the transformation of a relational system into a logistic system is simpler and may be more simply inverted than the inverse transformation. The former represents the logistic system as a part of the other and the theorems about the relational system are also valid for the logistic system. In fact the usual logistic systems will be more easily regarded as special cases of relational systems than inversely.[33] From this point of view relational systems are more general, whereas logistic systems give a more profound analysis in certain special cases.

Thus, we will begin with relational algebras.

---

[30]This is a part of the definition; see the introduction.

[31]See Heyting [5] and Tarski [8].

[32]Cf. Curry [1], Kap. I B Sect. 5.

[33]To put it another way, logistic algebras altogether comprise a part of a simple relational algebra by the transformation of I Sect. 5. This part is equivalent to a relational algebra by the simplest transformation (which does not use (13) and (14)). The derived rules of a relational algebra of this simple form are thus in principle equally valid for more general logistic algebras. But the inverse transformation leads to a very special relational algebra.

7 The Operations of Logical Algebras

We will not consider as logical algebras all conceivable algebras. The logical algebras require a restriction on the operations. We must consider that here.

In general the logical algebras have (at least) two operations which we call the sum or union and the product or intersection. We denote the sum, in general, by the sign "$\vee$", sometimes by "$+$". We designate the product by "$\wedge$", sometimes by simple juxtaposition. In the systems which we call logical groups there is only one operation; we designate it by juxtaposition.

All these operations are commutative and associative, as in ordinary algebra. But that which distinguishes logical algebras is that they are also idempotent, i.e., one has:

$$a \vee a = a, \qquad a \wedge a = a.$$

In a majority of cases one also has the following properties:

$$a \leq a \vee b, \qquad a \wedge b \leq a$$
$$b \leq a \vee b, \qquad a \wedge b \leq b$$
$$a \leq x \,\&\, b \leq x \to a \vee b \leq x, \qquad x \leq a \,\&\, x \leq b \to x \leq a \wedge b.$$

An algebra of this sort is called a *lattice*.[34] All the algebras which we will consider here, except for logical groups, are lattices. It is doubtful that any others have any logical importance.[35]

8 Interpretations of Logical Algebras

We consider some examples of interpretations of logical algebras in order to note their variety. Among the examples are some which belong to higher mathematics.

1^0 The oldest interpretation is that in which the obs are classes and "$\leq$" is the relation of inclusion. Here the product is the intersection, and $a \vee b$ is the class of objects which belong to a or to b (perhaps to both).[36] For this interpretation the relational form is the most natural.

2^0 One has also a similar interpretation in which the obs are relations. This may be regarded as a special case of 1^0 because relations are essentially classes of pairs. But the algebra of relations may have additional operations.

3^0 The obs are propositions and "$\leq$" is the relation of implication. This is a very important application because one may regard the explanation of implication as the main object of logic. The product $a \wedge b$ is *conjunction* (a and b) and the sum $a \vee b$ is *alternation* (a or b). The logistic form is natural, since "$\vdash$" is the affirmation of a proposition. As has already been noted, we must prevent the confusion of propositions which are stated in the theory of the system with the propositions which are the interpreted obs.

---

[34] Also called a structure.

[35] An idempotent algebra which is not a lattice has been considered by Moisil [7] from a purely structural point of view. I know of no other application.

[36] Translator's note: This is now called the union of the classes. I assume that this usage was not common in 1951.

There are many forms of this propositional interpretation according to the meaning which one gives to the word "proposition", i.e., to the obs. In the classical algebra implication is defined in terms of the truth values, true and false. The system is thus a system of matrices with two values (cf. 9^0 below). One may also have an intuitionistic propositional algebra. One is led to this system when one takes as obs certain propositions of another formal system, with certain slightly semantic definitions for the operators.[37]

4^0 The obs are functions of n real or natural numbers, and $\leq$ is the relation $f(x_1,\ldots,x_n) \leq g(x_1,\ldots x_n)$, where "$\leq$" has its usual meaning. One may have $n = 0$, in which case the obs are numbers, $a \wedge b$ is the smallest of the two numbers, $a \vee b$ the largest. The absolute value of f is $f \vee (-f)$.[38]

5^0 The obs are natural numbers, and $\leq$ is the relation of divisibility. Then $a \wedge b$ is the greatest common divisor, $a \vee b$ is the least common multiple.

6^0 A variety of interpretations that generalize 1^0 in the following manner: Suppose that $\leq$ is still the relation of inclusion between subclasses of a class K. Suppose also that we have a unary operation "$^-$" which associates to each subclass a another subclass $\bar{a}$ such that

$$a \leq \bar{a}$$
$$\bar{a} \leq \bar{\bar{a}} \qquad\qquad (21)$$
$$a \leq b \;\rightarrow\; \bar{a} \leq \bar{b}.$$

One calls an operation of this kind a *closure*, and the classes such that $a = \bar{a}$ closed classes. Then we take for obs the closed classes, and $\leq$ remains the inclusion relation. The produce is still the intersection, but $a \vee b$ is now the smallest closed class which includes a and b. There are many examples of this kind according to the nature of closure. Here are some of them (it is enough to say which are the closed classes):

a/ The sets closed in the sense of the general theory of sets of points (topology).

b/ Linear spaces (lines, planes, etc.). This example leads to a projective geometry.

c/ Convex sets of points.

d/ The subgroups of a mathematical group, or the subalgebras of an algebra.

e/ The classes of propositions which are closed according to certain given rules of deduction.

7^0 One may construct examples of openings by interchanging $\wedge$ and $\vee$.

8^0 Let K be a finite class of objects. One may indicate by a table or by some other means that the relation $\leq$ is true between certain pairs of things in K; the relation is defined as that generated from these axioms by the rule of transitivity. The indication may be given by a diagram, as in the following examples (Fig. 1).

In these examples the objects are circles and the primitive cases of the relation are indicated by lines with arrows. The examples g and h are not partially ordered and the arrows in them are necessary to indicate the sense of the relation; in the other cases, which do not contain cycles, and are consequently partially ordered, the sense may be identified by the sense of high and low on the paper.

---

[37] This is the principal thesis of Curry [2]. Cf. IV Sect. 6 below. (Translator's note: This section is not translated here.)

[38] For some generalizations, cf. IV, Sect. 4 below. (Translator's note: This section is not translated here.)

Fig. 1

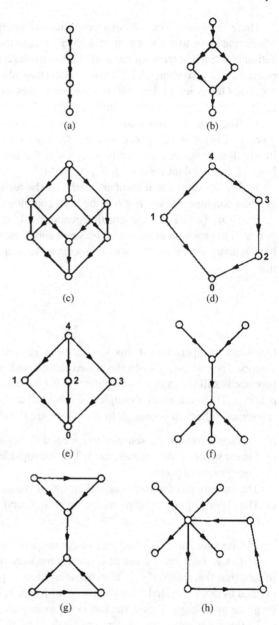

(a)　　　　　(b)

(c)　　　　　(d)

(e)　　　　　(f)

(g)　　　　　(h)

9^0 Let K be a class of a finite number of objects. One may define operations on K by tables as in the following examples (where K contains three objects 1, 2, 3).

x	x'
1	3
2	2
3	1

x/y	1	2	3
1	3	3	3
2	2	2	2
3	1	2	3

x/y	1	2	3
1	1	1	1
2	1	2	2
3	1	2	3

Let L be a subclass of K, and $F(x_1, \ldots, x_n)$ a function of the indeterminates $x_1, x_2, \ldots, x_n$. We interpret $\vdash F(x_1, \ldots, x_n)$ as saying that $F(x_1, \ldots, x_n)$ takes values in L no matter what are the values of $x_1, \ldots, x_n$. A set of conventions of this kind is called a *matrix*. A matrix is thus a description of a class of propositions, those which may be regarded as a formal system where all the propositions of this class are axioms and where there is no rule of deduction. This is a degenerate formal system of logistic form. Under certain restrictions a matrix gives us an algebra which satisfies the conditions of Sects. 6–7. Matrices are sometimes of capital importance in modern logic. They may be equally interesting in certain cases where the conditions of Sects. 6–7 are not all satisfied. The classical algebra of propositions appears equivalent to an algebra of matrices with two values.

References

1. Curry, H.B.: Grundlagen der kombinatorischen Logik. Am. J. Math., **52**, 509–536, 789–834 (1930). Inauguraldissertation
2. Curry, H.B.: A Theory of Formal Deducibility. Notre Dame Mathematical Lectures, vol. 6. University of Notre Dame, Notre Dame (1950). 2nd edn. (1957)
3. Curry, H.B.: Outlines of a Formalist Philosophy of Mathematics. North-Holland, Amsterdam (1951)
4. Dickson, L.E.: Definitions of a group and of a field by independent postulates. Trans. Am. Math. Soc. **6**, 198–204 (1905)
5. Heyting, A.: Die formalen Regeln der intuitionistischen Logik. In: Sitzungsberichte der preussischen Akademie der Wissenschaften, physikalisch-mathematische Klasse, pp. 42–56 (1930)
6. Kleene, S.C.: H. B. Curry, some aspects of the problem of mathematical rigor. Bull. Am. Math. Soc. **47**, 221–241 (1941) (review). J. Symb. Log. **6**, 100–102 (1941)
7. Moisil, Gr.C.: Recherches sur l'algèbre de la logique. Annales scientifiques de l'Université de Jassy **22**, 1–118 (1936)
8. Tarski, A.: Grundzüge der Systemenkalküls. Erster Teil. Fundam. Math. **25**, 503–526 (1935)
9. von Neumann, J.: Zur Hilbertschen Beweistheorie. Math. Z. **26**, 1–46 (1927)

Translator
J.P. Seldin
Department of Mathematics and Computer Science, University of Lethbridge, Lethbridge, Alberta, Canada
e-mail: jonathan.seldin@uleth.ca

Part 7
Jerzy Łos and Roman Suszko (1958)

Structural Consequence Operations and Logical Matrices Adequate for Them

Jan Zygmunt

Abstract The aim of this historical and expository essay is twofold. First, we describe the content and main features of Łoś and Suszko's paper entitled *Remarks on Sentential Logics*, 1958, emphasizing its novelty in a universal algebraic approach to the methodology of propositional logics. Second, we briefly discuss the impact of *Remarks* on research carried out in Poland up to the 1980s in the domain of metatheory of propositional calculi.

Keywords Sentential logic · Structural consequence operation · Rule of inference · Logical matrix · Adequacy problem · Methodology of propositional calculi

Mathematics Subject Classification (2000) Primary 01A60 · Secondary 03-03, 08-03—History of mathematics and mathematicians of 20th century (Mathematical logic and foundations; General algebraic systems) · Primary 03B22—General logic: Abstract deductive systems · Primary 03G25—Algebraic logic: Other algebras related to logic

1 Introduction

The paper, as regards its philosophical (or ideological) side and historical background, belongs to the methodology of deductive sciences in the spirit of Alfred Tarski; it is a distinctive piece of set-theoretical and algebraic metalogic. The basic concepts involved—sentential logic, consequence operation and logical matrix—have their origin in two classic works: Tarski 1930, and Łukasiewicz–Tarski 1930, and these concepts had already been somewhat investigated by the authors prior to *Remarks* (e.g. [18–22, 39–41]). In *Remarks*, however, the notions of 'sentential language', 'sentential logic', 'rule of inference' or 'matrix' are generalized and put in a wider universal-algebraic context, and the following new concepts are introduced: 'structural consequence', 'cardinality [or power] of consequence' and 'uniform consequence' (Point 2 of the paper), 'sequential basis of consequence' (Point 3), '[normal] extension of consequence' (Point 5), 'matrix consequence' and 'matrix [strongly] adequate for a consequence' (Points 7, 8). Generally speaking, the last two concepts concern semantics, while the former five are relevant to syntax.

The language of discourse and methods of reasoning applied in the paper are thoroughly algebraic with one excursion into general topology (in Point 8 in the proof of the theorem that the matrix consequence determined by a finite matrix is finitary, or finite to use the authors' original term); the elaboration of the Lindenbaum method in the proof of the main theorem (Point 9) is highly sophisticated.

Before Łoś and Suszko wrote *Remarks* they had collaborated in the field of model theory and jointly published two papers on the extension of models [25, 26]. These two papers are part of a five-instalment series on model theory developed under the supervision of Łoś (see [23, 24, 37]; see also an informative review of parts I–IV of the series by H.J. Keisler in JSL **27**, 93–95).

2 Main Features of *Remarks*

2.1 The starting point of the paper is the identification of sentential languages (or zero-order languages) with absolutely free algebras (or algebras of terms), so sentential variables are free generators, formulas are elements, and connectives are fundamental operations of such algebras. By means of this identification syntactical notions pertinent to a sentential logic are defined algebraically in an easy and precise way. Thus a sentential language $S = \langle S, F_1, \ldots, F_n \rangle$ is defined as a free algebra of some finitary type $\langle k_1, \ldots, k_n \rangle$, freely generated in the class of <u>all</u> similar algebras by a set of (sentential) variables p_t.[1] Hence, for every algebra $\mathcal{A} = \langle A, \ldots \rangle$ similar to S, any mapping from the variables p_t to A extends uniquely to a homomorphism from S to $\mathcal{A}$. In particular this is true when $\mathcal{A}$ is S: any endomorphism ε of S is uniquely determined by its values on the variables p_t, so endomorphisms are substitutions in algebraic disguise. Further, some subalgebras of S are freely generated by variables, and they are called *normal*. For any set of formulas $X \subseteq S$ there exists a least (minimal) normal subalgebra S_0 such that $X \subseteq S_0$, and the set of generators of S_0 is called the *support* of X, in symbols $s(X)$; intuitively, $s(X)$ is the set of all variables occurring in formulas of X.

2.2 Following Tarski, the concept of logical consequence is taken by the authors as a primitive fundamental metalogical notion. Axioms to be satisfied by any consequence operation should tell us, firstly, about the most general set-theoretical structure of a logical theory, and should encompass all possible inferences without specifying them. An operation Cn on sets of formulas is called a *consequence operation*—or, as the authors say, simply 'a consequence'—if it satisfies Tarski's axioms for a general theory of consequence (i.e., conditions (2.1) and (2.3) of Point 2). Moreover, any such operation Cn is uniquely determined by the family $\{X \subseteq S : Cn(X) = X\}$ of its fixed points, called *systems* of Cn.

Secondly, logical consequence is about forms of sentences, without regard to their content—by applying a substitution ε throughout a deduction of α from X one obtains a deduction of $\varepsilon\alpha$ from εX, and a consequence operation should take this fact into account. Hence the authors accept the following condition of *structurality* of Cn:

if $\alpha \in Cn(X)$ then $\varepsilon\alpha \in Cn(\varepsilon X)$ for every substitution (endomorphism) ε of S (where $\varepsilon X = \{\varepsilon\beta : \beta \in X\}$), or in short $\varepsilon Cn(X) \subseteq Cn(\varepsilon X)$

[i.e., (2.3) of the paper; *N.B.* The last inclusion may be replaced by the equality: $Cn(\varepsilon Cn(X)) = Cn(\varepsilon X)$]. Structurality is a kind of invariance. If Cn is structural then, obviously, the set $Cn(\emptyset)$ of tautologies (or logic) of Cn is *invariant*, i.e., closed under

---

[1]Łoś and Suszko do not specify the cardinality of the set of variables, nor the numbers $k_1, \ldots, k_n$. For general requirements as to the existence of free algebras see: [7] or [12]. From a logical point of view it is natural to assume that the set of variables is denumerable and at least one k_i is greater than 0.

substitutions. Moreover, Cn is structural if and only if the family of systems of Cn is closed with respect to the operation of taking <u>inverse</u> images of substitutions, i.e., if X is a system of Cn, so is $\{\beta \in S : \varepsilon\beta \in X\}$ for every substitution (see [6, p. 31, Theorem 1]).

Understanding 'consequence operation' as subject to the above pair of broad general constraints, in contemporary terminology one of the basis notions of *Remarks* might be stated as follows: that a sentential logic is any pair $\langle S, Cn \rangle$ where S is a sentential language and Cn is a structural consequence operation on S. The authors did not state it this way; the terminology has shifted slightly over fifty years.[2] But the basic notion was theirs, and is introduced in Point 2.

2.3 In Point 3, rules of inference are considered as set-theoretical relations $R(\alpha_0, \alpha_1, \alpha_2, \ldots)$ among formulas—α_0 is a conclusion while $\alpha_1, \alpha_2, \ldots$ are premises of R. Rules are said to be *finite* if the list of premises is finite. Rules are said to be *axiomatic* if the list of premises is of zero length, i.e., if the relation is unary. The notion of structurality is, in a natural way, extended to rules of inference: R is called *structural* if, for all formulas $\alpha_0, \alpha_1, \alpha_2, \ldots$ and every substitution ε, $R(\alpha_0, \alpha_1, \alpha_2, \ldots)$ implies $R(\varepsilon\alpha_0, \varepsilon\alpha_1, \varepsilon\alpha_2, \ldots)$. The well-known rule of modus ponens is structural but the rule of substitution r^*, defined by the clause: $r^*(\alpha_0, \alpha_1)$ iff $\alpha_0 = \varepsilon\alpha_1$ for some substitution ε, is <u>not</u>.[3] It is shown how rules generate consequences, and, for any consequence, how to find a set of rules which may generate it, or—as the authors say—be a basis for it. E.g., any structural consequence has a basis which consists of structural rules only, and—what is of some importance—any structural rule may be further replaced by a *sequential* rule, i.e., a rule determined by a schema of inference.[4]

2.4 Sentential languages, and sentential logics, are modelled by means of matrices, i.e., systems $\mathfrak{M}$ of the form $\langle A, f_1, f_2, \ldots, f_n, B \rangle$, where A is a non-empty set (the range of propositional variables), $f_1, f_2, \ldots, f_n$ are interpretations of the connectives $F_1, F_2, \ldots, F_n$, and $B \subseteq A$ is a set of designated "logical" values. Thus $\mathfrak{M}$ may be considered as a pair $\langle A, B \rangle$ where A is an algebra similar to the algebra S of formulas. Valuations of formulas in $\mathfrak{M}$ are identified with homomorphisms from S to A; let $Hom(S, A)$ be the set of all valuations. Each matrix $\mathfrak{M}$ determines on S the set of all its tautologies, $E(\mathfrak{M})$, that is the set of all formulas α such that $h(\alpha) \in B$ for every valuation $h \in Hom(S, A)$. Until the 1960s, and even later, there appeared many logical papers dealing with the *weak adequacy problem*: for a given logic L, understood as a set of formulas closed with respect to certain rules, find a matrix $\mathfrak{M}$ such that $L = E(\mathfrak{M})$ and,

---

[2] To be precise, this very definition and the term 'a sentential logic', do not occur in *Remarks*; the authors use 'sentential logics' to describe the domain of research or the variety of sentential calculi (cf. Point 7). They prefer to use only 'a consequence', thus stressing the second element of $\langle S, Cn \rangle$. As examples of how the terminology has shifted, our sentential logics are called by R. Wójcicki structural logics (in [49, p. 270]) or propositional logics (in [52]). J. Czelakowski sticks to 'a sentential logic' (cf. [9, Chapter 0]).

[3] Let Sb denote the consequence operation generated by the rule of substitution, i.e., $Sb(X)$ is the least set of formulas containing X and closed under r^*. Then Sb is not a structural consequence.

[4] Obviously, these are Hilbert-style rules. Also Genzen-style rules (4.1)–(4.5) are considered, and it is shown that they together with the rule of repetition will suffice to characterize completely the so-called full sequential basis of each finite and structural consequence (Point 4).

possibly, having certain desired properties prescribed in advance. The theorem of Lindenbaum (stated in [28, Theorem 3], and proved in a general case in [19, Theorem 10]) is a basic result pertinent to the weak adequacy problem, namely if $L \subseteq S$ is invariant, i.e., $L = Sb(L)$, then the matrix $\langle S, L \rangle$ is weakly adequate for L, i.e. $L = E(\langle S, L \rangle)$. Łoś and Suszko are critical about a logic research programme limited to the weak adequacy problem:

> "In the opinion of many logicians the theorem of Lindenbaum presents the nearest connection between sentential logics and interpretations by matrices. Indeed, every set of tautologies of a [structural] consequence $Cn(0) = T_0$ is an invariant set and therefore it may be represented in the form $T_0 = E(\mathfrak{M})$. If we find such a matrix $\mathfrak{M}$ we think that the problem of interpreting of this calculus is solved. But the calculus is given by means of the consequence and the matrix is adequate not for the whole consequence but only for its set of tautologies."

and instead they propose their meaning of adequacy:

> "In order to give a real connection between consequences and the matrices we shall define for every matrix $\mathfrak{M}$ a consequence $\mathfrak{M}(X)$ and we shall call $\mathfrak{M}$ adequate for Cn if $\mathfrak{M}(X) = Cn(X)$ for each $X \subseteq S$."

The definition of the operation $\mathfrak{M}(X)$—which we shall denote $\overrightarrow{\mathfrak{M}}$—can be stated as follows: $\overrightarrow{\mathfrak{M}}(X)$ is the set of all formulas that are verified by every valuation of S in $\mathfrak{M}$ that verifies every formula in X. It is easy to see that $\overrightarrow{\mathfrak{M}}$ is a structural consequence operation and $\overrightarrow{\mathfrak{M}}(\emptyset) = E(\mathfrak{M})$; $\overrightarrow{\mathfrak{M}}$ is called in *Remarks* the *matrix consequence* generated by $\mathfrak{M}$.

Now we may formulate the (**strong**) **adequacy problem**, which is the main question of *Remarks*: given a sentential logic $\langle S, Cn \rangle$, under what conditions on Cn, syntactical in nature, does there exist a single matrix $\mathfrak{M}$ such $Cn = \overrightarrow{\mathfrak{M}}$, i.e. a matrix strongly adequate for the logic?

2.5   The authors claim that two conditions are necessary and sufficient, namely structurality and *uniformity* of Cn. The latter means:

for every X, Y and α: if $s(X) \cap s(Y) = s(\{\alpha\}) \cap s(Y) = \emptyset$, $Cn(Y) \neq S$, and $\alpha \in Cn(X \cup Y)$, then $\alpha \in Cn(X)$.

In words: whenever a Cn-consistent set Y has no variable in common with a set X and a formula α, and α is a Cn-consequence of $X \cup Y$, then α is a Cn-consequence of X alone.

Since structurality is a part of our definition of a sentential logic, the main theorem (see Point 9) may be formulated in this way:

($\perp$) *For a sentential logic $\langle S, Cn \rangle$ there exists a matrix strongly adequate for it if, and only if, Cn is uniform.*

However—we are obliged to inform the reader right now—this general form of the theorem is not true; it is lacking a certain assumption, as we shall see this by analysing its proof.

The proof of the "only if" part (which is glossed over in *Remarks*) is by checking that every matrix consequence $\overrightarrow{\mathfrak{M}}$ is structural and uniform (on every propositional language that is similar to the underlying algebra of $\mathfrak{M}$).

The "if" part is proved by showing how the required matrix is constructed. This is a Lindenbaum type matrix $\langle S^*, B \rangle$ built up from the linguistic raw material: the sentential language S^* is a normal extension of S (obtained by adding new variables p_X corresponding to every Cn-consistent set $X \subseteq S$) and B is a system the consequence Cn^*, where Cn^* in an extension of Cn on S^* (Cn^* is defined according to the procedure described in Point 5, and Cn^* inherits from Cn all the desired properties).[5] More precisely,

$$B = Cn^* \left(\bigcup \{ \varepsilon X : X \subseteq S, \ Cn(X) \neq S, \ \varepsilon \in Aut^* \} \right)$$

where Aut^* is a set of 1–1 mappings of the variables of S into the variables of S^* such that any two different sets belonging to the family $\{ \varepsilon X : X \subseteq S, \ Cn(X) \neq S, \ \varepsilon \in Aut^* \}$ have no variable in common.

If B were Cn^*-inconsistent, that is, if B were equal to S^*, the matrix $\langle S^*, B \rangle$ would generate the inconsistent matrix consequence. Thus in order for the proof of the adequacy theorem to be correct one would have to prove that $B \neq S^*$. However this point is overlooked in the paper. Ryszard Wójcicki was the first to notice the lacuna (see [48] or [49], and also [52, p. 142 ff]).[6]

The most natural emendation to the formulation ($\bot$) is to require that the sentential logic $\langle S, Cn \rangle$ be *additive* in the following sense: whenever $\{X_i\}$ is a family of subsets of S such that:

1. $\{X_i\}$ is *separated*, i.e., every two different members of $\{X_i\}$ have no variable in common, in symbols: $s(X_i) \cap s(X_j) = \emptyset$, if $i \neq j$, and
2. there is a variable that does no occur in any formula of any member of $\{X_i\}$, and
3. every member of $\{X_i\}$ is a Cn-consistent set,

then the union $\bigcup \{X_i\}$ is also Cn-consistent.

Thus we have the following:

Łoś–Suszko–Wójcicki Adequacy Theorem *Given a sentential logic $\langle S, Cn \rangle$, there exists a matrix strongly adequate for $\langle S, Cn \rangle$ if, and only if, Cn is uniform and additive.*

Originally, Wójcicki [48] had himself discovered a necessary and sufficient condition by adding to uniformity the requirement that the members of any separated family $\{X_i\}$ of mutually uniform sets be mutually uniform with $\bigcup \{X_i\}$; where by the mutual uniformity

---

[5]This kind of extensions of sentential logics has been discussed in [16], where they are called normal extensions, and in [9, pp. 55–57], under the name of natural extensions.

[6]Neither H.B. Curry in his review of *Remarks* (see *Mathematical Reviews* **20** # 5125), nor A. Church (see *The Journal of Symbolic Logic* **40**(4), 603–604 (1975)), nor G.H. Müller (*Zentralblatt f. Mathematik u. ihre Grenzgebiete* **92**, 248) in their reviews have noticed this flaw. It is rather surprising that Church's review fails to mention it, as [51] was published in the same journal just the previous year. Moreover, Müller's review gives an incorrect statement of what it means to say a consequence is structural (in the sense that the authors use the term in *Remarks*).

of X and Y he means that $\alpha \in Cn(Z \cup X')$ iff $\alpha \in Cn(Z \cup Y')$, for all X' obtained from X by substitution of distinct variables for variables, all Y' obtained similarly from Y, and all $Z \subseteq S$, $\alpha \in S$ such that $Z \cup \{\alpha\}$ is separated from $X' \cup Y'$.

It turns out that Łoś and Suszko's original requirement of uniformity is sufficient in the case of finitary (*alias* compact) sentential logics. Hence we have, as a special case:

Łoś–Suszko Adequacy Theorem *For a finitary sentential logic $\langle S, Cn \rangle$ there exists a matrix strongly adequate for it if, and only if, Cn is uniform.*

2.6 The final part (Point 10) of *Remarks* contains a couple of brief comments on the classical propositional logic $\langle S, Cn_2 \rangle$, and in passing formulates two open problems.

The language S has a complete list of connectives: $\to$, $'$, $\vee$, $\wedge$ and $\equiv$, and the consequence Cn_2 is understood as generated by the sole rule of *modus ponens* and a suitable complete list of axiom schemata. The usual two-element Boolean matrix $\mathfrak{M}_2$ is strongly adequate for Cn_2 and $Cn_2(\emptyset)$ is the largest consistent and invariant system of Cn_2. The authors state, firstly, that there are three-element matrices $\mathfrak{M}$, with one designated element, such that $\overrightarrow{\mathfrak{M}}(\emptyset) = Cn_2(\emptyset)$ and the rule of *modus ponens* is not a derivable rule of $\overrightarrow{\mathfrak{M}}$ (it is not a normal rule of the matrix $\mathfrak{M}$), hence $\overrightarrow{\mathfrak{M}} \neq Cn_2$. The example of such an $\mathfrak{M}$ given in the original text is as it stands incorrect, and there are too many inaccuracies (misprints?) in the truth-tables to reconstruct the authors' original intent, so we have furnished another example suggested by P. Wojtylak, namely the matrix $\mathfrak{M}_D$ of [33, p. 135].[7] Secondly, it is indicated that the consequence Cn^* determined by the sole rule of *modus ponens* for equivalence $\equiv$ may be axiomatically strengthened to a consequence Cn_2^* with $Cn_2^*(\emptyset) = Cn_2(\emptyset)$ but in this case the set $Cn_2(\emptyset)$ is not the largest invariant and consistent system of Cn_2^*; the latter follows from a result of Słupecki to whom the authors refer, or we may prove it by considering the matrix $\mathfrak{M}_4 = \langle \mathcal{B}_4, \{0, 1\} \rangle$, where $\mathcal{B}_4$ is a four-element Boolean algebra and the zero, 0, and the unit, 1, of $\mathcal{B}_4$ are the designated elements. Indeed, the rule of *modus ponens* for $\equiv$ is a normal rule of this matrix, and the set of tautologies of $\mathfrak{M}_4$ is an invariant system properly containing $Cn_2(\emptyset)$ since $p \equiv p'$ is a tautology of $\mathfrak{M}_4$.

Furthermore, having written

> "We do not know any matrix adequate for the positive consequence or for the intuitionistic consequence"

the authors have delicately posed the problem of giving examples of such algebraic structures which might be naturally treated as logical matrices for Hilbert's positive logic and Heyting's intuitionistic propositional calculus (INT), or giving a mathematical characterization of adequate logical matrices for the two logics.[8]

---

[7]By the way, $\mathfrak{M}_D$ is isomorphic to the matrix described by Tables II in [8, p. 51]. Łoś and Suszko give no references to the literature at this point. It is of some historical interest that in [18] one finds a three-element (hence non-Boolean) matrix $\mathfrak{M}$ such that $\overrightarrow{\mathfrak{M}} = Cn_2$. This is probably the first published example of this type.

[8]We take it for granted that the authors knew that both logics are uniform. It is worth remarking, however, that the first proof of uniformity of Heyting's intuitionist propositional calculus and Hilbert's positive logic was published in [36, Theorem 5].

The problem for INT was solved by A. Wroński, and for the positive logic by J. Bernert and P. Wojtylak (see [1, 54] and our comments in 3.2 below).

3 The Impact of *Remarks* on the Methodology of Propositional Calculi

3.1 Before R. Wójcicki found a gap in Łoś and Suszko's adequacy theorem, somewhat earlier he considered a new type of semantic structure, namely generalized matrices, and making use of them he proved a result, later called the Wójcicki Adequacy Lemma, which extends the Lindenbaum theorem into all structural consequence operations [47]. Here are some details. If $\mathcal{A}$ is an algebra similar to the algebra S and $B_t \subseteq A$, for $t \in T$, then the system $\mathfrak{GM} = \langle \mathcal{A}, \{B_t : t \in T\}\rangle$ is called a *generalized* (or *ramified*) matrix for S.[9] Any such $\mathfrak{GM}$ induces on S a structural consequence operation according to the formula:

$$\alpha \in \overrightarrow{\mathfrak{GM}}(X) \text{ iff } h(\alpha) \in B_t \text{ whenever } hX \subseteq B_t \text{ for every } t \in T \text{ and every } h \in Hom(S, \mathcal{A}).$$

In particular, if Cn is a consequence on S, one can consider the generalized matrix of the Lindenbaum type $\mathfrak{GL} = \langle S, \{Cn(X) : X \subseteq S\}\rangle$, and obtain...

Wójcicki Adequacy Lemma *If $\langle S, Cn\rangle$ is a sentential logic (hence Cn is structural), then $\mathfrak{GL}$ is strongly adequate for it, i.e., $Cn = \overrightarrow{\mathfrak{GM}}$.*

To any ramified matrix $\mathfrak{GM} = \langle \mathcal{A}, \{B_t : t \in T\}\rangle$ there corresponds a class of usual matrices $\mathfrak{M}_t = \langle \mathcal{A}, B_t\rangle$, $t \in T$, and, as it is easily seen, $\overrightarrow{\mathfrak{GM}}(X) = \bigcap \{\overrightarrow{\mathfrak{M}_t}(X) : t \in T\}$ for all $X \subseteq S$. Hence, in other words, every structural consequence operation is a meet (infimum) of some family of matrix (i.e. structural, uniform and additive) consequences. Inspired by this simple result Wójcicki (see [48]), and also Suszko (see [3]), began investigating of the lattice of <u>all</u> structural consequences on a fixed S and its sublattices. That was the moment when Suszko and his American collaborators had started *abstract logics*,[10] and Wójcicki had taken first steps towards the theory of matrix semantics of logical calculi.

In light of the Wójcicki Adequacy Lemma one can pose one of the most natural problems:

P1. For a given particular sentential logic $\langle S, Cn\rangle$, is the matrix $\mathcal{L} = \langle S, Cn(\emptyset)\rangle$ strongly adequate for $\langle S, Cn\rangle$, or at least $\overrightarrow{\mathcal{L}}(X) = Cn(X)$ for every finite $X \subseteq S$?

This is a question about structural completeness of a logic, expressed in terms of consequence operations and not rules of inference as it was originally expressed by Pogorzelski

---

[9]Structures of this sort were earlier introduced in [38, p. 435]; they are called Smiley matrices in [50, p. 35], where it is shown that they are in a sense equivalent to generalized matrices.

[10]See this volume, pp. 257–260, where the paper *Some theorems on abstract logics* by Bloom, Brown and Suszko is reprinted and commented upon by R. Jansana.

[31]; structural completeness was widely discussed in the 1970s which contributed to a deeper understanding of the notion of an inference rule and completeness (see, e.g., [46]; and [32, 33] for a survey).

It follows from the Wójcicki Adequacy Lemma that every sentential logic $\langle S, Cn \rangle$ has a *complexity degree*, that is, the least cardinal number $\mathfrak{m}$ such that $Cn = \inf\{\overrightarrow{\mathfrak{M}_t} : t \in T\}$ for some matrices $\mathfrak{M}_t$, $t \in T$, with T of cardinality $\mathfrak{m}$. The following two problems then arise:

P2. For a fixed cardinal number $\mathfrak{m} > 0$, under what syntactic conditions is a sentential logic's complexity degree equal to or less than $\mathfrak{m}$?

P3. For a given sentential logic $\langle S, Cn \rangle$, what is its complexity degree?

Let us mention some early work relevant to these questions. The Łoś–Suszko–Wójcicki Adequacy Theorem gives necessary and sufficient conditions for a logic to have a complexity degree of 1. From the proof of this theorem as set out in [49] it follows that the complexity degree of a consistent logic $\langle S, Cn \rangle$ is not greater than its *degree of uniformity*, i.e., the cardinality of the quotient set $\{X : X \subseteq S,\ X = Cn(X) \neq S\}/_{\approx}$, where $\approx$ is the relation of mutual uniformity of sets of formulas (defined at the end of 2.5 above). A deeper analysis of the notion of complexity degree can be found in [15, 17] where it is proved that every $\mathfrak{m}$ (with $\mathfrak{m}$ being a natural number > 0 or $\mathfrak{m} = \aleph_0$, or $\mathfrak{m} = 2^{\aleph_0}$) may be the complexity degree of some logic, and that the complexity degree of Johansson's minimal logic is 2 and that of Rasiowa's logic with semi-negation (as defined in [34]) is $2^{\aleph_0}$. The complexity degree of both the relevant logics E and R is equal to 1 [29, 45].

3.2 R. Suszko was the first logician to look for natural examples of logics which have only uncountable strongly adequate matrices, and one of the first to find such a logic, in 1971[11]: the sentential calculus with identity (SCI)—the Suszko's basic system of non-Fregean logic. The proof that *SCI has no denumerable adequate matrix* is in [44] (and the conjecture was made in [4, p. 79]; see also [43]). Another example is EN-logic, the reduction of SCI to the two connectives: identity and negation (see [30]).

Suszko also conjectured that intuitionistic propositional logic and the so-called WT extension of SCI have no denumerable matrices. The first conjecture, which obviously corresponds to the problem left open in *Remarks* (see 2.6 above), was proved true by A. Wroński [54]), and the latter by S. Zachorowski [55]. Wroński's celebrated result that *there is no denumerable matrix for INT* is a corollary to another important theorem of his: a kind of embedding criterion for strong adequacy of a matrix for any intermediate logic, expressed in terms of denumerable strongly compact (i.e., subdirectly irreducible) Heyting algebras (see Theorem 1, in [54]). Wroński's criterion was subsequently used as the main tool in [11], where it is shown that *each of the three existing pre-finite intermediate logics has a denumerably infinite strongly adequate matrix*, and, moreover, these matrices are explicitly indicated.[12] Zachorowski [55] observed that Suszko's conjecture

---

[11]Shoesmith and Smiley were the first: in 1970 they already knew that any matrix strongly adequate for Łukasiewicz's logic $L_{\aleph_0}$ must be uncountable (see [36, Theorem 4]).

[12]An intermediate logic is said to be pre-finite if it is maximal in the set of intermediate logics having no finite matrix weakly adequate (see [29]).

about the WT extension of SCI is equivalent to a conjecture about the modal system S4, and using the McKinsey–Tarski mapping from INT into S4 and Wroński's result for INT, he proved that the consequence operation determined by the theses of S4 together with the rule of *modus ponens* has no denumerable strongly adequate matrix, though it does have one of the power of the continuum. J. Hawranek, at the suggestion of Suszko, considered two other modal consequence operations: the one determined by the theses of S5 and the sole rule of *modus ponens*, and its strengthening by the rule of necessitation. He showed, using the fact that every denumerable Boolean algebra is projective, that both of these consequence operations have denumerable strongly adequate matrices (see [13, 14]).

3.3 In [33], the matrix semantics is treated as a special case of a richer semantics in which logical matrices are replaced by more general set-theoretic entities called pre-ordered algebras, which are systems of the form $\mathfrak{A} = \langle \mathcal{A}, \leq \rangle$, where $\mathcal{A}$ is an abstract algebra and $\leq$ a preorder (reflexive and transitive relation) on A. Every such $\mathfrak{A}$ with $\mathcal{A}$ similar to S generates a structural consequence operation $\overrightarrow{\mathfrak{A}}$ on S in a natural way, and the following theorem holds: *if $\langle S, Cn \rangle$ is a finitary (compact) sentential logic with $Cn(\emptyset) \neq \emptyset$, then there is a preordered algebra $\mathfrak{A}$ such that $Cn = \overrightarrow{\mathfrak{A}}$ (op. cit, p. 71). This may be seen as a generalization of the above Łoś–Suszko adequacy theorem. Pogorzelski and Wojtylak (*op. cit*, pp. 72–74) look at the question of *Sb*-adequacy (called the "i-adequacy" in [6, p. 34]), that is, the existence of adequate matrices for invariant strengthening of structural consequences, and they obtain a quite general theorem corresponding to Łoś–Suszko–Wójcicki Adequacy Theorem: *For every sentential logic $\langle S, Cn \rangle$ (Cn is structural) there is a single logical matrix $\mathfrak{M}$ such that $Cn \circ Sb = \overrightarrow{\mathfrak{M}} \circ Sb$*, where $Cn \circ Sb$ is the composition of *Sb*, the consequence generated by the sole rule of substitution, and *Cn*. (See also Theorem XIII.4 in [6].)

3.4 Final Remarks   Outside Poland, *Remarks* remained unknown until the mid-1970s. We have not found any evidence that it was studied or referred to by somebody. It is not listed in either of the massive bibliographies of works on many-valued logic in [35] and [10]. Quite independently thirteen years later Shoesmith and Smiley published a paper whose central concern was identical to Łoś and Suszko's adequacy problem in *Remarks*. (See [36].) Consider their opening sentences (*op. cit*. p. 610)...

> Lindenbaum's construction of a matrix for a propositional calculus, in which the wffs themselves are taken as elements and the theorems as the designated elements, immediately establishes two general results: that every propositional calculus is many-valued, and that every many valued propositional calculus is also $\aleph_0$-valued. These results are however concerned exclusively with theoremhood, the inferential structure of the calculus being relevant only incidentally, in that it may serve to determine the set of theorems. We therefore ask what happens when deducibility is taken into consideration on a par with theoremhood. The answer is that in general the Lindenbaum construction is no longer adequate and both results fail. Instead we shall establish the conditions for a calculus to be many-valued, and show that every many-valued propositional calculus is also $2^{\aleph_0}$-valued.

Wójcicki [51] discusses the relation between the results contained in [36] and those contained in *Remarks* and some early papers of Wójcicki.

In Poland, *Remarks* was first carefully studied in the second half of the 1960s by R. Wójcicki and W.A. Pogorzelski (both of whom were then at the University of Wrocław). While Wójcicki concentrated more on the semantic issues of the paper, Pogorzelski focussed on its proof-theoretical side. In the 1970s each led research groups working on the methodology of propositional logic. There was close cooperation and a continual exchange of ideas between the groups, as well as the logicians from Jagiellonian University in Kraków. Results were presented at join seminars, conferences and season schools, and published in *Studia Logica*, *Reports on Mathematical Logic*, and *Bulletin of the Section of Logic*. The main achievements of this research, which was inspired through the 1970s and early 1980s by the ideas contained in *Remarks*, and by ideas handed down to a younger generation of logicians by Roman Suszko himself, are summarized in the two books by Pogorzelski and Wojtylak: *Elements of the Theory of Completeness in Propositional Logic* [32], and *Completeness Theory for Propositional Logic* [33]; two books by Wójcicki: *Lectures on Propositional Calculi* [52], and *Theory of Logical Calculi. Basic theory of consequence operations* [53]; and Czelakowski's monograph [9]. [The Reader is encouraged to read all the reviews (indicated in the bibliography) of the five books to acquaint himself with the reception of the work by the logic community.]

Acknowledgements   The author is very grateful to Robert Purdy for making many improvements concerning the English style of this paper, and to Piotr Wojtylak for conversations about *Remarks*.

References

1. Bernert, J.: A note on existing [existence] of matrices strongly adequate for some positive logics. Rep. Math. Log. **14**, 3–7 (1982) [MR671022 (84e:03021), reviewed by B. Lercher]
2. Bloom, S.L., Brown, D.J.: Classical abstract logics. Diss. Math. **102**, 43–52 (1973) [MR0446967 (56 # 5284), reviewed by G. Georgescu]
3. Bloom, S.L., Brown, D.J., Suszko, R.: A note on abstract logics. Bull. Acad. Pol. Sci., Sér. Sci. Math., Astr. Phys. **18**(3), 109–110 (1970) [MR0272606 (42 # 7487), reviewed by E. Mihailescu]
4. Bloom, S.L., Suszko, R.: Semantics for the sentential calculus with identity. Stud. Log. **28**, 77–82 (1971) [MR 45 # 6570 (from the author's introduction)]
5. Bloom, S.L., Suszko, R.: Investigations into the sentential calculus with identity. Notre Dame J. Form. Log. **13**, 289–308 (1972). Errata, Notre Dame J. Form. Log. **17**, 640 (1976) [MR0376300 (51 # 12478), reviewed by N.B. Cocciarella; Zbl 0188.01203, reviewed by the authors]
6. Brown, D.J., Suszko, R.: Abstract logics. Diss. Math. **102**, 9–41 (1973) [MR0446967 (56 # 5284), reviewed by G. Georgescu]
7. Burris, S., Sankappanavar, H.P.: A Course in Universal Algebra. Springer, New York (1981) [MR648287 (83k:08001), reviewed by R.S. Pierce; Zbl 0478.08001 reviewed by B.M. Schein]
8. Church, A.: Non-normal truth-tables for the propositional calculus. Bol. Soc. Mat. Mex. **10**, 41–52 (1953) [MR0058563 (15, 385d), reviewed by A. Rose; JSL **19**, 233–234 (1954), reviewed by G.F. Rose; Zbl 0053.34101, reviewed by G.H. Müller]
9. Czelakowski, J.: Protoalgebraic Logics. Trends in Logic-Studia Logica Library, vol. 10. Kluwer, Dordrecht (2001) [MR1828895 (2002f:03001), reviewed by S. Comer; Zbl 0984.03002, reviewed by J. Cirulis; Stud. Log. **74**, 313–342 (2003), reviewed by R. Elgueta]
10. Dunn, J.M., Epstein, G. (eds.): Modern Uses of Multiple-valued Logic. Invited Papers from the Fifth International Symposium on Multiple-Valued Logic, Indiana University, Bloomington, Ind., May 13–16, 1975. With a bibliography of many-valued logic by Robert G. Wolf. Edited by J. Michael Dunn and George Epstein. Episteme, Vol. 2. D. Reidel Publishing Co., Dordrecht–Boston (1977) [MR0469671 (57 # 9454); Zbl 0355.00008, registered]

11. Graczynska, E., Wroński, A.: Constructing denumerable matrices strongly adequate for pre-finite logics. Stud. Log. **33**, 417–423 (1974) [MR0387012 (52 # 7859), reviewed by M. Beeson; see also MR0396229 (53 # 97) for a review of Dov M. Gabbay of a shorter version of the paper; Zbl 0312.02023, reviewed by K. Schultz]

12. Grätzer, G.: Universal Algebra, 2nd edn. Springer, New York (1979); revised reprint (2008) [MR0538623 (80g:08001), reviewed by the Editors; Zbl 0412.08001, reviewed by B.M. Schein]

13. Hawranek, J.: A matrix adequate for S5 with MP and RN. Bull. Sect. Log. (Wrocław) **9**, 122–124 (1980) [Zbl 0459.03009]

14. Hawranek, J.: Macierze adekwatne dla logik stowarzyszonych z systemem modalnym S5 (Adequate matrices for logics associated with the modal system S5. Polish). Zesz. Naukowe Wyższej Szkoły Inżynierskiej Opolu, No 81 – Matematyka **4**, 87–90 (1983)

15. Hawranek, J., Zygmunt, J.: On the degree of complexity of sentential logics. A couple of examples. Stud. Log. **40**, 142–153 (1981) [MR0648574 (83i: 03052), reviewed by W. Rautenberg; Zbl 0484.03011 (summary)]

16. Hawranek, J., Zygmunt, J.: O normalnych rozszerzeniach logik zdaniowych (On normal extensions of sentential logics. Polish). Acta Univ. Wratislav. No 605 – Logika **10**, 21–29 (1983) [Zbl 0557.03006, reviewed by P. Materna]

17. Hawranek, J., Zygmunt, J.: On the degree of complexity of sentential logics. II. An example of the logic with semi-negation. Stud. Log. **43**, 406–413 (1984) [MR0803317(87c: 03069), reviewed by W. Rautenberg; Zbl 0569.03012, reviewed by the Editors]

18. Łoś, J.: Logiki wielowartościowe a formalizacja funkcji intensjonalnych (Many-valued logics and the formalization of intensional functions. Polish). Kwart. Filoz. **17**, 59–78 (1948) [MR0025416 (10, 1d), reviewed by H. Hiż; JSL **14**, 64–65 (1949), reviewed by R. Suszko]

19. Łoś, J.: O matrycach logicznych (On logical matrices. Polish). Pr. Wrocławskiego Tow. Nauk., Ser. B, **19**, 42 (1949) Wrocław [MR0089812 (19, 724b); JSL **16**, 59–61 (1951), reviewed by J. Kalicki]

20. Łoś, J.: An algebraic proof of completeness for the two-valued propositional calculus. Colloq. Math. **2**, 236–240 (1951) [MR0050541 (14, 345b), reviewed by H.B. Curry]

21. Łoś, J.: Sur le théorème de Gödel pour les théories indénombrables. Bull. Acad. Pol. Sci., Cl. III **2**, 319–320 (1954) [MR0063326 (16, 103f), reviewed by G. Kreisel]

22. Łoś, J.: The algebraic treatment of the methodology of elementary deductive systems. Stud. Log. **2**, 151–212 (1955) [JSL **21**, 193–194 (1956), reviewed by L. Henkin; MR0083958 (18, 785b), reviewed by L.N. Gál]

23. Łoś, J.: On the extending of models (I). Fundam. Math. **42**, 38–54 (1955) [MR0072071 (17, 224e), reviewed by G. Kreisel; JSL **27**, 93–95 (1962), reviewed by H.J. Keisler]

24. Łoś, J., Słomiński, J., Suszko, R.: On extending of models (V): Embedding theorems for relational models. Fundam. Math. **48**, 113–121 (1960) [MR0112825 (22 # 3676), reviewed by C.-C. Chang]

25. Łoś, J., Suszko, R.: On the extending of models (II): Common extensions. Fundam. Math. **42**, 343–347 (1955) [MR0075890 (17, 815d), reviewed by G. Kreisel; JSL **27**, 93–95 (1962), reviewed by H.J. Keisler]

26. Łoś, J., Suszko, R.: On the extending of models (IV): Infinite sums of models. Fundam. Math. **44**, 52–60 (1957) [MR0089813 (19, 724c), reviewed by G. Kreisel; JSL **27**, 93–95 (1962), reviewed by H.J. Keisler]

27. Łoś, J., Suszko, R.: Remarks on sentential logics, Proc. K. Ned. Akad. Wet., Ser. A **61** (= Indag. Math. **20**), 177–183 (1958). [Communicated by Prof. A. Heyting at the meeting of the Academy on November 30, 1957. MR 20 # 5125, reviewed by H.B. Curry; JSL **40**, 603–604 (1975) reviewed by A. Church; Zbl 0092.24802, reviewed by G.H. Müller]

28. Łukasiewicz, J., Tarski, A.: Untersuchungen über den Aussagenkalkül. Comptes Rendus des séances de la Société des Sciences et des lettres de Varsovie, classe III, **23**, 30–50 (1930) [JFM 57.1319.01, reviewed by W. Ackermann]

29. Maksimova, L.L.: The principle of separation of variables in propositional logics (Russian). Algebra Log. **15**, 168–184 (1976) 245. [English translation in Algebra Log. **15**, 105–114; MR0505212 (58 # 21417), reviewed by M. Rogava; Zbl 0354.02021, reviewed by K. Schultz]

30. Michaels, A., Suszko, R.: Sentential calculus of identity and negation. Rep. Math. Log. **7**, 87–106 (1976) [MR0505312 (58 # 21496); Zbl 0361.02018, reviewed by J.M. Plotkin]

31. Pogorzelski, W.A.: Structural completeness of the propositional calculus. Bull. Acad. Pol. Sci., Sér. Sci. Math. Astron. Phys. **19**, 349–351 (1971) [JSL **40**, 604–605 (1975), reviewed by M. Tírnoveanu]

32. Pogorzelski, W.A., Wojtylak, P.: Elements of the Theory of Completeness in Propositional Logic, Silesian University, Katowice (1982) [MR0681864 (84g:03001), reviewed by D. Ponasse; Zbl 0539.03002, reviewed by A. Rose]

33. Pogorzelski, W.A., Wojtylak, P.: Completeness Theory for Propositional Logic, Studies in Universal Logic. Birkhäuser, Basel (2008) [MR2410288 (2010c: 03001), reviewed by J.M. Czelakowski; Zbl 1143.03001, reviewed by V.V. Pambuccian]

34. Rasiowa, H.: An Algebraic Approach to Non-classical Logics, PWN, Warszawa. North-Holland, Amsterdam (1974) [MR0446968 (56 # 5285), reviewed by B. Poizat; JSL **42**, 432 (1977), reviewed by A. Preller]

35. Rescher, N.: Many-valued Logic. McGraw-Hill, New York (1969) [Zbl 0248.02023, reviewed by G.E. Minc; JSL **42**, 432–436 (1977), reviewed by G. Epstein]

36. Shoesmith, D.J., Smiley, T.J.: Deducibility and many-valuedness. J. Symb. Log. **36**, 610–622 (1971) [MR0300869 (46 # 29), reviewed by H.B. Curry; see also [51]]

37. Słomiński, J.: On the extending of models (III): Extensions in equationally definable classes of algebras. Fundam. Math. **43**, 69–76 (1956) [MR0078934 (18, 2e), reviewed by G. Kreisel; JSL **27**, 94–95 (1972), reviewed by H.J. Keisler]

38. Smiley, T.: The independence of connectives. J. Symb. Log. **27**, 426–436 (1962) [MR0172784 (30 #3003), reviewed by A. Heyting; JSL **40**, 250–251 (1975), reviewed by D. Ulrich]

39. Suszko, R.: W sprawie logiki bez aksjomatów (Concerning logic without axioms. Polish). Kwart. Filoz. **17**, 199–205 (1948) [MR0028254 (10, 421c), reviewed by H. Hiż; JSL **15**, 66 (1950), reviewed by A. Mostowski]

40. Suszko, R.: O analitycznych aksjomatach i logicznych regułach wnioskowania (Analytic axioms and logical rules of inference. Polish). Pr. Kom. Filoz. – Pozn. Tow. Przyj. Nauk **7**(5), 1–30 (1949) [JSL **15**, 223–224 (1950), reviewed by J. Kalicki]

41. Suszko, R.: Formalna teoria wartości logicznych. I (A formal theory of logical values, I. Polish). Stud. Log. **6**, 145–237 (1957) [MR0102477 (21 # 1270), reviewed by L. Rieger]

42. Suszko, R.: Concerning the method of logical schemes, the notion of logical calculus and the role of consequence relations. Stud. Log. **11**, 185–216 (1961) [MR0197279 (33 # 5450), reviewed by G. Kreisel]

43. Suszko, R.: A note on adequate models for non-Fregean sentential calculi. Bull. Sect. Log. **1**(4), 42–45 (1972) [MR0536411 (58 # 27348), reviewed by the editors]

44. Suszko, R.: Adequate models for the non-Fregean sentential calculus (SCI). In: Bogdan, R.J., Niiniluoto, I. (eds.): Logic, Language, and Probability. A Selection of Papers Contributed to Sections IV, VI, and XI of the Fourth International Congress for Logic, Methodology, and Philosophy of Science, Bucharest, September 1971, Synthese Library, vol. 51, pp. 49–54. Reidel, Dordrecht (1973) [MR0460068 (57 # 64), reviewed by S.L. Bloom]

45. Tokarz, M.: The existence of matrices strongly adequate for E, R and their fragments. Stud. Log. **38**, 75–85 (1979) [MR542420 (80h:03034), reviewed by L. Esakia; see also MR509687 (80a: 03031) for a review by C.F. Kielkopf of a shorter version of the paper]

46. Wojtylak, P.: On structural completeness of many-valued logics. Studia Logica **37**, 139–147 (1978) [MR503094 (80k:03027), reviewed by A. Rose]

47. Wójcicki, R.: Uogólnione pojęcie matrycy logicznej oraz konsekwencji matrycowej (A generalized notion of logical matrix and of matrix consequence. Polish). In: Rozprawy Filozoficzne, Prace Wydziału Filologiczno-Filozoficznego, vol. 21, fasc. 2, pp. 441–448. Towarzystwo Naukowe w Toruniu, Toruń (1969)

48. Wójcicki, R.: Logical matrices strongly adequate for structural sentential calculi. Bull. Acad. Pol. Sci., Sér. Sci. Math., Astr. et Phys. **17**, 333–335 (1969) [MR0258602 (41 # 3248), reviewed by Mihailescu]

49. Wójcicki, R.: Some remarks on the consequence operation in sentential logic. Fundam. Math. **68**, 270–279 (1970) [MR 42 # 1650, reviewed by B. Lercher; Zbl 0206.2741, reviewed by G.E. Minc]

50. Wójcicki, R.: Matrix approach in methodology of sentential calculi. Stud. Log. **32**, 7–39 (1973) [MR0381948 (52 # 2837), reviewed by S.L. Bloom]

51. Wójcicki, R.: Note on deducibility and many-valuedness. J. Symb. Log. **39**, 563–566 (1974) [MR0384481 (52 # 5358), reviewed by V.K. Finn; Zbl 0334.02007, reviewed by T. Hosoi]

52. Wójcicki, R.: Lectures on Propositional Calculi, Ossolineum, Wrocław (1984) [MR756633 (86h:03002), reviewed by J. Bernert; Zbl 0647.03019, reviewed by the Editors]

53. Wójcicki, R.: Theory of Logical Calculi: Basic Theory of Consequence Operations. Synthese Library, vol. 199, Kluwer Academic Publishers, Dordrecht (1988) [MR1009788 (90j:03001), reviewed by V. Verdú; Zbl 0682.03001, reviewed by C.F. Kielkopf; JSL **55**, 1324–1326 (1990), reviewed by S.L. Bloom; Stud. Log. **50**, 623–629 (1991), reviewed by R.A. Bull]

54. Wroński, A.: On cardinalities of matrices strongly adequate for the intuitionistic propositional logic. Rep. Math. Log. **3**, 67–72 (1974) [MR0387011 (52 # 7858), reviewed by M. Beeson; see also MR0396231 (53 # 99) for a review by B. van Rootselaar of a shorter version of the paper.; Zbl 0342.02011, reviewed by G. Kreisel]

55. Zachorowski, S.: Proof of a conjecture of Roman Suszko. Stud. Log. **34**, 253–256 (1975) [MR0411936 (54 # 65), reviewed by D. Makinson, see also MR 0409118 (53 # 12880) for a review by S.L. Bloom of a shorter version of the paper; Zbl 0325.02013, reviewed by M.J. Cresswell]

J. Zygmunt (✉)
Katedra Logiki i Metodologii Nauk, Uniwersytet Wrocławski, ul. Koszarowa 3/20,
51-149 Wrocław, Poland
e-mail: Logika@uni.wroc.pl

REMARKS ON SENTENTIAL LOGICS

BY

J. ŁOŚ AND R. SUSZKO

(Communicated by Prof. A. HEYTING at the meeting of November 30, 1957)

I. STRUCTURAL CONSEQUENCES

1. Let S be the set of all the formulas of a sentential logic formed by means of variables p_t and connectives $F_1, ..., F_n$, where F_i is a k_i-ary connective.

As known, S is a free algebra of the type $\langle k_1, ..., k_n \rangle$ and the variables p_t are free generators of it. Therefore, it follows that every function which maps the variables p_t in an algebra of the same type as S, may be extended to a homomorphism. Especially, if we assign to every variable p_t a formula $\varepsilon p_t = \alpha_t$ then we can extend this mapping to an endomorphism ε of S into itself. From the logical point of view the formula $\varepsilon \alpha$ is, for every α in S, a substitution of α, where α_t are substituted for p_t.

For formulas α in S a function $l(\alpha)$ with natural numbers as values, may be defined. It is called the *length* of the formula α. The recursive definition of this function is:

$$(1) \qquad l(p_t) = 1$$

$$(2) \qquad l(F_i(\alpha_1, ..., \alpha_{k_i})) = 1 + l(\alpha_1) + ... + l(\alpha_{k_i}).$$

Is A a subalgebra of S, then A is also a free algebra; the generators of A are those elements of it which are not values of F_i for formulas from A. The subalgebras generated by variables are called *normal-subalgebras*. For every subset X of S there is a least normal-subalgebra S_0 which contains X; the set of free generators of S_0 is called the *support* of X and is denoted by $s(X)$. The elements of $s(X)$ are those variables which appear in the formulas of X.

2. By a *consequence* in S we understand (see [1]) an operation Cn defined for every subset X of S and such that

$$(2.1) \qquad X \subset Cn(X) = Cn(Cn(X)) \subset S,$$

$$(2.2) \qquad X \subset Y \rightarrow Cn(X) \subset Cn(Y).$$

If for every endomorphism ε and subset X:

$$(2.3) \qquad \varepsilon Cn(X) \subset Cn(\varepsilon X)$$

12 Indagationes

J.-Y. Béziau (ed.), *Universal Logic: An Anthology*, 177–184
Studies in Universal Logic, DOI 10.1007/978-3-0346-0145-0_14, © Springer Basel AG 2012

then Cn is called a *structural* consequence. In future we shall be dealing with structural consequences only.

There are certainly such cardinals $\mathfrak{m}$ for which

$$(2.4) \qquad Cn(X) = \bigcup_{\substack{Y \in X \\ \overline{Y} < \mathfrak{m}}} Cn(Y).$$

The least of them is called the *cardinality* of Cn. If $\mathfrak{m} = \aleph_0$ then we say that Cn is a finite consequence.

A set X for which $Cn(X) = X$, is called a Cn-system or shortly a *system*. There are for every consequence two special systems: $T_0 = Cn(0)$ and $T_1 = \bigcap_{0 \neq X \subseteq S} Cn(X)$. If $T_0 = T_1 \neq 0$ then Cn is called an *axiomatic* consequence if $T_1 = 0$ it is called an *axiomless* consequence; if $T_0 \neq T_1$ it is called a *pseudoaxiomatic* consequence ([2, 3]). In future we shall not be concerned with the pseudoaxiomatic consequences. The set $T_0 = T_1$ is called the set of Cn-tautological formulas or briefly the set of tautologies.

A set $Y \subseteq S$ is Cn-consistent if $Cn(Y) \neq S$.

If for every subsets X and Y and for every α in S:

$$(2.5) \qquad \begin{cases} \text{if } s(X) \cap s(Y) = s(\alpha) \cap s(Y) = 0, \ Cn(Y) \neq S \\ \text{and } \alpha \in Cn(X \cup Y) \text{ then } \alpha \in Cn(X), \end{cases}$$

then Cn is called a *uniform* consequence.

The condition (2.3) yields that T_0 is invariant with respect to endomorphisms, i.e. $\varepsilon T_0 \subset T_0$ for every endomorphism ε of S.

3.   Each relation $R(\alpha_0, \alpha_1, \alpha_2, \ldots)$, finite or infinite, defined for formulas in S, is called a *rule of inference*. It is *structural* when for every endomorphism ε, from $R(\alpha_0, \alpha_1, \alpha_2, \ldots)$ it follows that $R(\varepsilon\alpha_0, \varepsilon\alpha_1, \varepsilon\alpha_2, \ldots)$. We observe that the so called rule of *substitution* (defined as follows: $R(\alpha, \beta)$ if and only if for an endomorphism $\varepsilon : \alpha = \varepsilon\beta$) is not structural in the above sense.

The consequence Cn is closed with respect to the rule R, if every Cn-system is closed with respect to this rule i.e. for every Cn-system X

$$(3.1) \qquad \alpha_1, \alpha_2, \ldots \in X \text{ and } R(\alpha_0, \alpha_1, \alpha_2, \ldots) \text{ involves } \alpha_0 \in X.$$

Let $\mathscr{R}$ be a set of rules. Is every Cn-system closed with respect to all rules in $\mathscr{R}$ and is every one of such a set a Cn-system, then $\mathscr{R}$ is said to be a *basis* of the consequence Cn. If we take into consideration the rules R_λ for $\lambda <$ cardinality of Cn, defined as follows

$$R_0(\alpha_0) =_{\mathrm{df}} \alpha_0 \in Cn(0)$$
$$R_1(\alpha_0, \alpha_2) =_{\mathrm{df}} \alpha_0 \in Cn(\{\alpha_1\})$$
$$\vdots$$
$$R_\lambda(\alpha_0, \alpha_1, \alpha_2, \ldots) =_{\mathrm{df}} \alpha_0 \in Cn(\{\alpha_1, \alpha_2, \ldots\})$$
$$\vdots$$

then we conclude that for every consequence there exists a basis and for structural consequences there are bases consisting of structural rules only.

For any sequence $\sigma = \langle \alpha_0, \alpha_1, \alpha_2, \ldots \rangle$ of formulas let R_σ denote such a rule that

(3.2) $\quad \begin{cases} R_\sigma(\beta_0, \beta_1, \beta_2, \ldots) \text{ if and only if for an endomorphism } \varepsilon: \\ \beta_0 = \varepsilon \alpha_0, \ \beta_1 = \varepsilon \alpha_1, \ \beta_2 = \varepsilon \alpha_2, \ \ldots . \end{cases}$

If $R = R_\sigma$ for some sequence σ then the rule R is said to be a *sequential* rule ([2, 3]). Of course, if R is structural then $R = \bigcup_\sigma R_\sigma$ where σ runs over all sequences $\langle \alpha_0, \alpha_1, \alpha_2, \ldots \rangle$ such that $R(\alpha_0, \alpha_1, \alpha_2, \ldots)$. Moreover, making use of the function $l(\alpha)$ one can prove that every structural rule may be decomposed into a union of sequential rules none of which contains [1] another. Therefore, for every structural consequence there is a basis which consists of sequential rules only none of which contains another.

The cardinality of a consequence is connected with the powers of arguments of the rules in the basis. Are all these powers $< \mathfrak{m}$, then the cardinality of the consequences is $\leqslant \mathfrak{m}$. Is the cardinality of Cn equal to $\mathfrak{m}$, then there exists a basis with the powers of arguments in their rules less than $\mathfrak{m}$. So e.g. for a finite consequence it may be assumed that its basis consists of finite rules only.

Is Cn of the cardinality $\mathfrak{m}$, then the set of all sequential rules with the powers of arguments less than $\mathfrak{m}$ and with respect to which Cn is closed, is called the *full sequential basis* of Cn.

4. There are such operations on sequential rules which if performed on the rules of a basis of some consequence, give the rules of the full basis of this consequence. In the finite case we have the following operations ([2, 3]).

(4.1) $\qquad \langle \alpha_0, \alpha_1, \ldots, \alpha_k \rangle \Rightarrow \langle \alpha_0, \alpha_{p_1}, \ldots, \alpha_{p_k} \rangle$
$\qquad\qquad$ (permutation of assumptions)

(4.2) $\qquad \langle \alpha_1, \alpha_1, \ldots, \alpha_k \rangle \Rightarrow \langle \alpha_0, \alpha_1, \ldots, \alpha_k, \beta \rangle$
$\qquad\qquad$ (addition of an assumption)

(4.3) $\qquad \langle \alpha_0, \alpha_1, \alpha_1, \alpha_2, \ldots, \alpha_k \rangle \Rightarrow \langle \alpha_0, \alpha_1, \alpha_2, \ldots, \alpha_k \rangle$
$\qquad\qquad$ (cancellation of a repeated assumption)

(4.4) $\qquad \begin{cases} \langle \alpha_0, \alpha_1, \alpha_2, \ldots, \alpha_k \rangle \\ \langle \alpha_1, \beta_1, \beta_2, \ldots, \beta_l \rangle \end{cases} \Rightarrow \langle \alpha_0, \beta_1, \beta_2, \ldots, \beta_l, \alpha_2, \ldots, \alpha_k \rangle$
$\qquad\qquad$ (joining of sequences)

(4.5) $\qquad \langle \alpha_0, \alpha_1, \ldots, \alpha_k \rangle \Rightarrow \langle \varepsilon \alpha_0, \varepsilon \alpha_1, \ldots, \varepsilon \alpha_k \rangle$
$\qquad\qquad$ (substitution, $\varepsilon = $ any endomorphism)

---

[1] Containing is meant here in the set-theoretical sense, because every rule is a subset of a suitable cartesian product

$$S \times S \times S \times \ldots$$

The following theorem is true for each finite and structural consequence: The full sequential basis of a consequence is the least set of sequential rules which is closed with respect to the operations (4.1)–(4.5) and contains some sequential basis of this consequence and the sequential rule of *repetition* (defined as follows: $R(\alpha, \beta)$ if and only if $\alpha = \beta$). This shows the *completeness* of the operations above.

Let Cn be a finite and structural consequence and let $\mathscr{R} = \{R_\sigma\}_{\sigma \in \Sigma}$ be a sequential basis of Cn, where Σ is a set of finite sequences of formulas. It is obvious that $\alpha \in Cn(X)$ if and only if α can be "proved" from formulas belonging to X in a finite number of steps according to the rules belonging to $\mathscr{R}$. Consider the least set Σ^* of finite sequences of formulas which is closed with respect to the operations (4.1)–(4.5) and contains Σ and some sequence $\langle p_t, p_t \rangle$. If $\mathscr{R}^* = \{R_\sigma\}_{\sigma \in \Sigma^*}$ then our theorem above states that $\mathscr{R}^*$ is the full sequential basis of Cn or, equivalently, that if $\sigma = \langle \alpha_0, \alpha_1, \ldots, \alpha_k \rangle$ then

(4.6)   if $\sigma \in \Sigma^*$ then $\alpha_0 \in Cn(\{\alpha_1, \ldots, \alpha_k\})$ and

(4.7)   if $\alpha_0 \in Cn(\{\alpha_1, \ldots, \alpha_k\})$ then $\sigma \in \Sigma^*$.

The implication (4.6) is obvious. One can prove the inverse implication (4.7) by considering the "proofs" of α_0 from the set $\{\alpha_1, \ldots, \alpha_k\}$.

5. Let S_0 be a normal subalgebra of the algebra S and let Cn_0 be a structural consequence defined in S_0. We assume that the cardinality of Cn_0 is not greater than that of the support of S_0. If Cn is a consequence in S, then Cn is called an *extension* of Cn_0 (and Cn_0 a restriction of Cn) if for every $X \subset S_0$ we have:

(5.1) $Cn_0(X) = S_0 \cap Cn(X)$.

Is Cn_0 a consequence in S_0 and Cn_1, Cn_2 two extensions of Cn_0 both structural and of the same cardinality as Cn_0, then Cn_1 and Cn_2 are equal. This means that a structural consequence Cn_0 has only one structural extension of a cardinality not greater than the cardinality of Cn_0. Is the extension Cn of Cn_0 structural and of the same cardinality as Cn_0 and, moreover, is Cn_0 uniform, then Cn is also uniform.

If we have a structural consequence Cn_0 in S_0, then for $X \subset S$ we can put

(5.2) $Cn(X) = \bigcup_Y \bigcup_\varepsilon \varepsilon^{-1} Cn_0(\varepsilon Y)$

where ε runs over all automorphism of S with $\varepsilon Y \subset S_0$, and Y runs over all subsets of X of the power less than the cardinality of Cn_0. The operation Cn defined in (5.2) is a structural consequence in S; it is an extension of Cn_0 and, obviously, it is of the same cardinality as Cn_0 [1]).

---

[1]) For the proof we must assume that the cardinality $\mathfrak{m}$ of Cn_0 fulfils the following condition:

if $\mathfrak{m}_\xi < \mathfrak{m}$ for $\xi \in \varXi$ and if $\overline{\overline{\varXi}} < \mathfrak{m}$, then $\sum_{\xi \in \varXi} \mathfrak{m}_\xi < \mathfrak{m}$.

II. Matrices

6. The set S of formulas is a free algebra. Therefore, every mapping of variables p_i in a similar algebra A may be extended to a homomorphism of S into A. Is A such an algebra and $B \neq 0$ a subset of A, then the pair $\langle A, B \rangle$ is called a *matrix* ([4]). Each homomorphism of S into A is called a *valuation*. A formula α is verified by the valuation h if $h\alpha \in B$. It is obvious that the verification of a formula α by a valuation h depends only on the values which h admits on variables belonging to the support of α.

The set of formulas verified by each valuation is called the set of *tautologies* of the matrix $\mathfrak{M}$ and denoted by $E(\mathfrak{M})$. It is invariant with respect to the endomorphism (substitutions).

Every invariant set $X \subset S$ may be represented as $E(\mathfrak{M}) = X$ with a suitable matrix $\mathfrak{M}$. This is the well known theorem of Lindenbaum ([4], theorem 3). For the purpose of its proof it is enough to consider the matrix $\langle S, X \rangle$. For this matrix the valuations are simply endomorphisms of S (see [5]).

7. In the opinion of many logicians the theorem of Lindenbaum presents the nearest connection between sentential logics and interpretations by matrices. Indeed, every set of tautologies of a consequence $Cn(0) = T_0$ is an invariant set and therefore, it may be represented in the form $T_0 = E(\mathfrak{M})$. If we find such a matrix $\mathfrak{M}$ we call it an adequate matrix for the calculus under consideration and we think that the problem of interpreting this calculus is solved. But the calculus is given by means of the consequence and the matrix is adequate not for the whole consequence but only for its set of tautologies.

In order to give a real connexion between consequences and matrices we shall define for every matrix $\mathfrak{M}$ a consequence $\mathfrak{M}(X)$ and we shall call $\mathfrak{M}$ *adequate* for Cn only if $\mathfrak{M}(X) = Cn(X)$ for each $X \subset S$.

8. Let $\mathfrak{M} = \langle A, B \rangle$ be a matrix and let X be a subset of S. By $\mathfrak{M}(X)$ we shall understand the set of all formulas verified by each such a valuation which verifies every formula in X. This definition may be written in signs as follows:

(8.1) $$\alpha \in \mathfrak{M}(X) \leftrightarrow \prod_h (hX \subset B \to h\alpha \in B).$$

For every matrix the operation defined in such a manner is a structural and uniform consequence; we call it the *matrix-consequence*.

It is difficult to say something about the cardinality of a matrix-consequence. We can present only the following theorem: Is the matrix $\mathfrak{M}$ finite (i.e. is the set A finite), then its matrix-consequence is finite.

Proof. Let $\alpha \in \mathfrak{M}(X)$. The task is to show the existence of such a finite set $X_0 \subset X$ that $\alpha \in \mathfrak{M}(X_0)$. We consider the product $A^{s(S)}$ as a bicompact space in the product-topology. The elements of this space, i.e.

the mappings of all p_t into A may be regarded as valuations of S. Let $V(\beta)$, for β in S, be the set of all valuations h with $h\beta \in B$. All these sets $V(\beta)$ are both closed and open in $A^{s(S)}$. The assumption that $\alpha \in \mathfrak{M}(X)$ is equivalent to the inclusion $\bigcap\limits_{\beta \in X} V(\beta) \subset V(\alpha)$ and this yields that $V(\alpha) \cup \bigcup\limits_{\beta \in X} V'(\beta) = A^{(s)S}$. All sets in this union are open and, because the space is bicompact, there is a finite set $X_0 \subset X$ such that $V(\alpha) \cup \bigcup\limits_{\beta \in X_0} V'(\beta) = A^{s(S)}$). The last equation is equivalent to the inclusion $\bigcap\limits_{\beta \in X_0} V(\beta) \subset V(\alpha)$ which involves $\alpha \in \mathfrak{M}(X_0)$.

9. We shall now give an outline of the proof of the main theorem:
For every structural and uniform consequence Cn there is an adequate matrix i.e. a matrix $\mathfrak{M}$ such $Cn(X) = \mathfrak{M}(X)$ for each $X \subset S$.

Proof. Let $\{X_\xi\}_{\xi \in \varXi}$ be the family of all Cn-consistent systems. It may be of a greater power than that of the support of S, but we can extend S by adding some new vaiables, to a set S^* the support of which equals in power the set $\varXi$. Further, we can extend Cn to a consequence Cn^* in S^*. Then, we can divide the support $s(S^*)$ into disjoint sets P_ξ ($\xi \in \varXi$) each of which is of a power not less than $s(S)$ and we can find such isomorphisms ε_ξ of S into S^* that $s(\varepsilon_\xi X_\xi) \subset P_\xi$. We put now

$$B = Cn^*(\bigcup\limits_{\xi \in \varXi} \varepsilon_\xi X_\xi) \qquad \mathfrak{M} = \,<S^*, B>$$

and we want to prove that $\mathfrak{M}$ is an adequate matrix for Cn.

Let X be a subset of S, α a formula such that $\alpha \in Cn(X)$ and h a valuation such that $hX \subset B$. By adding values for variables in $s(S^*) - s(S)$, the valuation h may be extended to an endomorphism of S^*. We have $h\alpha \in hCn(X) = h(Cn^*(X) \cap S) \subset Cn^*(hX) \subset Cn^*(B) = B$. This proves that $Cn(X) \subset \mathfrak{M}(X)$.

Let $\alpha \,\bar{\in}\, Cn(X)$. Then $Cn(X) = X_{\xi_0}$ for some suitable ξ_0 in $\varXi$. We choose $h = \varepsilon_{\xi_0}$ as a valuation. We have $hX \subset B$ as assumed. Suppose that

$$h\alpha \in B = Cn^*(hX_{\xi_0} \cup \bigcup\limits_{\substack{\xi \in \varXi \\ \xi \neq \xi_0}} \varepsilon_\xi X_\xi).$$

It follows that $h\alpha \in Cn^*(hX_{\xi_0})$ since Cn^* is uniform. The valuation h being an isomorphism, it may be extended to an automorphism h^* of the whole S^*. We have $h^*\alpha \in Cn^*(h^*X_{\xi_0})$, $h^{*-1}\,h^*\alpha = \alpha \in Cn^*(h^{*-1}\,h^*X_{\xi_0}) = Cn^*(X_{\xi_0})$ and, as $\alpha \in S$, $X_{\xi_0} \subset S$, finally $\alpha \in S \cap Cn^*(X_{\xi_0}) = Cn(X_{\xi_0}) = Cn(X)$, which contradicts the assumption. Therefore, we have proved that $\mathfrak{M}(X) \subset Cn(X)$ which, with the former inclusion, gives us that $\mathfrak{M}(X) = Cn(X)$ q.e.d.

10. Let S be now the set of formulas with the usual signs $\rightarrow$, $\vee$, $\wedge$, $'$, $\equiv$ and let Cn_0 be the consequence with a single rule of detachment only. This consequence is an axiomless one, but by adding suitable

axiomatic rules it can be made into a simple two-valued, an intuitionistic or a positive one. Let Cn_2 be the two-valued consequence and let T_2 be the set of its tautologies $Cn_2(0)$. This set is the largest invariant system of this consequence.

Is A a boolean algebra with the greatest element 1, then the matrix $\langle A, \{1\}\rangle$ is an adequate matrix for Cn_2. It is known that there are such matrices $\mathfrak{M}$ with $E(\mathfrak{M})=T_2$, which are not boolean algebras. As an example may be regarded the following matrix (where $A = \{0, 1/2, 1\}$ and $B = \{1\}$):

$\rightarrow$	0	1/2	1	$'$	$\vee$	0	1/2	1	$\wedge$	0	1/2	1	$\equiv$	0	1/2	1
0	1	1	1	0	0	0	0	1	0	0	0	0	0	1	1	0
1/2	1	1	1	0	1/2	0	0	1	1/2	0	0	0	1/2	1	1	0
1	0	1	1	1	1	1	1	1	1	0	0	1	1	0	0	1

This matrix is not adequate for Cn_2, since $p_1 \in \overline{\mathfrak{M}}(\{p_2 \rightarrow p_1, p_2\})$; to prove this it is enough to put $hp_1 = 1/2$ and $hp_2 = 1$.

We do not know any matrix adequate for the positive consequence or for the intuitionistic consequence. If we shall assume the rule of detachment not for implication but for equivalence, then by adding a finite number of axiomatic rules, we can make this consequence into a two-valued one i.e. into such Cn_2^* for which $T_2 = Cn_2^*(0)$, (see [6]). The set T_2 is not in this case the largest invariant set, because it may be enlarged by adding to it the formula $p_1 \equiv p_1'$ ([6], theorem IV). This shows that the rule of detachment for implication is not in the full basis of Cn_2^* and, therefore, Cn_2^* is not identical to Cn_2.

REFERENCES

1. TARSKI, A., Über einige fundamentalen Begriffe der Metamathematik, Comptes Rendus des séances de la Société des Sciences et des Lettres de Varsovie, 23, 22–29 (1930).
2. SUSZKO, R., O analitycznych aksjomatach i logicznych regułach wnioskowania (On analytic axioms and logical rules of inference), The Poznań Society of Friends of Sciences, Philosophical Section, 7, no. 5, 1–30 (1949).
3. ————, Formalna teoria wartości logicznych I (Formal theory of logical values I), Studia Logica IV, 145–236 (1957).
4. ŁUKASIEWICZ, J. and A. TARSKI, Untersuchungen über den Aussagenkalküls, Comptes Rendus des séances de la Société des Sciences et des Lettres de Varsovie, 23, 30–30 (1930).
5. ŁOŚ, J., O matrycach logicznych (On logical matrices), Travaux de la Société des Sciences et des Lettres de Wrocław, ser. B, no. 19 (1949).
6. SŁUPECKI, J., Über die Regeln des Aussagenkalküls, Studia Logica I, 19–40 (1953).

Mathematical Institute of the
Polish Academy of Sciences and
Institute of Philosophy and Sociology
of the Polish Academy of Sciences

184

Editorial Note

The above text of *Remarks on sentential logics* is a photographic copy of the original, and it requires the following corrigenda.

	Original text	Should be replaced by
p. 178_4	$R_1(\alpha_0, \alpha_2)$	$R_1(\alpha_0, \alpha_1)$
p. 181^{12}	endomorphism	endomorphisms
p. 183_{11}	Studia Logica IV	Studia Logica VI

The truth-tables of the original text on page 183, line 10 should be replaced by the following ones:

$\rightarrow$	0	½	1
0	1	1	1
½	0	1	1
1	0	1	1

$'$
1
0
0

$\vee$	0	½	1
0	0	1	1
½	1	1	1
1	1	1	1

$\wedge$	0	½	1
0	0	0	0
½	0	1	1
1	0	1	1

$\equiv$	0	½	1
0	1	0	0
½	0	1	1
1	0	1	1

Part 8
Saul Kripke (1963)

'Rolling Down the River':
Saul Kripke and the Course of Modal Logic

Johan van Benthem

Abstract We discuss Saul Kripke's seminal 1963 paper 'Semantical Considerations on Modal Logic', and sketch subsequent developments in modal logic with a view to their general logical thrust.

Keywords Modal logic · Kripke · Invariance · Translation · Complexity

1 Returning to the Sources: What a Famous Paper Contains

Kripke's classical paper that forms the point of departure for this brief essay[1] has a content that is easily described to the modern reader—especially, since much of it made its way soon into the widely used textbook Hughes & Cresswell 1968 [1], a major point of entry for my generation into the area.

The author first summarizes his own earlier work on relational possible worlds semantics of propositional modal languages, where truth of a formula $\Box\varphi$ at a world s means that is true at all accessible worlds t.[2] He notes the resulting completeness theorems for deduction in classical propositional modal logics such as T, $S4$ or $S5$.[3] But the key topic of the paper is modal predicate logic with object quantifiers, the vehicle for philosophical discourse about modality. This was an active area at the time. Semantic structures seem obvious: ordered families of possible worlds endowed with object domains, and an interpretation function for predicates with respect to objects and worlds. But the issue is the truth definition for the language. Kripke mentions earlier systems by Prior and Hintikka, taking different Fregean–Austinian views on atomic statements involving objects 'not of this world': these lack truth values.[4] While this diversity may reflect different legitimate conceptions of modal predication and quantification[5], a further fact hurts. Consistent logics for these semantics turn out to modify either the underlying modal propositional logics, or the standard inference rules of first-order predicate logic.

---

[1] Semantical Considerations on Modal Logic (Kripke 1963 [18]).

[2] The semantic structures in the paper are 'model structures' (nowadays, 'frames' with a distinguished world) and 'models' (adding a 'valuation'). Accessibility relations are to be reflexive: the paper has no 'minimal modal logic' K (named afterwards in honour of Kripke).

[3] In a footnote, Kripke cites Kanger and Hintikka as authors of related semantic ideas.

[4] Options spread to truth conditions for the modality, say, construed in a three-valued format.

[5] In this connection, Kripke later speaks of different *conventions*, all tenable in modal logic.

J.-Y. Béziau (ed.), *Universal Logic: An Anthology*, 187–195
Studies in Universal Logic, DOI 10.1007/978-3-0346-0145-0_15, © Springer Basel AG 2012

Kripke's own proposal takes an alternative Russelian line, giving truth values to modal formulas with any objects, inside or outside of the current world of evaluation. Boolean and modal clauses stay as for propositional systems, while quantification is only over objects existing at a world. He gives a proof system that is complete for the new semantics, summarizing his results as follows:

> The systems we have obtained have the following properties. They are a straightforward extension of the modal propositional logics, without the modifications of Prior's Q; the rule of substitution holds without restriction, unlike Hintikka's presentation; and nevertheless neither the Barcan formula nor its converse is derivable. Further, all laws of quantification theory—modified to admit the empty domain—hold. The semantic completeness theorem we gave for modal propositional logic can be extended to the new systems.

Finally, the paper discusses a number of mathematical 'provability interpretations' for the modalities, where possible worlds stand for models of arithmetic.

2 A First Reaction: 40 Years Later

One striking feature of this much-cited, but probably not much-read paper is how few results it contains! There is no new take on propositional modal systems, there is one new proposal for modal predicate logic—but not one that has eventually commanded allegiance – and there is no trace of the subsequent discovery of mathematical and philosophical problems with the 'straightforward' models of modal predicate logic, that have persisted until today (Brauner & Ghilardi 2006 [2], van Benthem 2009 [3, Chapters 11, 26]). Finally, the exploration of provability interpretations focuses precisely on a road not taken in the subsequent flowering of the field of 'Provability Logic' (Artemov 2006 [4]).

Now this is hindsight, and I do recall the liberating effect of this paper and others by Kripke, inspiring a young generation of logicians in the 1960s. Moreover, progress arises from seminal new ideas, as much as new theorems. And the paper definitely opens up a new semantic program that has kept growing since. Even so, I felt a bit like once when visiting the headwaters of the Mississippi in Itasca State Park, Minnesota. It is hard to imagine that the mighty river originates in this placid setting. And maybe this geographical analogy is apt. The Mississippi becomes a great river because of two further features. One is its teaming up with mighty allies, in particular the Missouri. Likewise, the later development of modal logic probably owes as much to the streams of powerful new ideas that were added to the paradigm by other strong logicians. Moreover, the Mississippi only becomes what it is thanks to the terrain on its path, with gradients driving it eventually to its delta. In a similar fashion, to me, it is the scientific community and its qualities of reception and nurture that determine whether ideas grow or die. The intellectual gradients of the time were right for modal logic.

3 The Historical Context: Backward, and Forward

The semantics for modal logic has a long history, and despite various authoritative surveys, it still has lots of mysteries to me. For instance, contrary to popular opinion, it just is not true that modal logic was in a 'syntax only' mode before the 1960s—except maybe in the hands of diehard formalists. Topological interpretations of the universal modality as an interior operation date back to Tarski in the 1930s (cf. van Benthem & Bezhanishvili 2007 [5]), and with hind-sight, we can see that this early topological semantics is even a generalization of the relational possible worlds semantics for $S4$ and its extensions, since Kripke models correspond to Alexandroff tree topologies. Indeed, the analogy extends from $S4$ to the minimal modal logic, and even below, once we move from topological to neighbourhood semantics. Topological methods were quite popular in the 1950s, and in particular, Beth 1957 [6] modeled intuitionistic logic in topological tree models. But somehow, relational models provided a simpler take that 'fired the imagination', perhaps also because of historical echoes to Leibnizian views of modality as truth in all possible worlds, that go back to the Middle Ages. But even their mathematical core content was around: it had been established in Jonsson & Tarski 1951 [7] on the representation theory of Boolean algebras with additional operators. Moreover, a further relevant line that was publicly known is Prior's extensive work in the 1950s on temporal logic, where the relational interpretation for the modalities in terms of 'earlier/later' is precisely the point. But in temporal logic, the relational semantics is so obvious that it seems no big deal, and its ideas remained underappreciated.

4 Intermezzo: Two Directions in Modal Logic

A distinction seems relevant here. One can connect modal languages with semantic structures in two ways. In one direction, the latter are given beforehand, and one looks for logical languages that *describe essential features*, and lead, hopefully, to axiomatizable, perhaps even decidable calculi of reasoning. For instance, one can view early topological semantics as some sort of 'proto-topology'. In this direction, no particular language or logic is sacrosanct, since one wants to get at relevant structure with whatever language is best suited. Thus, changing a modal language is quite acceptable on this approach, and indeed, temporal logics came in lots of different strengths. In the opposite direction of *giving a semantics*, however, one fixes a language and perhaps a deductive system, and asks for the design of a model class that makes the given logic come out 'complete'. The two modes suggest different questions. Of course, eventually, the two directions meet and interact, as is the reality of research in modal logic today.

In either direction, much of the subsequent history of modal logic could not have been predicted from its beginnings in the 1960s. The story that runs from the present paper to the extensive field described in the 2006 *Handbook of Modal Logic* [10] contains many further themes, and a much larger set of 'players', with applications and motivations coming not just from philosophy and mathematics, but also computer science, linguistics, and even economic game theory.

5 Why Was the Framework Attractive, and What Made it Stay?

One basic feature of Kripke's paper is its explanation of modal notions by means of essentially a classical model-theoretic picture. Intensionality is *extensionalized* via 'multiple reference' in sets of possible worlds, i.e., an extended notion of extension. This move 'de-mythologized' intensionality, and at the same time, it also *geometrized* it, providing appealing geometrical content to known modal axioms in terms of features of accessibility relations. I still recall how illuminating it was to match laborious modal syntactic deduction in a system like $S4$ with concrete pictures of reflexive transitive orders: it was as if one could suddenly see what one was doing. These semantic moves caught on fast, and proved illuminating across a wide range of philosophical themes beyond modality, such as time, knowledge, obligation, conditionals, and so on. Thus, modal logic became the 'calculus' of a booming area of philosophical logic. At the same time, technically, possible-worlds semantics brought modal logic much closer to extensional classical logic, and thus, methodologically, the unity of the field of logic became restored: insights and techniques could now flow freely between 'classical' and 'non-classical' areas. As a result, the distinction between 'mathematical' and 'philosophical' logic becomes pretty thin—as we shall see in more detail below.

Still there were criticisms from the start, saying that accessibility relations were ad-hoc formal devices, and that multiple extension was too coarse-grained for true intensionality. While these are valid points in many settings, the staying power of possible-worlds modeling has become ever clearer over time. And also, however justified the worries, the other crucial historical fact is that no convincing competitor has emerged with equal power and sweep. For instance, the onslaught of situation semantics in the 1980s as an alternative paradigm has failed—and later versions of situation theory even used modal logic to bring out their key features (cf. van Benthem & Martinez 2008 [8]).[6] Likewise, mathematical criticisms of the low content of modal logic have subsided, since the mathematics of modal logics has turned out much richer than what was imagined in the 1960s.

But over the years, modal logic has undergone some major changes, affecting its role in logic as a whole. We look at a few, and try to state their essence as a contribution to 'universal logic', i.e., the general thrust of the field.

6 Mathematical Changes in our Understanding of Modal Logic

While modalities have traditionally been viewed as expressions that enrich a classical system, while the matching semantics moves from single situations to complex families of worlds, the modern perspective has changed considerably. Modal logics are not about richer systems than classical logic, but poorer ones! The discussion to follow starts from propositional modal logic, that has become the dominant approach in the field, partly since it high-lights the modalities per se.[7]

Graph Structures   One shift is that 'possible worlds' in their original sense are no longer the ruling paradigm. Worlds can be as diverse as information states of an agent, states of a computer, points in time, board stages in games, linguistic parse trees, or just:

---

[6]This is not to say that all discussion is over. For instance, the appealing provability interpretation of the universal modality $\Box\varphi$ is an $\exists$-*type account* saying that there exists a proof for φ, rather than the above semantic $\forall$-type account that φ is true in all relevant possible worlds.

[7]Many modern insights have arisen from this move, letting the modalities 'speak for themselves' first. Later on in this paper, we will briefly discuss where *modal predicate logic* stands today.

points in a directed graph. The term 'possible world' is retained mainly for reasons of nostalgia and faded grandeur. Here is a better picture: modal logic is about *directed graphs*, and a reason for its broad sweep is the ubiquity of geometrical structures like this across a wide range of subjects.

Local Quantification   But structure is not all. Graphs suggest an 'internal' description language, where one views a large total situation from some current point via accessible neighbours. Typically then, modalities express *local quantification* over these accessible points. This is more restricted than the usual quantifiers of first-order logic, that give unbounded access to arbitrary points in a model.

Standard Translation and Tandem View   A powerful insight making modal logic more down-to-earth has been that Kripke's semantics drives a straightforward *translation* of the modal language into a first-order one of the right signature over possible worlds models, by sending modalities to bounded quantifiers. Thus, a modal formula $\Box \Diamond p$ describing a world w in a model M can be read equally well, using the truth conditions as translation clauses, as a first-order formula $\forall y(Rxy \rightarrow \exists z(Ryz \land Pz))$. This Gestalt Switch yields a *tandem approach* to reasoning with intensional notions without having to choose between modal or first-order logic, viewing modalities as bounded quantifiers ranging over the local environment of a world in an accessibility pattern, or abstractly, a point in a graph. Thus, from being an external 'challenger', modal logic gets integrated into classical logic.[8]

Fine-structure and Fragments   The above translation sends the basic modal language into a *fragment* of first-order logic. More generally, propositional modal languages tend to be fragments of classical logics, though not always first-order ones. What makes these fragments so insightful? For a start, the mini-language of basic modal logic has proved remarkably well-chosen for its combining various desirable features. One is the ease of modal deduction, without the tedious variable management needed for first-order proof in general. Modal core patterns of reasoning 'meet the eye' at once. But perhaps deeper are the following points:

Expressive Power and Invariance   The expressive power of the basic modal language turned out to match a natural structural *invariance relation* between graph models, viz. *bisimulation*, a natural back-and-forth 'process equivalence' between transitions from world to world that preserves atomic facts (Blackburn, de Rijke & Venema 2001 [9]). Bisimulation analyzes when two models are 'the same' from a modal point of view, a basic question that sets the semantic expressivity level for any well-designed logical language. And it has proven a natural level of identification for structures, not just in modal logic, but also in process theories, set theory, and other areas.[9]

---

[8]This co-existence was reinforced by other developments, omitted here, such as '*frame correspondence theory*' for modal axioms in terms of classical properties of the accessibility relation.

[9]The crucial issue of when two modal models represent 'the same structure' does not seem to have occurred much to philosophers, and it is still under-appreciated in philosophical logic.

Computational Complexity   Next, while the expressive power of the modal language is weaker than that of first-order logic, it shows much better behaviour in terms of the computational complexity of the core tasks a logic is used for: stating properties of structures, evaluating them, and reasoning about them. This computational viewpoint has often been dismissed as a matter of 'mere implementation', but by now, there may be more awareness that *procedural fine-structure* is as fundamental an issue as expressive fine-structure. In particular, testing for modal validity is decidable in polynomial space, model checking takes polynomial time: lower than for first-order logic. Thus expressive weakness can be computational strength, and the perspective shifts once more:

'The Balance'   Modal languages strike a balance between two competing forces: expressive power and computational complexity. This is a much broader theme in the field: first-order logic itself arose as a reasonably expressive fragment of second-order logic whose notion of consequence was axiomatizable (though undecidable).[10] But there are other natural compromises along these lines. In modern modal logic, many systems (cf. the 'hybrid logics' of Areces & ten Cate 2006 [11]) lie in between the basic language and full first-order logic, with extra modalities describing more graph structure.[11] This landscape of fragments of first-order logic has shown that 'small can be beautiful'. Hence, modal logics are also tools for exploring the fine-structure of complex classical systems, sometimes leading to the discovery of new classical logics in the process.[12]

The above features are my take on what makes modal logic general, and a source of perspectives and insights for the 'Universal Logic' in this volume. Of course, this is a personal stance, and more could be said. A large role in changing modal logic was also played by the mathematical studies of the 1970s, on frame definability, general completeness (and incompleteness) theorems, or the interplay of model theory and algebra. All these are well-documented in the *Handbook of Modal Logic* [10], and they are very much alive today, witness the lively current interactions of modal logic and Universal Algebra.

7 Descriptive Expansion: Modal Patterns and Transdisciplinary Migrations

But the development of modal logic since the 1960s is not just theoretical evolution of mathematical and computational perspectives. At the same time, its basic features changed through extension of descriptive coverage, and the study of modal patterns in a widening circle of fields. This was already visible in the late 1960s, when there was a

---

[10]Incidentally, our survey is by no means complete. There are many further viewpoints that capture important aspects of modal formalisms. One important more syntactic way to think of the above translation views modal languages as perspicuous *variable-free* formalisms for proof-theoretic purposes.

[11]The same is true for languages beyond first-order logic such as the *modal μ-calculus* (Bradfield & Stirling 2006 [12]). This is a decidable part of fixed-point logic with only local quantifiers.

[12]Modal logic influenced classical logic in the 'Guarded Fragment' of Andréka, van Benthem & Németi 1998 [13], a new large decidable part of first-order logic. See also the abstract model theory with Lindström theorems for weak languages in van Benthem, ten Cate & Väänänen 2009 [14].

wave of philosophical logics using modal ideas in innovative ways, including epistemic logic, doxastic logic, tense logic, deontic logic, conditional logic, and of course, Kripke's own studies of modality and quantification that reverberated for decades.

But also beyond philosophy, modal patterns are ubiquitous.[13] Linguists have used modal operators for describing modal, temporal and other expressions in natural language, making them important to communication and other cognitive functions. An even broader source of modal patterns is the study of information and agency, that crosses borders between philosophy and many other disciplines. In the 1970s, economists independently rediscovered epistemic logic as a convenient perspicuous formalism for the informational reasoning about rational agency that keeps players locked into game-theoretic equilibria. Around the same time, computer scientists started using modal, temporal, and epistemic logics for a wide variety of purposes in describing processes and information flow. The resulting dynamic and temporal logics of computation, information, knowledge, belief, preference, and other crucial features of social agency are coming together today in the study of what is sometimes called *intelligent interaction*.[14]

Besides all this, modal patterns have turned out crucial in ever more mathematical settings, witness non-well-founded set theory, provability logic, logics of space and space-time, or 'co-algebra' of non-well-founded infinite objects (Venema 2006 [16]). This wholly unplanned academic penetration of modal ideas since the 1960s is another reason for the remarkable staying power of the field.

8 Back to Philosophy and Mathematics

While philosophers may feel that all this newer activity is a farewell to the original philosophical motivations, there is really no reason why a prodigal son might not return to the old home, bringing travel tales far beyond what his parents ever imagined. Some modal logicians are returning to epistemology, philosophy of action, and other areas of the mother discipline these days. Examples are work on belief revision, interactive epistemology, and philosophy of action inspired by computational and game-theoretic influences. And examples can be multiplied.[15]

One might think of all this as 'applied modal logic', though I have argued elsewhere that the term 'applied logic' fails to do justice to what is happening today—as it suggests two falsehoods: that traditional ideas and logic systems suffice in the new areas, and that no new pure logical theory is at stake. But one striking feature of modal logics in the above areas has been the emergence of new fundamental notions and issues, beyond anything studied up to the 1980s:

One key example is the notion of what may be called *equilibrium*, in epistemic reflection or in actual behaviour: with examples like common knowledge, game-theoretic

---

[13]Freudental 1959 [15] proposed broadcasting reasoning patterns with basic modal dualities into outer space, as a way of making other intelligences aware of basics of thought on Planet Earth.

[14]Modal patterns also occur in knowledge representation, web languages, or spatially distributed computation, where again, they have often been rediscovered independently.

[15]Modal logics of *preference*, another crucial ingredient of intentional goal-driven rational agency, using evaluation orders of worlds, might well make their way back into deontic logic.

equilibria, or iterated action. This calls for modal logics beyond the usual ones, incorporating *recursion mechanisms* with 'fixed-points', whose theory, in full development, is rife with open problems. Another fundamental new theme is 'system architecture', and in particular, *system combination*. On a naive analytical view, one should just do modal logics for various features of agency, and then throw them together to get the whole picture. But it has become clear since around 1990, that things are much more delicate. Depending on the manner of combination, even decidable modal logics may give rise to highly undecidable combinations. Thus, the behaviour of complex modal systems is one more new fundamental challenge.

Both recursion and system combination make excellent sense for logic in general—and I would put them on a par with the 'universal' themes in Sect. 6.

Back to Modal Predicate Logic   An illustration of the preceding point takes us back to the modal predicate logic originally discussed here. One reason why the latter system has been so hard to define well, avoiding model-theoretic and proof-theoretic catastrophes, is the fact that is a system combination of two modal logics,[16] whose behaviour depends critically on the mode of combination. Even so, modal predicate logic is obviously important to old and new theory and applications—and in that sense, Kripke's paper is still highly relevant, decades later.

9 Conclusion

Is all well on the banks of the great river that Saul Kripke helped bring forth? Clearly, in its course, the river runs among diverse communities—and a common practical vision among them is hard to achieve. Also, major theoretical problems remain unresolved, from finding the right semantics for modal predicate logic to understanding general behaviour of system combination, the delicate balance of expressive power and computational complexity, and other issues raised here. And finally, different conceptions of modality are still alive beyond the possible worlds semantics of Kripke's paper. I mentioned alternative ∃-style proof views (Artemov 2006 [4])—and even more radically than that, non-operator *predicate views* of modality, dismissed in the 1960s partly under the influence of Montague, may still get their field day (Halbach, Leitgeb & Welch 2003 [17]).

However that may be, I feel that even in its present state, modal logic is not just an area of interdisciplinary application, but also an excellent conceptual laboratory for pure universal logic.

Acknowledgements   I would like to thank the anonymous reader for useful feedback, and Ştefan Minică for the LATEX apotheosis of this paper.

References

1. Hughes, G.E., Cresswell, M.J.: An Introduction to Modal Logic. Methuen, London (1968)

---

[16]One can profitably view first-order logic *itself* as a modal logic—or more precisely, as a dynamic logic of changing assignments in suitable computational state spaces.

2. Brauner, Th., Ghilardi, S.: First-order modal logic. In: Blackburn, P., van Benthem, J., Wolter, F. (eds.) Handbook of Modal Logic, pp. 549–620. Elsevier, Amsterdam (2006)
3. van Benthem, J.: Modal Logic for Open Minds. CSLI Publications, Stanford (2009)
4. Artemov, S.: Modal logic in mathematics. In: Blackburn, P., van Benthem, J., Wolter, F. (eds.) Handbook of Modal Logic, pp. 927–970. Elsevier, Amsterdam (2006)
5. van Benthem, J., Bezhanishvili, G.: Modal logics of space. In: Aiello, M., Pratt-Hartmann, I., van Benthem, J. (eds.) Handbook of Spatial Logics, pp. 217–298. Springer, Dordrecht (2007)
6. Beth, E.W.: Semantic construction of intuitionistic logic. Meded. K. Ned. Akad. Wet., New Ser. 19(11), 357–388 (1956) Amsterdam
7. Jonsson, B., Tarski, A.: Boolean algebras with operators. Am. J. Math. 73(4), 891–939 (1951)
8. van Benthem, J., Martinez, M.: The stories of logic and information. In: Adriaans, P., van Benthem, J. (eds.) Handbook of the Philosophy of Information, pp. 217–280. Elsevier, Amsterdam (2008)
9. Blackburn, P., de Rijke, M., Venema, Y.: Modal Logic. Cambridge University Press, Cambridge (2001)
10. Blackburn, P., van Benthem, J., Wolter, F. (eds.): Handbook of Modal Logic. Elsevier, Amsterdam (2006)
11. Areces, C., ten Cate, B.: Hybrid logics. In: Blackburn, P., van Benthem, J., Wolter, F. (eds.), Handbook of Modal Logic, pp. 821–868. Elsevier, Amsterdam (2006)
12. Bradfield, J., Stirling, C.: Modal μ-Calculi. In: Blackburn, P., van Benthem, J., Wolter, F. (eds.) Handbook of Modal Logic pp. 721–756. Elsevier, Amsterdam (2006)
13. Andréka, H., van Benthem, J., Németi, I.: Modal logics and bounded fragments of predicate logic. J. Philos. Log. 27, 217–274 (1998)
14. van Benthem, J.: ten Cate, B., Väänänen, J.: Lindström theorems for fragments of first-order logic. In: Proceedings of LICS, pp. 280–292 (2007). To appear in Logical Methods for Computer Science
15. Freudenthal, H.: Lincos: Design of a Language for Cosmic Intercourse. North-Holland, Amsterdam (1959)
16. Venema, Y.: Algebras and co-algebras. In: Blackburn, P., van Benthem, J., Wolter, F. (eds.) Handbook of Modal Logic, pp. 331–426. Elsevier, Amsterdam (2006)
17. Halbach, V., Leitgeb, H., Welch, Ph.: Possible worlds semantics for modal notions conceived as predicates. J. Philos. Log. 32, 179–223 (2003)
18. Kripke, S.A.: Semantical considerations on modal logic. Acta Philos. Fenn. 16, 83–94 (1963)

J. van Benthem (✉)
Institute for Logic, Language and Computation, University of Amsterdam, P.O. Box 94242, 1090 GE Amsterdam, The Netherlands
e-mail: johan.vanbenthem@uva.nl
url: http://staff.science.uva.nl/~johan
Department of Philosophy, Stanford University, Stanford, CA, USA

Semantical Considerations on Modal Logic

SAUL A. KRIPKE

This paper gives an exposition of some features of a semantical theory of modal logics [1]. For a certain quantified extension of S5, this theory was presented in [1], and it has been summarized in [2]. The present paper will concentrate on one aspect of the theory — the introduction of quantifiers — and it will restrict itself in the main to one method of achieving this end. The emphasis of the paper will be purely semantical, and hence it will omit the use of semantic tableaux, which is essential to a full presentation of the theory. (For these, see [1] and [11].) Proofs, also, will largely be suppressed.

We consider four modal systems. Formulae A, B, C, ... are built out of atomic formulae P, Q, R, ..., using the connectives $\wedge$, $\sim$, and $\square$. The system M has the following axiom schemes and rules:

A1. $\square A \supset A$
A2. $\square (A \supset B) \supset . \square A \supset \square B$
R1. $A, A \supset B \mathbin{/} B$
R2. $A \mathbin{/} \square A$

If we add the following axiom scheme, we get S4:

$$\square A \supset \square \square A$$

We get the *Brouwersche* system if we add to M:

$$A \supset \square \lozenge A$$

S5, if we add:

$$\lozenge A \supset \square \lozenge A$$

---

[1] The theory given here has points of contact with many authors: For lists of these, see [11] and Hintikka [6]. The authors closest to the present theory appear to be Hintikka and Kanger. The present treatment of quantification, however, is unique as far as I know, although it derives some inspiration from acquaintance with the very different methods of Prior and Hintikka.

J.-Y. Béziau (ed.), *Universal Logic: An Anthology*, 197–208
Studies in Universal Logic, DOI 10.1007/978-3-0346-0145-0_16, © Springer Basel AG 2012

Modal systems whose theorems are closed under the rules R1 and R2, and include all theorems of M, are called "normal". Although we have developed a theory which applies to such non-normal systems as Lewis's S2 and S3, we will restrict ourselves here to normal systems.

To get a semantics for modal logic, we introduce the notion of a (normal) *model structure*. A model structure (m.s.) is an ordered triple $(\mathbf{G}, \mathbf{K}, \mathbf{R})$ where $\mathbf{K}$ is a set, $\mathbf{R}$ is a reflexive relation on $\mathbf{K}$, and $\mathbf{G} \, \varepsilon \, \mathbf{K}$. Intuitively, we look at matters thus: $\mathbf{K}$ is the set of all "possible worlds;" $\mathbf{G}$ is the "real world." If $\mathbf{H}_1$ and $\mathbf{H}_2$ are two worlds, $\mathbf{H}_1 \, \mathbf{R} \, \mathbf{H}_2$ means intuitively that $\mathbf{H}_2$ is "possible relative to" $\mathbf{H}_1$; *i.e.*, that every proposition *true* in $\mathbf{H}_2$ is *possible* in $\mathbf{H}_1$. Clearly, then, the relation $\mathbf{R}$ should indeed be reflexive; every world $\mathbf{H}$ is *possible* relative to itself, since every proposition *true* in $\mathbf{H}$ is, *a fortiori*, possible in $\mathbf{H}$. Reflexivity is thus an intuitively natural requirement. We may impose additional requirements, corresponding to various "reduction axioms" of modal logic: If $\mathbf{R}$ is transitive, we call $(\mathbf{G}, \mathbf{K}, \mathbf{R})$ an S4-m.s.; if $\mathbf{R}$ is symmetric, $(\mathbf{G}, \mathbf{K}, \mathbf{R})$ is a *Brouwersche* m.s.; and if $\mathbf{R}$ is an equivalence relation, we call $(\mathbf{G}, \mathbf{K}, \mathbf{R})$ an S5-m.s. A model structure without restriction is also called an M-model structure.

To complete the picture, we need the notion of *model*. Given a model structure $(\mathbf{G}, \mathbf{K}, \mathbf{R})$, a *model* assigns to each atomic formula (propositional variable) P a truth-value $\mathbf{T}$ or $\mathbf{F}$ in each world $\mathbf{H} \, \varepsilon \, \mathbf{K}$. Formally, a *model* φ on a m.s. $(\mathbf{G}, \mathbf{K}, \mathbf{R})$ is a binary function $\varphi(P, \mathbf{H})$, where P varies over atomic formulae and $\mathbf{H}$ varies over elements of $\mathbf{K}$, whose range is the set $\{\mathbf{T}, \mathbf{F}\}$. Given a model, we can define the assignments of truth-values to non-atomic formulae by induction. Assume $\varphi(A, \mathbf{H})$ and $\varphi(B, \mathbf{H})$ have already been defined for all $\mathbf{H} \, \varepsilon \, \mathbf{K}$. Then if $\varphi(A, \mathbf{H}) = \varphi(B, \mathbf{H}) = \mathbf{T}$, define $\varphi(A \wedge B, \mathbf{H}) = \mathbf{T}$; otherwise, $\varphi(A \wedge B, \mathbf{H}) = \mathbf{F}$. $\varphi(\sim A, \mathbf{H})$ is defined to be $\mathbf{F}$ iff $\varphi(A, \mathbf{H}) = \mathbf{T}$; otherwise, $\varphi(\sim A, \mathbf{H}) = \mathbf{T}$. Finally, we define $\varphi(\square A, \mathbf{H}) = \mathbf{T}$ iff $\varphi(A, \mathbf{H}') = \mathbf{T}$ for every $\mathbf{H}' \, \varepsilon \, \mathbf{K}$ such that $\mathbf{H} \, \mathbf{R} \, \mathbf{H}'$; otherwise, $\varphi(\square A, \mathbf{H}) = \mathbf{F}$. Intuitively, this says that A is necessary in $\mathbf{H}$ iff A is true in all worlds $\mathbf{H}'$ possible relative to $\mathbf{H}$.

Completeness theorem. $\vdash A$ in M (S4, S5, the *Brouwersche* system) if and only if $\varphi(A, \mathbf{G}) = \mathbf{T}$ for every model φ on an M- (S4-, S5-, *Brouwersche*) model structure $(\mathbf{G}, \mathbf{K}, \mathbf{R})$.

(For a proof, see [11].)

This completeness theorem equates the syntactical notion of *provability* in a modal system with a semantical notion of *validity*.

The rest of this paper concerns, with the exception of some concluding remarks, the introduction of quantifiers. To do this, we must associate with each world a domain of individuals, the individuals that exist in that world. Formally, we define a *quantificational model structure* (q.m.s.) as a model structure (G, K, R), together with a function ψ which assigns to each $H \, \varepsilon \, K$ a set $\psi(H)$, called the *domain* of H. Intuitively $\psi(H)$ is the set of all individuals existing in H. Notice, of course, that $\psi(H)$ need not be the same set for different arguments H, just as, intuitively, in worlds other than the real one, some actually existing individuals may be absent, while new individuals, like Pegasus, may appear.

We may then add, to the symbols of modal logic, an infinite list of individual variables x, y, z, ..., and, for each nonnegative integer n, a list of n-adic predicate letters P^n, Q^n, ..., where the superscripts will sometimes be understood from the context. We count propositional variables (atomic formulae) as "0-adic" predicate letters. We then build up well-formed formulae in the usual manner, and can now prepare ourselves to define a quantificational *model*.

To define a quantificational model, we must extend the original notion, which assigned a truth-value to each atomic formula in each world. Analogously, we must suppose that in each world a given n-adic predicate letter determines a certain set of ordered n-tuples, its *extension* in that world. Consider, for example, the case of a monadic predicate letter $P(x)$. We would like to say that, in the world H, the predicate $P(x)$ is true of some individuals in $\psi(H)$ and false of others; formally, we would say that, relative to certain assignments of elements of $\psi(H)$ to x, $\varphi(P(x), H) = T$ and relative to others $\varphi(P(x), H) = F$. The set of all individuals of which P is true is called the *extension* of P in H. But there is a problem: should $\varphi(P(x), H)$ be given a truth-value when x is assigned a value in the domain of some *other* world H', and not in the domain of H? Intuitively, suppose $P(x)$ means "x is bald" — are we to assign a truth-value to the substitution instance "Sherlock Holmes is bald"? Holmes does not exist, but in other states of affairs, he would have existed. Should we assign a definite truth-value to the statement that he is bald, or not? Frege [3] and Strawson [4] would not assign the statement a truth-value; Russell [5] would [1]. For the purposes of modal logic we hold that different

---

[1] Russell, however, would conclude that "Sherlock Holmes" is therefore not a genuine name; and Frege would eliminate such empty names by an artifact.

answers to this question represent alternative *conventions*. All are tenable. The only existing discussions of this problem I have seen — those of Hintikka [6] and Prior [7] — adopt the Frege-Strawson view. This view necessarily must lead to some modification of the usual modal logic. The reason is that the semantics for modal propositional logic, which we have already given, assumed that every formula must take a truth-value in each world; and now, for a formula $A(x)$ containing a free variable x, the Frege-Strawson view requires that it not be given a truth-value in a world **H** when the variable x is assigned an individual not in the domain of that world. We thus can no longer expect that the original laws of modal propositional logic hold for statements containing free variables, and are faced with an option: either revise modal propositional logic or restrict the rule of substitution. Prior does the former, Hintikka the latter. There are further alternatives the Frege-Strawson choice involves: Should we take $\square \, A$ (in **H**) to mean that A is *true* in all possible worlds (relative to **H**), or just *not false* in any such world? The second alternative merely demands that A be either true or lack a truth-value in each world. Prior, in his system Q, in effect admits both types of necessity, one as "*L*" and the other as "*NMN*". A similar question arises for conjunction: if A is false and B has no truth-value, should we take $A \wedge B$ to be false or truth-valueless?

In a full statement of the semantical theory, we would explore all these variants of the Frege-Strawson view. Here we will take the other option, and assume that a statement containing free variables has a truth-value in each world for every assignment to its free variables [1]. Formally, we state the matter as follows: Let $U = \underset{H \, \varepsilon \, K}{\cup} \, \psi(H)$. U^n is the nth Cartesian product of **U** with itself. We define a quantificational *model* on a q.m.s. (**G, K, R**) as a binary

---

[1] It is natural to assume that an *atomic* predicate should be *false* in a world **H** of all those individuals not existing in that world; that is, that the extension of a predicate letter must consist of actually existing individuals. We can do this by requiring semantically that $\varphi \, (P^n, \, \mathbf{H})$ be a subset of $[\psi \, (\mathbf{H})]^n$; the semantical treatment below would otherwise suffice without change. We would have to add to the axiom system below all closures of formulae of the form $P^n \, (x_1, \ldots, x_n) \wedge (y)A(y) \, . \supset . \, A(x_i) \; (1 \leq i \leq n)$. We have chosen not to do this because the rule of substitution would no longer hold; theorems would hold for atomic formulae which would not hold when the atomic formulae are replaced by arbitrary formulae. (This answers a question of Putnam and Kalmar.)

function $\varphi(P^n, \mathbf{H})$, where the first variable ranges over n-adic predicate letters, for arbitrary n, and $\mathbf{H}$ ranges over elements of $\mathbf{K}$. If $n = 0$, $\varphi(P^n, \mathbf{H}) = \mathbf{T}$ or $\mathbf{F}$; if $n \geq 1$, $\varphi(P^n, \mathbf{H})$ is a subset of $\mathbf{U}^n$. We now define, inductively, for every formula A and $\mathbf{H} \varepsilon \mathbf{K}$, a truth-value $\varphi(A, \mathbf{H})$, relative to a given assignment of elements of $\mathbf{U}$ to the free variables of A. The case of a propositional variable is obvious. For an atomic formula $P^n(x_1, \ldots, x_n)$, where P^n is an n-adic predicate letter and $n \geq 1$, given an assignment of elements $a_1, \ldots, a_n$ of $\mathbf{U}$ to $x_1, \ldots, x_n$, we define $\varphi(P^n(x_1, \ldots, x_n), \mathbf{H}) = \mathbf{T}$ if the n-tuple $(a_1, \ldots, a_n)$ is a member of $\varphi(P^n, \mathbf{H})$; otherwise, $\varphi(P^n(x_1, \ldots, x_n), \mathbf{H}) = \mathbf{F}$, relative to the given assignment. Given these assignments for atomic formulae, we can build up the assignments for complex formulae by induction. The induction steps for the propositional connectives $\wedge$, $\sim$, $\square$, have already been given. Assume we have a formula $A(x, y_1, \ldots, y_n)$, where x and the y_i are the only free variables present, and that a truth-value $\varphi(A(x, y_1, \ldots, y_n), \mathbf{H})$ has been defined for each assignment to the free variables of $A(x, y_1, \ldots, y_n)$. Then we define $\varphi((x)A(x, y_1, \ldots, y_n), \mathbf{H}) = \mathbf{T}$ relative to an assignment of $b_1, \ldots, b_n$ to $y_1, \ldots, y_n$ (where the b_i are elements of $\mathbf{U}$), if $\varphi((A(x, y_1, \ldots, y_n), \mathbf{H}) = \mathbf{T}$ for every assignment of $a, b_1, \ldots, b_n$ to $x, y_1, \ldots, y_n$, respectively, where $a \varepsilon \psi(\mathbf{H})$; otherwise, $\varphi((x)A(x, y_1, \ldots, y_n), \mathbf{H}) = \mathbf{F}$ relative to the given assignment. Notice that the restriction $a \varepsilon \psi(\mathbf{H})$ means that, in $\mathbf{H}$, we quantify only over the objects actually existing in $\mathbf{H}$.

To illustrate the semantics, we give counterexamples to two familiar proposals for laws of modal quantification theory — the "Barcan formula" $(x) \square A(x) \supset \square (x)A(x)$ and its converse $\square (x)A(x) \supset (x) \square A(x)$. For each we consider a model structure $(\mathbf{G}, \mathbf{K}, \mathbf{R})$, where $\mathbf{K} = \{\mathbf{G}, \mathbf{H}\}$, $\mathbf{G} \neq \mathbf{H}$, and $\mathbf{R}$ is simply the Cartesian product $\mathbf{K}^2$. Clearly $\mathbf{R}$ is reflexive, transitive, and symmetric, so our considerations apply even to S5.

For the Barcan formula, we extend $(\mathbf{G}, \mathbf{K}, \mathbf{R})$ to a quantificational model structure by defining $\psi(\mathbf{G}) = \{a\}$, $\psi(\mathbf{H}) = \{a, b\}$, where a and b are distinct. We then define, for a monadic predicate letter P, a model φ in which $\varphi(P, \mathbf{G}) = \{a\}$, $\varphi(P, \mathbf{H}) = \{a\}$. Then clearly $\square P(x)$ is true in $\mathbf{G}$ when x is assigned a; and since a is the only object in the domain of $\mathbf{G}$, so is $(x)\square P(x)$. But, $(x)P(x)$ is clearly false in $\mathbf{H}$ (for $\varphi(P(x), \mathbf{H}) = \mathbf{F}$ when x is assigned b), and hence $\square(x)P(x)$ is false in $\mathbf{G}$. So we have a counterexample to the Barcan

formula. Notice that this counterexample is quite independent of
whether b is assigned a truth-value in **G** or not, so also it applies
to the systems of Hintikka and Prior. Such counterexamples can
be disallowed, and the Barcan formula reinstated, only if we require
a model structure to satisfy the condition that $\psi(\mathbf{H}') \subseteq \psi(\mathbf{H})$
whenever $\mathbf{H}\,R\,\mathbf{H}'$ ($\mathbf{H}, \mathbf{H}' \;\varepsilon\; \mathbf{K}$).

For the converse of the Barcan formula, set $\psi(\mathbf{G}) = \{a, b\}$,
$\psi(\mathbf{H}) = \{a\}$, where again $a \neq b$. Define $\varphi(P, \mathbf{G}) = \{a, b\}$, $\varphi(P, \mathbf{H})$
$= \{a\}$, where P is a given monadic predicate letter. Then clearly
$(x)P(x)$ holds in both **G** and **H**, so that $\varphi(\Box(x)P(x), \mathbf{G}) = \mathbf{T}$.
But $\varphi(P(x), \mathbf{H}) = \mathbf{F}$ when x is assigned b, so that, when x is assigned
$\varphi(\Box P(x), \mathbf{G}) = \mathbf{F}$. Hence $\varphi((x)\Box P(x), \mathbf{G}) = \mathbf{F}$, and we have the
desired counterexample to the converse of the Barcan formula.
This counterexample, however, depends on asserting that, in **H**,
$P(x)$ is actually *false* when x is assigned b; it might thus disappear
if, for this assignment, $P(x)$ were declared to lack truth-value in **H**.
In this case, we will still have a counterexample if we require a necessary statement to be *true* in all possible worlds (Prior's "*L*"),
but not if we merely require that it never be false (Prior's "*NMN*").
On our present convention, we can eliminate the counterexample
only by requiring, for each q.m.s., that $\psi(\mathbf{H}) \subseteq \psi(\mathbf{H}')$ whenever
$\mathbf{H}\,R\,\mathbf{H}'$.

These counterexamples lead to a peculiar difficulty: We have
given countermodels, in quantified S5, to both the Barcan formula
and its converse. Yet Prior appears to have shown in [8] that the
Barcan formula is derivable in quantified S5; and the converse
seems derivable even in quantified M by the following argument:

(A) $(x)A(x) \supset A(y)$ (by quantification theory)

(B) $\Box\,((x)A(x) \supset A(y))$ (by necessitation)

(C) $\Box\,((x)A(x) \supset A(y)) \supset \Box(x)A(x) \supset \Box\,A(y)$ (Axiom A2)

(D) $\Box\,(x)A(x) \supset \Box\,A(y)$ (from (B) and (C))

(E) $(y)\,(\Box\,(x)A(x) \supset \Box\,A(y))$ (generalizing on (D))

(F) $\Box\,(x)A(x) \supset (y)\,\Box\,A(y)$ (by quantification theory, and (E))

We seem to have derived the conclusion using principles that
should all be valid in the model-theory. Actually, the flaw lies in the
application of necessitation to (A). In a formula like (A), we give

the free variables the generality interpretation: [1] When (A) is asserted as a theorem, it abbreviates assertion of its ordinary universal closure

(A') $(y) ((x)A(x) \supset A(y))$

Now if we applied necessitation to (A'), we would get

(B') $\Box (y) ((x)A(x) \supset A(y))$

On the other hand, (B) itself is interpreted as asserting

(B") $(y) \Box ((x)A(x) \supset A(y))$

To infer (B") from (B'), we would need a law of the form $\Box (y)C(y) \supset (y)\Box C(y)$, which is just the converse Barcan formula that we are trying to prove. In fact, it is readily checked that (B") fails in the countermodel given above for the converse Barcan formula, if we replace $A(x)$ by $P(x)$.

We can avoid this sort of difficulty if, following Quine [15], we formulate quantification theory so that only *closed* formulae are asserted. Assertion of formulae containing free variables is at best a convenience; assertion of $A(x)$ with free x can always be replaced by assertion of $(x)A(x)$.

If A is a formula containing free variables, we define a *closure* of A to be any formula without free variables obtained by prefixing universal quantifiers and necessity signs, in any order, to A. We then define the axioms of quantified M to be the closures of the following schemata:

(0) Truth-functional tautologies

(1) $\Box A \supset A$

(2) $\Box (A \supset B). \supset . \Box A \supset \Box B$

(3) $A \supset (x)A$, where x is not free in A.

(4) $(x) (A \supset B). \supset . (x)A \supset (x)B$

(5) $(y) ((x)A(x) \supset A(y))$

---

[1] It is not asserted that the generality interpretation of theorems with free variables is the only possible one. One might wish a formula A to be provable iff, for each model φ, $\varphi(A, \mathbf{G}) = \mathbf{T}$ for every assignment to the free variables of A. But then $(x)A(x) \supset A(y)$ will not be a theorem; in fact, in the countermodel above to the Barcan formula, $\varphi ((x)P(x) \supset P(y), \mathbf{G}) = \mathbf{F}$ if y is assigned b. Thus quantification theory would have to be revised along the lines of [9] or [10]. This procedure has much to recommend it, but we have not adopted it since we wished to show that the difficulty can be solved without revising quantification theory or modal propositional logic.

The rule of inference is detachment for material implication. Necessitation can be obtained as a derived rule.

To obtain quantified extensions of S4, S5, the *Brouwersche* system, simply add to the axiom schemata all closures of the appropriate reduction axiom.

The systems we have obtained have the following properties: They are a straightforward extension of the modal propositional logics, without the modifications of Prior's Q; the rule of substitution holds without restriction, unlike Hintikka's presentation; and nevertheless neither the Barcan formula nor its converse is derivable. Further, all the laws of quantification theory — modified to admit the empty domain — hold. The semantical completeness theorem we gave for modal propositional logic can be extended to the new systems.

We can introduce *existence as a predicate* in the present system if we like. Semantically, existence is a monadic predicate $E(x)$ satisfying, for each model φ on a m.s. $(\mathbf{G}, \mathbf{K}, \mathbf{R})$, the identity $\varphi\,(E, \mathbf{H})$ $= \psi\,(\mathbf{H})$ for every $\mathbf{H}\,\varepsilon\,\mathbf{K}$. Axiomatically, we can introduce it through the postulation of closures of formulae of the form: $(x)A(x) \wedge E(y)$. $\supset . A(y)$, and $(x)E(x)$. The predicate P used above in the counterexample to the converse Barcan formula can now be recognized as simply existence. This fact shows how existence differs from the tautological predicate $A(x) \vee \sim A(x)$ even though $\Box(x)E(x)$ is provable. For although $(x)\,\Box\,(A(x) \vee \sim A(x))$ is valid, $(x)\Box E(x)$ is not; although it is necessary that every thing exists, it does not follow that everything has the property of necessary existence.

We can introduce identity semantically in the model theory by defining $x = y$ to be true in a world $\mathbf{H}$ when x and y are assigned the same value and otherwise false; existence could then be defined in terms of identity, by stipulating that $E(x)$ means $(\exists y)\,(x = y)$. For reasons not given here, a broader theory of identity could be obtained if we complicated the notion of quantificational model structure.

We conclude with some brief and sketchy remarks on the "provability" interpretations of modal logics, which we give in each case for propositional calculus only. The reader will have obtained the main point of this paper if he omits this section. Provability interpretations are based on a desire to adjoin a necessity operator to a formal system, say Peano arithmetic, in such a way that, for any formula A of the system, $\Box\,A$ will be interpreted as true iff A is provable in the system. It has been argued that such "provability" interpretations of a modal operator are dispensable in favor of a provability *predicate*, attaching to the Gödel number of A; but Professor Montague's contribution to the present volume casts at least some doubt on this viewpoint.

Let us consider the formal system **PA** of Peano arithmetic, as formalized in Kleene [12]. We adjoin to the formation rules operators $\wedge$, $\sim$, and $\square$ (the conjunction and negation adjoined are to be distinct from those of the original system), operating on closed formulae only. In the model theory we gave above, we took atomic formulae to be propositional variables, or predicate letters followed by parenthesized individual variables; here we take them to be simply the closed well-formed formulae of **PA** (*not* just the atomic formulae of **PA**). We define a model structure (**G**, **K**, **R**), where **K** is the set of all distinct (non-isomorphic) countable models of **PA**, **G** is the standard model in the natural numbers, and **R** is the Cartesian product **K**². We define a model φ by requiring that, for any atomic formula P and **H** ε **K**, $\varphi(P, \mathbf{H}) = \mathbf{T}$ (**F**) iff P is true (false) in the model **H**. (Remember, P is a wff of **PA**, and **H** is a countable model of **PA**.) We then build up the evaluation for compound formulae as before. [1] To say that A is true is to say it is true in the real world **G**; and, for any atomic P, $\varphi(\square P, \mathbf{G}) = \mathbf{T}$ iff P is provable in **PA**. (Notice that $\varphi(P, \mathbf{G}) = \mathbf{T}$ iff P is true in the intuitive sense.) Since (**G**, **K**, **R**) is an S5-m.s., all the laws of S5 will be valid on this interpretation; and we can show that *only* the laws of S5 are generally valid. (For example, if P is Gödel's undecidable formula, $\varphi(\square P \vee \square \sim P, \mathbf{G}) = \mathbf{F}$, which is a counterexample to the "law" $\square A \vee \square \sim A$.)

Another provability interpretation is the following: Again we take the atomic formulae to be the closed wffs of **PA**, and then build up new formulae using the adjoined connectives $\wedge$, $\sim$, and $\square$.

---

[1] It may be protested that **PA** already contain symbols for conjunction and negation, say "&" and "¬"; so why do we adjoin new symbols "$\wedge$" and "$\sim$"? The answer is that if P and Q are atomic formulae, then P & Q is *also* atomic in the present sense, since it is well-formed in **PA**; but $P \wedge Q$ is not. In order to be able to apply the previous theory, in which the conjunction of atomic formulae is not atomic, we need "$\wedge$". Nevertheless, for any **H** ε **K** and atomic P and Q, $\varphi(P\&Q, \mathbf{H}) = \varphi(P \wedge Q, \mathbf{H})$, so that confusion of "&" with "$\wedge$" causes no harm in practice. Similar remarks apply to negation, and to the provability interpretation of S4 in the next paragraph.

Let **K** be the set of all ordered pairs (**E**, α), where **E** is a consistent extension of **PA**, and α is a (countable) model of the system **E**. Let **G** = (**PA**, α_0), where α_0 is the standard model of **PA**. We say (**E**, α) **R** (**E'**, α'), where (**E**, α) and (**E'**, α') are in **K**, iff **E'** is an extension of **E**. For atomic P, define $\varphi(P, (\mathbf{E}, \alpha)) = \mathbf{T}$ (**F**) iff P is true (false) in α. Then we can show, for atomic P, that $\varphi(\Box P,$ (**E**, α)) = **T** iff P is provable in **E**; in particular, $\varphi(\Box P, \mathbf{G}) = \mathbf{T}$ iff P is provable in **PA**. Since (**G**, **K**, **R**) is an S4-m.s., all the laws of S4 hold. But not all the laws of S5 hold; if P is Gödel's undecidable formula, $\varphi((\sim \Box P \supset \Box \sim \Box P), \mathbf{G}) = \mathbf{F}$. But some laws are valid which are not provable in S4; in particular, we can prove for any A, $\varphi(\Box \sim \Box (\Diamond A \wedge \Diamond \sim A), \mathbf{G}) = \mathbf{T}$, which yields the theorems of McKinsey's S4.1 (cf. [13]). By suitable modifications this difficulty could be removed; but we do not go into the matter here.

Similar interpretations of M and the *Brouwersche* system could be stated; but, in the present writer's opinion, they have less interest than those given above. We mention one more class of provability interpretations, the "reflexive" extensions of **PA**. Let **E** be a formal system containing **PA**, and whose well-formed formulae are formed out of the closed formulae of **PA** by use of the connectives &, ¬, and $\Box$. (I say " & " and "¬" to indicate that I am using the same conjunction and negation as in **PA** itself, not introducing new ones. See footnote 1, p. 91.) Then **E** is called a reflexive extension of **PA** iff: (1) It is an inessential extension of **PA**; (2) $\Box A$ is provable in **E** iff A is; (3) there is a valuation α, mapping the closed formulae of **E** into the set $\{\mathbf{T}, \mathbf{F}\}$, such that conjunction and negation obey the usual truth tables, all the true closed formulae of **PA** get the value **T**, $\alpha(\Box A) = \mathbf{T}$ iff A is provable in **E**, and all the theorems of **E** get the value **T**. It can be shown that there are reflexive extensions of **PA** containing the axioms of S4 or even S4.1, but none containing S5.

Finally, we remark that, using the usual mapping of intuitionistic logic into S4, we can get a model theory for the intuitionistic predicate calculus. We will not give this model theory here, but instead will mention, for propositional calculus only, a particular useful interpretation of intuitionistic logic that results from the model theory. Let **E** be any consistent extension of **PA**. We say a formula P of **PA** is *verified* in **E** iff it is provable in **E**. We take the closed wffs $\dot{P}$ of **PA** as atomic, and build formulae out of them using the intuitionistic connectives $\wedge$, $\vee$, ¬, and $\supset$. We then stipulate inductively: $A \wedge B$ is verified in **E** iff A and B are; $A \vee B$ is verified in **E** iff *A* or *B* is; ¬ *A* is verified in **E** iff there is no consistent extension of **E** verifying *A*; *A* ⊃ *B* is verified in **E** iff every consistent extension **E'** of **E** verifying *A* also verifies *B*.

Then every instance of a law of intuitionistic logic is verified in **PA**; but, e.g., *A* v ¬ *A* is not, if *A* is the Gödel undecidable formula. In future work, we will extend this interpretation further, and show that using it we can find an interpretation for Kreisel's system FC of absolutely free choice sequences (cf. [14]). It is clear, incidentally, that **PA** can be replaced in the provability interpretations of S4 and S5 by any truth functional system (i.e., by any system whose models determine each closed formula as true or false); while the interpretation of intuitionism applies to any formal system whatsoever.

Harvard University.

References

[1] SAUL A. KRIPKE. *A completeness theorem in modal logic.* **The journal of symbolic logic,** vol. 24 (1959), pp. 1—15.

[2] SAUL A. KRIPKE. *Semantical analysis of modal logic* (abstract). **The journal of symbolic logic,** vol. 24 (1959), pp. 323—324.

[3] GOTTLOB FREGE. *Über Sinn und Bedeutung.* **Zeitschrift für Philosophie und philosophische Kritik,** vol. 100 (1892), pp. 25—50. English translations in P. Geach and M. Black, **Translations from the philosophical writings of Gottlob Frege,** Basil Blackwell, Oxford 1952, and in H. Feigl and W. Sellars (ed.), **Readings in philosophical analysis,** Appleton-Century-Crofts, Inc., New York 1949.

[4] P. F. STRAWSON. *On referring.* **Mind,** n. s., vol. 59 (1950), pp. 320—344.

[5] BERTRAND RUSSELL. *On denoting.* **Mind,** n. s., vol. 14 (1905), pp. 479—493.

[6] JAAKKO HINTIKKA. *Modality and quantification.* **Theoria** (Lund), vol. 27 (1961), pp. 119—128.

[7] A. N. PRIOR. **Time and modality.** Clarendon Press, Oxford 1957, VIII + 148 pp.

[8] A. N. PRIOR. *Modality and quantification in S5.* **The journal of symbolic logic,** vol. 21 (1956), pp. 60—62.

[9] JAAKKO HINTIKKA. *Existential presuppositions and existential commitments.* **The journal of philosophy,** vol. 56 (1959), pp. 125—137.

[10] HUGUES LEBLANC and THEODORE HAILPERIN. *Nondesignating singular terms.* **Philosophical review,** vol. 68 (1959), pp. 239—243.

[11] Saul A. Kripke. *Semantical analysis of modal logic I*. **Zeitschrift für mathematische Logik und Grundlagen der Mathematik**, vol. 9 (1963), pp. 67—96.

[12] Stephen C. Kleene. **Introduction to metamathematics.** D. Van Nostrand, New York 1952, x + 550 pp.

[13] J. C. C. McKinsey. *On the syntactical construction of systems of modal logic*. **The journal of symbolic logic**, vol. 10 (1945), pp. 83—94.

[14] G. Kreisel. *A remark on free choice sequences and the topological completeness proofs*. **The journal of symbolic logic**, vol. 23 (1958), pp. 369—388.

[15] W. Van O. Quine. **Mathematical logic.** Harvard University Press, Cambridge, Mass., 1940; second ed., revised, 1951, xii + 346 pp.

S.A. Kripke
Departments of Philosophy and Computer Science, CUNY Graduate Center, New York, USA

Part 9
Jean Porte (1965)

Formal Systems As Structures—A Bourbachic Architecture of Logic

Marcel Guillaume

Abstract First we situate Porte's work in the stream of creation of more and more abstract formal systems which began around the end of the nineteenth century until the time at which appeared this work taking formal systems themselves as subject as it had been undertaken since the 20s of the twentieth century, by referring to some landmarks quoted by Porte himself and of which the most directly influential on his thought is Tarski (1930) about 'the methodology of deductive sciences'. Second, we give a prefunctorial reading, hidden in and suggested just by a few Porte's notations, of some of the most abstract definitions and results of his thesis which he mentions without further explanations in its following programmatic first chapter and which pertain to the most general theoretical developments in this work. By the way, we call up Porte's main contribution: to base formal systems, viewed as combinatorial structures, on their assumed well-ordered alphabet.

Mathematics Subject Classification (2000) 01A60 · 03B22

In the following introductory first chapter of his doctoral thesis, Jean Porte defends the primordial thesis that the work he presents is entirely and uniquely devoted to pure mathematics, since, according to him, "[w]hether syntactic or semantic, or both, the theory of formal systems remains a branch of pure mathematics"—"a specialized part of 'combinatorics'". This view proceeds from the strong conviction that leads him, always preoccupied with not speaking for others, to announce: "the 'formal systems' here taken into account will be certain mathematical structures"[1] (accordingly, he sees the syntax as "the study of formal systems in themselves" and the semantics as "the study of the connections between the formal systems and other mathematical structures"). Hence, in particular, he claims the right to "use of all means to reason that the mathematicians are generally giving to themselves the right to use"—against empiricist or intuitionistic prohibitions, he quotes "the 'principle of the excluded third', the 'axiom of choice' or the 'class of all transfinite ordinal numbers'"—and appeals to the practice of "famous logicians, in particular Tarski".

So doing, he follows the trends towards greater and greater abstraction which did not cease spreading in mathematics since the beginning of the nineteenth century. According

---

[1]Porte specifies that he understands the term "structure" in a broader sense than [1], which comes down to admit transfinite products in place of finite products in the formation of any *scale of sets having a given basis*.

J.-Y. Béziau (ed.), *Universal Logic: An Anthology*, 211–222
Studies in Universal Logic, DOI 10.1007/978-3-0346-0145-0_17, © Springer Basel AG 2012

to the list of references having documented his work that he puts at the end of his thesis[2], he had before him, as main achieved sources of types of formal systems, besides the more fregean *Principia Mathematica* [68, 69] and Hilbert-style calculuses, a wide variety of other formal systems and of other *outlooks* about their generation: Markov's algorithms (in its English translation [44])[3], Curry's *combinatory logic* [9], Łoś and Suszko's *sentential logics* [35], Church's second *Introduction to mathematical logic* [5], Tarski's anthology of English translations of his main papers until 1938 [62], which includes a translation of his join work with Łukasiewicz on the sentential calculus [39], Gentzen's natural deduction and sequents calculuses (in its French translation [14]), Lorenzen's *operative Logik und Mathematik* [34], Rosser and Turquette's book on multivalued logics [54], Kleene's *Introduction to metamathematics* [28], Quine's *Mathematical logic* [53], von Wright's *Essay in modal logic* [66]—one of Porte's favourite own domain of research, Kemeny's *logical systems* [27], Post's systems of productions (in [52]) and *two-valued iterative systems* [51], Church's *calculi of λ-conversion* [3], Turing's machines (in [64]), Post's *finite combinatory process* [50], and Lewis and Langford's *Symbolic logic* [33]. Moreover to the formal systems displayed by these sources were yet added a crowd of systems akin to their own systems and introduced in commentaries and studies about these last ones in research papers.

In 1930 Tarski had devoted [58] and [59] to treat of *fundamental concepts of the methodology of deductive sciences*[4] such as deductive systems, freedom of contradiction, completeness and axiomatizability, in terms of the abstract notion of *consequence operation* that he denoted by Fl (from some grammatical form of the German word *Folgerung*[5]). This operation on the power set $\mathfrak{P}(S)$ of a set S of abstract but possibly left implicit ("meaningful") "propositions"[6] (in fact, *representations* thereof, according to intended interpretations possibly freed from inessential features) maps each set A of such "propositions", taken as "hypotheses"—added to some set of supposed but possibly left implicit "logical axioms"—to the set of their (syntactic) "consequences"—according to some set of supposed but possibly left implicit "rules of inference" taking "propositions" of S as "premisses" and "conclusion" (Porte speaks of these "rules" as "relations"). Then $Fl(A)$ is defined as the intersection of all subsets of S which include first the set of "logical axioms", second the set A, and which moreover are closed with respect to all operations represented by these "rules"[7].

In the *first part* of [59], no logical principle is taken in consideration, neither directly nor indirectly (what open the door to possible unintended interpretations of the notions of

---

[2]It his clear that this bibliography falls short of any historical preoccupation, in particular about earlier publications concerning concepts or theories of which he speaks. His reference to Frege's *Begriffsschrift* [11] constitutes the only exception.

[3]Although Porte refers also to the original edition of books which have not been published in French or in English, he specifies systematically that he perused them through their translations; in such cases we chosen to give *for the moment* only the reference to the translation.

[4]In the title of [58] the term 'metamathematics' is used in place of all the locution 'the methodology of deductive sciences', but in the text, he specifies that both denote the same discipline.

[5]*Consequence*, in English. The original version of the paper was written in German.

[6]Probably S is used as first letter of the German word *Satz*, even if the elements of S are called *Aussagen* (asserted propositions).

[7]I heard still in my mind the ironic tone that Jean Porte used to make this lot of inverted commas stand out during some of our conversations.

"propositions", "axioms" and "rules of inference" to which Porte alludes remotely in the text which follows), so that the generic consequence operation as described above satisfy four set-theoretical principles which can be taken as axiomatizing it in the frame of set theory. But in reality the first of them was quite irrelevant to the study of any consequence operation, since it stated only that the set S of "propositions" is countable[8] but nothing at all about the consequence operation; moreover it was doomed to become excessively limitative when more and more interesting results about uncountable systems will appear in contributions such as [16, 29, 40, 41, 70, 71] and further [20, 26]—and Porte, as many other contributors, did not retain it in his work.

Reference [59] is written in a notation in his time and country currently borrowed from Schröder's *Algebra der Logik* [56] taken in its set-theoretical interpretation: thus 'Σ' denotes union and '$\cdot$' intersection, '$A \subset B$' means that A is a proper or improper subset of B; and by $\mathfrak{E}$, Tarski denotes the class of finite[9] sets. Thus the three principles above—here written and labelled as in [59]—took the form

I.2 If $A \subset S$, then $A \subset Fl(A) \subset S$.

I.3 If $A \subset S$, then $Fl(Fl(A)) = Fl(A)$.

I.4 If $A \subset S$, then $Fl(A) = \sum_{X \in \mathfrak{P}(A) \cdot \mathfrak{E}} Fl(X)$.

Conversely, by taking these principles as axioms, the study of the notions definable in terms of the set-theoretical operations and of the operation Fl becomes completely independent of the inessential rules of construction of the "propositions" and choices of the "logical axioms" and "rules of inference" assumed to define the syntactic notion of "consequence". In particular, in order to refer to the parts of S which *include their consequences*, Tarski makes explicit his terminological shift from [23] "*closed* systems" to his own "deductive systems".

The level of abstraction at which stands thus the first part of [59] is the same as that at which stands the first attempts at combinatory logic [55]—in reality expounded in 1920 and farther taken over by [7]—at an abstract sequent calculus [21] to which [12] was the best contribution, and at a calculus of λ-conversion (included in the frame of a functional notation in two versions of a predicate calculus) in [2].

Reference [55] was indeed written at Hilbert's and Bernays' request by Heinrich Behmann which had jotted down some notes during Schönfinkel's talk. Starting from Frege's idea of replacing the predicates by characteristic functions out of [11] and from the definability of every boolean connective in terms of the single Sheffer's stroke established in [57]—it was still forgotten until 1933 that Charles Sanders Peirce had discovered this fact as early as in 1880, as can be seen in [46, pp. 13–18][10]—Schönfinkel proposed first merging universal quantifier with Sheffer's stroke, what led him to an "incompatibility functional"[11] U (Ufg would be taken as denoting $(x)[f(x)|g(x)]$, written with

---

[8]*I.e.* it is an image of some not necessarily proper initial segment of the sequence of natural numbers.

[9]*Endlich*, in German. By a "class", it is likely that Tarski understand here a *proper* class in von Neumann-Bernays' sense.

[10]This reference is borrowed from the footnote at the p. 48 of [25], which gives the title 'A boolean algebra with one constant'.

[11]Schönfinkel spoke in fact of a *function*; our rendering by the term "functional" which is here literally unfaithful but faithful to its real meaning as we can understand it today, aims just at underlining the facts that it is spoken about a function and that the sphere of thinking to which this function belongs is of a

a Sheffer's stroke), and then trying to eliminate the remaining fundamental concepts of proposition, propositional function and variable. In fact, every use of an individual bound variable can be removed by resorting to the incompatibility functor; and any proof where some individual variables are free can be transformed in another proof in which these variables are universally quantified, at the price, if necessary, of some renamings. As for functions, by understanding "$F(x, y)$ is the value of F at (x, y)" as meaning "$F(x, y)$ is the value at y of the function which is the value at x of F", every usual representation of a function can first be rewritten under the form of a string of operations of *application* to symbols of functions; by a further analysis of various universal processes of computation, such a string of *applications* can in turn be rewritten in terms of two "functionals" only, to wit a "merging functional" S (such that, according to Schönfinkel, every equation of the form $S\phi\chi f = (\phi f)(\chi f)$ holds[12]) and a "constancy[13] functional" C (such that the equations of the form $Cfg = f$ hold). Then, every formula of the predicate calculus can be written using only the functionals C, S, U, without referring to whatever domain. Schönfinkel was prevented from pursuing by illness and the search for combinatory logic has farther been successfully carried through by Curry.

Hilbert has had to face up to some mathematicians lacking totally of understanding on the occasion of the publication in *Mathematische Annalen* of one of [21–23]. Hertz's (representations of) propositions in these contributions were given as taking the form $(a_1, \ldots, a_n) \to b$, which according to him might be read "if $(a_1, \ldots, a_n)$ are together, then b is"; about the nature of the objects that the $a_1, \ldots, a_n$'s and the b represent, it is said only that they are among the "elements" of some given set. In [22], the inference rules take the form

$$\frac{(u_1, \ldots, u_n) \to a_1, \quad \ldots, \quad (u_1, \ldots, u_n) \to a_m, \quad (a_1, \ldots, a_m) \to b}{(u_1, \ldots, u_n) \to b}$$

and the "proofs", the tree form[14]. Hertz [23] says that a "complex" of elements *satisfies* a given proposition, if it does not contain the antecedent of this last one, or if it contains both its antecedent and its succedent, where the notions of antecedent and succedent apply to the sequences of elements according to their position before or after the arrow respectively. In this paper, Hertz accepts what we would call today the structural rule of thinning (in antecedent, of course). While studying the systems of propositions which are closed under the inference rules, he says that such a system $\mathfrak{T}$ is "axiomatized" by one of its subsets $\mathfrak{A}$ iff all members of $\mathfrak{T}$ are at the root of proofs in tree form of which all the leaves belong to $\mathfrak{A}$. His aim was to determine under which conditions a system of propositions could admit of an independent axiomatization—a problem that he did not solve—and to that end he reduced the proofs to some normal proofs. Gentzen [12] describes an infinite Hertz's system of propositions which do not admit of an independent axiomatization; he

---

nature which is quite different from the one to which the functions of the first order predicate calculus belong. The term "combinator", which was introduced further by Curry, would be anachronic in what concerns Schönfinkel.

[12] The right member represents the application of the function ϕf to the function χf.

[13] The corresponding German term is "*Konstanz*", to which Wahring [67, p. 557] attributes the terms "fixedness" and "invariablility" amongst its connotations.

[14] This form was inspired by the figures provided by Hilbert's method of "resolution in proofstrings" (*Beweisfäden* in German) of Hilbert's style sequential proofs.

splits up Hertz's complicated rules of inferences in a set of simple rules, to which he adds a "cut" rule. He introduces a notion of "consequence" which is *semantic* in essence: by referring to Hertz's notion of satisfaction, he defines the proposition q as a consequence of the propositions $p_1, \ldots, p_\nu$ iff every complex of elements which satisfies $p_1, \ldots, p_\nu$ satisfies q. Thus it is not a surprise that Hertz's papers are also well at their place in Tarski's [62] bibliography where Tarski sums up his references in his early works.

In which regards [2], its level of abstraction is well-known today, since it contains the first version of our λ-calculuses, which use the operations of "application" and of "abstraction" to construct terms which represents functions of which neither the domains nor the codomains are specified, in a way which is very similar to that which is used in combinatory logic.

As in his introductory chapter Porte understate his achievements sometimes up to not even name them otherwise that by presenting them under a common noun to which they answer, we will now add for the interested reader some explanations on the terms he use, on their meaning, and on the main features and properties of his theory of the beings in question in the more abstract part of his thesis, roughly one third of it after the first chapter. However one must know that quite a lot of important progresses made in his work, which rest on the theorems of that general theory, have been established by applying it to his "connective systems"—his locution for "sentential (or propositional) calculuses"—which fills more than the last half of the thesis; a general theory of these systems is developed by recapturing with a bigger precision a crowd of previous studies from various authors. By the way, we will say a few words about some of the most abstract pieces of his work that Porte mentions briefly at the end of its following first chapter, before he begins to present its first developments; these pieces form really some of Porte's contributions to the trend towards abstraction in his time.

This work stands at the same level of abstraction as those of Schönfinkel, Hertz, Tarski, Gentzen, Curry and Church just recalled. Porte defines his "logistic systems"[15] as structures $\mathcal{S} = \langle \mathcal{E}, \mathcal{F}, \mathcal{A}_1, \mathcal{A}_2, \ldots, \mathcal{R}_1, \mathcal{R}_2, \ldots, \mathcal{N} \rangle$, where the *language* $\mathcal{E}$—which Porte calls the "*alphabet*"—*is included* as *universe*—which Porte, following [1, p. 41], calls the *basis* (of the "scale of the sets having $\mathcal{E}$ as its basis"[16]), but absolutely no peculiarities of any language, such that rules of construction of significant strings of symbols, is taken in consideration; from the set $\mathcal{F}$ of "formulas", it is only said that this one is a subset of the set of the words on the alphabet $\mathcal{E}$; the (finite or transfinite) lists $\mathcal{A}_1, \mathcal{A}_2, \ldots$, and $\mathcal{R}_1, \mathcal{R}_2, \ldots$, enumerate, respectively, subsets of $\mathcal{F}$ called "schemes of axioms" and relations on $\mathcal{F}$,

---

[15]The adjective refers to the proposal made "without prior agreement nor previous communication by Messrs Itelson, Lalande and Couturat", to the "second Congress of philosophy" (in fact, the *Second International Congress of Philosophy, held at Geneva from 4 to 8 of September 1904*), "to give to the new logic the name of 'logistics'. This triple coincidence seems to justify the introduction of this new word, which is shorter and more exact than the usual locutions: symbolic, mathematical, algorithmic logic, algebra of logic", according to [6]. The quotes are borrowed from [30, p. 431], which comments: "the word itself is old. It referred during the Middle Ages to the *practical arithmetic* as opposed to the theoretical arithmetic". The scattered uses, as adjective and as substantive, of the word "logistic" during the first decades of the twentieth century have not been sufficient to supplant the use of the other locutions quoted by Couturat in his review.

[16]I explain, in [19], how the language of which a structure in the sense of Bourbaki is a set-theoretic interpretation, is hidden in the sets of the scale, where each constant symbol is interpreted—and represented—by its graph.

among which each has its own arity, "said to be *primitive rules of deduction*", without further addition on their natures and determinations; finally, "$\mathcal{N}$ is a bijective map from $\mathcal{E}$ onto an initial segment of the class of ordinals ... called 'numbering of the alphabet' ": so doing, Porte follows a recommendation of [60, 61], where it is followed—as almost universally since—and justified by evoking the complications which are inherent to other ways of proceeding to define the notion of satisfaction; Porte acknowledges that to count $\mathcal{N}$ as a determinative of the structure has the drawback that two logistic systems differing only in the numbering of their common alphabet will have to be counted as different. But this does not prevent the preservation of the level of abstraction as this one is secured according to the remarks above.

Porte was right when he chosen to place the alphabet at the basis. Since [27], in which the structures (called "semi-models" by Kemeny) were at long last distinguished from models, the logicians have realized little by little that structures were determined by the language only and models furthermore by the satisfaction of axioms written in the language.

We have now here, in order to continue describing Porte's three kinds of formal systems, to recall two definitions taken up by him—and this gives to us the opportunity of passing a *basic remark*, explaining what are the basic grounds of the presence in his theory of the lattices which he mentions at the end of his first chapter (and which all are complete). Of course on a logistic system the data of the axioms and of the deduction rules determine recursively "from below" the notion of *deduction* of *a formula from a finite[17] sequence of specific hypotheses[18]*, and *indirectly* the notion of *thesis* as *formal theorem*, that is, as formula which is *deducible without any specific hypotheses*, from basic axioms only—since indeed it is thus clear that the notion of thesis is already determined by the notion of deduction only. Porte did not, however, tell something about the finer notion of deductive system which one could view as tied to the notion of formula deducible from a finite sequence of specific hypotheses definable as the corresponding Tarski's notion of deductive system is tied to the notion of consequence. Porte's first worry in his chapter on logistic systems is establishing that the relation of deduction fulfils *universal conditions* in terms of formulas, finite sequences of formulas and constant operations between the second ones and the first ones; then, the conceivable as suggested above notion of deductive system is also determined "from above" as in the case of subspaces of linear spaces, *i.e.* any such deductive system is *uniformly uniquely determined by the determinatives of the logistic system* **and** *by an extra set of specific hypotheses*. Under such conditions, the set of these deductive systems, ordered "naturally", *i.e.* by inclusion, form a complete lattice, of which the smallest element is the set of theses, which moreover is by definition also *uniformly uniquely determined* just *by the determinatives of the logistic system*. Similarly, the notions of congruence considered by Porte fulfil definitions given by universal conditions, and hence, the congruences which satisfy these conditions form complete lattices, of which the finest element is the identity relation.

---

[17] The relation here taken in consideration is the one which holds from the hypotheses which are *used as terms* in a deduction towards the conclusion of that deduction, which is assumed to be of *finite* length.

[18] One has to contrast *specific* hypotheses with *basic* axioms, for example, mathematical hypotheses with logical axioms; but only by giving that as an example, for the boundary can very well be drawn elsewhere, according to what is in question. Porte did not fail to state a few propositions which apply to the *move* of the boundary.

Porte's notion of "derivational system"[19] is just derived from the notion of logistic system by replacing the lists of schemes of axioms and of primitive deduction rules by an abstract deduction relation, which is just a relation from the set $\mathfrak{S}(\mathcal{F})$ of finite sequences of formulas towards the set $\mathcal{F}$ of formulas which fulfils, among the *universal* structural properties established before of a deduction relation on a logistic system, those which do not refer to the notions of axioms and of primitive deduction rules. Then the 'basic remark' above still holds, and there is still, as smallest element of a complete lattice of sets of sequences of specific hypotheses which are stable under the taking of the formulas deducible from these hypotheses, a set of formulas which, being *deducible from the void sequence* of formulas (taken as specific hypotheses), may be called, as in the case of logistic systems, the *theses* of the system. Moreover, for similar reasons, the graphs of abstract deduction relations over the same datum of an alphabet, a numerotation of it, and a set of formulas on it, form also a complete lattice.

Then it is clear that each logistic system is so mapped to a single derivational system that Porte denotes as a derivative by $S' = \langle \mathcal{E}, \mathcal{F}, \mathcal{H}, \mathcal{N} \rangle$, where $\mathcal{E}, \mathcal{F}, \mathcal{N}$ are as in the definition of a logistic system and $\mathcal{H}$ is a deduction relation from $\mathfrak{S}(\mathcal{F})$ towards $\mathcal{F}$ as it was just described. This is how this definition of a derivational system can be read if the functional notation made up of one sign "prime" is forgotten; otherwise, $\mathcal{E}, \mathcal{F}, \mathcal{N}$ are the corresponding determinatives of the logistic system S, of which $\mathcal{H}$ denotes the deduction relation, and S' is *the* derivational system of which Porte says that it is *generated* by S. As for the theses in this case, note that their set remains unaltered by this mapping from S to the derivational system generated by him.

It was tempting to regard this mapping as the side turned towards objects of a functor between suitable categories. Porte did not take the plunge, probably owing to the lack of obviousness as for defining in a sufficiently natural way the notion of morphism between two logistic systems to state. It is in the order of things that Porte did call every logistic system which belong to the inverse image of any derivational system under this mapping an *axiomatization* of the derivational system; a little experience of formal systems shows that each derivational system can possess many axiomatizations. Then, Porte describes another potentially functorial mapping from derivational systems towards logistic systems, each of which is a canonical axiomatization of the derivational system from which it is the image under this mapping, and which is *uniformly uniquely determined by the determinative* of the derivational system—before all, by its relation of deduction.

Finally, commenting the relationship between the notions of deduction (under hypotheses) and of consequence, Porte notes that each is definable, in the frame of the other determinatives, in terms of the other (however, the definition of the relation of deduction in terms of the operation of consequence is universal, while the definition of the second in

---

[19]By the French locution "système déductionnel", Porte intended to shorten a more long locution like "sytèmes déterminés par leur relation de déductibilité". He coined the adjective "déductionnel" because he could not use the adjective "déductif" ("deductive", in English): this last was no longer at disposal in order to form the locution "deductive systems", of which we recalled above the central position that it had already taken up in Tarski's papers under a quite different meaning; for Tarski himself, and also for Porte, Tarski's notion of deductive systems referred to Pieri's notion of *hypothetico-deductive* (*axiomatic*) *systems* [48], to Padoa's notion of *deductive* (*axiomatic*) *theories* [45], and to the *deductive sciences* of Polish logicians of the first decades of the twentieth century, the methodology of which was the subject of some of Tarski's contributions. Faced with the same problem in order to translate Porte's adjective "déductionnel" in English, [8, p. 80], coined the adjective "derivational" here employed.

terms of the first is existential). He notes that the role of the derivational systems could thus have been played by structures in which an operation of consequence will replace the relation of deduction of the derivational structures.

Porte's notion of "thetic system" too is derived from the notion of derivational system, in a way quite similar to the way through which the notion of derivational system is derived from the notion of logistic system, by just replacing the relation of deduction by a subset of the set $\mathcal{F}$, said to be *the set of theses* of the system, and which can this time be quite arbitrary.

Then each derivational system is mapped to a single thetic system that Porte denotes by $S'' = \langle \mathcal{E}, \mathcal{F}, \mathcal{T}, \mathcal{N} \rangle$, where $\mathcal{E}, \mathcal{F}, \mathcal{N}$ are as in the definition of a logistic system and $\mathcal{T}$ is an arbitrary subset of $\mathcal{F}$. This is how this definition of a derivational system can be read if the functional notation made up of the *two* signs "prime" is completely forgotten; but one can also forget only the part of the functional notation made up of the *first* sign "prime": in this case, $\mathcal{E}, \mathcal{F}, \mathcal{N}$ are the corresponding determinatives of the *derivational* system S', of which $\mathcal{T}$ denotes the set of theses, and S'' is *the* thetic system of which Porte says that it is *generated* by S'; otherwise, the entire functional notation is then taken in account, S'' denotes the thetic system generated by the derivational system S' generated by the logistic system S, so that $\mathcal{E}, \mathcal{F}, \mathcal{N}$ are the corresponding determinatives of S, of which $\mathcal{T}$ denotes the set of theses, and S'' is also *the* thetic system of which Porte says that it is *generated* by S.

So after having introduced a mapping from the logistic systems onto the derivational systems, Porte successively introduced another mapping from the derivational systems onto the thetic systems, and then the composite of these mappings. The "values" taken by each of these mappings are *uniformly uniquely determined* by the determinatives of the structures to which they have been applied. We have noted above that usually a *given* derivational system possess many axiomatizations; Porte extends this terminology to the thetic systems, to which the same remark apply, all the more since the 'basic remark' above applies to the derivational systems which generate a *given* thetic systems, which then form a complete lattice[20], as Porte notes it after having defined *two* other mappings, of which both present the same prefunctorial feature, from every thetic system to a derivational system which generates it, and under which the thetic system has *two* different images provided the thetic system at which they are applied possesses at least two formulas which are not theses. What these propositions show is that the amount of information afforded by a thetic system is lesser (strictly in the great majority of cases) than that which is provided by a derivational system which generates it; in the same way, the amount of information afforded by a derivational system is lesser (strictly in the great majority of cases) than that which is provided by a logistic system which generates it: to write axioms and rules of inference is the more precise of the three ways taken in account by Porte for formalizing a scientific discipline.

Then, Porte introduces two equivalence relations between logistic systems: first, to generate the same derivational system, *i.e.* to present the same abstract relation of deduction, and second, to generate the same thetic system, *i.e.* to present the same set of

---

[20]The determination of the logistic systems by axiom schemes and deduction rules is too much complicated for the inverse image of a given derivational system under the aforesaid first mapping to have a structure as simple as that of lattice; the same holds for the inverse image of thetic systems under the aforesaid *composite* mapping.

theses. If we consider the inverse images of the three mappings from structures affording the highest amount of information to structures providing a lesser amount, it is clear that the two equivalence relations on logistic systems are comparable, and that the finest is the first, which corresponds to the lesser loss of information. Porte shows that it is strictly finer in many cases; nevertheless, he proves also that if two logistic systems present the same set of theses and obey the same list of primitive rules of inference, they generate also the same derivational system. This gave to him the opportunity first of observing that under these assumptions the independence of axiom schemes can be evaluated indiscriminately on the theses or on the deduction relation, while no corresponding similar fact holds about deduction rules, and second of refusing to regard axiom schemes as 0-ary deduction rules because this discrepancy "establish a dissymmetry between the axiom schemes and the deduction rules of a logistic system".

Now, usually, in Porte's time, the notions of independence and of acceptability for deduction rules were defined in terms of "equivalence of formal systems", and Porte notices that by analyzing which is the one of the two aforesaid natural equivalence relations between logistic systems to which it was referred in various works published before his own, the answer was sometimes the first, sometimes the second. Then, from the existence of logistic systems which determine the same set of theses without determining the same abstract deduction relation, he concluded to the necessity of distinguishing the two corresponding notions of independence for deduction rules to which he alludes in is footnote $1,1$[21] and provides examples of primitive deduction rules which are independent with regard to the deduction relation but not with regard to the theses. The case of acceptability, which can be defined in terms of independence and vice versa, is similar.

Elementary as they are, these distinctions threw much light already within the theory of connective systems as Porte began to develop it.

References

1. Bourbaki, N.: Éléments de Mathématique. Première partie: Les structures fondamentales de l'Analyse. Livre I, Théorie des Ensembles (Fascicule de résultats). Actualités scientifiques et industrielles, vol. 846. Hermann, Paris (1939)
2. Church, A.: A set of postulates for the foundations of logic. Ann. Math., Ser. 2 **33**, 346–366 (1932–1933); **34**, 839–864
3. Church, A.: The Calculi of λ-Conversion. Annals of Mathematical Studies, vol. 6. Princeton University Press, Princeton (1941)
4. Church, A.: Introduction to Mathematical Logic, Part I. Annals of Mathematical Studies, vol. 13. Princeton University Press, Princeton (1944)
5. Church, A.: Introduction to Mathematical Logic I. Princeton University Press, Princeton (1956). Revised and much enlarged edition of Church [4]
6. Couturat, L.: Compte rendu du deuxième Congrès de Philosophie. Rev. Métaphys. Morale **12**, 1042 (1904)
7. Curry, H.B.: Grundlagen der kombinatorischen Logik. Am. J. Math. **52**, 551–558 (1930)
8. Curry, H.B.: Foundations of Mathematical Logic. McGraw-Hill, New York (1963)

---

[21] Indeed, the work from which the text which follow is taken is divided in sections numbered from the beginning of this work until its end independently of its division in chapters; in this work, Porte uses of footnote references of the form x, y, where x is the number of the section, and y the number of the note among the notes of the section.

9. Curry, H.B., Feys, R.: Combinatory Logic I. North-Holland, Amsterdam (1958)
10. Davis, M. (ed.): The Undecidable. Basic Papers on Undecidable Propositions, Unsolvable Problems, and Computable Functions. Raven Press, Hewlitt (1965)
11. Frege, G.: Begriffsschrift. Eine der arithmetischen nachgebildete Formelsprache des reinen Denkens. Nebert, Halle (1879). Partial English translation in [65, pp. 1–82]
12. Gentzen, G.C.E.: Über die Existenz unhabhängiger Axiomensysteme zu unendlichen Satzsystemen. Math. Ann. **107**, 329–350 (1932). English translation in [15, pp. 29–52]
13. Gentzen, G.C.E.: Untersuchungen über das logische Schließen. Math. Z. **39**, 176–210 (1934); 405–431. English translation in [15, pp. 69–131]
14. Gentzen, G.C.E.: Recherches sur la déduction logique, traduction française de [13] par J. Ladrière (avec notes de R. Feys et J. Ladrière). Presses Universitaires de France, Paris (1955)
15. Gentzen, G.C.E.: The Collected Papers of Gerhardt Gentzen, Szabo, M.E. (ed.) North-Holland, Amsterdam (1955)
16. Gödel, K.: Eine Eigenschaft der Realisierungen des Aussagenkalküls. Ergebnisse eines mathematischen Kolloquiums **3**, 20–21 (1932). English translation in [17, pp. 239–241]
17. Gödel, K.: Collected Works. Volume I, Publications 1929–1936 Feferman, S., Dawson, J.W. Jr, Kleene, S.C., Moore, G.H., Solovay, R.M., van Heijenoort, J. (eds.) Oxford University Press, New York (1986)
18. Guillaume, M.: La Logique Mathématique en sa jeunesse. [47, pp. 185–365] (1994)
19. Guillaume, M.: La logique mathématique en France entre les deux guerres mondiales: quelques repères. Rev. d'Hisitoire Sci. **62**(1), 177–220 (2009)
20. Henkin, L.: Some remarks on infinitely long formulas. In: Infinitistic Methods. Proceedings of the Symposium on Foundations of Mathematics, 2–9 September 1959, pp. 167–183. Pergamon Press, New York (1961)
21. Hertz, P.: Über Axiomensysteme für beliebige Satzsystem. I Sätze ersten Grades. Math. Ann. **87**, 246–269 (1922)
22. Hertz, P.: Über Axiomensysteme für beliebige Satzsystem. II Sätze höheren Grades. Math. Ann. **89**, 76–102 (1923)
23. Hertz, P.: Über Axiomensysteme für beliebige Satzsystem. Math. Ann. **101**, 457–514 (1929)
24. Hilbert, D., Bernays, P.: Grundlagen der Mathematik I. Springer, Berlin (1934)
25. Hilbert, D., Bernays, P.: Grundlagen der Mathematik I. Springer, Berlin (1968). 2nd edn. revised and extended, of [24]
26. Karp, Carol Ruth: Languages with Expressions of Infinite Length. North-Holland, Amsterdam (1964)
27. Kemeny, J.G.: Models of logical systems. J. Symb. Log. **13**, 16–30 (1948)
28. Kleene, S.C.: Introduction to Metamathematics. North-Holland, Amsterdam (1952)
29. Krasner, M.: Une généralisation de la notion de corps. J. Math. Pures Appl., Sér. 9 **17**, 367–385 (1938)
30. Lalande, A.: Vocabulaire technique et critique de la philosophie, 4th edn. Notablement augmentée, vol. I. Alcan, Paris (1938)
31. Lewis, C.I.: A Survey of Symbolic Logic. University of California Press, Berkeley (1918). Reprinted by Dover, New York (1960)
32. Lewis, C.I., Langford, C.H.: Symbolic Logic. Century, New York (1932)
33. Lewis, C.I., Langford, C.H.: Symbolic Logic. Dover, New York (1959). 2nd revised edn. of [32]
34. Lorenzen, P.: Einführung in die operative Logik und Mathematik. Springer, Berlin (1955)
35. Łoś, J., Suszko, R.: Remarks on Sentential Logics. Proc. K. Ned. Akad. Wet., Ser. A **61**, 177–183 (1958)
36. Łukasiewicz, J.: Tresœ wykladu pozegnalnego wygloszonego w auli Uniwersytetu Warszawskiego 7 marca 1918 (Warszawa) (1918). English translation: Wojtasiewicz, O.: Farewell Lecture, Warsaw University Lecture Hall, March 7, 1918, [38, pp. 84–86]
37. Łukasiewicz, J.: O logice trójwartościowej. Ruch Filoz., **5**, 170–171 (1920). English translation in [38, pp. 87–88]
38. Łukasiewicz, J.: Selected Works. Borkowski, L. (ed.) Polish Scientific Publishers, Warszawa (1970)
39. Łukasiewicz, J., Tarski, A.: Untersuchungen über den Aussagenkalkül, Sprawozdania z Posiedzeń Towarzystwa Naukowego Warszawskiego, Wydział III, Nauk-Matematyczno-fizycznych (= Comptes-Rendus des Séances de la Société des Sciences et des Lettres de Varsovie, Classe III Sciences mathématiques et physiques) **23**, 30–50 (1930). Revised English translation in [62, pp. 38–59], reprinted in [38, pp. 131–152]

40. Maltsev, A.I.: Untersuchungen aus dem Gebiet der mathematischen Logik, Mat. Sb. N.S. **1**, 323–336 (1936). English translation in [42, pp. 1–14]

41. Maltsev, A.I.: Ob odnom obščém métodé polučéniá lokal'nyh téorém téorii grupp, Uč. Zap. Fiz.-Mat. Nauki, Ivanovsk. Gos. Pédagog. Inst., Fiz.-Mat. Fak. **1**, 3–9 (1941). English translation in [42, pp. 15–21]

42. Maltsev, A.I.: The Metamathematics of Algebraic Systems: Collected Papers 1936–1967, Wells III, B.F. (ed. and transl.) North-Holland, Amsterdam (1971)

43. Markov, A.A.: Teoriia Algorifmov. Trudy Mathemetischeskogo Instituta imeni V.A. Steklova, vol. 42. Akad. Nauk. S.S.S.R., Moscow (1954)

44. Markov, A.A.: Theory of Algorithms. Israel Program for Scientific Translations, Jerusalem (1961). English translation of Markov [43]

45. Padoa, A.: Essai d'une théorie algébrique des nombres entiers, précédé d'une introduction logique à une théorie déductive quelconque. In: Bibliothèque du Congrès International de Philosophie. T. 3, pp. 309–365. Armand Colin, Paris (1901). Partial English translation in [65, pp. 118–123]

46. Peirce, C.S.: Collected Papers of Charles Sanders Peirce, vol. III: Exact Logic. Hartshorne, C., Weiss, P. (eds.) Harvard University Press, Cambridge (1933)

47. Pier, J.-P.: Development of Mathematics 1900–1950, Pier, J.-P. (ed.) Birkhäuser, Basel (1994)

48. Pieri, M.: I principii delle geometria di positione composti in sistema logico-dedutivo. Mem. R. Accad. Sci. Torino, Ser. 2a **48**, 1–62 (1899)

49. Post, E.L.: Introduction to a general theory of elementary propositions. Am. J. Math. **43**, 163–185 (1921). Reprinted in [65, pp. 264–283]

50. Post, E.L.: Finite combinatory processes – formulation I. J. Symb. Log. **1**, 103–105 (1936). Reprinted in [10, pp. 288–291]

51. Post, E.L.: The Two-Valued Iterative Systems of Mathematical Logic. Annals of Mathematical Studies, vol. 5. Princeton University Press, Princeton (1941)

52. Post, E.L.: Formal reductions of the general combinatorial decision problem, Am. J. Math., **65**, 197–215 (1943)

53. Quine, W.v.O.: Mathematical Logic, revised edn. Harvard University Press, Cambridge (1951)

54. Rosser, J.B., Turquette, A.R.: Many-Valued Logics. North-Holland, Amsterdam (1952)

55. Schönfinkel, M.: Über die Bausteine der mathematischen Logik. Math. Ann. **92**, 305–316 (1924). English translation in [65, pp. 355–366]

56. Schröder, F.W.K.E.: Vorlesungen über die Algebra der Logik (exakte Logik) 1 (1890); 2, Erster Teil (1891); 3, Erster Teil (1895); 2, Zweiter Teil (1905), Teubner, Leipzig (1890–1905)

57. Sheffer, H.M.: A set of five independent postulates for Boolean algebras, with application to logical constants. Trans. Am. Math. Soc. **14**, 481–488 (1913)

58. Tarski, A.: Über einige fundamentale Begriffe der Metamathematik. Sprawozdania z Posiedzeń Towarzystwa Naukowego Warszawskiego, Wydział III, Nauk-Matematyczno-fizycznych (= Comptes-Rendus des Séances de la Société des Sciences et des Lettres de Varsovie, Classe III, Sciences Mathématiques et Physiques) **23**, 22–29 (1930). Reprinted in [63, I, pp. 313–320]; revised English translation in [62, pp. 30–37]

59. Tarski, A.: Fundamentale Begriffe der Methodologie der deduktiven Wissenschaften I. Monatshefte Math. Phys. **37**, 361–404 (1930). Reprinted in [63, I, pp. 347–390]; revised English translation in [62, pp. 60–109]

60. Tarski, A.: Pojęcie prawdy językach nauk dedukcyjnych. Prace Towarzystwa Naukowego Warszawskiego, Wydział III, Nauk Matematyczno-fizycznych (= Travaux de la Société des Sciences et des Lettres de Varsovie, Classe III, Sciences Matématiques et Physiques), vol. 34. Warszawskie Towarzystwo Naukowe, Warszawa (1933)

61. Tarski, A.: Der Wahrheitsbegriff in die formalisierten Sprachen. Stud. Philos. **1**, 261–405 (1935). German translation of Tarski [60] by L. Blaustein, reprinted in [63, II, pp. 53–198]; revised English translation in [62, pp. 152–378]

62. Tarski, A.: Logic, Semantics, Metamathematics. Papers from 1923 to 1938. Clarendon Press, Oxford (1956)

63. Tarski, A.. In: Givant, S.R., McKenzie, R.N. (eds.) Collected Papers, vol. 4. Birkhäuser, Basel (1986)

64. Turing, A.M.: On computable numbers, with an application to the Entscheidungsproblem. Proc. London Math. Soc., Ser. 2 **42**, 230–265 (1937); A correction, Proc. London Math. Soc., Ser. 2 **43**, 544–546. Reprinted in [10, pp. 115–153]

65. van Heijenoort, J.: From Frege to Gödel. A Source Book in Mathematical Logic 1879–1931. Cambridge University Press, Cambridge (1967)
66. von Wright, G.H.: An Essay in Modal Logic. North-Holland, Amsterdam (1951)
67. Wahrig, G.: Deutsches Wörterbuch, neu herausgeben von Dr. Renate Wahrig-Burfeind. Bertelsmann, Gütersloh (1997)
68. Whitehead, A.N., Russell, B.: Principia Mathematica. I, 1910; II, 1912; III, 1913. Cambridge University Press, Cambridge (1910–1913)
69. Whitehead, A.N., Russell, B.: Principia Mathematica. Cambridge University Press, Cambridge (1925–1927). 2nd revised edn. of [68]
70. Zermelo, E.: Über mathematische Systeme und die Logik des Unendlichen. Forsch. Forstschritte **8**, 6–7 (1932)
71. Zermelo, E.: Grundlagen einer allgemeinen Theorie der mathematischen Satzsysteme. Fundam. Math. **25**, 136–146 (1935)

M. Guillaume (✉)
33, Rue des Prés Hauts, 63100 Clermont-Ferrand, France
e-mail: mguil@wanadoo.fr

Researches in the General Theory of Formal Systems and in Connective Systems

Chapter 1
Aim of the Present Work

Jean Porte

Abstract Jean Porte devoted his thesis to the development of an abstract theory of formal systems, viewed as mathematical structures based on their assumed well-ordered alphabet and belonging to combinatorial analysis rather than to metamathematics. In the text which follows, the first chapter of his thesis, he presents his asserted purely mathematical programme of systematization of many previous studies, splitting formal systems into three classes: the logistic ones, the derivational ones, and the thetic ones. He announces some results which were new at this time, more particularly about the links between these three species and about the illustration of his general theory by applying it to many previous versions of the propositional parts of several logics, and sometimes to the relations between different logics, if necessary regardless of any applicability of the systems introduced for the sake of theoretical studies.

Mathematics Subject Classification (2000) 03B22

1 General Points

Most of contemporary mathematical logic consists in studies of various "formal systems". The present work aims to specify, and sometimes to generalize, some of the most general notions used (implicitly or explicitly) by contemporary logicians.

The notions used here being very general, there will not be many strong theorems. But some classical results, proved in the literature for various particular "formal systems", will here be obtained as easy consequences of the same general definitions. At last, the general definitions and the results which follow from them will become clearer when they are isolated, released from the features which are proper to one or another such "formal system". Such an undertaking of generalization is vindicated when it leads to new methods: now, it will be seen that it led me, on one hand, to give precise definitions of various notions which are used in the works of contemporary logicians, but not much studied for themselves—and that ends up sometimes with results which are easy to prove from the general definitions, but surprising in their appearance[1,1]. On the other hand, I

Translated by M. Guillaume, with kind permission of his daughter in law, Helene W. Blanc, from: Porte, J., "Recherches sur la théorie général des systèmes formels et sur les systèmes connectifs". Collection de Logique Mathématique, Série A, No. 18. Gauthier-Villars, 1965

[1,1] See, in particular, the distinction between two kinds of "independences" for deduction rules (§ 19), and the traps which are waiting the logician who wants to "change primitive notions".

was motivated to state (and sometimes to solve) general problems of which I do not know any equivalent in the literature[1,2].

2 Foundations of Mathematics, or a Particular Branch of Mathematics?

Many regard mathematical logic as the study of a philosophical problem said to be that of the "foundations of mathematics". That famous problem will not be studied here; I shall not even discuss the question whether that philosophical problem has a scientific meaning or not.

The "formal systems" here taken into account will be some mathematical structures[2,1], neither more nor less "fundamental" than the class of algebraic structures, for example. In fact, I shall define *three classes* of mathematical structures: the "logistic systems", the "derivational systems", and the "thetic systems" (see Chaps. 3, 4 and 5) and the expression "formal systems" will designate the union of these three classes. It has to be remarked that these three classes of structures have already been defined in the literature, but often in a not very precise way, above all with respect to the distinction between these three classes of structures[2,2]. On the other hand, my most interesting results come from the study of the links between these three classes of formal systems (a study which has never before been made, to my knowledge).

The theory of formal systems is thus a particular branch of pure mathematics[2,3].

The "connective systems" are a particular class of formal (logistic, derivational and thetic) systems, which possess some distinctive features that make them into generalizations of the classical propositional calculus[2,4]. In fact, even in the first part of the present work, the examples of formal systems of which it has been made use are most often connective calculi (see *e.g.* Theorems 18.2, 20.4, *etc.*).

3 Mathematics or Metamathematics?

Modern mathematical logic was born from the study, by mathematical methods, of mathematical reasoning itself; and one can defend the contention that the whole of nowadays

---

[1,2]For open problems, see in particular §§ 27 and 31.

[2,1]The word "structure" is taken here in a meaning akin to, but slightly different from the one given to it by Bourbaki.

[2,2]This imprecision is besides due in great part to the lack of a precise vocabulary for referring to the notions about which it is spoken.

[2,3]If one would like to insert this theory in Bourbaki's *Elements of Mathematics*, this is easy, provided it is set as a "Book I *bis*" (after Set Theory) or, perhaps better, as a "Book II *bis*" (between Algebra and general Topology). The only fragment of Bourbaki's *Elements* which treats (very partly) of the theory of formal systems is a part of the Appendix to Chap. I of the Book I (see [1], Chap. I, Appendix, §§ 1, 2 and 3).

[2,4]Insofar as these systems have been studied in a rather general way in the literature, the authors have called them "propositional calculi" or "sentential calculi": see for example [3] or [6]. [*Translator's note*: *The words "calculs propositionnels" of the French original have been replaced by the content of the parenthesis in English which followed them in Porte's note.*]

existing mathematics constitutes the development of a particular logistic system. But in the theory of formal systems, one is brought to study a great number of other systems; some of these systems can abstractly represent some aspects (or some parts) of mathematical reasoning (I shall say that they are mathematical beings which are *adequate* for the concrete reasonings that they are able to represent)[3,1]; other systems are adequate for methods of calculation rather than for reasonings; lastly, other systems are adequate for nothing concrete (this is even the general case)—some of which will be explicitly defined and studied in a somewhat detailed way, for example in order to serve as counterexamples in the proofs of various general theorems—but most of the formal systems included in the general definitions are (until now) of no use[3,2].

The theory of formal systems is sometimes called "metamathematics"[3,3] (see *e.g.* [5]) by authors who are above all interested in formal systems which are more or less adequate for certain aspects of mathematical reasoning. I shall carefully avoid making use of this word which could suggest false ideas: in reality, the relationship between "mathematics" and "metamathematics" has nothing in common with the relationship between "physics" and "metaphysics". Moreover, the present work is not particularly concerned with the formal systems which are adequate for mathematical reasoning.

It can be said that the theory of formal systems, as it is studied here, is a specialized part of "combinatorial analysis": it is a question of studying certain classes of arrangements with repetitions.

4 Syntax and Semantics

The preceding considerations apply immediately only to that part of modern mathematical logic that it has been agreed will be called "syntax". In contrast, one calls *"semantics"* the study of the links between the formal systems "which have some meaning" and "what they mean"—in other words, the study of the question whether certain formal systems are adequate for certain mathematical reasoning. But this study has been put into a form abstract enough (see *e.g.* [4], followed by many other authors) for the general considerations on syntax to apply to semantics as well. It can be said that one calls "syntax" the study of formal systems in themselves, and "semantics" the study of the connections between the formal systems and other mathematical structures. There are practically no memoirs devoted to "pure syntax" nor to "pure semantics", but "rather syntactic" or "rather semantic" studies. The present work is "rather syntactic"; however, all that concerns the connections between the connective systems and their "evaluation matrices" (see Chaps. 11 and *sqq.*) is semantic.

Whether syntactic or semantic, or both, the theory of formal systems remains a branch of pure mathematics. Then, in the present work, I shall use all means of reasonning that

---

[3,1] Such is the case, in particular, of different formal systems called "formulations of the classical propositional calculus", see Chap. 13.

[3,2] In the same way, in general topology, one studies "spaces" which are adequate for physical space or for fragments of physical space (for example the three- or two-dimensional euclidean spaces), but one studies also many other spaces.

[3,3] It seems that this word (or rather the German word "Metamathematik") was coined by Hilbert.

mathematicians generally give to themselves the right to use[4,1] (including for example, possibly, the "principle of the excluded third", the "axiom of choice" or the "class of all transfinite ordinal numbers"). If the reader is anxious to get more precision, he can regard the present work as a chapter of Bernays' set theory, as it was modified by Gödel (see [2])[4,2].

5 Original Features of the Present Work

About one third of this work consists in reminders of definitions and of already known (and even classical) results; these reminders were necessary, on one hand for making the text continuous, on another hand (and before all) for specifying the vocabulary and the notation. Another third consists in notions and problems which have already been mentioned in the literature but have here been "renovated", that is, presented in more general, more simple or more precise ways. The last third consists in notions and results which are new[5,1].

Among the notions and results which are new (or "renovated" in an important way), the attention of the reader is directed to the points which follow:

– Distinction between two kinds of "independences" for deduction rules (see Chap. 6).
– Lattice of deduction relations* which are compatible with one and the same thetic system (see Chap. 6). Lattice of deduction relations which extend one and the same deduction relation (see Chap. 6).
– Congruences of connectives systems: lattice of the congruences; connectively determined congruences (see Chap. 8).
– Classical propositional calculus: simple systems, where the rule of detachment is not valid; systems with a very weak deduction relation; method of construction of evaluation matrices where such or such rule is not valid (see Chap. 13).
– Modal calculi: study of a very weak system (see Chap. 14).
– Precise syntactic definition of the "equipollence" between two connective systems, of the "formal definition" of the connectives, and applications to the classical propositional calculus and to the intuitionistic propositional calculus (Chaps. 12, 13 and 14).

References

1. Bourbaki, N.: Théorie des ensembles, Chaps. I et II, Actualités scientifiques et industrielles, vol. 1212. Hermann, Paris (1954)

[4,1] In that, I am only following the example given by famous logicians, in particular by Tarski.

[4,2] Personally, I do not know any account of pure mathematics which could not be regarded as a chapter of Bernays-Gödel's set theory, or of a rather simple extension of this theory.

[5,1] Some of these results have already been announced in abstracts published, whether for congresses or colloquia, or as notes at the Academy of Sciences.

Translator's note: Porte used the French term "déductibilité" (as an abbreviation of the locution "relation de déductibilité"), but according to its own definition, he speaks of relations from the specific hypotheses of a deduction to its conclusion.

2. Gödel, K.: The consistency of the continuum hypothesis, Princeton Ann. Math. Stud. **3** (1940)
3. Harrop, R.: On the existence of finite models and decision procedures for propositional calculi. Proc. Camb. Philos. Soc. **54**, 1–15 (1958)
4. Kemeny, J.G.: Models of logical systems. J. Symb. Log. **13**, 16–30 (1948)
5. Kleene, S.C.: Introduction to Metamathematics, Van Nostrand, New York (1952)
6. Łoś, J., Suszko, R.: Remarks on sentential logics, Proc. K. Ned. Akad. Wet., Ser. A **61**, 177–183 (1958)

Translator
M. Guillaume
33, Rue des Prés Hauts, 63100 Clermont-Ferrand, France
e-mail: mguil@wanadoo.fr

Part 10
Per Lindström (1969)

Lindström's Theorem

Jouko Väänänen

Abstract Lindström's Theorem is an important fact about first order logic and cannot be overlooked by anybody interested in the question, why is first order logic so useful and so widely used. Are there some deeper reasons for this or is it just a coincidence. I give an overview of Lindström's theorem and a sketch of its proof in modern notation.

Mathematics Subject Classification (2000) 03C95

Lindström's Theorem is a model-theoretic characterization of first order logic. It says in effect that any formal language that goes beyond first order logic has to distinguish between some infinite cardinalities in the sense that some sentence has a model of some infinite cardinality but not of all infinite cardinalities. Loosely speaking Lindström's Theorem tells us that any proper extension of first order logic has to detect something non-trivial about the set-theoretic universe. The equivalent original formulation says that first order logic is a maximal logic which satisfies the Downward Löwenheim–Skolem Property and the Countable Compactness Property. I will give an introduction to this result. The background and history of the result is told by Lindström himself in [13].

When we characterize first order logic we have to fix the domain where it is characterized. To this end we introduce the concept of abstract logic. We refer the reader to [2, Chapter II] for complete definitions. What we present here is merely a sketch. This concept was introduced by Lindström [9] and Mostowski [15] independently of each other. The concept was further developed by Barwise [1].

As a first approximation, an *abstract logic* is a pair $L = (S, T)$, where S is a set and T is a relation between arbitrary structures and elements of the set S. Intuitively, S is the set of sentences of the abstract logic L and T is the truth predicate. If τ is a vocabulary, let $\mathrm{Str}(\tau)$ denote the class of structures of vocabulary τ. Obviously we make some assumptions about S and T. They are listed below after we introduce some new concepts:

For $\varphi \in S$ we write $\mathrm{Mod}_{L,\tau}(\varphi) = \{ \mathfrak{M} \in \mathrm{Str}(\tau) : T(\mathfrak{M}, \varphi) \}$. An abstract logic L is said to be *closed under negation*, if for all vocabularies τ and all $\varphi \in S$ there is $\neg\varphi \in S$ such that $\mathrm{Mod}_{L,\tau}(\neg\varphi) = \mathrm{Str}(\tau) \setminus \mathrm{Mod}_{L,\tau}(\varphi)$. We say L is *closed under conjunction* if for all vocabularies τ and all $\varphi, \psi \in S$ there is $\varphi \wedge \psi \in S$ such that $\mathrm{Mod}_{L,\tau}(\varphi \wedge \psi) = \mathrm{Mod}_{L,\tau}(\varphi) \cap \mathrm{Mod}_{L,\tau}(\psi)$. We say L is *closed under existential quantification*, if for all vocabularies τ, for all constant symbols c in τ and for all $\varphi \in S$, there is $\varphi' \in S$ such that:

$$\mathrm{Mod}_{L,\tau \setminus \{c\}}(\varphi') = \{ \mathfrak{M} : (\mathfrak{M}, c^{\mathfrak{M}}) \in \mathrm{Mod}_{L,\tau}(\varphi) \text{ for some } c^{\mathfrak{M}} \in M \}.$$

Research partially supported by grant 40734 of the Academy of Finland, and by the ESF EUROCORES LogICCC program LINT.

We say that L is *closed under renaming* if whenever $\pi : \tau \to \tau'$ is a permutation which respects arity, and we extend π in a canonical way to $\hat{\pi} : \text{Str}(\tau) \to \text{Str}(\tau')$, then for all and $\varphi \in S$, there is $\varphi' \in S$ such that $\{\hat{\pi}(\mathfrak{M}) : \mathfrak{M} \in \text{Mod}_{L,\tau}(\varphi)\} = \text{Mod}_{L,\tau'}(\varphi')$. We say that L is *closed under free expansions* if whenever $\tau \subseteq \tau'$ and $\varphi \in S$, there is $\varphi' \in S$ such that $\text{Mod}_{L,\tau}(\varphi) = \text{Mod}_{L,\tau'}(\varphi')$. Finally, we say that L is *closed under isomorphisms*, if whenever $\varphi \in S$, $\mathfrak{M} \in \text{Mod}_{L,\tau}(\varphi)$ and $f : \mathfrak{M} \cong \mathfrak{N}$, then also $\mathfrak{N} \in \text{Mod}_{L,\tau}(\varphi)$.

We get an abstract logic $L_{\omega\omega} = (S_0, T_0)$ satisfying the above closure properties by letting S_0 be the set of all first order sentences and T_0 the usual truth predicate of first order logic:

$$T_0(\mathfrak{M}, \varphi) \iff \mathfrak{M} \models \varphi.$$

For example, the closure under free expansions can be satisfied simply by choosing $\varphi' = \varphi$. Other abstract logics arise from infinitary languages, generalized quantifiers, higher order logic and combinations of such.

An abstract logic $L = (S, T)$ is a *sublogic* of another abstract logic $L' = (S', T')$, in symbols $L \leq L'$, if for all $\varphi \in S$ there is $\varphi' \in S'$ such that for all τ $\text{Mod}_{L,\tau}(\varphi) = \text{Mod}_{L',\tau}(\varphi')$. If $L \leq L'$ and $L' \leq L$, we say that L and L' are *equivalent*, $L \equiv L'$.

Now we are ready to give the real definition:

Definition An *abstract logic*[1] is a pair $L = (S, T)$, where S is a set and T is a relation between structures and elements of S, such that L is closed under isomorphisms, renaming, free expansions, negation, conjunction, and existential quantification.

An abstract logic $L = (S, T)$ satisfies the (*Countable*) *Compactness Property* if for any (countable) $\Sigma \subseteq S$, if $\bigcap\{\text{Mod}_{M,\tau}(\varphi) : \varphi \in \Sigma\} = \emptyset$, then $\bigcap\{\text{Mod}_{M,\tau}(\varphi) : \varphi \in \Sigma_0\} = \emptyset$ for some finite $\Sigma_0 \subseteq \Sigma$. An abstract logic $L = (S, T)$ satisfies the *Downward Löwenheim–Skolem Property* if for every countable τ every non-empty $\text{Mod}_{L,\tau}(\varphi)$, $\varphi \in S$, contains a countable model. Now we are ready to state Lindström's Theorem:

Lindström's Theorem *Suppose L is an abstract logic such that $FO \leq L$. Then the following conditions are equivalent:*

(1) *L has the Countable Compactness Property and the Downward Löwenheim–Skolem Property.*

(2) *$L \equiv L_{\omega\omega}$.*

There are several other characterizations of first order logic, all due to Lindström. An abstract logic $L = (S, T)$ satisfies the *Upward Löwenheim–Skolem Property* if every $\text{Mod}_{L,\tau}(\varphi)$, $\varphi \in S$, which contains an infinite model, contains an uncountable model. We can replace condition (1) above by

(1)' *L has the Upward and Downward Löwenheim–Skolem Properties*[2].

---

[1] Lindström called abstract logics *extended first order logics*.

[2] Lindström makes here an additional assumption that he calls "strongness". It suffices that L is closed under substitution of (first order) formulas into atomic formulas.

In this form Lindström's characterization of first order logic among all abstract logics extends Mostowski's characterization [14] of $L_{\omega\omega}$ among logics obtained by adding simple unary generalized quantifiers to $L_{\omega\omega}$.

For another characterization, we fix some notation. We use $\mathrm{Th}_L(\mathfrak{M})$ to denote $\{\varphi \in S : T(\mathfrak{M}, \varphi)\}$. If $\mathfrak{M} \subseteq \mathfrak{N}$ and $\mathrm{Th}_L((\mathfrak{M}, a)_{a \in M}) = \mathrm{Th}_L((\mathfrak{N}, a)_{a \in N})$, we write $\mathfrak{M} <_L \mathfrak{N}$. An abstract logic $L = (S, T)$ satisfies the *Tarski Union Property* if $\mathfrak{M}_0 <_L \mathfrak{M}_1 <_L \cdots$ implies $\mathfrak{M}_n <_L \bigcup_n \mathfrak{M}_n$. We can replace condition (1) above by

(1)″ L has the Compactness Property and the Tarski Union Property [10].

We can also replace (1) by a condition derived from the Omitting-Types Theorem of first order logic [12]. By assuming a little more effectiveness about the abstract logics, (1) can be replaced by a combination of the Downward Löwenheim–Skolem Property and either a property derived from the Beth Definability Theorem of first order logic [10], extending a result of Mostowski [15], or a property derived from the Completeness Theorem of first order logic [10]. Lindström has himself written a very readable survey of his results in [11].

Barwise [1] characterized $L_{\kappa\omega}$ for $\kappa = \beth_\kappa$. Lindström's Theorem was rediscovered later by H. Friedman.

Proof of Lindström's Theorem[3] Suppose there were an abstract logic $L = (S, T)$ that satisfies both the Downward Löwenheim Property and the Countable Compactness Property, but some $\varphi \in S$ is not first order definable, i.e. $\mathrm{Mod}_{L,\tau}(\varphi)$ is not of the form $\mathrm{Mod}_{L_{\omega\omega},\tau}(\psi)$ for any first order ψ. We assume w.l.o.g. that τ is finite and relational.

For every n there are only finitely many (logically non-equivalent) first order sentences $\psi_i^n, i = 1, \ldots, k_n$, of vocabulary τ and of quantifier rank at most n. Let us call two L-structures n-equivalent if they satisfy the same ψ_i^n. There are only $\leq 2^{k_n}$ different n-equivalence classes, and each class is first order definable. Since φ is not definable in first order logic, we can find for any n L-structures $\mathfrak{M}_n$ and $\mathfrak{N}_n$ such that:

$$T(\mathfrak{M}_n, \varphi)$$
$$T(\mathfrak{N}_n, \neg\varphi) \tag{1}$$
$$\mathfrak{M}_n \text{ and } \mathfrak{N}_n \text{ are } n\text{-equivalent.}$$

Lindström uses then a characterization of n-equivalence in terms of back-and-forth sequences. Ehrenfeucht [5] and Fraisse [6] showed that two models are n-equivalent if and only if there are relations $I_i, i < n$, such that

- If $(a_1, \ldots, a_i)I_i(b_1, \ldots, b_i)$, then $a_1, \ldots, a_i \in M$ and $b_1, \ldots, b_i \in N$.
- $()I_0()$.
- If $(a_1, \ldots, a_i)I_i(b_1, \ldots, b_i)$ then for all $a_{i+1} \in M$ ($b_{i+1} \in N$) there is $b_{i+1} \in N$ ($a_{i+1} \in M$) such that $(a_1, \ldots, a_{i+1})I_{i+1}(b_1, \ldots, b_{i+1})$.
- If $(a_1, \ldots, a_{i-1})I_i(b_1, \ldots, b_{i-1})$, then for all atomic formulas $\varphi(v_1, \ldots, v_{i-1})$ we have $\mathfrak{M} \models \varphi(a_1, \ldots, a_{i-1})$ if and only if $\mathfrak{N} \models \varphi(b_1, \ldots, b_{i-1})$.

---

[3]This proof is from [9]. Lindström proved a very similar result for logics with generalized quantifiers already in [8].

If there are such relations I_i, $i < \omega$, then we say that that $\mathfrak{M}$ and $\mathfrak{N}$ are ω-equivalent. Note:

$$\text{If } \mathfrak{M} \text{ and } \mathfrak{N} \text{ are countable and } \omega\text{-equivalent,} \quad \text{then } \mathfrak{M} \cong \mathfrak{N}, \tag{2}$$

as we can go "back-and-forth" between the countable models generating infinite sequences $(a_1, \ldots, a_i, \ldots)$ and $(b_1, \ldots, b_i, \ldots)$ such that for all i we have $(a_1, \ldots, a_i)I_i(b_1, \ldots, b_i)$, and moreover, $M = \{a_i : i < \omega\}$ and $N = \{b_i : i < \omega\}$.

Lindström writes (1), supplemented with a little bit of arithmetic, into a sentence $\psi(n)$ in S, using the above back-and-forth characterization of n-equivalence. By the Countable Compactness Property there is a model of $\psi(n)$ in which n is non-standard. Due to the coding used, this model yields two other models $\mathfrak{M}$ and $\mathfrak{N}$ such that $T(\mathfrak{M}, \varphi)$, $T(\mathfrak{N}, \neg\varphi)$ and $\mathfrak{M}$ and $\mathfrak{N}$ are ω-equivalent. By the Downward Löwenheim–Skolem Property, we may assume $\mathfrak{M}$ and $\mathfrak{N}$ are countable. But then they are isomorphic by (2). Thus L cannot be closed under isomorphisms, contrary to assumption. This ends the proof.   □

The setup of Lindström's Theorem can be modified in many interesting ways. We can consider monadic structures [17], Banach spaces [7], topological structures [18], or modal logic [3] and prove similar characterizations. Also, the assumption $FO \leq L$ can be relaxed [4]. It is still open whether the Compactness Property and the Craig Interpolation Theorem together characterize first order logic. However, there is an extension of $L_{\omega\omega}$ with the Compactness Property that satisfies the Beth Definability Theorem [16].

An overview of model theoretic properties of various extensions of first order logic can be found in [2].

References

1. Barwise, J.: Axioms for abstract model theory. Ann. Math. Logic **7**, 221–265 (1974)
2. Barwise, J., Feferman, S. (eds.): Model-Theoretic Logics. Perspectives in Mathematical Logic. Springer-Verlag, New York (1985)
3. van Benthem, J.: A new modal Lindström theorem. Logica Universalis **1**(1), 125–138 (2007)
4. van Benthem, J., ten Cate, B., Väänänen, J.: Lindström theorems for fragments of first-order logic. Log. Methods Comput. Sci. **5**(3), 27 (2009)
5. Ehrenfeucht, A.: An application of games to the completeness problem for formalized theories. Fund. Math. **49**, 129–141 (1960/1961)
6. Fraïssé, R.: Etude de certains opérateurs dans les classes de relations, définis à partir d'isomorphismes restreints. Z. Math. Logik Grundlagen Math. **2**, 59–75 (1956)
7. Iovino, J.: On the maximality of logics with approximations. J. Symbol. Log. **66**(4), 1909–1918 (2001)
8. Lindström, P.: First order predicate logic with generalized quantifiers. Theoria **32**, 186–195 (1966)
9. Lindström, P.: On extensions of elementary logic. Theoria **35**, 1–11 (1969)
10. Lindström, P.: A characterization of elementary logic. In: Modality, Morality and Other Problems of Sense and Nonsense, pp. 189–191. CWK Gleerup Bokförlag, Lund (1973)
11. Lindström, P.: On characterizing elementary logic. In: Logical Theory and Semantic Analysis. Synthese Library, vol. 63, pp. 129–146. Reidel, Dordrecht (1974)
12. Lindström, P.: Omitting uncountable types and extensions of elementary logic. Theoria **44**(3), 152–156 (1978)
13. Lindström, P.: Prologue. In: Krynicki, M., Mostowski, M., Szczerba, L. (eds.) Quantifiers, Logics, Models and Computation, vol. 248, pp. 21–24. Kluwer Academic, Dordrecht (1995)
14. Mostowski, A.: On a generalization of quantifiers. Fund. Math. **44**, 12–36 (1957)

15. Mostowski, A.: Craig's interpolation theorem in some extended systems of logic. In: Logic, Methodology and Philos. Sci. III, Proc. Third Internat. Congr., Amsterdam, 1967, pp. 87–103. North-Holland, Amsterdam (1968)
16. Shelah, S.: Remarks in abstract model theory. Ann. Pure Appl. Log. **29**(3), 255–288 (1985)
17. Tharp, L.H.: The characterization of monadic logic. J. Symb. Log. **38**(3), 481–488 (1973)
18. Ziegler, M.: A language for topological structures which satisfies a Lindström-theorem. Bull. Amer. Math. Soc. **82**(4), 568–570 (1976)

Added in proof: A recent important paper on the Lindström's Theorem is "Nice infinitary logics" by Saharon Shelah, J. Amer. Math. Soc. **25**, 395–427 (2012).

J. Väänänen (✉)
Department of Mathematics and Statistics, University of Helsinki, Helsinki, Finland
e-mail: jouko.vaananen@helsinki.fi
Institute for Logic, Language and Computation, University of Amsterdam, Amsterdam, The Netherlands

On Extensions of Elementary Logic

by

PER LINDSTRÖM

(University of Gothenburg)

In this paper we present some results to the effect that certain combinations of theorems from the theory of models of elementary logic (EL) cannot be generalized to proper extensions of EL satisfying various conditions depending on the theorems in question. For example, we prove (Theorem 2) that Löwenheim's theorem together with the compactness theorem for denumerable sets of sentences cannot be extended to any generalized first order logic (defined below) which properly extends EL. Some of our results are improvements of earlier theorems of Mostowski [6] and Lindström [5].

We use the following notation and terminology. By a type we understand a sequence $t = \langle t_0, \ldots, t_{m-1} \rangle$ of positive integers; $Dt = m = \{0, \ldots, m-1\}$. A structure of type t, $\mathfrak{A} = \langle |\mathfrak{A}|, R_k^{\mathfrak{A}} \rangle_{k < m}$, consists of a non-empty set $|\mathfrak{A}|$ and relations $R_k^{\mathfrak{A}} \subseteq |\mathfrak{A}|^{t_k}$ for $k < m$; $t^{\mathfrak{A}} = t$. Structures are said to be similar if they are of the same type. By the cardinality of $\mathfrak{A}$ we understand that of $|\mathfrak{A}|$. $\mathfrak{B}$ is an expansion of $\mathfrak{A}$ if $Dt^{\mathfrak{A}} \leq Dt^{\mathfrak{B}}$ and $R_k^{\mathfrak{B}} = R_k^{\mathfrak{A}}$ for $k < Dt^{\mathfrak{A}}$. If π is a permutation of $Dt^{\mathfrak{A}}$, then $\mathfrak{A}_\pi = \langle |\mathfrak{A}|, R_{\pi(k)}^{\mathfrak{A}} \rangle_{k < m}$, where $m = Dt^{\mathfrak{A}}$. If $t_k^{\mathfrak{A}} \geq 2$ for $k < m = Dt^{\mathfrak{A}}$ and $a \in |\mathfrak{A}|$, then $\mathfrak{A}^{(a)}$ is the structure $\langle |\mathfrak{A}|, S_k \rangle_{k < m}$ such that for any $k < m$ and any $b_0, \ldots, b_{n-2} \in |\mathfrak{A}|$, where $n = t_k^{\mathfrak{A}}$, $\langle b_0, \ldots, b_{n-2} \rangle \in S_k$ iff $\langle b_0, \ldots, b_{n-2}, a \rangle \in R_k^{\mathfrak{A}}$. K, M, N are classes of pairwise similar structures. K is said to be a free expansion of M if for some type t, $\mathfrak{A} \in K$ iff $\mathfrak{A}$ is of type t and $\mathfrak{A}$ is an expansion of a member of M. $K_\pi = \{\mathfrak{A}_\pi \colon \mathfrak{A} \in K\}$ and $K^+ = \{\mathfrak{A} \colon \mathfrak{A}^{(a)} \in K \text{ for every } a \in |\mathfrak{A}|\}$. $\overline{K}$ is the complement of K with respect to the class of structures similar to the members of K. K is said to characterize the class M of

1 – Theoria, 1: 1969

J.-Y. Béziau (ed.), *Universal Logic: An Anthology*, 237–247
Studies in Universal Logic, DOI 10.1007/978-3-0346-0145-0_20, © Springer Basel AG 2012

structures of type t if for every $\mathfrak{A}$ of type t, $\mathfrak{A}$ is isomorphic to a member of M iff some expansion of $\mathfrak{A}$ is a member of K. K characterizes the structure $\mathfrak{B}$ if K characterizes $\{\mathfrak{B}\}$. We write $K \in F$ ($K \in F_\omega$) to mean that K is a class of structures of type $\langle 1 \rangle$ such that (K has no finite member and) if $\langle A, R \rangle \in K$, then R is finite and $\neq 0$ and for every n, there is a countable member $\langle B, S \rangle$ of K such that S is of power $n + 1$.

Let $L = (\Sigma, T)$, where Σ is an arbitrary non-empty set and T is a binary relation between members of Σ on the one hand and structures on the other; $\Sigma_L = \Sigma$, $T_L = T$. Members of Σ_L will be called L-sentences. If φ is an L-sentence, then $\mathrm{Mod}_{t,L}(\varphi)$ is the class of structures of type t, $\mathfrak{A}$, such that $T_L(\varphi, \mathfrak{A})$. $K \in C_L$ means that there are $\varphi \in \Sigma_L$ and t such that $K = \mathrm{Mod}_{t,L}(\varphi)$. K ($\mathfrak{A}$) is said to be L-characterizable if some $M \in C_L$ characterizes K ($\mathfrak{A}$). We say that L is a generalized first order logic if L satisfies the following conditions:

(i) If $K \in C_L$, then K is closed under isomorphism.
(ii) If $K \in C_L$, the members of K are of type t, and π is a permutation of Dt, then $K_\pi \in C_L$.
(iii) If $K \in C_L$ and M is a free expansion of K, then $M \in C_L$.
(iv) If $K \in C_L$, then $\overline{K} \in C_L$.
(v) If K, $M \in C_L$, then $K \cap M \in C_L$.

From now on we assume that L and L' are generalized first order logics. L is a strong generalized first order logic if L satisfies the following condition:

(vi) If $K \in C_L$, then K^+ is L-characterizable.

Let t be any type. By a t-formula (t-sentence) we understand an elementary formula (sentence) with no non-logical symbols other than the predicates P_k for $k < Dt$, where P_k is t_k-ary. Let Γ be the union of the sets of t-sentences for arbitrary t and let T_0 be the relation such that $T_0(\varphi, \mathfrak{A})$ iff $\varphi \in \Gamma$ and φ holds in $\mathfrak{A}$. Set $EL = (\Gamma, T_0)$. Thus $K \in C_{EL}$ iff K is elementarily definable. Next let $Q_0, \ldots, Q_{k-1}$ be arbitrary (generalized) quantifiers and let Q be the set of the corresponding quantifier symbols. (For definitions of the relevant notions pertaining to logic with generalized quantifiers see [5]. We assume here that Q-formulas may contain the ordinary quantifier symbols and sentential connectives.) Let Δ be the set of Q-sentences and let T_1 be the relation such that $T_1(\varphi, \mathfrak{A})$ iff $\varphi \in \Delta$ and φ holds in $\mathfrak{A}$. Set $L(Q) = (\Delta, T_1)$. Clearly EL and $L(Q)$ are strong generalized first order logics. If φ is a Q-formula with no free variables other than $v_0, \ldots, v_{m-1}$ and $x \in |\mathfrak{A}|^m$, then $\mathfrak{A} \models \varphi[x]$ means that x satisfies φ in $\mathfrak{A}$.

L' is an extension of L, in symbols $L \subseteq L'$, if for every L-sentence φ and every type t, there is an L'-sentence ψ such that $\text{Mod}_{t,L}(\varphi) = \text{Mod}_{t,L'}(\psi)$. L is equivalent to L', $L \equiv L'$, if $L \subseteq L'$ and $L' \subseteq L$. $L \subseteq_{\text{inf}} L'$ means that for every L-sentence φ and every type t, there is an L'-sentence ψ such that $\text{Mod}_{t,L}(\varphi)$ and $\text{Mod}_{t,L'}(\psi)$ have the same infinite members. $L \equiv_{\text{inf}} L'$ iff $L \subseteq_{\text{inf}} L'$ and $L' \subseteq_{\text{inf}} L$.

Consider now the following conditions on L:

(I) If $K_n \in C_L$ for every n and $\bigcap_{n<\omega} K_n = 0$, then $\bigcap_{n \leq m} K_n = 0$ for some m.

(II) If $K \in C_L$ and K has an infinite member, then K has a denumerable member.

(III) If $K \in C_L$ and K has a denumerable member, then K has an uncountable member.

As is well-known, EL satisfies these conditions and certain much stronger conditions as well. In what follows we shall obtain some results in the converse direction. These results are simple consequences of the following

THEOREM 1. If L satisfies (II), $EL \subseteq_{\text{inf}} L$, and $L \not\subseteq_{\text{inf}} EL$, then some member of F_ω is L-characterizable.

For the proof of Theorem 1 we require the following two lemmas the first of which is due to R. Fraïssé [4] and A. Ehrenfeucht [2] and the second to R. Fraïssé [3]. To state these lemmas we need the following additional notation and terminology. If $x = \langle x_0, \ldots, x_{m-1} \rangle$, then $x \widehat{\ } a = \langle x_0, \ldots, x_{m-1}, a \rangle$. By an I-sequence of length $m+1$ for $\langle \mathfrak{A}, \mathfrak{B} \rangle$, where $\mathfrak{A}$ and $\mathfrak{B}$ are of type t, we understand a sequence $\langle I_k \rangle_{k \leq m}$ of relations such that

(1) $I_k \subseteq |\mathfrak{A}|^k \times |\mathfrak{B}|^k$ for $k \leq m$,

(2) $\langle \rangle \ I_0 \ \langle \rangle$,

(3)   if $k < m$ and $x I_k y$, then for every $a \in |\mathfrak{A}|(b \in |\mathfrak{B}|)$, there is a $b \in |\mathfrak{B}| \ (a \in |\mathfrak{A}|)$ such that $x \widehat{\ } a I_{k+1} y \widehat{\ } b$,

(4)   if $x I_m y$, then for any atomic t-formula φ with no variables othen than $v_0, \ldots, v_{m-1}$, $\mathfrak{A} \models \varphi[x]$ iff $\mathfrak{B} \models \varphi[y]$.

$\langle I_k \rangle_{k < \omega}$ is an I-sequence of length ω for $\langle \mathfrak{A}, \mathfrak{B} \rangle$ if for every m, $\langle I_k \rangle_{k \leq m}$ is an I-sequence of length $m + 1$ for $\langle \mathfrak{A}, \mathfrak{B} \rangle$.

LEMMA 1. Let $\varkappa$ be any cardinal. If K does not have the same members of power $\varkappa$ as any member of C_{EL}, then for every m, there are structures $\mathfrak{A}, \mathfrak{B}$ of power $\varkappa$ such that $\mathfrak{A} \in K$, $\mathfrak{B} \in \overline{K}$ and an I-sequence of length $m + 1$ for $\langle \mathfrak{A}, \mathfrak{B} \rangle$.

PROOF. Suppose the members of K are of type t. For $n \leq m$ we define the notions of an (m, n)-condition and of a complete (m, n)-condition as follows. An (m, m)-condition is an atomic t-formula with no variables other than $v_0, \ldots, v_{m-1}$. Let $\varphi^{(1)}$ be φ or $\neg \varphi$ according as $i = 0$ or $i = 1$. Next let $\varphi_0, \ldots, \varphi_k$ be all (m, n)-conditions. Then for any $i_0, \ldots, i_k$, $\varphi_0^{(i_0)} \wedge \ldots \wedge \varphi_k^{(i_k)}$ is a complete (m, n)-condition. Finally, if $n > 0$ and φ is a complete (m, n)-condition, then $\exists v_{n-1} \varphi$ is an $(m, n-1)$-condition. No other formulas are (complete) (m, n)-conditions. Note that the free variables of an (m, n)-condition are among $v_0, \ldots, v_{n-1}$.

By hypothesis, the class of members of K of power $\varkappa$ does not coincide with the class of models of power $\varkappa$ of any disjunction of complete $(m, 0)$-conditions. It follows that there are $\mathfrak{A}$ and $\mathfrak{B}$ of power $\varkappa$ such that $\mathfrak{A} \in K$, $\mathfrak{B} \in \overline{K}$, and the same $(m, 0)$-conditions hold in $\mathfrak{A}$ and $\mathfrak{B}$. Now for $k \leq m$, define the relation I_k thus: $x I_k y$ iff $x \in |\mathfrak{A}|^k$, $y \in |\mathfrak{B}|^k$, and for every (m, k)-condition φ, $\mathfrak{A} \models \varphi[x]$ iff $\mathfrak{B} \models \varphi[y]$. Obviously, the sequence $\langle I_k \rangle_{k \leq m}$ satisfies conditions (1), (2), and (4). To show that it also satisfies (3) suppose $k < m$, $x I_k y$, and $a \in |\mathfrak{A}|$. Let φ be the complete $(m, k+1)$-condition such that $\mathfrak{A} \models \varphi[x \widehat{\ } a]$. Then $\psi = \exists v_k \varphi$ is an (m, k)-condition and $\mathfrak{A} \models \psi[x]$. Hence $\mathfrak{B} \models \psi[y]$, whence there is a $b \in |\mathfrak{B}|$ such that $\mathfrak{B} \models \varphi[y \widehat{\ } b]$. Clearly $x \widehat{\ } a I_{k+1} y \widehat{\ } b$. This proves one half of (3). The proof of the other half is the same. Thus

$\langle I_k \rangle_{k \leq m}$ is an I-sequence of length $m+1$ for $\langle \mathfrak{A}, \mathfrak{B} \rangle$ and Lemma 1 is proved.

LEMMA 2. If $\mathfrak{A}$ and $\mathfrak{B}$ are denumerable and there is an I-sequence of length ω for $\langle \mathfrak{A}, \mathfrak{B} \rangle$, then $\mathfrak{A}$ is isomorphic to $\mathfrak{B}$.

PROOF. Let $|\mathfrak{A}| = \{a_n : n \in \omega\}$, $|\mathfrak{B}| = \{b_n : n \in \omega\}$, and let $\langle I_k \rangle_{k < \omega}$ be an I-sequence for $\langle \mathfrak{A}, \mathfrak{B} \rangle$. We define sequences $\langle c_n \rangle_{n < \omega}$ and $\langle d_n \rangle_{n < \omega}$ such that for every n,

(5) $c_{2n} = a_n$,
(6) $d_{2n+1} = b_n$,
(7) $\langle c_0, \ldots, c_{n-1} \rangle I_n \langle d_0, \ldots, d_{n-1} \rangle$

as follows. Suppose c_n and d_n have been defined for $n < k$. If k is even, $k = 2r$, set $c_k = a_r$. By (3), there is a least index s such that $\langle c_0, \ldots, c_k \rangle I_{k+1} \langle d_0, \ldots, d_{k-1}, b_s \rangle$. Set $d_k = b_s$. If, on the other hand, k is odd, $k = 2r + 1$, set $d_k = b_r$. Again by (3), and (2) if $r = 0$, there is a least index s such that $\langle c_0, \ldots, c_{k-1}, a_s \rangle I_{k+1} \langle d_0, \ldots, d_k \rangle$. Set $c_k = a_s$.

Now let $f = \{\langle c_n, d_n \rangle : n \in \omega\}$. Then, by (4)—(7), f is an isomorphism on $\mathfrak{A}$ onto $\mathfrak{B}$.

PROOF OF THEOREM I. Let $K_0 \in C_L$ be such that there is no $M \in C_{EL}$ such that K_0 and M have the same infinite members. Then

(8) there is no $M \in C_{EL}$ such that K_0 and M have the same denumerable members.

Indeed, Suppose $M \in C_{EL}$. There is then a class $M_0 \in C_L$ such that M and M_0 have the same infinite members. Let $M_1 = (\overline{K_0} \cap M_0) \cup (K_0 \cap \overline{M_0})$. By (iv) and (v), $M_1 \in C_L$. Clearly M_1 has an infinite member. Hence, by (II), M_1 has a denumerable member and so (8) follows.

Suppose, for simplicity, that the members of K_0 are of type $\langle 2 \rangle$. Let K_1 be the class of structures $\langle A, R_k \rangle_{k < 7}$ of type $t = = \langle 1, 2, 2, 2, 2, 3, 3 \rangle$ such that

(9) R_0 is non-empty,
(10) R_3 is a one-one function on A into a proper subset of A,

(11) R_4 is a linear ordering of R_0 such that R_0 has an R_4-first member and every member of R_0 which has an R_4-successor has an immediate R_4-successor,

(12)   for every $a \in A$, the relation $f_a = \{\langle x, y \rangle : \langle a, x, y \rangle \in R_5\}$ is a function on R_0 into A,

(13)   if x is the R_4-first member of R_0, then there are a, b such that $\langle x, a, b \rangle \in R_6$,

(14)   if $\langle x, a, b \rangle \in R_6$, $x \in R_0$, y is the immediate R_4-successor of x, and z is any member of A, then there are c, d, u such that $\langle y, c, d \rangle \in R_6$, $f_c(y) = z$, $f_d(y) = u$, and for every $v \in R_0$, if $v \neq y$, then $f_c(v) = f_a(v)$ and $f_d(v) = f_b(v)$,

(15)   (Like (14) except that a and b are interchanged.),

(16)   if $\langle x, a, b \rangle \in R_6$ and y, z are R_4-predecessors of x, then $\langle f_a(y), f_a(z) \rangle \in R_1$ iff $\langle f_b(y), f_b(z) \rangle \in R_2$.

It is easily seen that $K_1 \in C_{EL}$ and that it has no finite members, whence $K_1 \in C_L$. Next, it follows at once from (ii)—(v) that there is a class $K_2 \in C_L$ of structures of type t such that $\mathfrak{A} \in K_2$ iff $\langle |\mathfrak{A}|, R_1^{\mathfrak{A}} \rangle \in K_0$ and $\langle |\mathfrak{A}|, R_2^{\mathfrak{A}} \rangle \in \overline{K}_0$. Now set $K = K_1 \cap K_2$. Then, by (v), $K \in C_L$. We propose to show that K characterizes a member of F_ω.

Let n be any natural number $\neq 0$. By (8) and Lemma 1, there are relations $R_1, R_2 \subseteq \omega^2$ such that $\mathfrak{A} = \langle \omega, R_1 \rangle \in K_0$, $\mathfrak{B} = \langle \omega, R_2 \rangle \in \overline{K}_0$, and an I-sequence $\langle I_k \rangle_{k < n}$ of length n for $\langle \mathfrak{A}, \mathfrak{B} \rangle$. Let $R_0 = n$ and let R_4 be the $<$-relation restricted to R_0. Let g be a one-one function on ω onto the set of functions on R_0 into ω. We write g_x for g(x). Next let $R_5 = \{\langle m, x, y \rangle : x \in R_0 \text{ and } g_m(x) = y\}$. Let $R_6 = \{\langle m, x, y \rangle : m \in R_0 \text{ and } \langle g_x(0), \ldots, g_x(m-1) \rangle \ I_m \ \langle g_y(0), \ldots, g_y(m-1) \rangle\}$. Finally, let R_3 be a one-one function on ω into a proper subset of ω. The verification that $\langle \omega, R_k \rangle_{k < 7}$ is a member of K presents no difficulties.

Suppose now, for reductio ad absurdum, that there is a structure $\mathfrak{A} \in K$ such that $R_0^{\mathfrak{A}}$ is infinite. Let M_1 be the free expansion of K whose members are of type $t' = t^\frown 2$. Next let M_2 be the class of structures $\mathfrak{B}$ of type t' such that $R_7^{\mathfrak{B}}$ is a one-one function on $R_0^{\mathfrak{B}}$ into a proper subset of $R_0^{\mathfrak{B}}$. Clearly $M_0 = M_1 \cap M_2 \in C_L$ and M_0 has an infinite member. Hence, by (II), M_0 has a denumerable member $\mathfrak{C}$. Clearly $R_0^{\mathfrak{C}}$ is infinite. Hence, by (11), for every n, there is a member c_n of $R_0^{\mathfrak{C}}$ which has exactly n $R_4^{\mathfrak{C}}$-predecessors. For $a \in |\mathfrak{C}|$ let $f_a = \{\langle x, y \rangle : \langle a, x, y \rangle \in R_5^{\mathfrak{C}}\}$. Now, for each n, define the relation I_n as follows: $\langle a_0, \ldots, a_{n-1} \rangle \, I_n \, \langle b_0, \ldots, b_{n-1} \rangle$ iff there are $a, b \in |\mathfrak{C}|$ such that $\langle c_n, a, b \rangle \in R_6^{\mathfrak{C}}$ and $f_a(c_k) = a_k$ and $f_b(c_k) = b_k$ for $k < n$. It is then easily checked, using (12)—(16), that $\langle I_n \rangle_{n < \omega}$ is an I-sequence of length ω for $\langle \mathfrak{C}_1, \mathfrak{C}_2 \rangle$, where $\mathfrak{C}_m = \langle |\mathfrak{C}|, R_m^{\mathfrak{C}} \rangle$, $m = 1, 2$. Hence, by Lemma 2, $\mathfrak{C}_1$ is isomorphic to $\mathfrak{C}_2$. But $\mathfrak{C}_1 \in K_0$ and $\mathfrak{C}_2 \in \overline{K}_0$. Hence, by (i), $\mathfrak{C}_1$ is not isomorphic to $\mathfrak{C}_2$. A contradiction from which it follows that if $\mathfrak{A} \in K$, then $R_0^{\mathfrak{A}}$ is finite. Since, finally, in view of (9) and (10), if $\mathfrak{A} \in K$, then $\mathfrak{A}$ is infinite and $R_0^{\mathfrak{A}} \neq 0$, this concludes our proof that K characterizes a member of F_ω.

COROLLARY 1. *If L satisfies (II), $EL \subseteq L$, and $L \not\subseteq EL$, then some member of F is L-characterizable.*

PROOF. If $L \not\subseteq {}_{int}EL$, then the conclusion follows from Theorem 1. Suppose then $L \subseteq {}_{int}EL$. There are then classes K, M such that $K \in C_L$, $K \notin C_{EL}$, $M \in C_{EL}$, and K and M have the same infinite members. It follows that for every m, there is an $n > m$ such that K and M does not have the same members of power n. Indeed, otherwise, as is easily seen, we would have $K \in C_{EL}$. Suppose the members of K are of type $t = \langle t_0, \ldots, t_{m-1} \rangle$. Set $t' = \langle 1, t_0, \ldots, t_{m-1} \rangle$. Let N be the class of structures such that $\mathfrak{A} \in N$ iff $\mathfrak{A}$ is of type t', $R_0^{\mathfrak{A}} \neq 0$ and $\langle |\mathfrak{A}|, R_{k+1}^{\mathfrak{A}} \rangle_{k < Dt} \in (\overline{K} \cap M) \cup (K \cap \overline{M})$. Then, clearly, $N \in C_L$ and N characterizes a member of F.

COROLLARY 2. *If L is strong, L satisfies (II), $EL \subseteq {}_{int}L$, and $L \not\subseteq {}_{int}EL$, then $\langle \omega, \leq \rangle$ is L-characterizable.*

PROOF. By Theorem 1, there is a class $K \in C_L$ such that K characterizes a member of F_ω. Since L is strong, it follows that there is a class $M \in C_L$ which characterizes K^+. Let N be the class of structures of the same type as the members of M, $\mathfrak{A}$, such that $R_0^{\mathfrak{A}}$ is a reflexive linear ordering of $|\mathfrak{A}|$ such that every member of $|\mathfrak{A}|$ has an $R_0^{\mathfrak{A}}$-successor. Then $M \cap N \in C_L$ and characterizes $\langle \omega, \leq \rangle$.

From Corollary 1 we can now easily derive the following

THEOREM 2. *If L satisfies (I) and (II) and* $EL \subseteq L$, *then* $L \equiv EL$.

PROOF. Suppose L satisfies (II), $EL \subseteq L$, and $L \not\subseteq EL$. Then, by Corollary 1, there is a class $K_0 \in C_L$ which characterizes a member of F. Let t be the type of the members of K_0. For $n > 0$ let K_n be the class of structures $\mathfrak{A}$ of type t such that $R_0^{\mathfrak{A}}$ has at least n members. Then clearly $K_n \in C_L$ for every n, $\bigcap_{n < \omega} K_n = 0$, and $\bigcap_{n \leq m} K_n \neq 0$ for every m. Thus L does not satisfy (I).

If K characterizes $\langle \omega, \leq \rangle$, then, obviously, K has a denumerable member but no uncountable member. Hence, in view of Corollary 2, we have the following

THEOREM 3. *If L is strong, L satisfies (II) and (III), and* $EL \subseteq_{int} L$, *then* $L \equiv_{int} EL$.

Theorems 2 and 3 are improvements of Theorems 5.1 and 5.6 [5]. The present proofs are, however, almost the same as those given in [5].

In order to be able to apply the concepts of recursive function theory we now assume that the members of Σ_L and $\Sigma_{L'}$ are finite configurations of symbols from certain given finite sets or, equivalently, natural numbers. We write $L \subseteq_{eff} L'$ to mean that there is an effective method whereby, given any type t and any L-sentence φ, an L'-sentence ψ can be found such that $\text{Mod}_{t,L}(\varphi) = \text{Mod}_{t,L}(\psi)$. $L \equiv_{eff} L'$ iff $L \subseteq_{eff} L'$ and $L' \subseteq_{eff} L$. L is n.c.-effective if (a) there is an effective method by means of which for any type t and any L-sentence φ, an L-sentence ψ can be found such that $\text{Mod}_{t,L}(\psi) = \overline{\text{Mod}_{t,L}(\varphi)}$ and (b) there is an effective method by means of which for any type t and any L-sentences φ, ψ, an L-sentence θ can be found such that $\text{Mod}_{t,L}(\theta) = \text{Mod}_{t,L}(\varphi) \cap \text{Mod}_{t,L}(\psi)$. Clearly $EL \subseteq_{eff} L(Q)$ and EL and L(Q) are both n.c.-effective. Set $V_L = \{\langle \varphi, t \rangle : t$ is any type, $\varphi \in \Sigma_L$, and $T_L(\varphi, \mathfrak{A})$ for every $\mathfrak{A}$ of type t$\}$. L is said to be axiomatizable if V_L is recursively enumerable. As is well-known, EL is axiomatizable. In the converse direction we have the following

THEOREM 4. *If L is n.c.-effective and axiomatizable, L satisfies (II), and* $EL \subseteq_{eff} L$, *then* $L \equiv_{eff} EL$.

In the proof of Theorem 4 we use the following result due to Trakhtenbrot [7]. A t-sentence is said to be finitely valid if it holds in all finite structures of type t.

LEMMA 3. There is a type t such that the set of finitely valid t-sentences is not recursively enumerable.

PROOF OF THEOREM 4. We first prove that $L \subseteq EL$. Suppose not. Then, by Corollary 1, there is an L-sentence θ and a type $t' = = \langle t'_0, \ldots, t'_{m'-1} \rangle$ such that $Mod_{t',L}(\theta)$ characterizes a member of F. Let $t = \langle t_0, \ldots, t_{m-1} \rangle$ be as in Lemma 3, set $t^+ = \langle t'_0, \ldots, t'_{m'-1}, t_0, \ldots, t_{m-1} \rangle$, and let θ^+ be an L-sentence such that $Mod_{t^+,L}(\theta^+)$ is a free expansion of $Mod_{t',L}(\theta)$. Now let φ be any t-sentence. For every r, replace P_r everywhere in φ by $P_{m'+r}$ and then relativize all quantifier expressions to P_0, i.e. replace $\exists v_k \psi$ by $\exists v_k (P_0 v_k \wedge \psi)$ and $\forall v_k \psi$ by $\forall v_k (P_0 v_k \rightarrow \psi)$. Let φ_0 be the sentence thus obtained. Since $EL \subseteq_{eff} L$, we can now effectively find an L-sentence ψ such that $Mod_{t^+,L}(\psi) = Mod_{t^+,EL}(\varphi_0)$. Next, since L is n.c.-effective, we can find an L-sentence η such that

$$Mod_{t^+,L}(\eta) = \overline{Mod_{t^+,L}(\theta^+)} \cup Mod_{t^+,L}(\psi).$$

Clearly, $\langle \eta, t^+ \rangle \in V_L$ iff φ is finitely valid. Since η was found effectively from φ and since V_L is recursively enumerable, we may now conclude that the set of finitely valid t-sentences is recursively enumerable. But this contradicts Lemma 3. Thus the assumption that $L \not\subseteq EL$ was false and so $L \subseteq EL$.

That $L \subseteq_{eff} EL$ can now be shown as follows. Let φ be any L-sentence and t any type. Given t we can obviously find an effective enumeration $\psi_0, \psi_1, \psi_2, \ldots$ of all t-sentences. Next, for each n, an L-sentence η_n can be found such that $Mod_{t,L}(\eta_n) = = Mod_{t,EL}(\psi_n)$. Finally, we can find L-sentences ξ_n such that for every n,

$$Mod_{t,L}(\xi_n) = (Mod_{t,L}(\varphi) \cap Mod_{t,L}(\eta_n)) \cup$$
$$\cup \overline{(Mod_{t,L}(\varphi) \cap \overline{Mod_{t,L}(\eta_n)})}.$$

Clearly $\langle \xi_n, t \rangle \in V_L$ iff $Mod_{t,L}(\varphi) = Mod_{t,EL}(\psi_n)$. But, since $L \subseteq EL$, there is an n for which this equation holds. Hence, since V_L is recursively enumerable, we can find an n such that $\langle \xi_n, t \rangle \in$

$\in V_L$. Thus, given a type t and an L-sentence φ a t-sentence ψ_n can be found effectively such that $Mod_{t,EL}(\psi_n) = Mod_{t,L}(\varphi)$; in other words, $L \subseteq {}_{eff}EL$, as was to be proved.

Theorem 4 is a generalization of a theorem of Mostowski (Theorem 4 [6]).

We conclude this paper by discussing the possibility of extending Beth's theorem on definability together with (II) to proper extensions of EL of the form L(Q). Let φ be an L(Q)-sentence which contains the q-ary predicate P_s. φ is said to define P_s implicitly if for any two structures $\mathfrak{A}$ and $\mathfrak{B}$, if $t^{\mathfrak{A}} = t^{\mathfrak{B}}$, $\mathfrak{A} \models \varphi$, $\mathfrak{B} \models \varphi$, $|\mathfrak{A}| = |\mathfrak{B}|$, and $R_k^{\mathfrak{A}} = R_k^{\mathfrak{B}}$ for $k < Dt^{\mathfrak{A}}$ and $k \neq s$, then $R_s^{\mathfrak{A}} = R_s^{\mathfrak{B}}$. φ is said to L(Q)-define P_s explicitly if there is a Q-formula ψ which contains no predicates other than those occurring in φ, does not contain P_s, contains no free variables othen than $v_0, \ldots, v_{q-1}$, and is such that

$$\mathfrak{A} \models \varphi \rightarrow \forall v_0 \ldots v_{q-1}(P_s v_0 \ldots v_{q-1} \leftrightarrow \psi),$$

for every structure $\mathfrak{A}$ of suitable type. We say that L(Q) has Beth's property if for every L(Q)-sentence φ and every predicate P_s, if φ defines P_s implicitly, then φ L(Q)-defines P_s explicitly. As is well-known, EL (Q empty) has Beth's property. In the converse direction we have the following

THEOREM 5. If L(Q) has Beth's property and satisfies (II), then $L(Q) \equiv {}_{int}EL$.

PROOF IN OUTLINE. Let $L = L(Q)$. As noted above, L is a strong generalized first order logic. Suppose L satisfies (II) and $L \not\subseteq {}_{int}EL$. Then, by Corollary 2, there is an L-sentence θ and a type t such that $Mod_{t,L}(\theta)$ characterizes $\langle \omega, \leq \rangle$. Suppose now we have defined a Gödel numbering of the Q-formulas satisfying the usual effectiveness conditions. Then, by a suitable modification of the construction used in the proof of Theorem 2.1[1] and using the sentence θ, we can find a type t' and an L-sentence φ such that $Mod_{t',L}(\varphi)$ is non-empty and if $\mathfrak{A} \in Mod_{t',L}(\varphi)$, then there is a structure $\mathfrak{B}$ isomorphic to $\mathfrak{A}$ such that $|\mathfrak{B}| = \omega$, $R_0^{\mathfrak{B}} = \{\langle m, n \rangle : m \leq n\}$, $R_1^{\mathfrak{B}} = \{\langle k, m, n \rangle : k + m = n\}$, $R_2^{\mathfrak{B}} = \{k, m, n\rangle : k \cdot m = n\}$, and $R_3^{\mathfrak{B}} = \{\langle m, n \rangle : n$ is the Gödel number of a formula ψ such that ψ does not contain P_3 and for some r, ψ contains no free variables other than $v_0, \ldots, v_{r-1}$, $m = 2^{m_0} \cdot \ldots \cdot p_{r-1}^{m_{r-1}}$, where p_{k-1} is the kth prime, and $\mathfrak{B} \models \psi[m_0, \ldots, m_{r-1}]\}$. Clearly φ defines P_3 implicitly. Suppose now L has Beth's property and let $\mathfrak{B}$ be as described. There is then a Q-formula η not containing P_3 such that

$$\mathfrak{B} \models \forall v_0 v_1 (P_3 v_0 v_1 \leftrightarrow \eta).$$

It follows that there is a Q-formula ξ with no free variable other than v_0 and not containing P_3 such that for every n, $\mathfrak{B} \models \xi[n]$ iff $\langle 2^n, n \rangle \in R_3^{\mathfrak{B}}$. Let p be the Gödel number of $\neg\xi$. Then $\mathfrak{B} \models \xi[p]$ iff $\langle 2^p, p \rangle \in R_3^{\mathfrak{B}}$ iff $\mathfrak{B} \models \neg\xi[p]$; a contradiction. Thus L does not have Beth's property and our proof is finished.

BIBLIOGRAPHY

[1] W. Craig and R. L. Vaught. Finite axiomatizability using additional predicates. J. Symb. Logic, vol. 23 (1958), pp. 289—308.

[2] A. Ehrenfeucht. An application of games to the completeness problem for formalized theories. Fund. Math., vol. 49 (1961), pp. 129—141.

[3] R. Fraïssé. Sur l'extension aux relations de quelques propriétés des ordres. Ann. Sci. École Norm. Sup., 3ième série, vol. 71 (1954), pp. 363—388.

[4] R. Fraïssé. Sur quelques classifications des relations, basées sur des isomorphismes restreints. Alger-Mathématiques, vol. 2 (1955), pp. 16—60, 273—295.

[5] P. Lindström. First order predicate logic with generalized quantifiers. Theoria, vol. 32 (1966), pp. 186—195.

[6] A. Mostowski. On a generalization of quantifiers. Fund. Math., vol. 44 (1957), pp. 12—36.

[7] B. A. Trakhtenbrot. On recursive separability. Dodkl. Akad. Nauk SSSR, vol. 88 (1953), pp. 953—956.

Received on November 30, 1968

The author has himself pointed out the following error: The assumption $EL \subseteq_{INF} L$ of Theorem 1 and its Corollaries should be replaced by the stronger assumption $EL \subseteq L$.

Part 11
Stephen L. Bloom, Donald J. Brown
and Roman Suszko (1970)

Bloom, Brown and Suszko's Work on Abstract Logics

Ramon Jansana

Abstract In this short note that serves as an introduction to the 1970 paper of Bloom, S.L., Brown, D.J. and Suszko, R. entitled *Some theorems on abstract logics* I briefly explain the main ideas of the period that shaped the context where the notion of abstract logic was introduced by these authors, the fundamental ideas of the theory of abstract logics, the content of the paper and finally I refer to some related work.

Mathematics Subject Classification (2010) 03G10 · 03G27

In 1969 Donald Jerome Brown defended his PhD. dissertation *Abstract Logics*, the principal advisor of which was Roman Suszko. Part of the material was published in 1973 in [3]. Previously, in 1970, Brown and Suszko published [1] and [2], together with Stephen L. Bloom. The second of these papers is reprinted in this anthology. The first one is only a note that presents without proofs the main results contained in the second. As an introduction to [2], I first expound some of the main ideas of the period that shape the framework in which the research was done, I then briefly describe the main ideas of the theory of abstract logics introduced in [3] and [4], point to references on further developments, and summarize the contents of the paper. Finally I will refer to some studies in which the idea of abstract logic was rediscovered (under perhaps different guises) without the rediscoverers being aware of Bloom, Brown and Suszko's research.

During the 1920's and 1930's, Alfred Tarski worked on the methodology of the deductive sciences [15–18]. The goal was to study the axiomatic method abstractly, and to do so he introduced the abstract concept of consequence operation in the study of logics. Tarski mainly had the mathematical axiomatic theories in mind and set the framework for the study of the most general properties of the operation that assigns to a set of axioms the formulas that are provable from them in a given logical calculus (like Frege's or Russell's).

Given a set of axioms and rules—known today as a Hilbert-style calculus—a formula ϕ is provable from a set of formulas X if there is a finite sequence of formulas, each element of which belongs to X, it is an axiom, or it is obtained from previous formulas in the sequence by one of the inference rules. Let $Cn(X)$ be the set of formulas provable from X. Then Cn is an operation that applies to sets of formulas and outputs sets of formulas. The operation Cn has the following properties:

(1) $X \subseteq Cn(X)$.
(2) $Cn(Cn(X)) = Cn(X)$.
(3) $Cn(X) = \bigcup \{Cn(Y) : Y \subseteq X, X \text{ finite}\}$.

Tarski took these properties to define the notion of consequence operation. In fact, he added that there is a formula ϕ such that $Cn(\{\phi\})$ is the set of all formulas. He also assumed that the set of formulas is finite or of the cardinality of the set of the natural numbers. Condition (3) implies the weaker, and important, condition of monotonicity:

(4) if $X \subseteq Y$, then $Cn(X) \subseteq Cn(Y)$.

Let τ be a set of connectives and consider it also as an algebraic similarity type. The algebra of formulas $\mathbf{Fm}_\tau$ built from the set of sentential variables and the connectives in τ is, in algebraic terminology, the absolutely free algebra of type τ with the set of sentential variables as generators. In 1958 Łoś and Suszko considered in [12] a new condition on consequence operations in order to obtain their notion of sentential logic, namely structurality (nowadays also known as invariance under substitutions). A *structural consequence operation* on the algebra of formulas of type τ is an operation Cn on the set Fm_τ of the formulas of type τ that satisfies (1), (2) and (4) above and also that for every substitution σ, i.e. an endomorphism of $\mathbf{Fm}_\tau$,

(5) $\sigma[Cn(X)] \subseteq Cn(\sigma[X])$.

A *sentential logic*[1] of type τ is a pair $L = \langle \mathbf{Fm}_\tau, Cn_L \rangle$ where Cn_L is a structural consequence operation on Fm_τ. If in addition Cn_L satisfies condition (3) above, then L is said to be finitary.

Sentential logics may enjoy a diversity of models. A particular kind of model which is suitable to the study of any sentential logic is the logical matrix. Tarski defined the general concept of logical matrix in the 1920's but the concept was already implicit in previous work by Łukasiewicz, Bernays, Post and others, who used truth-tables, either in independence proofs or to define logics different from classical logic. A *logical matrix* is a pair $\langle A, D \rangle$ where A is an algebra and D is a subset of the domain of A, called the set of designated elements. The general theory of logical matrices was developed mainly by Polish logicians, starting with Łoś [11] and continuing in Łoś and Suszko [12], building on previous work by Lindenbaum. In [12] matrices are used for the first time both as models of arbitrary sentential logics and to define objects of this kind. The interested reader is addressed to [22] for information on the theory of logical matrices and sentential logics.

Let $L = \langle \mathbf{Fm}_\tau, Cn_L \rangle$ be a sentential logic. A logical matrix $\mathcal{M} = \langle A, D \rangle$, with A an algebra of type τ, is a *model* of L if whenever $\phi \in Cn_L(\Gamma)$ then for every valuation v on A (i.e. a homomorphism from $\mathbf{Fm}_\tau$ to A) if $v[\Gamma] \subseteq D$, then $v(\phi) \in D$. A class M of logical matrices is an adequate semantics for a logic L if

$$\phi \in Cn_L(\Gamma) \text{ iff for every } \langle A, D \rangle \in \text{M and every valuation } v \text{ on } A,$$

$$\text{if } v[\Gamma] \subseteq D, \text{ then } v(\phi) \in D.$$

Obviously this definition applies to single matrices by taking M to be a singleton. The standard way to define a sentential logic from a logical matrix or from a class of them is to take the clause on the right hand side of the 'iff' above as the defining condition.

---

[1] Sentential logics as defined here are also known as deductive systems or Hilbert systems.

Wojcicki [20] proved that every sentential logic L has a matrix semantics, which consists of the matrices $\langle \mathbf{Fm}_\tau, T \rangle$ where T is a theory of L, i.e. a set of formulas such that $Cn_L(T) = T$. These matrices are known as the *Lindenbaum matrices* of L.

1 Abstract Logics

The concept of abstract logic generalizes the concept of sentential logic by forgetting (i) that the algebra is a free algebra and (ii) the condition of structurality. Many concepts applicable to sentential logics do not depend on these conditions, so they admit a generalization to a more abstract setting. The theory of abstract logics is precisely this setting.

Although the theory of abstract logics on full is developed by Brown and Suszko, there is a precedent of the concept in Smiley [14] where a new notion of matrix, more general than the notion described above, is proposed and which in fact is the notion of abstract logic.

Let A be a set. A map C from the power set of A to itself is called a consequence operation or a closure operation if it satisfies conditions (1), (2) and (4) above. It is said to be finitary if in addition it satisfies condition (3). The set of closure operations on A forms a complete lattice under the ordering given by $C_1 \le C_2$ if and only if for every $X \subseteq A$, $C_1(X) \subseteq C_2(X)$. The infimum of a family $\{C_i : i \in I\}$ of closure operations on A is the closure operation C defined by $C(X) = \bigcap_{i \in I} C_i(X)$, for every $X \subseteq A$. This induces a complete lattice on the set of abstract logics over an algebra A with domain A.

Let τ be an algebraic similarity type. An *abstract logic* of type τ is a pair $\langle A, C \rangle$ where A is an algebra of type τ and C is a closure operation on the domain A of A. When A is the absolutely free algebra and C is structural, then we have a sentential logic of type τ. Note that if τ is empty, then an algebra of type τ is a set, so a pair $\langle A, C \rangle$ where C is a closure operation on A is an abstract logic of the empty algebraic similarity type. This has the consequence that the results of the theory of abstract logics apply to closure operations on sets.

Let $\langle A, C \rangle$ be an abstract logic. A set $X \subseteq A$ is C-closed if $C(X) = X$. The family of C-closed sets contains A and is closed under arbitrary intersections of nonempty families and so is a complete lattice when it is ordered by inclusion. The families of subsets of A that enjoy these two properties are known as closure systems or closed set systems. Due to the duality between closure operations and closure systems on a set, in [3] abstract logics are also presented as an algebra plus a closure system on its domain, and the authors move freely from one presentation to the other according to convenience[2]. This alternative way of looking at abstract logics establishes a similarity between abstract logics and topological spaces, which inspires the notion of morphism between abstract logics, introduced in [3]. Let $\langle A_1, C_1 \rangle$, $\langle A_2, C_2 \rangle$ be abstract logics of type τ. A map $f : A_1 \to A_2$ is a *logical morphism* if (i) f is a homomorphism from A_1 to A_2 and (ii) it is continuous, i.e. for every C_2-closed set $Y \subseteq A_2$, $f^{-1}[Y]$ is a C_1-closed set. The collection of abstract logics of type τ, taken as objects, and the collection of logical morphisms between them, taken as morphisms, constitute a category, the category of abstract logic of type τ.

---

[2]We will be systematic and will define all the concepts using the closure operation presentation, though this is not done in [3].

Important concepts introduced in [3] to study the categories of abstract logics are logical congruence and bilogical morphism. Let $\mathcal{A} = \langle A, C \rangle$ be an abstract logic. A *logical congruence* of $\mathcal{A}$ is a congruence θ of A such that for every $a, b \in A$, if $a\theta b$, then $C(a) = C(b)$. The family of logical congruences of $\mathcal{A}$ ordered by inclusion has a greatest element[3], known as the Tarski congruence of $\mathcal{A}$, and it is closed under intersections of arbitrary nonempty families; therefore it is a complete lattice. Let θ be a logical congruence of $\mathcal{A}$; the *quotient abstract logic* of $\mathcal{A}$ by θ is the abstract logic $\mathcal{A}/\theta = \langle A/\theta, C_\theta \rangle$ where A/θ is the quotient algebra of A modulo θ and C_θ is the closure operation on A/θ defined by $C_\theta(X) = \{a/\theta : a \in C(\{b : b/\theta \in X\})\}$. If θ is the greatest logical congruence of $\mathcal{A}$, $\mathcal{A}/\theta$ is called the *reduction* of $\mathcal{A}$ and has only one logical congruence, the identity relation. The abstract logics with this property are nowadays called *reduced* and they play a prominent role in the theory of abstract logics. Let $\mathcal{A}_1 = \langle A_1, C_2 \rangle$, $\mathcal{A}_2 = \langle A_2, C_2 \rangle$ be two abstract logics of the same similarity type. There is a subset of the set of logical morphisms from $\mathcal{A}_1$ to $\mathcal{A}_2$ of special importance. A logical morphism h from $\mathcal{A}_1$ to $\mathcal{A}_2$ is a *bilogical morphism* if it is onto A_2 and the set of C_1-closed elements of $\mathcal{A}_1$ is the set $\{h^{-1}[X] : X \text{ is a } C_2\text{-closed element of } \mathcal{A}_2\}$. If there is a bilogical morphism from A_1 to A_2, then their reductions are isomorphic.

In [4] Bloom and Brown study the abstract logics that enjoy the properties of classical propositional logic. Let us consider the similarity type given by a binary connective $\vee$ and a unary connective $\neg$. An abstract logic $\langle A, C \rangle$ of this type is classical if C is finitary, and the following two conditions hold

- $a \in C(X)$ iff $C(X, \neg a) = A$.
- $C(X, a) \cap C(X, b) = C(X, a \vee b)$.

Examples of classical abstract logics are classical propositional logic and for every Boolean algebra B, presented in the type $\{\vee, \neg\}$, the abstract logic $\langle B, \text{Fi} \rangle$, where Fi is the closure operation that maps every set $X \subseteq B$ to the filter of B generated by X. These logics are called Boolean logics in [4], where it is shown that they are exactly the reduced classical abstract logics. One of the main results of [4] is that the category of classical abstract logics is dually equivalent to the category of Boolean spaces and continuous maps. The well-known duality theorem of Stone for Boolean algebras implies that the category of classical abstract logics is equivalent to the category of Boolean algebras.

1.1 Further Developments

Several authors in Barcelona continued the research in [3, 4] by studying categories of abstract logics that correspond to some non-classical logics (intuitionistic, modal, etc.) in the same manner as the category of classical abstract logics correspond to classical propositional logic. Those abstract logics were defined by conditions like the two above, that is, conditions that establish the behavior of the operations of the algebra with respect to the consequence operation. These conditions are often called Tarski-style conditions. Later on, the research on abstract logics was integrated with the research started by Blok, Czelakowski and Pigozzi that lead to the creation of the field Abstract Algebraic Logic.

---

[3]This fact was proved in Font, Jansana [8], where the name Tarski congruence is also introduced.

For information on this field see [6, 9], and for the research done in Barcelona on abstract logics see [8] and the references therein.

2 The Content of the Paper

In the paper, and in [3], some concepts applicable to sentential logics are generalized to arbitrary abstract logics. One is structural consequence operation. An abstract logic $\langle A, C \rangle$ is *structural* if for every endomorphism e of A, $e[C(X)] \subseteq C(e[X])$ for every $X \subseteq A$. Moreover the notion of invariant abstract logic is introduced. An abstract logic $\langle A, C \rangle$ is *invariant* if for every endomorphism e of A, $e[C(X)] \subseteq C(X)$ for every $X \subseteq A$.

A logical matrix or a class of matrices can be used to define an abstract logic over a given algebra in an analogous way as they are used to define a sentential logic. Let $\mathcal{M} = \langle B, D \rangle$ be a logical matrix and let A be an algebra of the same type as B. Let $\mathrm{Hom}(A, B)$ denote the set of homomorphisms from A to B. The closure operation $Cn_{\mathcal{M}}$ on A is defined in [2] by:

$$a \in Cn_{\mathcal{M}}(X) \text{ iff for every } h \in \mathrm{Hom}(A, B), \text{ if } h[X] \subseteq D, \text{ then } h(a) \in D$$

for every $a \in A$ and every $X \subseteq A$. Then a class of matrices M over algebras of the type of A gives the closure operation $C_{\mathsf{M}} = \mathrm{Inf}_{\mathcal{M} \in \mathsf{M}} C_{\mathcal{M}}$ on A. The abstract logics $\langle A, C_{\mathcal{M}} \rangle$ and $\langle A, C_{\mathsf{M}} \rangle$ are structural.

An abstract logic $\langle A, C \rangle$ is said to be M-adequate if C is the closure operation C_{M} defined on A by M as above. The following theorem is one of the main results of the paper: an abstract logic is structural if and only if it is M-adequate for some class of matrices M.

The paper discusses a different abstract logic on A that can be defined using a logica matrix $\mathcal{M}$. The closure operation is now denoted by $Cn_{\mathcal{M}}^{+}$ and is defined by

$$a \in Cn_{\mathcal{M}}^{+}(X) \text{ iff for every } h \in \mathrm{Hom}(A, B) \ h[X] \subseteq D,$$

$$\text{then for every } h \in \mathrm{Hom}(A, B) \ h(a) \in D.$$

As above, a class of matrices M over algebras of the type of A gives the closure operation $C_{\mathsf{M}}^{+} = \mathrm{Inf}_{\mathcal{M} \in \mathsf{M}} C_{\mathcal{M}}^{+}$ on A. The abstract logics $\langle A, C_{\mathcal{M}}^{+} \rangle$ and $\langle A, C_{\mathsf{M}}^{+} \rangle$ are invariant. Moreover, an abstract logic $\langle A, C \rangle$ is said to be M^{+}-adequate if C is the closure operation C_{M}^{+} defined on A by M. The following theorem is proved in the paper: an abstract logic is invariant if and only if it is M^{+}-adequate for some class of matrices M.

3 Related Work

Abstract logics have been rediscovered again and again by several authors, each one working independently of the others. The book [5] gives exactly the notion of abstract logic in Definition 3.7. The book [13] discusses closure spaces, which are abstract logics over the empty similarity type, and considers functions on their universes with logical behavior w.r.t. the closure operation; therefore, it implicitly considers algebras with a closure operation, that is, abstract logics of nonempty types. The concept of atlas in [7], which is the same as the concept of generalized matrix in [21], is also closely related to that of abstract logic. Finally, abstract logics are closely related to the implication relations considered by A. Koslow in [10] in his abstract approach to logics.

References

1. Bloom, S.L., Brown, D.J., Suszko, R.: A note on abstract logics. Bull. Acad. Polon. Sci. Sér. Sci. Math. Astronom. Phys. **18**, 109–110 (1970)
2. Bloom, S.L., Brown, D.J., Suszko, R.: Some theorems on abstract logics. Algebra Log. **9**, 274–280 (1970)
3. Brown, D.J., Suszko, R.: Abstract logics. Diss. Math. (Rozprawy Mat.) **102** (1973), 52 pp.
4. Bloom, S.L., Brown, D.J.: Classical abstract logics. Diss. Math. **102**, 43–51 (1973)
5. Cleave, J.P.: A Study of Logics. Oxford University Press, Oxford (1991)
6. Czelakowski, J.: Protoalgebraic Logics. Kluwer Academic, Dordrecht (2001)
7. Dunn, J.M., Hardegree, G.M.: Algebraic Methods in Philosophical Logic. Oxford University Press, Oxford (2001)
8. Font, J.M., Jansana, R.: A General Algebraic Semantics for Sentential Logics. Lecture Notes in Logic, vol. 7. Springer-Verlag, Berlin (1996), vi+135 pp. Second revised edition, edited by Association for Symbolic Logic, 2009. Electronic version freely available through Project Euclid at http://projecteuclid.org/euclid.lnl/1235416965
9. Font, J.M., Jansana, R., Pigozzi, D.: A survey of abstract algebraic logic. Abstract algebraic logic, Part II (Barcelona, 1997). Stud. Log. **74**(1–2), 13–97 (2003)
10. Koslow, A.: A Structuralist Theory of Logic. Cambridge University Press, Cambridge (1992)
11. Łoś, J.: On matrycach ligicznych (On logical matrices). Trav. Soc. Sci. Lettres Wrocław, Ser. B **19** (1949)
12. Łoś, J., Suszko, R.: Remarks on sentential logics. Indag. Math. **20**, 177–183 (1958)
13. Martin, V., Pollard, S.: Closure Spaces and Logic. Kluwer, Dordrecht (1996)
14. Smiley, T.: The independence of the connectives. J. Symbol. Log. **27**, 426–436 (1962)
15. Tarski, A.: Über einige fundamentale Begriffe der Metamathematik, C. R. Soc. Sci. Lettr. Varsovie, Cl. III **23**, 22–29 (1930). English translation [19]: On some fundamental concepts of metamathematics, 30–37
16. Tarski, A.: Fundamentale Begriffe der Methodologie der deduktiven Wissenschaften, I. Monatshefte Math. Phys. **37**, 361–404 (1930). English translation in [19]: Fundamental concepts of the methodology of the deductive sciences, 60–109
17. Tarski, A.: Grundzüge der Systemenkalküls. Erster Teil. Fundam. Math. **25**, 503–526 (1935). English translation in [19]: Foundations of the calculus of systems, 342–383
18. Tarski, A.: Grundzüge der Systemenkalküls. Zweiter Teil. Fundam. Math. **26**, 283–301 (1936). English translation in [19]: Foundations of the calculus of systems, 342–383
19. Tarski, A.: Logic, Semantics, Metamathematics. Papers from 1923 to 1938. Hackett Pub., Indianapolis; Corcoran J. (ed.) 2nd edn. (1983)
20. Wójcicki, R.: Logical matrices strongly adequate for structural sentential calculi. Bull. Acad. Polon. Sci. (Ser. Sci. Math. Astronom. Phys.) **17**, 333–335 (1969)
21. Wójcicki, R.: Matrix approach in the methodology of sentential calculi. Stud. Log. **32**, 7–37 (1973)
22. Wójcicki, R.: Theory of Logical Calculi. Basic Theory of Consequence Operations. Kluwer, Dordrecht (1988)

R. Jansana (✉)

Department of Logic, History and Philosophy of Science, University of Barcelona, Barcelona, Spain

e-mail: jansana@ub.edu

SOME THEOREMS ON ABSTRACT LOGICS*

S. L. Bloom, D. J. Brown, and
R. Suszko

It was an idea of A. Lindenbaum that the language of a logical calculus provides an adequate semantical interpretation for this calculus; see Lindenbaum theorem in [2]. In this paper the Lindenbaum idea is applied to structural and invariant logical calculi.

1. Closure Systems and Closure Operations (E. H. Moore, A. Tarski, G. Birkhoff, O. Ore). An operation Cn on the powerset $P(S)$ of the set S is called <u>closure operator</u> (or consequence operation) on S if for all $X, Y \subseteq S$:

$$X \subseteq Cn(X) = Cn(Cn(X)) \text{ and } Cn(X) \subseteq Cn(Y) \quad \text{if } X \subseteq Y.$$

A family $CL \subseteq P(S)$ is said to be a closure system on S if CL is closed under arbitrary intersections, i.e., $\cap F$ is in CL is $F \subseteq CL$. Clearly, $S \in CL$.

(1.1) Every closure system CL on S constitutes a complete lattice under set inclusion where $\text{Inf } X_i \cap X_i$ and

$$\text{Sup}_i X_i = \text{Inf} \{ Z \in CL \mid X_i \subseteq Z \text{ for all } i \}.$$

(1.2) Every closure operator Cn on S defines the closure system $CL^{(Cn)} = \{ X \subseteq S \mid Cn(X) = X \} =$ the family of all Cn closed subsets of S. Conversely, every closure system CL on S determines the closure operator Cn_{CL} defined as follows:

$$Cn_{CL}(X) = \cap \{ Z \in CL \mid X \subseteq Z \}.$$

This correspondence between closure systems and closure operators is one-one, i.e., $CL^{(Cn_{CL})} = CL$ and $Cn_{CL}(Cn) = Cn$.

(1.3) The family of all closure operators on S is a complete lattice under the ordering:

$$Cn_1 \leq Cn_2 \text{ if-if } Cn_1(X) \subseteq Cn_2(X) \quad \text{for all } X \subseteq S.$$

Here, $Cn = \text{Inf}_i Cn_i$ means that $Cn(X) = \cap Cn_i(X)$ for all $X \subseteq S$ and $\text{Sup}_i Cn_i = \text{Inf} \{ Cn \mid Cn_i \leq Cn \text{ for all } i \}$.

(1.4) The family of all closure systems on S is a complete lattice under set inclusion where $\text{Inf}_i CL_i = \{ X \subseteq S \mid X \in CL_i \text{ for all } i \}$ and $\text{Sup}_i CL_i = \text{Inf} \{ CL \mid CL_i \subseteq CL \text{ for all } i \}$.

(1.5) The lattice of all closure systems on S and the lattice of all closure operators on S are dually isomorphic under the correspondence defined in (1.2).

The closure operator Cn on S is called <u>proper</u> if $Cn(a) \neq Cn(b)$ for some $a, b \in S$. Clearly, Cn is proper if and only if $Cn(x) \neq S$ for some $x \in S$.

---

*Presented by the third author at the tenth algebraic colloquium held in memory of A. I. Maltsev at Novosibirsk, September 20-27, 1969. Original article published in English.

Stevens Institute of Technology, U.S.A. Polish Academy of Sciences, Poland. Published in Algebra i Logika, Vol. 9, No. 3, pp. 274-280, May-June, 1970. Original article submitted November 26, 1969.

2. Abstract Logics. Suppose κ is the class of all similar algebras of a given similarity tube τ. Let be the carrier of the algebra A. If A is in κ and Cn is a closure operator on $|A|$ then the system $<A,Cn>$ is called an abstract logic of type τ.[†] The logic $<A,Cn>$ is free, if A is free in κ. It is proper, if Cn is a proper closure operator on $|A|$.

Logical calculi studied in mathematical logic are free logics in our sense. If $L=<A,Cn>$ is a free logic, then Cn is the consequence operation in L, the Cn-closed sets are called Cn-theories, and the least Cn-theory $Cn(\emptyset)$, where $\emptyset$ is the empty set, is the set of tautologies in L; compare [4].

3. Structurality and Invariance. Given an algebra A of type τ, the operation Sb is defined as follows where ε runs over all endomorphisms of A and $x \subseteq |A|$: $Sb(x) = \bigcup_{\varepsilon} \varepsilon(x)$. Clearly, Sb is a closure operator on $|A|$. A set $x \subseteq |A|$ is said to be invariant (with respect to endomorphisms), if $x = Sb(x)$, i.e., x is Sb-closed. If Cn is any closure operator on $|A|$ then the operation $Cn Sb$ on $P(|A|)$ is defined as $Cn Sb(x) = Cn(Sb(x))$ for all $x \subseteq A$. Obviously, $Cn Sb(x) = Cn(x)$ if x is invariant.

A logic $L=<A,Cn>$ is called structural [1] if for every endomorphism ε of A and all $x \subseteq |A|$: $\varepsilon(Cn(x)) \subseteq Cn(\varepsilon(x))$. The logic $L = <A,Cn>$ is called invariant, if for every endomorphism ε of A and all $x \subseteq |A|$: $\varepsilon(Cn(x)) \subseteq Cn(x)$. Obviously, invariance of $L=<A,Cn>$ means that every Cn-closed set is invariant, or by duality, $Sb \leqslant Cn$. One may easily prove the following

LEMMA 1. If the logic $L_i=<A,Cn_i>$ is structural or invariant for every $i \in I$ and $Cn = \underset{i}{Inf} Cn_i$ then $L = <A,Cn>$ is structural or invariant, respectively.

LEMMA 2. If $L = <A,Cn>$ is a structural logic then $Sb(Cn(x)) \subseteq Cn(Sb(x))$ for all $x \subseteq |A|$.

Proof. If $a \in Sb(Cn(x))$ then $a \in \varepsilon(Cn(x))$ for some ε. By structurality, $a \in Cn(\varepsilon(x))$. Hence, $a \in Cn(Sb(x))$.

COROLLARY. If the logic $L=<A,Cn>$ is structural then the set $Cn(\emptyset)$ is invariant.

THEOREM 1. If $L = <A,Cn>$ is a structural logic, then $L' = <A, Cn Sb>$ is an invariant logic and $Cn Sb = Sup(Cn, Sb)$.

Proof. Obviously, $x \subseteq Cn Sb(x)$ and $Cn Sb(x) \subseteq Cn Sb(Y)$ if $x \subseteq Y$. Moreover, $Cn Sb(Cn Sb(x)) = Cn Sb(x)$ since, by the above lemma:

$$Cn(Sb(Cn(Sb(x)))) \subseteq Cn(Cn(Sb(Sb(x)))) = Cn(Sb(x)).$$

Thus $Cn Sb$ is a closure operator. By the lemma again: $Sb(Cn(Sb(x))) \subseteq Cn(Sb(Sb(x))) = Cn(Sb(x))$. Hence, $L'=<A,Cn Sb>$ is an invariant logic. Now we show that

$$(\cdot)\, Cn Sb(x) = x \text{ if and only if } Cn(x) = x = Sb(x).$$

Suppose that $Cn Sb(x) = x$. Then, $Cn(x) = Cn(Cn(Sb(x))) = Cn(Sb(x)) = x$. Moreover, if $a \in Sb(x)$ then $a \in \varepsilon(x)$ for some ε. Hence, $a \in Cn(x)$ and, by structurality, $a \in Cn(\varepsilon(x)) \subseteq Cn(Sb(x)) = x$. Thus $Sb(x) = x$. On the other hand, if $Cn(x) = x = Sb(x)$, then obviously $Cn(Sb(x)) = x$. The statement (•) is proved. It means that: $Cn Sb$-closed sets = invariant Cn-closed sets. We conclude, by duality, that $Cn Sb = Sup(Cn, Sb)$.

4. Logical Matrices and Closure Operators. A system $M=<B,T>$ is called a logical matrix [2] of type τ if B is in κ and $T \subseteq |B|$. The matrix $M=<B,T>$ of type τ generates two closure operators Cn_M and Cn_M^+ on $|A|$ where A is any algebra in κ : (1) $a \in Cn_M(x)$ if and only if for all $h \in Hom(A,B)$: if $h(x) \subseteq T$ then $h(a) \in T$; see [1]. (2) $a \in Cn_M^+(x)$ if and only if: if $h(x) \subseteq T$ for all $h \in Hom(A,B)$ then $g(a) \in T$ also for all $g \in Hom(A,B)$; see [3], p. 204. Clearly, $Cn_M \leqslant Cn_M^+$ and $Cn_M(\emptyset) = Cn_M^+(\emptyset)$. Moreover, one may easily prove the following

LEMMA 3. The logic $<A,Cn_M>$ is structural. The logic $<A,Cn_M^+>$ is invariant.

5. Matrix Adequacy and Lindenbaum Matrices. Let $N = \{M_i\}$ be a class of matrices of type τ and A be an algebra of type τ. Define two closure operators on $|A|$:

$$M\text{-}Cn = \underset{i}{Inf}\, Cn_{M_i} \quad \text{and} \quad M\text{-}Cn^+ = \underset{i}{Inf}\, Cn_{M_i}^+.$$

By Lemmas 1 and 3: $<A,M\text{-}Cn>$ is structural and $<A, N\text{-}Cn^+>$ is invariant.

[†] An extensive paper on abstract logics, by the second and third authors, is under preparation.

A logic $L = <A, Cn>$ of τ is said to be M-adequate or M^+-adequate if correspondingly:

$$Cn = M \sim Cn \quad \text{or} \quad Cn = M \sim Cn^+.$$

If $L = <A, Cn>$ is a logic then every matrix $M_T = <A, T>$ where T is any Cn-closed set, is called **Lindenbaum matrix** for L. The logic L and all the matrices M_T are of the same type. The family $M_L = \{M_T\}_{T \neq |A|}$ is said to be the **canonical family** of matrices for L.

THEOREM 2 (Wojcicki [5]). If $L = <A, Cn>$ is a structural logic, then L is M_L-adequate.

Proof. Wojcicki formulated his theorem for free structural logics. His proof, however, applies to any structural abstract logic. We repeat his proof. Suppose that $L = <A, Cn>$ is structural. We have to show that: $a \in Cn(x)$ if and only if for every Cn-closed set $T \subset |A|$ and each endomorphism ε of A if $\varepsilon(x) \subseteq T$ then $\varepsilon(a) \in T$. If $a \in Cn(x)$ then $\varepsilon(a) \in Cn(\varepsilon(x))$ by structurality. Hence, if $\varepsilon(x) \subseteq T = Cn(T)$ then $Cn(\varepsilon(x)) \subseteq T$ and, finally, $\varepsilon(a) \in T$. Conversely, if $a \notin Cn(x)$, then set $T = Cn(x) \subset |A|$ and ε = the identity map. Obviously, $\varepsilon(x) \subseteq T$ and $\varepsilon(a) \notin T$.

THEOREM 3. If $L = <A, Cn>$ is an invariant logic, then L is M_L^+-adequate.

Proof. We have to show that: $a \in Cn(x)$ if and only if for every Cn-closed set $T \subset |A|$ if $Sb(x) \subseteq T$ then $Sb(a) \subseteq T$. If $a \in Cn(x)$ then $Sb(a) \subseteq Cn(x)$ by invariance. Hence, if $Sb(x) \subseteq T = Cn(T)$ then $Cn(x) \subseteq Cn(Sb(x)) \subseteq Cn(T) = T$ and finally, $Sb(a) \subseteq T$. Conversely, if $a \notin Cn(x)$ then set $T = Cn(x) \subset |A|$. Obviously, it is not true that $Sb(a) \subseteq T$. However, by invariance, $Sb(x) \subseteq T$.

THEOREM 4. A logic L is a structural if and only if L is M-adequate for some family of matrices M.

THEOREM 5. A logic L is invariant if and only if L is M^+-adequate for some family of matrices M.

6. **Free Logics.** The last theorems when applied to free logics (i.e., logical calculi) have quite clear semantical meaning. If $L = <A, Cn>$ is a free logic of type τ, then every matrix $M = <B, T>$ of type τ is an interpretation of L and the homomorphisms of A and B are so called valuations of L in M. If h is in $\text{Hom}(A, B)$, then the inverse image $h(T) = STF_h(M) =$ the set of all elements (formulas) in $|A|$ satisfied by h in M and the intersection $\bigcap_h STF_h(M) = TR(M) =$ the set of all elements (formulas) in $|A|$ which are true in M.

To say that the free logic $L = <A, Cn>$ is M-adequate means that: $a \notin Cn(x)$ if and only if there is a matrix $M = <B, T>$ in M and h in $\text{Hom}(A, B)$ such that $x \subseteq STF_h(M)$ and $a \notin STF_h(M)$ On the other hand, to say that the free logic $L = <A, Cn>$ is M-adequate means that: $a \notin Cn(x)$ if and only if there is a matrix $M = <B, T>$ in M such that $x \subseteq TR(M)$ and $a \notin TR(M)$.

It is clear that M-adequacy or M^+-adequacy of a free logic express a **semantical completeness** property of this logic. Observe that if $L = <A, Cn>$ is M-adequate or M^+-adequate then the set of tautologies $= Cn(\emptyset) = \bigcap_{M \text{ in } M} TR(M) =$ the set of all elements (formulas) in $|A|$ which are **valid** in M. The converse is not true, in general.

LEMMA 4. If the logic $L = <A, Cn>$ is structural, invariant, and free, with at least two generators for A, then L is not proper.

Proof. Let g_1, g_2 be two different generators for A. If a_1, a_2 are in $|A|$ then, since A is free in K, there is an endomorphism ε of A such that $\varepsilon(g_1) = a_1$ and $\varepsilon(g_2) = a_2$. Clearly, $g_2 \in Sb(g_1)$. By invariance, $g_2 \in Cn(g_1)$ and, by structurality, $\varepsilon(g_2) \in Cn(\varepsilon(g_1))$. Thus, $a_2 \in Cn(a_1)$. But a_1, a_2 are arbitrary elements in $|A|$. Hence, Cn is not proper.

MAIN THEOREM. Every proper free logic $L = <A, Cn>$, with at least two generators for A, has one of the following mutually exclusive properties:

(1) L is structural and M-adequate for some family M;

(2) L is invariant and M^+-adequate for some family M;

(3) L is neither structural nor invariant and there does not exist a family M such that L is M-adequate or M^+-adequate.

LITERATURE CITED

1.   J. Los and R. Suszko, "Remarks on Sentential Logics," Proc. Kon. Neth. Akad. van Weten., Ser. A, 67 (1958), pp. 177-183.
2.   J. Lukasiewicz and A. Tarski, Untersuchungen über den Aussagenkalkül, Comptes Rendus Soc. Sci. Varsovie, cl. III, 23 (1930), pp. 30-50.
3.   R. Suszko, "Concerning the method of logical schemes, the notion of logical calculus, and the role of consequence relations," Studia Logica, 11, 185-214 (1961).
4.   A. Tarski, Über einige fundamentalen Begriffe der Metamathematik, Comptes Rendus Soc. Sci. Varsovie, cl. III, 23 (1930), pp. 22-29.
5.   R. Wohcicki, "LogicalMatrices strongly adequate for structural sentential calculi," Bull. Acad. Polon. Sci. (ser. Sci Math. Astronom. Phys.), 17, 333-335 (1969).

S.L. Bloom (✉)

Computer Science Department, Stevens Institute of Technology, Hoboken, NJ, USA
e-mail: bloom@cs.stevens.edu

Part 12
Dana Scott (1974)

Dana Scott's Work with Generalized Consequence Relations

Lloyd Humberstone

Abstract Scott's elegant abstract approach to logical matters, summarized in the opening section of 'Completeness and Axiomatizability in Many-Valued Logic'—as well as in other publications of his from around the same period (the 1970s)—is presented and discussed here without too much attention to the treatment of many-valued logic suggested in that work. However, we do touch on this and its relations to some later work, after the general discussion.

Mathematics Subject Classification (2000) 03B05 · 03B50

After its opening historical and autobiographical remarks, Dana Scott's 'Completeness and Axiomatizability in Many-Valued Logic'—from now on: C & A—outlines in Sect. 1 an elegant approach to logical issues in general and then applies this to the particular field of many-valued logic *à la* Łukasiewicz (though without negation) in Sects. 2–5. The present introductory companion concentrates on Sect. 1 of C & A, in which this general approach is expounded, though some brief remarks on the business of the later sections are included at the end (i.e., before 'Minor Qualms and Further Reading'). The general approach in question, further elaborated in various other publications of Scott's, such as [30, 31], and [33], has found its way into several monographs by those who had come under Scott's influence—such as Gabbay [15, 16], Segerberg [36]—to say nothing of numerous individual papers by these and other authors. For reasons that will shortly become clear, our title, though convenient, is not quite accurate, since Scott does not actually work with generalized (sometimes called "multiple-conclusion") consequence relations proper, but with a certain 'finitized' version of these.

To reduce novelty, Scott's notation will be followed: italic capitals for formulas (or sentences) and script capitals for sets thereof. But for consequence relations and variations thereof it will be helpful to employ a different turnstile notation, however, reserving "$\Vdash$" (variously decorated) for use in the case in which multiple and empty right-hand sides are permitted and "$\vdash$" (again perhaps with decorations) when only a single formula appears to the right. Scott, by contrast, uses both of these for the former case, and discusses the latter only, in terms of consequence *operations* ("Cn") rather than consequence relations. A contrast between the relational versions will make for greater clarity.[1] So let us define

---

[1] The connection between the two formulations is given in (1) of §1 of C & A. At p. 287 of the paper—all such page references being to the reprinting of the paper in present anthology—Scott says that he feels it would be wrong to call a relation satisfying (in our notation below) the conditions (R)$_{\Vdash}$, (M)$_{\Vdash}$ and

the relevant notions for this case first and then proceed to the generalization allowing a set of formulas instead of a single formula as the second 'relatum'.

For a language $\mathscr{L}$ (here identified with its set of formulas), a *consequence relation* on $\mathscr{L}$ is a relation $\vdash\, \subseteq \wp(\mathscr{L}) \times \mathscr{L}$ satisfying the following three conditions,[2] for all $\mathscr{A}, \mathscr{A}' \subseteq \mathscr{L}, A, B \in \mathscr{L}$:

(R)$_\vdash$ $A \vdash A$;

(M)$_\vdash$ If $\mathscr{A} \vdash B$ then $\mathscr{A}, \mathscr{A}' \vdash B$;

(T)$_\vdash^+$ For every $\mathscr{C} \subseteq \mathscr{L}$: If $\mathscr{A} \vdash C$ for each $C \in \mathscr{C}$ and $\mathscr{A}, \mathscr{C} \vdash B$, then $\mathscr{A} \vdash B$.

Note the conventions above, common in logical work: for (R)$_\vdash$, "$A \vdash A$" has been written, meaning "$\{A\} \vdash A$"; "$\mathscr{A}, \mathscr{A}'$" means "$\mathscr{A} \cup \mathscr{A}'$"; and so on.

Similarly, a relation $\Vdash\, \subseteq \wp(\mathscr{L}) \times \wp(\mathscr{L})$ is a *generalized consequence relation* (on $\mathscr{L}$) when the three conditions below[3]—stated using the same notational conventions— are met, for all $\mathscr{A}, \mathscr{A}', \mathscr{B}, \mathscr{B}' \subseteq \mathscr{L}, A \in \mathscr{L}$:

(R)$_\Vdash$ $A \Vdash A$;

(M)$_\Vdash$ If $\mathscr{A} \Vdash \mathscr{B}$ then $\mathscr{A}, \mathscr{A}' \Vdash \mathscr{B}, \mathscr{B}'$;

(T)$_\Vdash^+$ For every $\mathscr{C} \subseteq \mathscr{L}$: If $\mathscr{A}, \mathscr{C}_0 \Vdash \mathscr{C}_1, \mathscr{B}$ for every $\mathscr{C}_0, \mathscr{C}_1$, for which $\mathscr{C}_0 \cup \mathscr{C}_1 = \mathscr{C}$ and $\mathscr{C}_0 \cap \mathscr{C}_1 = \varnothing$, then $\mathscr{A} \Vdash \mathscr{B}$.

A consequence relation $\vdash$ (generalized consequence relation $\Vdash$) is said to be *finitary* if whenever $\mathscr{A} \vdash B$, we have $\mathscr{A}_0 \vdash B$, for some finite $\mathscr{A}_0 \subseteq \mathscr{A}$ (resp. whenever $\mathscr{A} \Vdash \mathscr{B}$, we have $\mathscr{A}_0 \Vdash \mathscr{B}_0$ for some finite $\mathscr{A}_0 \subseteq \mathscr{A}, \mathscr{B}_0 \subseteq \mathscr{B}$). If we were simply defining finitary consequence or generalized consequence relations, the defining conditions (T)$_\vdash^+$ and (T)$_\Vdash^+$ could be simplified, in effect restricting attention to unit sets playing the role indicated by "$\mathscr{C}$" in those conditions; we call the corresponding simplified conditions (T)$_\vdash$ and (T)$_\Vdash$, and write them in "rule notation", i.e., interpreted as saying that if the conditions above the horizontal line are satisfied so is that below the line:

$$(T)_\vdash \quad \frac{\mathscr{A} \vdash C \quad\quad \mathscr{A}, C \vdash B}{\mathscr{A} \vdash B} \quad\quad\quad (T)_\Vdash \quad \frac{\mathscr{A} \Vdash C, \mathscr{B} \quad\quad \mathscr{A}, C \Vdash \mathscr{B}}{\mathscr{A} \Vdash \mathscr{B}}.$$

(T)$_\Vdash$ a consequence relation because it is "only in special cases that we can realize" such a relation "by formal manipulation". However, the terminology objected to here is not "deducibility relation" or "derivability relation"—which do indeed have such syntactic connotations, but "consequence relation"—and that terminology has no such connotations (semantic consequence, etc.). The best reason for not calling such relations consequence relations is that a consequence relation relates a set of formulas to a formula, not to another set of formulas. Hence "generalized consequence relation"—generalizing from unit sets of formulas to arbitrary sets of formulas (though in Scott's own case, these are specifically arbitrary *finite* sets). Note in particular that a generalized consequence relation is not a (kind of) consequence relation, any more than, for example, a generalized Boolean algebra is a (kind of) Boolean algebra. A more subtle point: although generalized consequence relations are more general than consequence relations in not restricting the number of formulas 'on the right' to 1, by contrast with the generalized Boolean algebra example, consequence relations themselves are not a kind of generalized consequence relation.

[2]The etymology of these labels is given after their first appearance (p. 284) in C & A. Scott applies the labels only in the case of generalized consequence relations, as defined below, and gives a formulation of (R)—dropping the subscript $\Vdash$ to match C & A's notation—which, unlike that given here, incorporates some aspects of (M).

[3]The third condition is conspicuously missing from p. 37 of Segerberg [36], where it is replaced by two conditions shown in Theorems 2.6, 2.7, of Shoesmith and Smiley [37] to be strictly weaker, when taken together, than that condition.

As Scott says, these are essentially versions of the Cut rule—with C as the 'cut formula' (the formula cut out, on passage from the premises to the conclusion)[4]—encountered in sequent calculus proof systems of the kind originated by Gentzen, for the latter's intuitionistic and classical systems respectively (*LJ* and *LK*, as Gentzen called them).[5] In fact Scott works not with finitary generalized consequence relations but with a variation thereon (under names like "entailment relation" and "conditional assertion relation") in which the sets of formulas on the left and right are themselves required to be *finite*;[6] it is for this reason that the title 'Scott's Work with Generalized Consequence Relations' is (as remarked earlier) not accurate. However, the main role played by this restriction is to secure that the associated generalized consequence relations—which Scott also touches on in passing—are *finitary*, as defined above; since we can always restrict attention to the finitary case when desired, it is better not to insist that the sets related by ⊩ themselves be finite. (Later, we shall encounter a general theoretical consideration for not imposing this restriction: see the paragraph with inset (a) and (b) below.[7])

Adapting Scott's terminology, a *valuation* for a language is a function assigning to each formula of the language one or other of the truth-values $\mathbf{t}$, $\mathbf{f}$, presumed distinct, and we say that a valuation v is *consistent with* a consequence relation ⊢ just in case $v(A) = \mathbf{t}$ for each $A \in \mathscr{A}$ implies $v(B) = \mathbf{t}$, whenever $\mathscr{A} \vdash B$, and that v is *consistent with* a generalized consequence relation ⊩ just in case $v(A) = \mathbf{t}$ for each $A \in \mathscr{A}$ implies that for some $B \in \mathscr{B}$, $v(B) = \mathbf{t}$, whenever $\mathscr{A} \Vdash \mathscr{B}$. In C & A, and other publications of the same vintage and orientation [30, 31, 33] Scott stresses this notion in connection with generalized consequence relations rather than consequence relations, in which setting there is an elegant even-handedness (duality) between $\mathbf{t}$ and $\mathbf{f}$. To see this in action let us rehearse the Lindenbaum style proof of the result usually called Lindenbaum's Lemma as it applies to finitary generalized consequence relations ⊩ on a language with countably many formulas, of which we take $C_1, \ldots, C_n, \ldots$ to be some fixed enumeration.[8] According to

---

[4]We have done some re-lettering here so that "C" can be used to suggest "Cut", whereas in C & A, "B" is used in the corresponding formulation. Note that a similar 'rule' formulation of the condition $(T)^+_{\Vdash}$ would be highly contentious, since it would have not only infinitely many 'premises' but (in the case of a language of the most commonly encountered kind, with countably many formulas) *uncountably* many of them.

[5]One should replace the metalinguistic predicate symbols "⊢" and "⊩" by proper sequent-separators to turn the conditions here into schematic figures for sequent-to-sequent rules, and note that the conception of sequent involved does not attend to the order in which formulas are written (on a given side of the separator) or to the number of times a formula occurs—a departure from Gentzen's practice. Nor is the association alluded to quite exact, since while Gentzen's framework for classical logic allows empty and multiple—rather than just singleton—succedents, his treatment for intuitionistic logic allows empty as well as singleton succedents (whereas "$\mathscr{A} \vdash \varnothing$" does not make sense for ⊢ a consequence relation).

[6]In this case we could put the above requirements in the form of $(R)_{\Vdash}$ and a simple combination of (M) and (T) style conditions—for all C: $\mathscr{A} \Vdash \mathscr{B}$ iff $\mathscr{A}, C \Vdash \mathscr{B}$ and $\mathscr{A} \Vdash C, \mathscr{B}$.

[7]There are also *specific* reasons arising from semantic considerations when certain logics are treated as consequence relations or generalized consequence relations for which the finitariness condition is not satisfied. To illustrate this from Scott's own work, let us cite [35] and [27].

[8]Scott imposes no such cardinality restrictions and for his proof of the result accordingly uses instead an appeal to Zorn's Lemma—as well as putting matters in terms of extensions of ⊩ rather than of $\mathscr{A}$ and $\mathscr{B}$ (on which see below); the present version for countable languages is common in the literature, though, appearing for example in the proof of Lemma 1.4 of Gabbay [15] (where, however, there is a slight complication caused by the fact that predicate logic rather than propositional logic is being treated).

this result, given any such $\Vdash$, whenever for sets of formulas $\mathscr{A}$, $\mathscr{B}$ of the language of $\Vdash$, we have $\mathscr{A} \not\Vdash \mathscr{B}$, then there is some valuation v for that language with (1) $v(A) = \mathbf{t}$ for all $A \in \mathscr{A}$, (2) $v(B) = \mathbf{f}$ for all $B \in \mathscr{B}$, and (3) v is consistent with $\Vdash$. One puts $\mathscr{A}_0 = \mathscr{A}$, $\mathscr{B}_0 = \mathscr{B}$, and, beginning with the pair $\langle \mathscr{A}_0, \mathscr{B}_0 \rangle$, defines the sequence of pairs:

$$\langle \mathscr{A}_{i+1}, \mathscr{B}_{i+1} \rangle = \begin{cases} \langle \mathscr{A}_i \cup \{C_{i+1}\}, \mathscr{B}_i \rangle & \text{if } \mathscr{A}_i, C_{i+1} \not\Vdash \mathscr{B}_i \\ \langle \mathscr{A}_i, \mathscr{B}_i \cup \{C_{i+1}\} \rangle & \text{if } \mathscr{A}_i, C_{i+1} \Vdash \mathscr{B}_i. \end{cases}$$

Now we can define a valuation v by putting $v(C) = \mathbf{t}$ if $C \in \mathscr{A}^+ = \bigcup_{i \in \mathbb{N}} \mathscr{A}_i$ and $v(C) = \mathbf{f}$ if $C \in \mathscr{B}^+ = \bigcup_{i \in \mathbb{N}} \mathscr{B}_i$. This is indeed a valuation, since each formula gets into one or other of the unions just mentioned, as C is C_{i+1} for some $i \in \mathbb{N}$, and so was added either to $\mathscr{A}_i$ to obtain $\mathscr{A}_{i+1}$ or else to $\mathscr{B}_i$ to obtain $\mathscr{B}_{i+1}$,[9] and no formula gets added to both sides for the following reason: this would contradict (given conditions $(R)_{\Vdash}$ and $(M)_{\Vdash}$) the fact that $\mathscr{A}^+ \not\Vdash \mathscr{B}^+$.

But how do we know that this *is* a fact? The explanation is facilitated by adopting another of Scott's uses of the term "consistent": call a pair $\langle \mathscr{C}, \mathscr{D} \rangle$ of sets of formulas *consistent* (more explicitly: $\Vdash$-*consistent*) when $\mathscr{C} \not\Vdash \mathscr{D}$. Then by induction on n we can see that all the pairs $\mathscr{A}_n, \mathscr{B}_n$ are consistent; the basis ($n = 0$) case is given by the starting assumption that $\mathscr{A} \not\Vdash \mathscr{B}$. For the inductive step we check that consistency is preserved on passing from $\langle \mathscr{A}_i, \mathscr{B}_i \rangle$ to $\langle \mathscr{A}_{i+1}, \mathscr{B}_{i+1} \rangle$ in accordance with the instructions inset above. If the first instruction was followed, C_{i+1} was added to $\mathscr{A}_i$, then the condition securing the consistency of $\langle \mathscr{A}_{i+1}, \mathscr{B}_{i+1} \rangle$ had to have been satisfied. If the second instruction was followed, and C_{i+1} was added to $\mathscr{B}_i$, this was because (a) $\mathscr{A}_i, C_{i+1} \Vdash \mathscr{B}_i$. In this case, for $\langle \mathscr{A}_{i+1}, \mathscr{B}_{i+1} \rangle$ to fail to be consistent is for it to be the case that (b) $\mathscr{A}_i \Vdash C_{i+1}, \mathscr{B}_i$. But then by $(T)_{\Vdash}$, (a) and (b) yield the conclusion that $\mathscr{A}_i \Vdash \mathscr{B}_i$, contradicting the inductive hypothesis (i.e., the supposition that $\langle \mathscr{A}_i, \mathscr{B}_i \rangle$ was consistent).

It still remains to be checked that the 'limit pair' $\langle \mathscr{A}^+, \mathscr{B}^+ \rangle$ is consistent, and this is where the assumption of finitariness for $\Vdash$ comes in.[10] Had it been the case that $\mathscr{A}^+ \Vdash \mathscr{B}^+$, then by that condition there would be finite subsets $\mathscr{A}^-, \mathscr{B}^-$, of $\mathscr{A}^+, \mathscr{B}^+$ (respectively), with $\mathscr{A}^- \Vdash \mathscr{B}^-$. But since only finitely formulas are involved here, all are included in some initial segment $C_1, \ldots, C_n$, say, of the enumeration of all formulas, in which case we would have $\mathscr{A}_n \Vdash \mathscr{B}_n$, contradicting the finding above that each of these pairs, including $\langle \mathscr{A}_n, \mathscr{B}_n \rangle$, is consistent.[11] The consistency of $\langle \mathscr{A}^+, \mathscr{B}^+ \rangle$, just established, amounts to the consistency of the valuation v (defined above) with $\Vdash$. And since, recalling that the original $\mathscr{A}$ and $\mathscr{B}$ (for which we supposed $\mathscr{A} \not\Vdash \mathscr{B}$) are subsets of $\mathscr{A}^+$ and $\mathscr{B}^+$, we have been led to a valuation consistent with $\Vdash$ which verifies all of $\mathscr{A}$ and falsifies all of $\mathscr{B}$. Since beyond the supposition that $\mathscr{A} \not\Vdash \mathscr{B}$, $\mathscr{A}$ and $\mathscr{B}$ were arbitrary, this shows that $\Vdash$ is determined by the class of all valuations consistent with $\Vdash$, where to say that a generalized consequence relation $\Vdash$ is *determined by* a class of valuations $\mathscr{V}$ is

---

[9]The term "added", here, is to be taken with something of a pinch of salt since C_{i+1} may already belong to $\mathscr{A}_i$ or to $\mathscr{B}_i$, respectively, in which case no change is made. We continue to use this term for convenience, though.

[10]The condition of finitariness can be weakened (see Example 0.1.1 of [19]) though it cannot simply be dropped.

[11]Note that this remains the case even if the initial $\mathscr{A}$ and $\mathscr{B}$ (alias $\mathscr{A}_0$ and $\mathscr{B}_0$) were themselves infinite.

to say that for all $\mathscr{A}$, $\mathscr{B}$:

$$\mathscr{A} \Vdash \mathscr{B} \Leftrightarrow \forall v \in \mathscr{V}, \quad \text{if } \forall A \in \mathscr{A}, v(A) = \mathbf{t} \quad \text{then } \exists B \in \mathscr{B}, v(B) = \mathbf{t}.$$

Especially if $\Vdash$ has been presented syntactically (i.e., as the least generalized consequence relation satisfying certain formal conditions) we may think of the $\Rightarrow$-direction of this $\Leftrightarrow$-statement as asserting the *soundness* and the $\Leftarrow$-direction as asserting the *completeness* of the syntactic presentation (proof-system[12]) with respect to a semantics provided by $\mathscr{V}$. When we take $\mathscr{V}$ as the class of all valuations consistent with $\Vdash$ we have secured the soundness half by definition, and the above Lindenbaum argument (or the streamlined version in C & A) establishes the completeness half.

There are numerous variations on the above Lindenbaum construction which would have served equally well and remained within the same general approach. For example, we could have 'biased' the addition of the C_i in the opposite way, defining the sequence of pairs $\langle \mathscr{A}_i, \mathscr{B}_i \rangle$ with the same starting point but proceeding via the instructions:

$$\langle \mathscr{A}_{i+1}, \mathscr{B}_{i+1} \rangle = \begin{cases} \langle \mathscr{A}_i, \mathscr{B}_i \cup \{C_{i+1}\} \rangle & \text{if } \mathscr{A}_i \nVdash \mathscr{B}_i, C_{i+1} \\ \langle \mathscr{A}_i \cup \{C_{i+1}\}, \mathscr{B}_i \rangle & \text{if } \mathscr{A}_i \Vdash \mathscr{B}_i, C_{i+1}. \end{cases}$$

Essentially the same reasoning as in the case worked through would apply here, too. Or we could have gone for something more even-handed:

$$\langle \mathscr{A}_{i+1}, \mathscr{B}_{i+1} \rangle = \begin{cases} \langle \mathscr{A}_i \cup \{C_{i+1}\}, \mathscr{B}_i \rangle & \text{if } \mathscr{A}_i, C_{i+1} \nVdash \mathscr{B}_i \\ \langle \mathscr{A}_i, \mathscr{B}_i \cup \{C_{i+1}\} \rangle & \text{if } \mathscr{A}_i \nVdash \mathscr{B}_i, C_{i+1}. \end{cases}$$

For this version the division of labor amongst $(R)_{\Vdash}$, $(M)_{\Vdash}$ and $(T)_{\Vdash}$ would be rather different. For example, $(T)_{\Vdash}$ now needs to be appealed to show that when its turn comes up—i.e. in passing from $\langle \mathscr{A}_i, \mathscr{B}_i \rangle$ to $\langle \mathscr{A}_{i+1}, \mathscr{B}_{i+1} \rangle$—the formula C_{i+1} gets added to the $\mathscr{A}_i$ side or else to the $\mathscr{B}_i$ side: something that was automatic on the earlier versions.

What would definitely *not* be consilient with Scott's approach, by contrast, would be to reach for $\neg$ (negation) and do the Lindenbaum construction by adding $\neg C_{i+1}$ to $\mathscr{A}_i$ instead of adding C_{i+1} itself to $\mathscr{B}_i$ (or vice versa). The procedures we have been describing abstract from the issue of what connectives the set of formulas of $\mathscr{L}$ is closed under, or what logical behavior the given $\Vdash$ prescribes for them.[13] If we want classical negation, we can add suitable conditions on $\Vdash$, which in their simplest form would be the following:

$$A, \neg A \Vdash \quad (\text{i.e. } A, \neg A \Vdash \varnothing) \quad \text{and} \quad \Vdash A, \neg A.$$

---

[12]Should one use the term *axiomatization* in this context? This is what "axiomatizability" alludes to in the title of C & A, though for me it remains too closely associated with the Frege–Hilbert 'logics as sets of formulas' approach to be suitable.

[13]Indeed in C & A, Scott emphasizes (with examples like that at the bottom of p. 287) that the pairs of sets related by $\Vdash$ need not even be sets of formulas from anything normally regarded as a language at all. For such reasons, the phrase "generalized closure relation" is used in place of "generalized consequence relation" in the present author's [19]. On the desirability of presenting as much as possible in abstraction from the details of various connectives, such as $\neg$, Scott remarks in [31], p. 255f.: "If, however, the rôles of the various connectives are to be investigated, it seems better to analyze out the connecting glue from the connective patchwork; and that is what $\vdash$ [$= \Vdash$ here] neatly accomplishes." See also the paragraph in C & A after the proof of Proposition 1.3, beginning "It is clear…"

These conditions are clearly satisfied by the generalized consequence relation determined by any class of valuations v satisfying the 'Boolean' condition that for all $A \in \mathscr{L}$, $v(\neg A) = \mathbf{t}$ if and only if $v(A) = \mathbf{f}$, and the valuation v extracted from the Lindenbaum construction (or Scott's high-speed version using Zorn's Lemma) will in turn satisfy this condition, which amounts to saying that $\neg A \in \mathscr{A}^+$ if and only if $A \in \mathscr{B}^+$.[14] Indeed, *every* valuation consistent with a $\Vdash$ satisfying the above $\neg$ principles must similarly associate the usual negation truth-function with $\neg$.[15] Analogous conditions for $\wedge$ and $\vee$ are as follows, understood again as holding for all $A, B \in \mathscr{L}$:

$$A, B \Vdash A \wedge B, \quad A \wedge B \Vdash A, \quad A \wedge B \Vdash B, \quad A \vee B \Vdash A, B, \quad A \Vdash A \vee B, \quad B \Vdash A \vee B.$$

These six conditions are equivalent (for generalized consequence relations $\Vdash$) to the two two-way *conditional* conditions formulated as rules ($\wedge$) and ($\vee$) on p. 293, but the present formulation is more convenient (and often used by Scott elsewhere—e.g., [31], p. 254) for seeing that any $\Vdash$ satisfying them must associate the usual conjunction and disjunction truth-functions with $\wedge$ and $\vee$ respectively, and also that any $\Vdash' \supseteq \Vdash$ must satisfy them if $\Vdash$ does. A common observation, which goes back to Carnap,[16] is that for consequence relations (i.e., putting "$\vdash$" for "$\Vdash$") the fourth of the above conditions does not make sense (since we have more than one formula represented on the right—except for applications in which $A = B$), and neither does either of the two conditions given above for $\neg$ (in the case of the first, because there are too *few* formulas on the right, rather than too many). There are no similarly syntactic conditions—conditional or otherwise—on consequence relations which *force* valuations consistent with them to associate the standard disjunction and negation truth-functions with $\vee$ and $\neg$, and so simply providing a valuation consistent with $\vdash_{CL}$, the consequence relation of classical propositional logic, verifying all of $\mathscr{A}$ and falsifying B for arbitrary $\mathscr{A}, B$, for which $\mathscr{A} \nvdash_{CL} B$, will not suffice to show the completeness of $\vdash_{CL}$ w.r.t. the class of Boolean valuations. (For example, the valuation $v_{\mathscr{A}}$ defined by: $v_{\mathscr{A}}(C) = \mathbf{t}$ iff $\mathscr{A} \vdash C$ is always consistent with $\vdash$, verifies all formulas in $\mathscr{A}$ but not C, but will typically not associate the disjunction and negation truth-functions with $\vee$ and $\neg$, even when $\vdash$ is $\vdash_{CL}$.) Of course for this to constitute something in need of being shown, we have to think of $\vdash_{CL}$ as being defined syntactically in the first place— as opposed to be being defined as the consequence relation determined by the class of Boolean valuations;[17] to make up for the loss of the "$A \vee B \Vdash A, B$" condition, we will

---

[14] A and $\neg A$ cannot both lie in $\mathscr{A}^+$, or the first inset condition on $\neg$, together with $(M)_{\Vdash}$, would imply $\mathscr{A}^+ \Vdash \mathscr{B}^+$, and they cannot both lie in $\mathscr{B}^+$ for similar reasons, but invoking the second of the two conditions.

[15] A valuation v (for $\mathscr{L}$) associates the n-ary truth-function f with an n-ary connective # (of $\mathscr{L}$) just in case for all formulas $A_1, \ldots, A_n$ (of $\mathscr{L}$), $v(\#(A_1, \ldots, A_n)) = f(v(A_1), \ldots, v(A_n))$.

[16] References and fuller discussion of this issue may be found in §2 of the author's [21].

[17] For an example of such a syntactic characterization, we can take $\vdash_{CL}$ in connectives $\wedge$, $\vee$, $\neg$, to be the least consequence relation satisfying the $\vdash$ analogues of the first, second, third, fifth and sixth $\Vdash$-conditions on conjunction and disjunction, with the fourth replaced by the conditional condition given below, together with conditions $A, \neg A \vdash B$ and: if $\mathscr{A}, B \vdash C$ and $\mathscr{A}, \neg B \vdash C$, then $\mathscr{A} \vdash C$. Of course this last condition could be replaced by the unconditional form $\vdash B \vee \neg B$, but we have given conditions governing a single connective each, so as to make for easier isolation of fragments, and so on.

need to impose the condition one would express in the 'rule' notation as:

$$\frac{\mathscr{A}, A \vdash C \qquad \mathscr{A}, B \vdash C}{\mathscr{A}, A \vee B \vdash C}$$

which will allow a Lindenbaum style argument for $\vdash$ (adapting that for $\Vdash$ above) to extend any $\mathscr{B}$ for which $\mathscr{B} \nvdash D$ for some formula D to a $\mathscr{B}^+$ which contains at least one disjunct of every disjunction it contains ("primeness", as this property is often called) and which is still such that $\mathscr{B}^+ \nvdash D$.[18] Because, however, the above condition is itself conditional in form, it is not guaranteed to be inherited by consequence relations extending a given $\vdash$ satisfying it (by contrast with the "$A \vee B \Vdash A, B$" condition and generalized consequence relations).

The above contrast between $\wedge$ on the one hand and $\vee$ (along with $\rightarrow$ and $\neg$) on the other is put in Gabbay [15] in terms of strong and weak classicality, though instead of defining strong classicality as a matter of a given truth-function's being associated with the connective on every $\vdash$-consistent valuation and weak classicality as being determined by some class of valuations on each of which some one such association obtains, Gabbay's definition—which will not be reproduced here (see [15], 12–14)—quantifies over Scott-style 'finitized' generalized consequence relations agreeing with the consequence relation $\vdash$ under consideration, thereby invoking another idea from C & A (p. 285, though this terminology is from Gabbay): for any choice of $\mathscr{L}$, a generalized consequence relation $\Vdash \subseteq \wp(\mathscr{L}) \times \wp(\mathscr{L})$ *agrees with* a consequence relation $\vdash \subseteq \wp(\mathscr{L}) \times \mathscr{L}$ just in case for all $\mathscr{A} \subseteq \mathscr{L}, B \in \mathscr{L}$:

$$\mathscr{A} \Vdash B \quad \text{if and only if } \mathscr{A} \vdash B.$$

When attention is restricted to finitary generalized consequence relations *à la* C & A (roughly speaking), Scott shows how to define, given $\vdash$, a minimum and maximum $\Vdash$ agreeing with $\vdash$; that is, calling these $\Vdash_{min}$ and $\Vdash_{max}$, respectively, we have $\Vdash_{min} \subseteq \Vdash \subseteq \Vdash_{max}$ for all $\Vdash$ agreeing with the given $\vdash$.[19] An intriguing point: according to Shoesmith and Smiley [37], if the restriction to finitary (or as they say, compact) generalized consequence relations is lifted, there is still, for any consequence relation, a *minimum* agreeing generalized consequence relation but not necessarily a *maximum* such relation (a maximum 'counterpart', in the terminology of [37]: see Theorem 5.7 therein).

As mentioned in note 8, for the Lindenbaum-related completeness argument reviewed above, Scott himself considers in C & A, for the proof of Proposition 1.3 there, a sequence of extensions of the generalized consequence relation $\Vdash$[20] for which we are supposing that

---

[18] The idea is that the envisaged $\mathscr{B}^+$ will be maximal in not having D as a $\vdash$-consequence, and that the valuation assigning **t** to precisely its elements will associate with $\vee$ the standard (inclusive) disjunction truth-function since the primeness of $\mathscr{B}^+$ will guarantee that this valuation verifies a disjunction only if it verifies at least one disjunct. (The corresponding "if" version of this condition will be secured by the more straightforward route of having disjunctions as $\vdash$-consequences of their disjuncts—an *unconditional* condition on $\vdash$.)

[19] Reminder: this is not Scott's notation in Theorem 1.2 of C & A, where "$\vdash$" is used alongside "$\Vdash$" for generalized consequence relations, and all information about consequence relations is packaged in terms of the corresponding consequence operation Cn.

[20] Since Scott uses not only "$\Vdash$", variously decorated, for (his finite version of) generalized consequence relations, but also "$\vdash$", reserved here for consequence relations themselves, his formulation is that each

$\mathscr{A} \not\Vdash \mathscr{B}$, rather than, as above, a sequence of pairs $\langle \mathscr{A}_i, \mathscr{B}_i \rangle$, though from the way these extensions are defined, the effect is much the same, since the 'limit' is a generalized consequence relation $\Vdash^+ \supseteq \Vdash$ for which we still have $\mathscr{A} \not\Vdash^+ \mathscr{B}$ while for every formula C, we have exactly one of $\Vdash^+ C$ and $C \Vdash^+$. Generalized consequence relations extending $\Vdash$ with this property (on the same language) are, as Scott notes, in one-to-one correspondence with the valuations consistent with $\Vdash$. (See the bottom of p. 286 of C & A.) Each such generalized consequence relation is of the form $\Vdash_v$, following Scott's notation, where this means $\Vdash_{\{v\}}$, and in general $\Vdash_{\mathscr{V}}$ is the generalized consequence relation determined by $\mathscr{V}$.[21] The generalized consequence relation determined by (the unit set of) a valuation is already finitary, indeed, without the need for artificially finitarizing it as in the preceding note, as is also the case for any finite set of valuations. (When $\mathscr{V}$ is finite, having n elements, say, then $\Vdash_{\mathscr{V}}$ is finitary in a particularly dramatic manner: $\mathscr{A} \Vdash_{\mathscr{V}} \mathscr{B}$ implies that there are $\mathscr{A}_0 \subseteq \mathscr{A}$, $\mathscr{B}_0 \subseteq \mathscr{B}$ for which $\mathscr{A}_0 \Vdash_{\mathscr{V}} \mathscr{B}_0$ where $\mathscr{A}_0 \cup \mathscr{B}_0$ has cardinality at most n.) So the argument issues in a v consistent with the original $\Vdash$, just as before, verifying all of $\mathscr{A}$ and none of $\mathscr{B}$, showing the completeness of $\Vdash$ with respect to the class of its consistent valuations.[22]

While being perfectly respectable generalized consequence relations, those such as $\Vdash_v$ (amongst others—typically the various left and right extensions mentioned in note 20) figuring in the proof of Proposition 1.3 of C & A lack a feature one might expect of consequence relations and generalized consequence relations when the latter are thought of as embodying *logics*, namely *substitution-invariance* (sometimes called structurality), i.e., that for all substitutions s (of arbitrary formulas for sentence letters):

$$\mathscr{A} \Vdash \mathscr{B} \quad \text{implies} \quad \{s(A) \mid A \in \mathscr{A}\} \Vdash \{s(B) \mid B \in \mathscr{B}\}.$$

(finitary) such relation $\vdash$ is the intersection of all consistent and complete $\Vdash \supseteq \vdash$, and he appeals to Zorn's Lemma to obtain a maximal such relation for which $\mathscr{A}_0 \not\Vdash \mathscr{B}_0$, where (in his notation) $\mathscr{A}_0 \not\vdash \mathscr{B}_0$. This means the "sequence of extensions" is not visibly appealed to in the proof of Proposition 1.3. (Note that in calling $\Vdash$ consistent, Scott means that $\varnothing \not\Vdash \varnothing$—or equivalently, that some pair of sets of formulas of the language of $\Vdash$ is $\Vdash$-consistent.) With $\mathscr{L}$ as earlier, the same fixed enumeration of its elements, a step-by-step Lindenbaum construction of the kind currently envisaged starts with $\Vdash_0 = \Vdash$ and puts $\Vdash_{i+1}$ as the left extension by C_i or else the right extension by C_i of $\Vdash_i$, whichever of these is the first w.r.t. which $\langle \mathscr{A}, \mathscr{B} \rangle$ is a consistent pair; here the *left extension* of $\Vdash$ by a formula C is defined to be that $\Vdash'$ such that for all $\mathscr{C}, \mathscr{D}$, we have $\mathscr{C} \Vdash' \mathscr{D}$ iff $\mathscr{C}, C \Vdash' \mathscr{D}$ and the *right extension* of $\Vdash$ by C as that $\Vdash'$ for which we have $\mathscr{C} \Vdash' \mathscr{D}$ iff $\mathscr{C} \Vdash' C, \mathscr{D}$.

[21] For accuracy—given the latitude we have already allowed ourselves in thinking of Scott's relations $\Vdash$ as finitary generalized consequence relations—we should say that for Scott $\Vdash_{\mathscr{V}}$ is the generalized consequence relation *finitarily determined* by $\mathscr{V}$, meaning that it is the set of all pairs $\langle \mathscr{A}, \mathscr{B} \rangle$ for which there are finite subsets $\mathscr{A}^-$, $\mathscr{B}^-$, of $\mathscr{A}$, $\mathscr{B}$, respectively, such that every valuation in V verifying all formulas in $\mathscr{A}^-$, verifies some formula in $\mathscr{B}^-$.

[22] Proposition 1.3 should have the word "consistent" inserted before the first occurrence of "$\vdash$", as is clear from the first sentence in of the proof. And the second sentence of the proof should read "Suppose $\mathscr{A}_0 \not\Vdash \mathscr{B}_0$" rather than "Suppose $\mathscr{A}_0 \Vdash \mathscr{B}_0$". This typo was corrected by Scott in the page proofs—as the present author can attest, having been given a copy of the proofs by Scott as a graduate student of his in Oxford at the time the paper was in press—but the correction was not acted on by the copy editor. This is one of several such overlooked corrections, but the others are inconsequential or readily correctable, as is also the case for the unnoticed typos (such as "probability" for "provability" between (11) and (12) on p. 298, '1973' for '1972' as the date on item 5 in C & A's bibliography, which, incidentally, appeared—see our reference [31]—under the title 'Background to Formalization', not 'Background to Formalization, I', as the bibliography of C & A says). Let me take this opportunity to thank Steve Gardner for the detection of numerous typographical mishaps in an earlier draft of the present introductory essay.

Of course, such conditions would not be in keeping with the level of abstraction at which Scott is working in C & A (see note 13 above), in particular because the 'languages' $\mathscr{L}$ involved are not being required to have the structure of absolutely free algebras (with sentence letters as the free generators and primitive connectives as fundamental operations).[23] In particular, even if the $\Vdash$ we start with is substitution-invariant the further generalized consequence relations appealed to in Scott's version of the proof will typically not be. Anyone uncomfortable with this aspect of the discussion can always use one of the earlier versions mentioned above of Lindenbaum's Lemma, in which $\Vdash$ is fixed and the maximization strategy is applied to the sets $\mathscr{A}$ and $\mathscr{B}$. Thus either way, we have the $\supseteq$-direction of the Scott-style equation:

$$\Vdash \, = \, \Vdash_{\mathscr{V}_{\Vdash}}$$

for any finitary generalized consequence relation $\Vdash$, the converse direction being an immediate consequence of the definition of consistent valuations.[24] (What about for arbitrary generalized consequence relations? Here the non-trivial—"completeness" or "$\supseteq$"—direction is even simpler, since if $\mathscr{A} \nVdash \mathscr{B}$, the condition $(T)_{\Vdash}^+$, taking $\mathscr{C} = \mathscr{L}$, instantly gives a $\Vdash$-consistent pair $\langle \mathscr{C}_0, \mathscr{C}_1 \rangle$ of mutually exclusive jointly exhaustive sets of formulas, for which by $(M)_{\Vdash}$ and $(R)_{\Vdash}$ we have $\mathscr{A} \subseteq \mathscr{C}_0$ and $\mathscr{B} \subseteq \mathscr{C}_1$, inducing the desired valuation v as the characteristic function of $\mathscr{C}_0$.) An immediate consequence of the inset equation above is that

$$\mathscr{V}_{\Vdash_0} = \mathscr{V}_{\Vdash_1} \Rightarrow \Vdash_0 \, = \, \Vdash_1,$$

since we may apply (the function) $\Vdash$ to both sides of the antecedent here.

Scott gives a more refined formulation at (11) on p. 287 of C & A, with the first and second occurrences of "$=$" replaced by "$\supseteq$" and "$\subseteq$" respectively, and the whole thing then written in biconditional form, and the remark on the same page, after (11) and (12), is his way of noting (in topological terms) that the corresponding claims with the functions $\Vdash$ and $\mathscr{V}$—see again note 24—interchanged, namely (a) and (b):

$$(a) \quad \mathscr{V} = \mathscr{V}_{\Vdash_{\mathscr{V}}}, \qquad (b) \quad \Vdash_{\mathscr{V}_0} = \Vdash_{\mathscr{V}_1} \Rightarrow \mathscr{V}_0 = \mathscr{V}_1,$$

do *not* hold for his own favored $\Vdash$-relations—or our presentation of them as finitary generalized consequence relations (for which $\Vdash_{\mathscr{V}}$ should be understood as the generalized consequence relation finitarily determined by $\mathscr{V}$, as defined in note 21 above). For example, the generalized consequence relations on a countable language $\mathscr{L}$ finitarily determined by the class $\mathscr{V}$ of all valuations for $\mathscr{L}$ and by the class $\mathscr{V} \setminus \{v_{\mathbf{f}}\}$ coincide, where $v_{\mathbf{f}}$ is the valuation assigning $\mathbf{f}$ to every formula of $\mathscr{L}$.[25] This is somewhat awkward for those

---

[23] By contrast, when discussing modal predicate logic in [29], and thinking of logics as sets of formulas, Scott wrote (p. 156): "When we know that a formula is *logically valid* (. . .) then it should remain valid when *formulas* are substituted for *predicate letters*. There is no other convention that is reasonable."

[24] Note that the leftmost and rightmost occurrences of "$\Vdash$" in this equation refer to an arbitrary but fixed generalized consequence relation, whereas the middle occurrence is a function symbol taking us from the set of valuations consistent with that relation to the generalized consequence relation determined thereby. This notational practice is followed here despite that confusing aspect of it, simply because it was Scott's practice in his publications in the 1970s, C & A included.

[25] This example is inspired by Bencivenga [3]. We could equally well have considered $\mathscr{V} \setminus \{v_{\mathbf{t}}\}$, where $v_{\mathbf{t}}$ is the valuation assigning $\mathbf{t}$ to every formula, a valuation well known to be consistent with every

motivated to pass from consequence relations to generalized consequence relations by the Carnapian considerations alluded to earlier, from the perspective of which it is more elegant not to impose the finitariness (or, as in Scott's case, finiteness) restriction. Without such a restriction, (a), and therefore (b), does hold. To see this, consider the non-trivial direction of (a): $\mathscr{V}_{\Vdash_\mathscr{V}} \subseteq \mathscr{V}$. Arguing this contrapositively, take $v \notin \mathscr{V}$—to show that $v \notin \mathscr{V}_{\Vdash_\mathscr{V}}$. Since $v \notin \mathscr{V}$, for each $u \in \mathscr{V}$ there is a formula C_u, say, with $v(C_u) \neq u(C_u)$. Put $\mathscr{A} = \{C_u \mid u(C_u) = \mathbf{f}\}$, $\mathscr{B} = \{C_u \mid u(C_u) = \mathbf{t}\}$, and note that v verifies every formula in $\mathscr{A}$ but no formula in $\mathscr{B}$, whereas $\mathscr{A} \Vdash_\mathscr{V} \mathscr{B}$. Thus $v \notin \mathscr{V}_{\Vdash_\mathscr{V}}$, as desired. Accordingly, in this setting, we have the neat Carnap-inspired contrast between consequence relations and generalized consequence relations: for the latter we always have $\mathscr{V} = \mathscr{V}_{\Vdash_\mathscr{V}}$, but for the former, understanding the notation analogously, we do not in general have $\mathscr{V} = \mathscr{V}_{\vdash_\mathscr{V}}$ (even though again in this case we have $\vdash = \vdash_{\mathscr{V}_\vdash}$).[26]

That concludes our introduction to Scott's general perspective on logical matters and some variations on that perspective. Since the present discussion serves as something of an introductory companion to C & A, however, it is appropriate to include also a few comments on the main business of the paper (Sects. 2–5), in which Scott treats the many-valued logics of Łukasiewicz; a passing familiarity with these sections is assumed, in that even the main points are not summarized here. This treatment is also mentioned in less detail in Scott [31] and [34], which give the general idea: replace many values with many (bivalent) valuations. The idea of not taking fully seriously there being more than two genuine truth-values is famously associated with Michael Dummett and Roman Suszko, who each had their own reasons; Scott joins them, coming at it from a slightly different angle.[27]

Since Łukasiewicz worked with logics as sets of formulas rather than as consequence relations (let alone generalized consequence relations) there is a great degree of freedom in associating a generalized consequence relation which agrees with a consequence relation which 'agrees with' a logic in the set-of-formulas sense, meaning that the formulas in the logic are precisely the consequences of the empty set by the consequence relation in question.[28] Taking the consequence relation which preserves the property of taking the designated value—more generally taking a designated value, though for the current matrices there is only one such value—on any assignment of values in a Łukasiewicz

---

consequence relation—and a special case of the fact that the 'conjunctive combination', as it is put in [21], *q.v.* for this point, of any set of valuations consistent with a consequence relation is itself consistent with that consequence relation.

[26] We can usefully combine the notations also, and make the observation—provable by appeal to the fact mentioned at the end of note 25—that for the minimum generalized consequence relation $\Vdash_{min}$ agreeing with a consequence relation $\vdash$, we have $\Vdash_{min} = \Vdash_{\mathscr{V}_\vdash}$.

[27] Essentially Dummett's perspective is conveyed in one of the publications in our bibliography: p. 86*f.* of Smiley [38]. In fact Scott's sentiments are very close to Dummett's: consider the remark at p. 289 of C & A: "Is not the division of statement types into the designated and the undesignated just a truth valuation? Of course. So why not call it one?" For an implementation of Dummett's approach, which recognizes both the assignments of matrix values to formulas—there (and below) called matrix evaluations—and the induced bivalent valuations, see §§4–5 of my [20], where the relevant Dummett references can also be found. For information and references on Suszko's approach, see §3 of [9], and the more recent [47] and [11], especially Sect. 1, and, for remarks on Scott, Sect. 2.2.

[28] In Scott's terminology (from Theorem 3.3 of C & A), combining these two steps gives different entailment relations with the same absolute assertions.

matrix, gives one such agreeing candidate, while taking what we might call the $\leq$-based consequence relation (where $\leq$ is the linear ordering of values associated with the matrix) gives another: the consequence relation, which in the terminology of Font *et al.* [12] preserves "degrees of truth".[29] It is the latter consequence relations (one for each matrix) that Scott's generalized consequence relations should be thought of as generalizing.[30] There is no reason to do this only for the case of linear orderings $\leq$, and modal logic presents a case in which the order is only a partial—the inclusion relation amongst the truth-sets of formulas in a Kripke model, which leads to the so-called local as opposed to global consequence relations in modal logic (where in its matrix incarnation taking the designated value amounts to being true throughout the model). Scott [33] made the same move of urging that attention be paid to the $\leq$-based ('local truth-preservation') semantical idea in this context too, criticizing those (such as H.B. Curry) who had tacitly opted for designation-preservation ('global truth-preservation').[31]

The modal case has long been familiar—see Massey [24] and references therein—but it will do no harm to illustrate it here in an especially simple incarnation, the modal logic of the two-point Kripke frame with universal accessibility relation. Picking on one of the points which for the labels to follow may be thought of as the actual world, classify a formula as *contingently true* if it is true at the chosen point only, *contingently false* if it is true at the other point only, *necessary* if it is true at both points and *impossible* if it is true at neither. Then we can construct a four-element matrix with these 'values' and obvious tables for the Boolean connectives and for $\Box$ (the necessity operator).[32] (We can do the same for more than two points, getting an expansion by the $\Box$ operation of the appropriate direct product of the 2-element Boolean algebra with itself—only now there are no similarly convenient shorthand labels for the different subcases of contingency.) So far only the algebra of the matrix has been described. We can choose to designate only the value *necessary*, in which case we get the (generalized or plain) consequence relation determined by the ("Henle") matrix being the ($\pm$ generalized) consequence relation determined by the frame, or the values *contingently true* and *necessary*, for the local case. In the latter case, we have the $\leq$-formulation (taking $\leq$ as $\subseteq$, with $\|A\|^{\mathcal{M}}$ being the set of points at which A is true in a model $\mathcal{M}$ on the frame in question) $\mathscr{A} \Vdash \mathscr{B}$ if and only if:

$$\bigcap_{A \in \mathscr{A}} \|A\|^{\mathcal{M}} \subseteq \bigcup_{B \in \mathscr{B}} \|B\|^{\mathcal{M}}$$

---

[29] [12] shows how to rescue this consequence relation from the shadows into which it is cast by early work—Blok and Pigozzi [7]—in contemporary ('abstract') algebraic logic. (Since Scott conducts the discussion in C & A in terms of degrees of error rather than degrees of truth, he has "$\geq$" for "$\leq$".) See also Font [11].

[30] Smiley [38] distinguishes these two and adds a third to the mix, which is not quite a consequence relation since it relates multisets rather than sets of formulas to individual formulas. Scott also has an additional generalized consequence relation symbolized by a boldface turnstile, in C & A: see Theorem 3.3.

[31] Font and Jansana [14] show how to give this idea its due in the framework of abstract algebraic logic, rescuing it from the second-class status in which it appeared in [7]. The term "designation-preservation" in the sentence to which this note is appended refers to preservation of a *single* designated value.

[32] A confusing point: of course, Łukasiewicz himself (in)famously devised a four-valued modal logic, but with a different treatment of $\Box$ from that in play here. This 'Ł-modal logic' is not amongst those we have in mind throughout this discussion in alluding to Łukasiewicz's many-valued logics. For much information about the system in question, see Font and Hájek [13].

for every model $\mathcal{M}$ (on the two-point frame). Simplifying for expository convenience to the case in which $\mathscr{A} = \{A\}$ and $\mathscr{B} = \{B\}$, we can think of this—now $\|A\|^{\mathcal{M}} \subseteq \|B\|^{\mathcal{M}}$— as telling us that for every matrix evaluation h (every homomorphism from the algebra of formulas to the algebra of the matrix, that is), $h(A) \le h(B)$, giving the $\le$-based reformulation of the (non-Henle) matrix semantics.[33] Alternatively, noting that what $\mathcal{M}$ tells us about our two points—call them 1 and 2—is completely given by bivalent valuations v_1 and v_2, we can, instead of saying $h(A) \le h(B)$ for all h, give the straightforward truth-preservation account: However these valuations are chosen, for $i \in \{1, 2\}$, $v_i(A) = \mathbf{t}$ implies $v_i(B) = \mathbf{t}$. In this sense, to echo Scott's slogan in this banal setting, many $(= 4)$ values have been replaced by many $(= 2)$ valuations; the cost is that, by contrast with the matrix evaluations h, the bivalent valuations v_i do not provide a compositional semantics when taken one by one, since whether $v_i(\square A)$ is $\mathbf{t}$ or $\mathbf{f}$ depends on $v_j(A)$ for $i \ne j$ (in the manner dictated by the Kripke semantics).

The model-theoretic semantics just parenthetically alluded to—compare also the Kripke semantics for intuitionistic and intermediate logics as opposed to the use of Heyting algebras[34]—helps to motivate the logics concerned, as Scott urges at p. 283 of C & A,[35] and to explain why the algebras and matrices based on them are of any interest.[36] The discussion throughout §§2–3 of C & A in particular contains no explicit algebraic semantics, but instead contains model-theoretic semantics—in valuational dress (the subscripts on "v" representing the model elements)—in which the models are based on algebras, namely Abelian groups.[37] This is the material which, as Scott notes (C & A, p. 291, n. 6), was independently developed by Urquhart [41, 42] for Łukasiewicz logics, in which this aspect of the situation is perhaps especially clear. (Scott also mentions— C & A, p. 289 —Emil Post as having had similar ideas, but supplies no definite reference. For this, see Scott [31], p. 266, n. 31.) This distinction may appear over-subtle in view of the ease of reformulating matters illustrated above in the modal case to ($\le$-based) algebraic semantics, but it is worth observing in view of the contrast with the modal and intuitionistic cases in which the models are based on (i.e., have as their frames) relational structures rather than algebras.

Minor Qualms and Further Reading   Scott begins C & A with historical remarks about Alfred Tarski and Jan Kalicki; additional historical information on these figures can be obtained from Feferman and Feferman [10] and Zygmunt [48], respectively. (See

---

[33]In "$h(A) \le h(B)$", the particular $\le$ is the lattice ordering associated with the four-element Boolean algebra, with *impossible $\le$ contingently true $\le$ necessary* and *impossible $\le$ contingently false $\le$ necessary*; *necessary* and *impossible* would normally be called $\mathbf{1}$ and $\mathbf{0}$ in this context.

[34]This comparison is suggested in Urquhart [43].

[35]Scott writes "On the other side I would also like to question whether the original matrix method is actually the most intuitively satisfying way of setting up new 'logics'. (...) (i)f there is supposed to be *philosophical* interest in the system it is quite important to present it in the right light."

[36]This is not to deny the greater theoretical utility of modal algebras over Kripke frames arising from the fact that every normal modal logic is determined by a class of the former, which is not so in the case of the latter. See, for example, §6 of Venema [44], and, for an interesting reaction to the discovery—by K. Fine and S.K. Thomason at around the time that C & A was written—of normal modal logics determined by no class of frames, van Benthem [4].

[37]Well, not quite algebras in the strict sense since, in Scott's exposition, we are dealing with *ordered* Abelian groups.

also the several retrospective articles devoted to Tarski's work in Volume 51—the 1986 volume—of the *Journal of Symbolic Logic*.) The richest source of information on generalized consequence relations (under the name "multiple-conclusion logics") is probably still Shoesmith and Smiley [37], which contains several striking points of contrast with the single-conclusion case (i.e. consequence relations proper). A Scott-inspired approach to modal logic, in which structural conditions are used to distinguish different logics rather than connective-specific principles, can be found in Blamey and Humberstone [6];[38] likewise for what its author calls 'partial logic' in Blamey [5]. (Scott's attitude to modal logic, after helping the subject along with the advances in propositional modal logic recorded in Lemmon and Scott [23] and numerous recommendations on—amongst other things— quantified modal logic in [29], came to take on a distinctly jaundiced tone: see the reference to the "Coca-Cola of the modal logicians" in [31], p. 245. One wonders what he would make of the dramatic subsequent developments in the field.) The theme of minimum/maximum generalized consequence relations agreeing with a given consequence relation (all finitary) is taken up in Gabbay [16], where it is shown that if we start with the standard consequence relation of intuitionistic propositional logic, the minimum and maximum agreeing generalized consequence relations are those associated (in the obvious way) with the Beth semantics and the Kripke semantics, respectively, for intuitionistic logic. (See Chapter 3 of [16], Sect. 2, Theorem 5, and Sect. 1, Theorem 6, respectively.)

A philosophical problem for Scott's (admittedly tentative) motivation in terms of degrees of error is raised in §1 of Smiley [38]; actually Smiley raises two such problems, one of which assumes that Scott's "degrees of error" idea is to be applied in connection with negated statements—whereas in fact negation is studiously avoided in C & A. (The objection would stand in a modified form: there is no way of extending Scott's motivating discussion to accommodate it. Oddly, Scott does not comment on his decision to omit negation.) A technical criticism of one detail in the treatment in C & A of Łukasiewicz's infinite-valued logic appears in Harris and Fitelson [17], note 17.

This passage from p. 284 of C & A deserves comment:

> In many formalizations a great deal of effort is expended to eliminate cut as a primitive rule; but it has to be proved as a derived rule. In general cut is *not* eliminable; it is an essential property of the kind of relations we have in mind.

The formalizations Scott has in mind are sequent calculi (also called: Gentzen systems). There is a confusing usage of the phrases such as *derived rule*, *derivable rule*, according to which, rather than contrasting with it, these mean the same as *admissible rule*: evidently this is Scott's use here. In the stricter sense the Cut rule is not derivable but merely admissible on the basis of the remaining rules.[39] And no-one has ever meant by Cut Elimination anything in the vicinity of rendering this rule not admissible. In fact Scott was perfectly

---

[38] It would be interesting to see if something could be done for the quantifiers ($\forall$ and $\exists$) along similar lines—or perhaps along other lines reviewed in Wansing [46] for the treatment of modal operators.

[39] See Wang [45]. The same confusion between the primitive/derived distinction and the derivable/admissible distinction appears in Segerberg [36], p. 83. Further examples of authors conflating these distinctions and some discussion of the damage done thereby can be found in [22]. Scott himself gives a proper explanation of derivability (as opposed to mere admissibility) in [33]. A proper discussion of sequent-to-sequent rules naturally requires explicit isolation of the notion of a sequent—see note 5— something not done in C & A; for information on the analysis of such rules from the perspective of Scott's valuational semantics see [18] and [21], and further references given in the latter source.

well aware of this (*cf.* [28], p. 244, also [27], p. 336*f.*) and is best construed here simply as expressing irritation at the way Cut Elimination had become such an obsession for proof theorists. The irritation may have caused a certain amount of tone deafness on Scott's part, however; he refers to his own rules as supplying a "Gentzen-style formalization" (C & A, p. 297) but this is very far from being the case, if it is to mean anything more than that they are sequent-to-sequent rules. For a Gentzen system (or sequent calculus) we want more than this: we want the rules governing particular connectives to insert a single occurrence of the connective in question as the main connective of a formula on the left or the right of the sequent-separator in the conclusion sequent of the rule—a feature conspicuously missing from Scott's rules in C & A, not only for $\wedge$ and $\vee$, where the familiar Gentzen rules could serve equally well, but also for $\rightarrow$, which is governed in Scott's presentation by five rules (($\rightarrow_1$) and ($\rightarrow_2$) from p. 294 counted as two rules each, ($\rightarrow_3$) as one) which do not come close to being as described above.[40] I speak of tone deafness here because since, even setting aside the Cut rule (a rule version of the condition $(T)_{\Vdash}$) Scott's other rules do not have the subformula property, so the issue of 'Cut Elimination' loses its rationale.

Another note of impatience is sounded by Scott, first as a summary of Tarski's attitude to non-classical logics by the 1950s: "Just how many calculi do you want anyway?" (C & A, p. 282), and further (p. 283), capturing and also endorsing that attitude:

> Logic is logic, and if certain algebras arise, they must be investigated. As he [= Tarski] found, this attitude leads to many difficult problems. There is no need to manufacture strange "logics" and their algebras out of thin air.

What might no doubt seem strange to many readers is the very idea of manufacturing or inventing—as opposed to isolating or discovering—various logics; Scott is plainly not antipathetic (see [28]) to intuitionistic logic as a worthwhile object of investigation, even by those not willing to endorse it as *the*—or even[41] as *a*—correct logic for reasoning with, perhaps reserving only classical logic for the latter status. But the question then immediately arises, given a clarification as to exactly what counts as a logic, as to which (and how many, etc.) logics lie between these two. Such—as they are called—*intermediate* logics (taken here only as one range of examples to make the point), are not 'manufactured' by anyone, but objects waiting to be described and investigated by those interested.[42] Per-

---

[40]In fact the best way to treat the Łukasiewicz systems in the spirit of Gentzen has been to have the sequent-separator separate not sets but multisets of formulas, since these are what have, from about 15 years after the publication of C & A, come to be called substructural logics, with the structural rule of Contraction being absent (from such Gentzen systems). Further, certain distinctive features of these logics associated with the linear ordering of matrix elements have meant that arguably the best realization of Gentzen's desiderata involves the use of *hypersequents* rather than just such multiset-based sequents; see [1, 8, 25].

[41]This, we add for logical pluralists in the style of Beall and Restall [2].

[42]Indeed, Scott himself has taken such an interest, and his name has come to be associated with one particular intermediate logic he isolated (though not in published work). This logic has received considerable attention, being discussed for example in the papers summarized in Minari [26] which are numbered there as [3, 72, 89].

haps Scott is simply expressing the view that such investigations will be fruitless and the interest that motivates them is an indicator of poor taste.[43]

Scott introduces the relations we have been discussing under the rubric of finitary generalized consequence relations (though recall that he uses the notation "⊢", reserved in our discussion for consequence relations) with the words: "let us agree that ⊢ is a relation between finite sets."[44] Since two such sets are involved, ⊢ is a *binary* relation between (finite) sets. But on the same page, we have a casual reference to "a multi-ary relation like ⊢", presumably meaning a relation of no fixed arity (so that "⊢" is a multigrade—rather than dyadic—predicate of the metalanguage). There is a way of reconciling these conflicting ideas, worked out by Barry Taylor in the 1970s and eventually published much later in the form of Taylor and Hazen [40], where the predicates concerned are described as *flexible*: they take a fixed number of lists (of no fixed length) of terms, as in "Tom, Dick, and Harry were fighting with Tim, Dave, Elroy and Horace", where the flanking lists can be expanded or reduced at will. Whether one should bother with such refinements in the present case, or just say that ⊢ is a binary relation between sets, as in Scott's original formulation, is not so clear. Note, however, that when Scott writes, to quote the last cited remark more fully, that "(i)t would not be misleading to regard these laws [his (R), (M) and (T)] as generalizations of the laws of a partial ordering to those appropriate to a multi-ary relation like ⊢", he overdoes things even at this impressionistic level, and—as Matthew Spinks has pointed out to me—would have done better to speak of pre-orders (alias quasi-orders) rather than partial orders, since there should be no suggestion of antisymmetry here. One might think that this is too 'syntactic' a reaction, especially since, on the opening page of C & A, Scott remarks that "the general theme is algebraic logic", and syntactically distinct formulas, such as an A and a B for which $A \vdash B$ and $B \vdash A$ can still correspond to the same element of the Lindenbaum algebra. But although this can happen, it need not, since the language of ⊢ may also provide connectives, for example a 1-ary connective #, for which the ⊢-equivalence of A and B (to summarize the situation just described) does not guarantee the ⊢-equivalence of #A and #B. This point just made has a somewhat different realization for algebraic logic in the style of Blok and Pigozzi [7], which we do not go into here; it is implicit in Spinks and Veroff [39]. Scott (C & A, p. 301) himself defines the relevant equivalence relation in terms of each of $A \to B$ and $B \to A$ being consequences of the empty set—which amounts to the same thing as the definition just given (in view of Theorem 3.1(i) of C & A), and notes that this is indeed a congruence relation on the algebra of formulas, for the Łukasiewicz logics; the present point is simply that this need not always be the case.

References

1. Avron, A.: The method of hypersequents in the proof theory of propositional non-classical logics. In: Hodges, W., Hyland, M., Steinhorn, C., Truss, J. (eds.) Logic: From Foundations to Applications, pp. 1–32 Clarendon Press, Oxford (1996)

---

[43]There are several passages in C & A in which Scott goes out of his way to tell us what not to think about. From p. 292: "There are many more mathematical structures available than applications"; from p. 301: "The *use* of one algebra is not enough justification for a *theory* of special algebras."

[44]C & A, p. 284; in the source, Scott has italics for emphasis on the word "sets".

2. Beall, J.C., Restall, G.: Logical Pluralism. Oxford University Press, Oxford (2006)
3. Bencivenga, E.: Dropping a few worlds. Log. Anal. **26**, 241–246 (1983)
4. van Benthem, J.: Possible worlds semantics: A research program that cannot fail? Stud. Log. **43**, 379–393 (1984)
5. Blamey, S.: Partial logic. In: Gabbay, D.M., Guenthner, F. (eds.) Handbook of Philosophical Logic, vol. 5, 2nd edn. pp. 261–354. Kluwer, Dordrecht (2002)
6. Blamey, S., Humberstone, L.: A perspective on modal sequent logic. Publ. Res. Inst. Math. Sci., Kyoto Univ. **27**, 763–782 (1991)
7. Blok, W.J., Pigozzi, D.: Algebraizable Logics. Mem. Am. Math. Soc. **77**, #396 (1989)
8. Ciabattoni, A., Metcalfe, G.: Bounded Łukasiewicz logics. In: Cialdea Mayer, M., Pirri, F. (eds.) TABLEAUX 2003. Lecture Notes in Artificial Intelligence, vol. 2796, pp. 32–47. Springer-Verlag, Berlin (2003)
9. da Costa, N., Béziau, J.-Y., Bueno, O.: Malinowski and Suszko on many-valued logics: On the reduction of many-valuedness to two-valuedness. Mod. Log. **6**, 272–299 (1996)
10. Burdman Feferman, A., Feferman, S.: Alfred Tarski: Life and Work. Cambridge University Press, Cambridge (2004)
11. Font, J.M.: Taking degrees of truth seriously. Stud. Log. **91**, 383–406 (2008)
12. Font, J.M., Gil, A.J., Torrens, A., Verdú, V.: On the infinite-valued Łukasiewicz logic that preserves degrees of truth. Arch. Math. Log. **45**, 839–868 (2006)
13. Font, J.M., Hájek, P.: On Łukasiewicz's four-valued modal logic. Stud. Log. **70**, 157–182 (2002)
14. Font, J.M., Jansana, R.: Leibniz filters and the strong version of a protoalgebraic logic. Arch. Math. Log. **40**, 437–465 (2001)
15. Gabbay, D.M.: Investigations in Modal and Tense Logics with Applications to Problems in Philosophy and Linguistics. Reidel, Dordrecht (1976)
16. Gabbay, D.M.: Semantical Investigations in Heyting's Intuitionistic Logic. Reidel, Dordrecht (1981)
17. Harris, K., Fitelson, B.: Distributivity in $Ł_{\aleph_0}$ and other sentential logics. J. Autom. Reason. **27**, 141–156 (2001)
18. Humberstone, L.: Valuational semantics of rule derivability. J. Philos. Log. **25**, 451–461 (1996)
19. Humberstone, L.: Classes of valuations closed under operations galois-dual to boolean sentence connectives. Publ. Res. Inst. Math. Sci., Kyoto Univ. **32**, 9–84 (1996)
20. Humberstone, L.: Many-valued logics, philosophical issues In: Craig, E. (ed.) Routledge Encyclopedia of Philosophy, vol. 6, pp. 84–91. Routledge, London (1998)
21. Humberstone, L.: Sentence connectives in formal logic (2010, in press), as an entry in the Stanford Encyclopedia of Philosophy, http://plato.stanford.edu/
22. Humberstone, L.: Smiley's distinction between rules of inference and rules of proof. In: Lear, J., Oliver, A. (eds.) The Force of Argument, Routledge, in press
23. Lemmon, E.J., Scott, D.S., Segerberg, K. (eds.) An Introduction to Modal Logic American Philosophical Quarterly Monograph Series. Basil Blackwell, Oxford (1977) (Originally circulated 1966)
24. Massey, G.J.: The modal structure of the Prior–Rescher family of infinite product systems. Notre Dame J. Form. Log. **13**, 219–233 (1972)
25. Metcalfe, G., Olivetti, N., Gabbay, D.M.: Sequent and hypersequent calculi for abelian and Łukasiewicz logics. ACM Trans. Comput. Log. **6**, 578–613 (2005)
26. Minari, P.: Intermediate Logics. An Historical Outline and a Guided Bibliography, Rapporto Matematico #79, Dept. of Mathematics, University of Siena (1983)
27. Scott, D.: Logic with denumerably long formulas and finite strings of quantifiers. In: Addison, J.W., Henkin, L., Tarski, A. (eds.) Proceedings of the International Symposium on the Theory of Models, Berkeley, 1963, pp. 329–341 North-Holland, Amsterdam (1965)
28. Scott, D.: Constructive validity. In: Laudet, M., Lacombe, D., Nolin, L., Schützenberger, M. (ed.) Symposium on Automatic Demonstration. Lecture Notes in Mathematics, vol. 125, pp. 237–275. Springer-Verlag, Berlin (1970)
29. Scott, D.: Advice on modal logic. In: Lambert, K. (ed.) Philosophical Problems in Logic, pp. 143–173. Reidel, Dordrecht (1970)
30. Scott, D.: On engendering an illusion of understanding. J. Philos. **68**, 787–807 (1971)
31. Scott, D.: Background to formalization. In: Leblanc, H. (ed.) Truth, Syntax and Modality, pp. 244–273. North-Holland, Amsterdam (1973)

32. (= "C & A") Scott, D.: Completeness and axiomatizability in many-valued logic. In: Henkin, L. et al. (eds.), Proc. of the Tarski Symposium, pp. 411–435. American Math. Society, Providence (1974) (Reprinted above)
33. Scott, D.: Rules and derived rules. In: Stenlund, S. (ed.) Logical Theory and Semantic Analysis, pp. 147–161. Reidel, Dordrecht (1974)
34. Scott, D.: Does many-valued logic have any use? In: Körner, S. (ed.) Philosophy of Logic, pp. 64–74. Basil Blackwell, Oxford (1976)
35. Scott, D., Tarski, A.: The sentential calculus with infinitely long expressions. Colloq. Math. **6**, 166–176 (1958)
36. Segerberg, K.: Classical Propositional Operators. Clarendon Press, Oxford (1982)
37. Shoesmith, D.J., Smiley, T.J.: Multiple-Conclusion Logic. Cambridge University Press, Cambridge (1978)
38. Smiley, T.J.: Comment on Scott [34]. In: Körner, S. (ed.) Philosophy of Logic, pp. 74–88. Basil Blackwell, Oxford (1976)
39. Spinks, M., Veroff, R.: Constructive logic with strong negation is a substructural logic: I, II. Stud. Log. **88**, 325–348 (2008) and **89**, 401–425
40. Taylor, B., Hazen, A.P.: Flexibly structured predication. Log. Anal. **35**, 375–393 (1992)
41. Urquhart, A.: A general theory of implication (abstract). J. Symbol. Log. **37**, 443 (1972)
42. Urquhart, A.: An interpretation of many-valued logic. Z. Math. Log. Grundl. Math., **19**, 111–114 (1974)
43. Urquhart, A.: Review of Scott [32]. In: Mathematical Reviews **51** (1976), Review No. 55 (at MathSciNet as MR0363802)
44. Venema, Y.: Algebras and coalgebras. In: Blackburn, P., van Benthem, J., Wolter, F. (eds.) Handbook of Modal Logic, pp. 331–426. Elsevier, Amsterdam (2007)
45. Wang, H.: Note on rules of inference, Z. Math. Log. Grundl. Math. **11**, 193–196 (1965)
46. Wansing, H.: Displaying Modal Logic. Kluwer, Dordrecht (1998)
47. Wansing, H., Shramko, Y.: Suszko's thesis, inferential many-valuedness, and the notion of a logical system. Stud. Log. **88**, 405–429 (2008)
48. Zygmunt, J.: The logical investigations of Jan Kalicki. Hist. Philos. Logic **2**, 41–53 (1981)

L. Humberstone (✉)
Monash University, Victoria, Australia
e-mail: Lloyd.Humberstone@arts.monash.edu.au

COMPLETENESS AND AXIOMATIZABILITY
IN MANY-VALUED LOGIC

DANA SCOTT[1]

The Polish school of logic has enjoyed a wide influence on the field due in large measure to the excellent writings and teachings of Alfred Tarski spanning a period of some 50 years. The many hours I have spent in lectures, seminars and private conversations with Professor Tarski were a real privilege, and it gives me great pleasure to be able to record my gratitude on this occasion. I only wish I had learned more—had been able to serve the subject better. But I have not always had the strength of purpose to meet the high standards set for us by Tarski's example. It is one of his special virtues that he has opened up so many intellectual vistas for us and has made it possible to see how to distinguish the important from the trivial. This last is one of the many lessons I hope also to be able to pass on in some measure. I was especially happy to find at this symposium that Tarski himself approaches the decade of the '70's with characteristic vigor, continuing his teaching with his usual interest and excitement in new ideas.

The purpose of the present essay is retrospective. There are no new ideas here; rather I wish to examine how certain concepts that have long been a concern of Tarski's fit together. The specific subject is many-valued logic (particularly Łukasiewicz'), but the general theme is *algebraic* logic.

Since Boole, algebra has been an important method in logical investigations, and Tarski has made innumerable contributions in this area, carrying on the early work of people like Schröder and of later writers like M. H. Stone. One direction of Tarski's thought can be traced[2] through *Der Wahrheitsbegriff* to the

---

[1] Support of the National Science Foundation is gratefully acknowledged.

[2] [2] and [6] may be consulted for the work of Tarski, while [3] has a wide ranging general bibliography.

very recent *Cylindric algebras*. Another direction leads to universal algebra and model theory. Other reports at this symposium are devoted to the first track; I wish to discuss the second from a special point of view.

My first teacher in "Polish logic" was the late Jan Kalicki, whom I was lucky enough to have as an instructor in a sophomore class at Berkeley. (I hope I can be forgiven some biographical remarks, as there is a definite point I wish to make.) Prior to taking this course I had a job in the Periodicals Room of the main library. While putting some journals away one day, I ran across the *Journal of Symbolic Logic*, became fascinated, and checked a few issues out. Nearly everything was 'way over my head, but one paper I could understand was on many-valued truth tables by Kalicki. The next term I was delighted to find that the Kalicki teaching *Theory of equations* was the same as Kalicki the author, and we soon became good friends. He was pleased I understood his paper so well, but—under the influence of Tarski—one of the first things he did was to divert my attention from *logical matrices* to *abstract algebras*. The result of this tutelage was my initial piece of published research on equational completeness of theories. The move was a good one for me, since I was thereby well prepared for further studies of model theory under Tarski. But let me ask whether this was more than an historical accident, or whether there was some deeper reason for discouraging further thought on the truth tables.

The influential paper by Łukasiewicz and Tarski on sentential calculus not only summarized several years of work but also initiated a long series of studies of logical matrices by innumerable authors. Many-valued logic was given its main impetus around 1920 independently by Łukasiewicz and Post. There are important earlier anticipations, but it is not my purpose here to write a history.[3] Tarski seems, however, to have been the first to view matrix formation as a general method of constructing systems of calculi by regarding matrices as abstract algebras provided with designated elements. Aside from its intrinsic interest, the method had been previously employed mainly in independence proofs, notably by Bernays in special cases.

By the early '50's I think Tarski judged rightly that the theme was to a great extent played out. Just how many calculi do you want anyway? Multiple-valued logics had not found much of a foundational role, and there did not seen much point in creating new ones. Of course they could be studied abstractly, as he himself suggested, but the study seemed rather specialized. The emphasis on designated elements came from a desire to interpret systems of *assertions*, but the main structure was built into the operations (connectives). A shift of interest to algebras (systems of operations) thus seemed natural. This shift made it possible to regard the algebras (matrices) as models of various sets of equations or other first-order axioms. Structural questions (about subalgebras, direct products, homomorphic images) are thereby converted to standard questions of model theory.

---

[3] Cf. Rescher [3] for the history of the subject.

Whatever cleverness one possessed in making up logical matrices is now channeled into the mainstream of first-order logic and its model theory.

On the other hand, we must note that Tarski himself continued his work in algebraic logic (relation algebras, closure and Brouwerian algebras, and cylindric algebras) aided by many collaborators. There is a difference, however, for all these "algebras of logic" are closely related to Boolean algebra: Some of them are extensions by operators, some are derived as substructures. I think it is fair to say that Tarski's main concern has been with *classical logic*; interpretations of modal or intuitionistic logic were side products of other work concerning, especially, Boolean algebras (as say, for example, his study of the lattice of theories or of ideals of Boolean algebras). Logic is logic, and if certain algebras arise, they must be investigated. As he found, this attitude leads to many difficult problems. There is no need to manufacture strange "logics" and their algebras out of thin air.

As outlined here, that attitude seems perfectly reasonable. Aside from the work on cylindric algebras, Tarski's papers influenced a whole group of modal logicians (I think of Lemmon and Bull in particular) who cast their studies in algebraic form and made considerable progress in sorting out the structures of various systems. But I would like to question whether the algebraic method is really the most *efficient* way to approach these problems. On the other side I would also like to question whether the original matrix method is actually the most *intuitively satisfying* way of setting up new "logics". Certainly it is a generalization of the two-valued method, but is it the correct generalization? If all that is wanted is independence proofs, it does not matter what type of structure is employed for the argument. But if there is supposed to be *philosophical* interest in the system, it is quite important to present it in the right light.

In another paper [5] I have begun a philosophical debate on these and related issues. In this paper I shall confine attention to points growing more directly out of the many-valued systems of Łukasiewicz which I was not able to present in detail in the more wide ranging discussion.

1. Consequence and valuation. One of Tarski's early and much quoted papers presents the axioms for an abstract *consequence relation*, and idea that underwent an intensive development in later work. More precisely, **Cn** is an operator from sets of sentences to the set of all their (joint) consequences. There are many interpretations for such a notion: closure under logical deduction, algebraic closure (as with subalgebras), and to a certain extent topological closure (though this is more infinitistic). At more or less the same time Hertz studied abstract sentential logic in terms of a relation ⊢ between finite sets (sequences) of "sentences" and their consequences. The connection between the two presentations is obvious:

(1) $\qquad A_0, A_1, \ldots, A_{n-1} \vdash B \quad iff \quad B \in \mathbf{Cn}(\{A_0, A_1, \ldots, A_{n-1}\}).$

Axioms can be translated from one form to the other; there are advantages in clarity that can be claimed for each approach. Hertz' method has a slight gain in convenience in stating explicit, formal rules, and, as is known, his work directly influenced Gentzen.

In the later work of Gentzen the new idea was introduced of making the right-hand side of the $\vdash$ multiple as well as the left. When we write

$$(2) \qquad A_0, A_1, \ldots, A_{n-1} \vdash B_0, B_1, \ldots, B_{m-1}$$

we mean roughly that the *conjunction* of the sentences A_i has the *disjunction* of the B_j as a logical consequence. That is rough; the precise meaning will be discussed below. But, even with a very vague understanding of the notion, certain properties of the relation can be recognized as correct.

To simplify writing and to obviate certain trivial rules, let us agree that $\vdash$ is a relation between finite *sets*; thus

$$(3) \qquad \mathscr{A} \vdash \mathscr{B},$$

where $\mathscr{A}, \mathscr{B} \subseteq \mathscr{S}$. Here $\mathscr{S}$ is the (abstract) set of all "sentences". In this way (2) is a shorthand version of (3) in case $\mathscr{A} = \{A_0, A_1, \ldots, A_{n-1}\}$ and $\mathscr{B} = \{B_0, B_1, \ldots, B_{m-1}\}$. A related shorthand allows

$$(4) \qquad \mathscr{A}, \mathscr{B} = \mathscr{A} \cup \mathscr{B},$$
$$(5) \qquad \mathscr{A}, B = \mathscr{A} \cup \{B\}.$$

Using this notation, the basic properties of $\vdash$ can be stated as follows:

$$(R) \qquad \mathscr{A} \vdash \mathscr{B} \text{ if } \mathscr{A} \cap \mathscr{B} \neq \varnothing,$$

$$(M) \qquad \frac{\mathscr{A} \vdash \mathscr{B}}{\mathscr{A}, \mathscr{A}' \vdash \mathscr{B}, \mathscr{B}'},$$

$$(T) \qquad \frac{\mathscr{A} \vdash B, \mathscr{C} \quad \mathscr{A}, B \vdash \mathscr{C}}{\mathscr{A} \vdash \mathscr{C}}.$$

We have stated (M) and (T) as inference rules, which means simply that the relationships above the line imply those below. It would not be misleading to regard these laws as generalizations of the laws of a *partial ordering* to those appropriate for a multi-ary relation like $\vdash$. First (R) is the *reflexive* property; then (M) is a *monotonic* law; finally (T) is the principle of *transitivity* or the well-known *cut rule*. In many formalizations a great deal of effort is expended to eliminate cut as a primitive rule; but it has to be proved as a derived rule. In general, cut is *not* eliminable; it is an essential property of the kind of relations we have in mind. It is only for some very special relations on special sets $\mathscr{S}$ obtained by a special means of generation that the rule can be avoided as an assumption. The reason why we expect it as a true property of $\vdash$ will appear shortly.

There are many relations satisfying (R), (M), and (T), and every one determines a consequence relation in the sense of Tarski.

PROPOSITION 1.1. *Let* $\vdash$ *satisfy* (R), (M), *and* (T). *Define a mapping on subsets* $\mathscr{X} \subseteq \mathscr{S}$ *by the formula*

(i) $\mathbf{Cn}_{\vdash}(\mathscr{X}) = \{B \in \mathscr{S} : \mathscr{A} \vdash B,$ *some finite* $\mathscr{A} \subseteq \mathscr{X}\}.$

Then $\mathbf{Cn}_{\vdash}$ *satisfies Tarski's axioms*[4] 2, 3, 4.

The proof is completely straightforward. Conversely, every $\mathbf{Cn}$ determines a $\vdash$, but the correspondence is not in general unique.

THEOREM 1.2. *Let* $\mathbf{Cn}$ *satisfy Tarski's axioms. Define* $\vdash_{\min}$ *and* $\vdash_{\max}$ *by the formulas*

(i) $\mathscr{A} \vdash_{\min} \mathscr{B}$ *iff* $\mathbf{Cn}(\mathscr{A}) \cap \mathscr{B} \neq \varnothing,$

(ii) $\mathscr{A} \vdash_{\max} \mathscr{B}$ *iff* $\bigcap_{B \in \mathscr{B}} \mathbf{Cn}(\mathscr{A}' \cup \{B\}) \subseteq \mathbf{Cn}(\mathscr{A}'),$ *all* $\mathscr{A}' \supseteq \mathscr{A}.$

Then $\vdash_{\min}$ *and* $\vdash_{\max}$ *satisfy* (R), (M), *and* (T). *Furthermore, for any other* $\vdash$, *we have*

(iii) $\mathbf{Cn} = \mathbf{Cn}_{\vdash}$ *iff* $\vdash_{\min} \subseteq \vdash \subseteq \vdash_{\max}.$

PROOF. That $\vdash_{\min}$ satisfies (R), (M), and (T) and that $\vdash_{\min} \subseteq \vdash$ provided $\mathbf{Cn} = \mathbf{Cn}_{\vdash}$ is obvious. It is also easy to check that $\vdash_{\max}$ satisfies (R) and (M) from (ii) and the properties of $\mathbf{Cn}$.

To check (T), suppose that

$$\mathbf{Cn}(\mathscr{A}', B) \cap \bigcap_{C \in \mathscr{C}} \mathbf{Cn}(\mathscr{A}', C) \subseteq \mathbf{Cn}(\mathscr{A}') \quad \text{and} \quad \bigcap_{C \in \mathscr{C}} \mathbf{Cn}(\mathscr{A}', C) \subseteq \mathbf{Cn}(\mathscr{A}', B),$$

for all $\mathscr{A}' \supseteq \mathscr{A}$. Therefore, at once, $\mathscr{A} \vdash_{\max} \mathscr{C}$ follows.

Next, suppose that $\mathbf{Cn} = \mathbf{Cn}_{\vdash}$ and that $\mathscr{A} \vdash \mathscr{B}$ holds. We wish to show $\mathscr{A} \vdash_{\max} \mathscr{B}$ holds. Suppose

$$C \in \bigcap_{B \in \mathscr{B}} \mathbf{Cn}(\mathscr{A}', B);$$

we must show $C \in \mathbf{Cn}(\mathscr{A}')$, where $\mathscr{A}' \supseteq \mathscr{A}$. Now $\mathscr{A}' \vdash \mathscr{B}, C$ holds by virtue of (M). If $\mathscr{B}$ were $\varnothing$, we would be done. Proceed by induction. Suppose $\mathscr{B} = \mathscr{B}', B$. We would have

$$\mathscr{A}', B \vdash C \quad \text{and} \quad \mathscr{A}' \vdash B, \mathscr{B}', C,$$

whence

$$\mathscr{A}' \vdash \mathscr{B}', C.$$

Continuing in this way, $\mathscr{A}' \vdash C$ does indeed hold, as was to be shown.

Finally, suppose $\vdash_{\min} \subseteq \vdash \subseteq \vdash_{\max}$. Clearly $\mathbf{Cn} \subseteq \mathbf{Cn}_{\vdash_{\min}} \subseteq \mathbf{Cn}_{\vdash} \subseteq \mathbf{Cn}_{\vdash_{\max}}$. We need to show that $\mathbf{Cn}_{\vdash_{\max}} \subseteq \mathbf{Cn}$. Thus suppose

$$\mathbf{Cn}(\mathscr{A}', B) \subseteq \mathbf{Cn}(\mathscr{A}')$$

for all $\mathscr{A}' \supseteq \mathscr{A}$. Obviously

$$B \in \mathbf{Cn}(\mathscr{A}, B) \subseteq \mathbf{Cn}(\mathscr{A}).$$

The proof is complete. $\square$

From Proposition 1.1 and Theorem 1.2 we can see that there is no loss in passing from the $\mathbf{Cn}$ theory to the $\vdash$ theory. In fact, the $\vdash$ relation has a chance of being

---

[4] Tarski's axioms are those in [6, p. 31].

slightly more subtle. Before seeing just how subtle, we recast Lindenbaum's Theorem. Call a relation $\Vdash$ *consistent* if there is no $A \in \mathscr{S}$ such that both $\Vdash A$ and $A \Vdash$ (that is, $\varnothing \Vdash A$ and $A \Vdash \varnothing$). Since we always assume $\mathscr{S}$ is nonempty, this is the same as supposing that $\varnothing \Vdash \varnothing$. Call $\Vdash$ *complete* if for all $A \in \mathscr{S}$, either $\Vdash A$ or $A \Vdash$.

PROPOSITION 1.3 (LINDENBAUM). *Every relation $\vdash$ satisfying* (R), (M), *and* (T) *is the intersection of all such consistent and complete relations* $\Vdash$ *containing* $\vdash$.

PROOF. The relation $\vdash$ is contained in the intersection of the $\Vdash$. Suppose $\mathscr{A}_0 \vdash \mathscr{B}_0$. Let $\Vdash$ be a maximal relation such that $\vdash \subseteq \Vdash$ and $\mathscr{A}_0 \nVdash \mathscr{B}_0$. Such a relation exists by Zorn's Lemma. The relation $\Vdash$ is clearly consistent. Suppose it is not complete. Let $A \in \mathscr{S}$ be such that $\nVdash A$ and $A \nVdash$. Define new relations such that

$$\mathscr{A} \Vdash_0 \mathscr{B} \quad iff \quad \mathscr{A} \Vdash A, \mathscr{B},$$
$$\mathscr{A} \Vdash_1 \mathscr{B} \quad iff \quad \mathscr{A}, A \Vdash \mathscr{B},$$

for all $\mathscr{A}, \mathscr{B} \subseteq \mathscr{S}$. By choice of A, both $\Vdash_0$ and $\Vdash_1$ are proper consistent extensions of $\Vdash$; and they both satisfy (R), (M) and (T), as is easily checked. It follows that $\mathscr{A}_0 \Vdash_0 \mathscr{B}_0$ and $\mathscr{A}_0 \Vdash_1 \mathscr{B}_0$ both hold by the maximality of $\Vdash$. But then by (T), we have $\mathscr{A}_0 \Vdash \mathscr{B}_0$, which is a contradiction. $\square$

It is clear that this proof is the same as the original proof of Lindenbaum's Theorem[5]—but the difference here lies in the fact that we have assumed absolutely nothing about the closure under sentential connectives for the set $\mathscr{S}$. Whatever sentential "calculus" was needed has been absorbed into the properties (R), (M), and (T) of $\vdash$. The advantage of doing so lies in the fact that these properties hold in many systems having quite diverse selections of sentential connectives.

Consistent and complete relations $\Vdash$ are obviously truth valuations in disguise. Given $\Vdash$, satisfying (R), (M), and (T), of course, and complete and consistent, we can define $v_{\Vdash}: \mathscr{S} \rightarrow \{\mathbf{t}, \mathbf{f}\}$ by the formula

(6) $$v_{\Vdash}(A) = \mathbf{t} \quad iff \quad \Vdash A.$$

Conversely, given any $v: \mathscr{S} \rightarrow \{\mathbf{t}, \mathbf{f}\}$, define

(7) $$\mathscr{A} \Vdash_v \mathscr{B} \quad iff \quad whenever\ v(A) = \mathbf{t}\ for\ all\ A \in \mathscr{A},$$
$$then\ v(B) = \mathbf{t}\ for\ some\ B \in \mathscr{B}.$$

It is easily checked that

(8) $$\Vdash_{v_{\Vdash}} = \Vdash \quad and \quad v_{\vdash_v} = v;$$

hence the v and the $\Vdash$ are in a one-one correspondence.

Given an arbitrary $\vdash$, we can think of all complete and consistent extensions $\Vdash$ and the corresponding valuations $v_{\Vdash}$. These form a set $\mathscr{V}_{\vdash}$ of all valuations v

---

[5] Cf. [6, pp. 34, 98].

consistent with $\vdash$. Independently defined:

(9) $\quad v \in \mathscr{V}\vdash \quad$ *iff whenever* $\mathscr{A} \vdash \mathscr{B}$ *and* $v(A) = t$ *for all* $A \in \mathscr{A}$, *then* $v(B) = t$ *for some* $B \in \mathscr{B}$.

Lindenbaum's Theorem then reads

(10) $\qquad\qquad \mathscr{A} \vdash \mathscr{B} \quad$ iff $\quad \mathscr{A} \Vdash_v \mathscr{B}$ *for all* $v \in \mathscr{V}\vdash$.

We see further that, for two such relations $\vdash_0$ and $\vdash_1$,

(11) $\qquad\qquad \vdash_0 \subseteq \vdash_1 \quad$ iff $\quad \mathscr{V}\vdash_0 \supseteq \mathscr{V}\vdash_1$

and

(12) $\qquad\qquad \mathscr{V}\vdash_0 \cap \vdash_1 = \mathscr{V}\vdash_0 \cup \mathscr{V}\vdash_1.$

In fact, we have a lattice anti-isomorphism between the set of all $\vdash$-relations and the family of all closed sets of the product space $\{t, f\}^{\mathscr{S}}$, where the two-point set $\{t, f\}$ has the discrete topology.

All that has been said here is well known in some form or another (with the possible exception of Theorem 1.2). The point I wish to make emphasizes the simplicity and generality of the results and in addition concerns the intuitive interpretation of a $\vdash$-relation.

I now feel that it would be wrong to call a relation $\vdash$ satisfying (R), (M), and (T) a *consequence relation*. It is only in special cases that we can realize $\vdash$ by formal manipulation, by giving *derivations* of conclusions from hypotheses. It would be better to read $\mathscr{A} \vdash \mathscr{B}$ as $\mathscr{A}$ *entails* $\mathscr{B}$, though the word "entail" also involves extraneous meanings. I would prefer to call $\mathscr{A} \vdash \mathscr{B}$ a *conditional assertion*. This is not to imply that someone has gone to the trouble of carrying out an *act* of assertion; rather when we consider the whole system $(\mathscr{S}, \vdash)$, then the relationships $\mathscr{A} \vdash \mathscr{B}$ are those allowed as correctly (or coherently) assert*able*. This relationship must be taken in a conditional sense, since, as we have seen, $\mathscr{A} \vdash \mathscr{B}$ means that *whenever* all the statements in $\mathscr{A}$ are true under a consistent valuation, *then* at least one in $\mathscr{B}$ must be true also. The terminology of a *relation of conditional assertion* is very cumbersome, however; and we shall speak simply of *entailment relations* $\vdash$ as those satisfying (R), (M), and (T).

As examples of entailment relations, we may mention the case of $\mathscr{S}$ as a Boolean algebra of sets. Then it is natural to take

(13) $\qquad\qquad \mathscr{A} \vdash \mathscr{B} \quad$ iff $\quad \bigcap \mathscr{A} \subseteq \bigcup \mathscr{B}.$

In this case the consistent valuations correspond to the ultrafilters. If, more abstractly, $\mathscr{S}$ were just a distributive lattice, then $\mathscr{V}\vdash$ would correspond to the prime ideals. Thus Proposition 1.3 includes, not surprisingly, the well-known representation theorems. Note, by the way, that in the lattice case if we use (13) as a definition, then (R) and (M) are immediate, but (T) is *equivalent* to the distributive law. This too is well known, but I feel it useful to put all these facts together in this context.

Another example of why the approach through entailment relations is pleasant is afforded by the Compactness Theorem. The version we shall have occasion to quote concerns the generalization of the entailment relation from finite to arbitrary subsets of $\mathscr{S}$. Suppose $\vdash$ is an entailment relation and $\mathscr{X}, \mathscr{Y} \subseteq \mathscr{S}$ are possibly infinite subsets. Define

(14) $\mathscr{X} \vdash \mathscr{Y}$ iff $\mathscr{A} \vdash \mathscr{B}$ for some finite pair $\mathscr{A} \subseteq \mathscr{X}$ and $\mathscr{B} \subseteq \mathscr{Y}$.

The negation of this relationship, $\mathscr{X} \nvdash \mathscr{Y}$, is a kind of *consistency condition*. (Simple consistency of the set $\mathscr{X}$ corresponds to $\mathscr{X} \nvdash \varnothing$.) We can now prove

PROPOSITION 1.4.   *Suppose, for an entailment relation $\vdash$, that $\mathscr{X} \nvdash \mathscr{Y}$. Then for some $v \in \mathscr{V}_\vdash$ we have*
  (i) $v(A) = \mathbf{t}$ *for all $A \in \mathscr{X}$, and*
  (ii) $v(B) = \mathbf{f}$ *for all $B \in \mathscr{Y}$.*

PROOF.   Given that $\mathscr{X} \nvdash \mathscr{Y}$, define a new relation $\Vdash$ by the formula

$$\mathscr{A} \Vdash \mathscr{B} \quad \text{iff} \quad \mathscr{X}, \mathscr{A} \vdash \mathscr{Y}, \mathscr{B}.$$

Obviously $\Vdash$ extends $\vdash$. Also $\Vdash$ is consistent by assumption on $\mathscr{X}$ and $\mathscr{Y}$. By Lindenbaum's Theorem (cf. (10)) we know $\mathscr{V}_\Vdash \neq \varnothing$. Let $v \in \mathscr{V}_\Vdash$; then $v \in \mathscr{V}_\vdash$. But note by construction that

$$\Vdash A \quad \text{for all } A \in \mathscr{X}, \text{ and}$$
$$B \Vdash \quad \text{for all } B \in \mathscr{Y}.$$

Hence (i) and (ii) follow as desired.   □

The reader can think of many variations on this same type of argument.

2. Many values vs. many valuations.   Everyone can understand $\{\mathbf{t}, \mathbf{f}\}$-valuations, but few—even the creators of the subject—can understand many-valued truth tables. They were constructed by abstract *analogy* and heve never been given, in my opinion, adequate motivation. Consider for example Łukasiewicz' 6-valued implication:

$\rightarrow$	1	$\frac{4}{5}$	$\frac{3}{5}$	$\frac{2}{5}$	$\frac{1}{5}$	0
*1	1	$\frac{4}{5}$	$\frac{3}{5}$	$\frac{2}{5}$	$\frac{1}{5}$	0
$\frac{4}{5}$	1	1	$\frac{4}{5}$	$\frac{3}{5}$	$\frac{2}{5}$	$\frac{1}{5}$
$\frac{3}{5}$	1	1	1	$\frac{4}{5}$	$\frac{3}{5}$	$\frac{2}{5}$
$\frac{2}{5}$	1	1	1	1	$\frac{4}{5}$	$\frac{3}{5}$
$\frac{1}{5}$	1	1	1	1	1	$\frac{4}{5}$
0	1	1	1	1	1	1

Numerically we can write

(1) $\begin{aligned}(p \rightarrow q) &= 1 && \text{if } p \leq q, \\ &= 1 - p + q && \text{if } p \geq q.\end{aligned}$

Similarly conjunction and disjunction can be defined by

(2) $(p \wedge q) = \min(p, q),$

(3) $(p \vee q) = \max(p, q).$

A rough interpretation makes p and q "probabilities", but this is a *very rough* idea of what it means to be "partly" true. The main trouble with the probabalistic approach is that the probability of a compound proposition (a conjunction, say) is hardly ever a *function* of the probabilities of the parts (the propositions must be independent or something). Thus the computation rules (1)–(3) are not to be motivated by probabilities.

   This is the first puzzle, and one I feel that Łukasiewicz never faced squarely. Another puzzle concerns the "designated" elements. Usually the value 1 is designated (as indicated by *1 in the table). Some authors have tried designating several elements, with interesting formal results—particularly in predicate calculus. Rescher has tried designation and *anti*designation, but the argument for doing this does not seem very conclusive to me. We shall try a different plan, which I was surprised and pleased to see had already occurred to Post. (It is not so easy to have new ideas.)

   First let us *renumber* the values in Łukasiewicz' table to make generalization easier.

$\rightarrow$	0	1	2	3	4	5
0	0	1	2	3	4	5
1	0	0	1	2	3	4
2	0	0	0	1	2	3
3	0	0	0	0	1	2
4	0	0	0	0	0	1
5	0	0	0	0	0	0

Now 0 corresponds to *true* and 5 to *false*, but for the moment do not give the numerology any metaphysical significance. Indeed, do not call the elements of $\{0, 1, 2, 3, 4, 5\}$ *values*; take them rather as standing for *types* of sentences. Abstractly we could set $\mathscr{S} = \{0, 1, 2, 3, 4, 5\}$ or we could map the sentences to their respective types. The *assumption* of Łukasiewicz logic is that this mapping is *functional* on the connectives. This may be quite mad or it may be a discovery— there is simply no way to telling at this stage of analysis.

   The next step is to reconsider the process of designation. Is not the division of statement types into the designated and undesignated just a truth valuation? Of course. So why not call it one? Indeed, there are many valuations that it would be natural to consider in view of the vague intuitions suggested by Łukasiewicz.

We can put them in a table:

$\mathcal{V}$	v_0	v_1	v_2	v_3	v_4
0	t	t	t	t	t
1	f	t	t	t	t
2	f	f	t	t	t
3	f	f	f	t	t
4	f	f	f	f	t
5	f	f	f	f	f

Here we find $\mathcal{V} = \{v_0, v_1, v_2, v_3, v_4\}$ consists of 5 valuations making it possible to distinguish 6 types of sentences. Each one of the valuations considered represents *one* scheme of designating the elements—it takes all of them together to make the full range of distinctions, however.

Now given any $\mathcal{S}$ and any $\mathcal{V} \subseteq \{t, f\}^{\mathcal{S}}$, we can always define an entailment relation

(4) $\qquad\qquad \mathcal{A} \vdash_{\mathcal{V}} \mathcal{B} \quad iff \quad \mathcal{A} \Vdash_v \mathcal{B} \text{ for all } v \in \mathcal{V}.$

In our special case above, we find that

(5) $\qquad\qquad\qquad \vdash_{\mathcal{V}} A \quad iff \quad v_0(A) = t.$

This circumstance makes it *seem* as though the usual scheme of designation is sufficient, but this is a mistake. The other valuations are "built in" to the $\rightarrow$-connective. I shall argue that the only way to really "understand" $\rightarrow$ is to make all of $\mathcal{V}$ explicit.

Keeping in mind the numerical transformation from $\{1, \frac{4}{5}, \frac{3}{5}, \frac{2}{5}, \frac{1}{5}, 0\}$ to $\{0, 1, 2, 3, 4, 5\}$ and using the more abstract notation of $A, B \in \mathcal{S}$, we may first rewrite the numerical formulas (2) and (3) as follows:

(6) $\qquad\qquad v_i(A \wedge B) = t \quad iff \quad v_i(A) = t \text{ and } v_i(B) = t,$

(7) $\qquad\qquad v_i(A \vee B) = t \quad iff \quad v_i(A) = t \text{ or } v_i(B) = t.$

Here $0 \leq i < 5$; in other words the numbers enter only in the subscripts to the v_i. So far they do not enter in a very interesting way; the main difficulty lies with $\rightarrow$. This is not such a difficult exercise, though. What we find is this.

(8) $\quad v_i(A \rightarrow B) = t \quad iff \quad$ whenever $i + j \leq k$ and $v_j(A) = t$, then $v_k(B) = t.$

For example $1 \rightarrow 4 = 3$, and it is easy to see why rule (8) works. In checking such rules, the reader should keep in mind that

(9) $\qquad\qquad$ whenever $i \leq j$ and $v_i(A) = t$, then $v_j(A) = t.$

Whether (8) exactly fits Łukasiewicz' table or not is a bit beside the point (except for historical interest). Anyone can see at once that (8) as a definition provides *some kind* of an implication. The main question is whether anyone should be interested in this notion. It is simple enough (and to my taste better looking than formula (1)), but there may be better reasons for engaging our attention. We note again that the numbers are restricted to the subscripts, though in quantifying over them in (8) a certain amount of "numerology" takes place in the $i + j \leqq k$ relationship.[6] What could all this mean?

Instead of imprecise (and inappropriate) talk of probabilities, let us think of the equation $v_i(A) = t$ in the following way. In case i is small, it means that A is very true; when i is larger, A is not so true. The extreme cases (in the example) are $i = 0$ and $i = 5$. For 0, the sentence A must be perfectly true; for 5, all but false is allowed. Note that we have now completely eliminated numbers as "values" for sentences. This explains the title of this section: It seems much better to consider a variety of *valuations* rather than a variety of "truth" *values*. Valuations use the ordinary truth values, t and f, and sentential "values" could creep in again as types of sentences as distinguished by the valuations (or as sets of valuations); but it is not necessary to directly interpret these "values".

It remains, however, to interpret the indices $0 \leqq i < 5$. Here is my suggestion: Call them *degrees of error*. The index 0 means *no* error; while 5 indicates *maximum* error. (Even with maximum allowable error, there is one type of sentence which remains *false*.) Now error is as vague an idea as probability; maybe it is even vaguer. The big assumption here is that errors *add arithmetically*. Suppose they do. Then we can give an intuitive reading to (8). *We ask*: When is it reasonable to say that an implication $A \rightarrow B$ is true to within degree i? *Answer*: Just in case i represents an additive *shift* of error between hypothesis and conclusion. More precisely, it must be the case that whenever A is true to degree j, then B will also be true at least in degree $i + j$. Shifty implication. Well, why not? There is an idea here, but has it any application?

So far the only nice example I can think of involves the ideas of *distance* and *approximate equality* of points. Suppose P is a space of points with a real-valued metric defined on it where $|a - b|$ denotes the distance between a and b. An equality statement $a = b$ may be true or only imperfectly true. We might find it natural to say that $a = b$ is true to within degree i if and only if

(10) $$|a - b| \leqq i.$$

Condition (10) is often used for an approximate notion of equality. The only trouble with it is that, as a relationship, it is *not* transitive; whereas perfect equality is. Formally this means that the conditional assertion,

(11) $$a = b, \quad b = c \vdash a = c,$$

does *not* hold in general. This is a place where Łukasiewicz' implication could step in, for the conditional assertion

(12) $$a = b \vdash [b = c \rightarrow a = c]$$

---

[6] This reading of Łukasiewicz' implication was independently discovered by Alasdair I. F. Urquhart; cf. [7].

is valid (using our valuations and our interpretation of →). The reader will find that (12) is nothing more than an implicit statement of the *triangle inequality* in metric spaces. In Łukasiewicz' logic we are noting here the difference between statements of the form

$$[[A \wedge B] \rightarrow C] \quad \text{and} \quad [A \rightarrow [B \rightarrow C]].$$

What is mildly remarkable is that such distinctions might even be pertinent to a mathematically interesting theory.

It would be pleasant to conclude then that Łukasiewicz' logic is a *calculus of error* where, however, reference to the errors are conveniently left implicit. We need more examples before we can tell whether this is a significant idea. In particular, we must note that an attempted interpretation has to avoid any involvement with the notion of *probability of error*; since such relationships as (6) and (7) can only subsist when the propositions A and B are *independent*—an unrealistic assumption when probabilities are invoked. What needs argument is the hypothesis that there is an intuitive notion of—what should we call it?—*cumulative error* that roughly fits rules (6), (7), (8), and (9). Without such an argument this is all an academic exercise. There are many more mathematical structures available than there are applications.

3. Rules and axioms. Having shown in the last section just how *many valuations* replace *many values* and how the connectives interact with these valuations, we turn now to the problem of giving an elementary axiomatization to such a "logic". It does not seem to be an automatic phenomenon for the desired kind of axiomatization to exist; though a translation from the proportional calculus into a first-order predicate calculus (as suggested by rules (6)–(9) of §2) would show that the valid entailments ought to be *recursively enumerable*. Of course, much depends on the intended range of i and the meaning of the relationship $i + j \leq k$. For a suitable construal of I (the range of i, j, k), of $+$, and of $\leq$, it would even be possible to argue for the *decidability* of the entailment relation (in the case that $\mathcal{S}$ consists of the usual kind of propositional formulas) by virtue of well-known results about theories of ordered Abelian groups. We shall, however, pursue a more direct approach here which allows us to *exhibit* the axioms and rules.

In the background it is assumed that $\mathcal{S}$ is a *nonempty* set of "sentences"; while in the foreground we assume above all that ⊢ is an entailment relation between finite subsets of $\mathcal{S}$ satisfying the basic inference rules (R), (M), and (T). These are uniform assumptions for *all* logical systems founded on the notion of bivalent valuation, which we have shown appropriate even to the Łukasiewicz system. As more specialized assumptions we take first (again more or less in the background) the condition that $\mathcal{S}$ is *closed* under these connectives: $[A \wedge B]$, $[A \vee B]$, and $[A \rightarrow B]$. As in algebraic logic, we can imagine $\mathcal{S}$ as an abstract set closed under three binary operations; but, as distinguished from the algebraic approach, we are not supposing that the equality relation in $\mathcal{S}$ has "logical" significance. What we shall find is that there is generally a whole spectrum of equivalence relations over $\mathscr{S}$ that it will be technically useful to isolate and to employ. (Though by virtue of the axioms we shall assume a certain *minimal* equivalence relation could be used to produce an algebra.)

Generalizing the discussion of the last section, we could say that what we are interested in are the properties of entailment relations of the form $\Vdash_{v_i}$, where $\mathscr{V} = \{v_i : i \in I\}$ is a system of valuations on $\mathscr{S}$ indexed by an interval $I = [0, e)$ of an ordered Abelian group, and where the v_i satisfy conditions (6)–(9) of §2. Here, of course, the $+$ is the addition operation of the group, and $\leq$ is the ordering. For "pure" logic, we might even specialize $\mathscr{S}$ to a syntactical class of formulas, but the completeness proof of the next section does not require this assumption in the main part of the argument. We could also ask whether we should worry about special groups (integers, reals, rationals) as contrasted with general groups. This question will be discussed at the end of the next section. Here we shall try to state what is common to all such interpretations.

In generalizing to abstract groups, however, we must take care. The finite segments I of the group of integers are obviously such that to each A, for which $v_i(A) = t$ holds for *some* i, we can associate a *minimal* such i. In view of condition (9) of the last section, this minimal i determines all the other places at which the valuations make A true. Such a situation would not at all be necessary in infinite groups such as the reals or rationals. As we shall see, this minimum condition (call it (min)) has a strong effect on the valid statements of the logic. *We shall assume it* (to be able to have exactly Łukasiewicz' system), but later we shall give reasons why the assumption seems unwarranted—especially in the presence of quantifiers in predicate logic.

The connectives $\wedge$ and $\vee$ give us no trouble, since we can take over the usual classical rules.

$$(\wedge) \qquad \frac{\mathscr{A}, A, B \vdash \mathscr{B}}{\mathscr{A}, [A \wedge B] \vdash \mathscr{B}},$$

$$(\vee) \qquad \frac{\mathscr{A} \vdash A, B, \mathscr{B}}{\mathscr{A} \vdash [A \vee B], \mathscr{B}}.$$

These inference rules are each a *pair* of rules, because they can be used in both directions: up and down. Their validity is obviously guaranteed by principles (6) and (7), where the semantics is the same as for ordinary logic. Implication, however, is not so obvious in its properties.

In the case of *classical* material implication, we have as valid the well-known principle

$$(\supset) \qquad \frac{\mathscr{A}, A \vdash B, \mathscr{B}}{\mathscr{A} \vdash [A \supset B], \mathscr{B}}.$$

We also know that $(\wedge)$, $(\vee)$, $(\supset)$ are complete for classical logic. However, the rule $(\supset)$ fails *in both directions* for Łukasiewicz' implication $\rightarrow$ in place of $\supset$ — even in the three-valued case. The rule $(\supset)$ is too closely tied to the ordinary truth tables. What is required is a pair of weaker principles.

$$(\rightarrow_1) \qquad \frac{A, \mathscr{B} \vdash \mathscr{C}}{A \rightarrow \mathscr{B} \vdash A \rightarrow \mathscr{C}},$$

$$(\rightarrow_2) \qquad \frac{\mathscr{C} \vdash \mathscr{B}, A}{\mathscr{B} \rightarrow A \vdash \mathscr{C} \rightarrow A}.$$

In both of these (double!) rules we require the side condition

$$(*) \qquad \mathscr{C} \neq \varnothing,$$

and we have used the shorthand

$$A \rightarrow \mathscr{B} = \{[A \rightarrow B] : B \in \mathscr{B}\},$$
$$\mathscr{C} \rightarrow A = \{[C \rightarrow A] : C \in \mathscr{C}\}.$$

But even these two rules are not enough; the *Abelian* character of the assumed underlying group finds expression in this axiom:

$$(\rightarrow_3) \qquad [A \rightarrow [B \rightarrow C]] \vdash [B \rightarrow [A \rightarrow C]].$$

Now the list is closed.[7] We leave to the reader the verification of the failure of $(\supset)$, but we shall check the correctness of the others.

THEOREM 3.1. *The rules* $(\rightarrow_1)$, $(\rightarrow_2)$, $(\rightarrow_3)$ *are valid in any interpretation in any ordered Abelian group.*

PROOF. Actually all we need to know about I is that $0 = \min(I)$ and that I is *convex* (i.e., it contains everything between any two of its elements). The existence of the $\sup(I)$ is irrelevant.

The easiest question is that of Axiom $(\rightarrow_3)$. The verification can be laid out in a table which is just like a proof by natural deduction. In the table $\sqrt{}$ means *true* and $\times$, *false*. The argument is by contradiction. The first two lines are assumed and the others follow, in this case, by the rule for valuations of $\rightarrow$-formulas. A line of the proof of the form

$$i \, |\sqrt{}| \, A$$

simply means the same as

$$v_i(A) = \mathbf{t};$$

similarly for falsehoods. The convexity of I makes the use of $k \geq i + j$ unnecessary in line (4). This saves many variables over the indices (= degrees of

---

[7] The rules for implication embody improvements and corrections found since the presentation of the lecture.

error). From *this* example, it looks as simple as ordinary logic; but not all problems are so straightforward. Note that assumption (min) has not been used here.

The Verification of Axiom ($\rightarrow_3$).

(1)	$i \|\sqrt{}\| A \rightarrow [B \rightarrow C]$	*assume*
(2)	$i \|\times\| B \rightarrow [A \rightarrow C]$	*assume*
(3)	$j \|\sqrt{}\| B$	*by (2), for some j*
(4)	$i + j \|\times\| A \rightarrow C$	*by (2), the same j*
(5)	$k \|\sqrt{}\| A$	*by (4), for some k*
(6)	$(i + j) + k \|\times\| C$	*by (4), the same k*
(7)	$i + k \|\sqrt{}\| B \rightarrow C$	*by (1) and (5)*
(8)	$(i + k) + j \|\sqrt{}\| C$	*by (7) and (3)*
(9)	contradiction	*by (6) and (8),*
		the associative
		and commutative
		laws

Let us now turn to the inference rule ($\rightarrow_1$). There are two inferences to verify. Assume first that $A, \mathscr{B} \vdash \mathscr{C}$ holds. It will be sufficiently general if we take

$$\mathscr{B} = \{B\} \quad \text{and} \quad \mathscr{C} = \{C, D\}.$$

Suppose then that $A \rightarrow B \vdash A \rightarrow C, A \rightarrow D$ does *not* hold. For some i we would then have the following situation:

$$i \|\sqrt{}\| A \rightarrow B$$
$$i \|\times\| A \rightarrow C$$
$$i \|\times\| A \rightarrow D$$

It follows for suitable j, j' that

$$j \|\sqrt{}\| A$$
$$i + j \|\times\| C$$
$$j' \|\sqrt{}\| A$$
$$i + j' \|\times\| D.$$

Let $j'' = \min(j, j')$; we can then set j'' into the last four statements. (This is the step where the side condition (∗) comes into play.) But by our original assumption on i we see that

$$i + j'' \|\sqrt{}\| B$$

and of course

$$i + j'' \|\sqrt{}\| A.$$

But then since $A, B \vdash C, D$ is assumed,

$$i + j'' \|\sqrt{}\| C, D$$

follows (i.e., either C or D is *true* under $v_{i+j''}$). And now we have a contradiction. Note again that assumption (min) was not employed.

For the converse direction of the rule, assume $A \to \mathcal{B} \vdash A \to \mathcal{C}$, or in our special case $A \to B \vdash A \to C$, $A \to D$. In case $A, B \vdash C, D$ failed, we would have the following situation:

$$i \, |\sqrt{}| \, A$$
$$i \, |\sqrt{}| \, B$$
$$i \, |\times| \, C$$
$$i \, |\times| \, D$$

Let j be the minimum degree such that

$$j \, |\sqrt{}| \, A.$$

In view of the choice of i and of j, we see that

$$i - j \, |\sqrt{}| \, A \to B$$

must hold. On the other hand, we find that

$$i - j \, |\times| \, A \to C$$
$$i - j \, |\times| \, A \to D.$$

We are thus in contradiction to our basic assumption. $\square$

(In this last argument (min) is essential. Taking rational (or real) degrees with $I = [0, \infty)$ (or a finite interval), we can try

$$\{i \in I : v_i(A) = \mathbf{t}\} = (0, \infty),$$
$$\{i \in I : v_i(B) = \mathbf{t}\} = [1, \infty),$$
$$\{i \in I : v_i(C) = \mathbf{t}\} = (1, \infty).$$

Neither A nor C has a minimum; however, the minimum for both $A \to B$ and $A \to C$ is the same: namely, 1. Thus $A \to B \vdash A \to C$ is correct, but alas $A, B \vdash C$ is not. This is a pity, since the inference is a very useful one for making neat proofs.)

THEOREM 3.2. *The following three rules are provable from* $(\to_1)$–$(\to_3)$:

(i) $$\frac{A \vdash B}{\vdash A \to B},$$

(ii) $$\frac{A \vdash B \to C}{B \vdash A \to C},$$

(iii) $$\vdash A \to B, \quad B \to A.$$

PROOF. Rule (i) is a special case of $(\to_1)$ where $\mathcal{B}$ is empty and $\mathcal{C} = \{B\}$. Rule (ii) follows easily from (i) in view of $(\to_3)$. The proof of (iii) takes a bit longer. First we derive from $(\vee)$

(1) $$A \vee B \vdash A, B.$$

Then by $(\to_1)$, since $\vdash [A \vee B] \to [A \vee B]$,

(2) $\vdash [A \vee B] \to A, \quad [A \vee B] \to B.$

Next again by $(\vee)$

(3) $A \vdash A \vee B;$

whence by $(\to_2)$

(4) $[A \vee B] \to B \vdash A \to B.$

Similarly we have

(5) $[A \vee B] \to A \vdash B \to A.$

We then conclude (iii) from (2), (4), (5) by two applications of (T). □

We may note that the use of $(\vee)$ in the proof of (iii) was inessential, because the combination $[A \to B] \to B$ can be used in place of $[A \vee B]$, for which purpose the rules $(\to_1)$–$(\to_3)$ are sufficient. Next note that in our formulation the whole Łukasiewicz system can be condensed into assertions of the form $\vdash A$ by repeated applications of the basic rules and those of Theorem 3.2. This shows the connection between the Gentzen-style formalization and the usual axiomatic systems. It seems to me, however, that the former has many advantages. For example, there *are* systems of many-valued logic *without* tautologies. For these, absolute assertions $\vdash A$ are trivial, but conditional assertions $A \vdash B$ are not. This, however, is a topic for another essay. In the meantime it is hoped that enough evidence is given here to show that conditionals and rules of inference are easy to use. In this connection, the next theorem is of interest.

THEOREM 3.3.   *Given a consistent entailment relation* $\vdash$ *satisfying the Łukasiewicz rules, the relation* $\vdash$ *is also an entailment relation which has the same absolute assertions, where we define*

$$\mathscr{A} \vdash \mathscr{B} \quad iff \quad \vdash (\bigwedge \mathscr{A})^n \to \bigvee \mathscr{B} \text{ for some } n > 0.$$

Here $\mathscr{B} \neq \varnothing$ *and* $\bigwedge$ *and* $\bigvee$ *stand for repeated conjunction and disjunction; further* $A^n \to B$ *means*

$$A \to [A \to \cdots [A \to B] \cdots].$$
$$\underbrace{\qquad\qquad\qquad\qquad}_{n\text{-times}}$$

PROOF.   We have avoided the empty disjunction since we have no *false* statements $\perp$ such that $\perp \vdash$. However, as there are *true* statements T where $\vdash \mathsf{T}$, we can take an empty conjunction to be any such T. It is of course obvious that $\vdash \mathscr{B}$ holds if and only if $\vdash \mathscr{B}$ holds (since $\vdash$ is consistent, $\mathscr{B} \neq \varnothing$ is necessary). What is not so obvious is that $\vdash$ satisfies (R), (M), and (T).

Actually the rules (R) and (M) are very easy to verify, so we shall concentrate on (T). A special case will illustrate the method:

$$\text{(6)} \qquad \frac{\begin{array}{c} \vdash A^n \to [B \vee C] \\ \vdash [A \wedge B]^m \to C \end{array}}{\vdash A^{n \cdot m} \to C} .$$

If rule (6) holds for $\vdash$, it is then clear that $\vdash$ satisfies (T). Whether the estimate $n \cdot m$ in the conclusion is the best possible, I do not know; but it makes for an easy inductive proof.

In case $n = 1$, since $m > 0$, several applications of (T) will yield the conclusion $\vdash A \to C$. Of course such rules as those in Theorem 3.2 are also required to move the parts around.

Assume then that rule (6) holds for n for *all* A, B, C. We wish to prove it for $n + 1$. Assume that

$$\text{(7)} \qquad\qquad \vdash A^n \to [A \to [B \vee C]],$$

$$\text{(8)} \qquad\qquad \vdash [A \wedge B]^m \to C.$$

Now from the general rules we can easily prove

$$\text{(9)} \qquad\qquad A \to [B \vee C] \vdash A \to B, \quad A \to C;$$

so from (7) and (9) we derive

$$\text{(10)} \qquad\qquad \vdash A^n \to [A \to B] \vee [A \to C].$$

Note too the provability of

$$\text{(11)} \qquad\qquad A \to C \vdash A^m \to C.$$

Thus if we let $B' = [A \to B]$ and $C' = A^m \to C$ we find

$$\text{(12)} \qquad\qquad \vdash A^n \to B' \vee C'.$$

Next note the probability of

$$\text{(13)} \qquad\qquad [A \wedge B] \to C \vdash [A \to B] \to [A \to C].$$

(Why? Because it is equivalent to

$$\text{(14)} \qquad\qquad [A \to B] \vdash [A \to [[A \wedge B] \to C] \to C],$$

which by $(\to_1)$ is equivalent to

$$\text{(15)} \qquad\qquad A, B \vdash [[A \wedge B] \to C] \to C.$$

In turn (15) is equivalent to

$$\text{(16)} \qquad\qquad A \wedge B \vdash [[A \wedge B] \to C] \to C,$$

and to

(17) $$[A \wedge B] \to C \vdash [A \wedge B] \to C.$$

Since (17) *is* provable, so is (13).)

From (8), by repeated applications of (13) changing the choice of C each time, we then obtain

(18) $$\vdash (A \to B)^m \to (A^m \to C).$$

This can be weakened to

(19) $$\vdash [A \wedge [A \to B]]^m \to [A^m \to C],$$

or in other words

(20) $$\vdash [A \wedge B']^m \to C'.$$

By (12) and (20) and rule (6) for B' and C', we derive

(21) $$\vdash A^{n \cdot m} \to [A^m \to C],$$

which is the same as

(22) $$\vdash A^{(n+1) \cdot m} \to C.$$

Thus rule (6) holds for $n + 1$. $\square$

Theorem 3.3 is "the same as" the crucial Lemma 2 of Chang, which was stated in terms of prime ideals of MV-algebras.[8] The new form seems to me to make more "sense", however, when formulated as a property of Łukasiewicz' implication. In particular we can see ways in which $\vdash$ is "better than" $\vdash$. For example, under $\vdash$ implication becomes *transitive* in a conditional form

(23) $$A \to B, \quad B \to C \vdash A \to C.$$

This holds in view of

(24) $$\vdash [[A \to B] \wedge [B \to C]]^2 \to [A \to C].$$

It might be interesting to try to axiomatize $\vdash$, but I have not done so here because the *semantical* interpretation of $\vdash$ does not seem very "natural". We shall only use $\vdash$ as an aid for the completeness proof (in the same way that Chang used his prime ideals).

The quibble about the empty disjunction could have been avoided in Theorem 3.3 in the case where there is an $\perp \in \mathscr{S}$ with $\perp \vdash$. When this is so $\bigvee \varnothing$ could be defined as this $\perp$. Then $\mathscr{A} \vdash \varnothing$ would be allowed to hold (for the appropriate $\mathscr{A}$). With this change in the definition of $\vdash$, we could say without exception that $\vdash$ is always an *extension* of $\vdash$ with the same absolute assertions. This does not mean, as we already noted, that $\vdash$ and $\vdash$ satisfy the same inference *rules*.

---

[8] Cf. Chang [1] and references there to earlier papers.

4. Completeness. In Theorem 3.1 we showed the soundness of the rules of Łukasiewicz' logic for interpretations in ordered Abelian groups. We emphasize again that the valuations were assumed to satisfy not only (6)–(9) of §2, but also the condition (min). We wish to sketch now the converse.

THEOREM 4.1. *Given an entailment relation* $\vdash$ *satisfying the Łukasiewicz rules, if* $\mathscr{A} \vdash \mathscr{B}$ *does not hold, then some system of valuations* $\{v_i : i \in I\} \subseteq \mathscr{V}_{\vdash}$ *provides an interpretation in an ordered Abelian group where* $\mathscr{A} \Vdash_{v_j} \mathscr{B}$ *fails for some* $j \in I$.

To prove the existence of the system of valuations and the ordered Abelian group, we require as a lemma a general separation theorem for sets in a vector space.[9]

LEMMA 4.2. *Let* Q *be the field of rational numbers, and let* V *be a linear vector space over* Q. *Suppose* $\mathscr{P}$ *and* $\mathscr{N}$ *are two subsets of* V *such that no nonempty sum of elements of* $\mathscr{P}$ *is ever equal to a sum of elements of* $\mathscr{N}$. *Then there is an ultrapower* Q^* *of* Q *and a linear functional* $\varepsilon : V \to Q^*$ *such that*

(i) $\varepsilon(A) > 0$ *for all* $A \in \mathscr{P}$;

(ii) $\varepsilon(B) \leqq 0$ *for all* $B \in \mathscr{N}$.

PROOF IN OUTLINE. Finite subsets $\mathscr{F} \subseteq V$ generate finite-dimensional subspaces with $\mathscr{F} \subseteq V_{\mathscr{F}} \subseteq V$. The condition about sums tells us that, in $V_{\mathscr{F}}$, the convex hull of $\mathscr{P} \cap V_{\mathscr{F}}$ is disjoint from the cone generated by $\mathscr{N} \cap V_{\mathscr{F}}$. Hence by the separation theorem in finite-dimensional spaces[10] we can find a linear functional $\varepsilon_{\mathscr{F}} : V_{\mathscr{F}} \to Q$ which satisfies (i) and (ii) for all $A, B \in \mathscr{F}$. Let F be the family of all finite subsets of V and let $\mathfrak{U}$ be an ultrafilter in the algebra of all subsets of F such that

$$\{\mathscr{X} \in \mathbf{F} : \mathscr{F} \subseteq \mathscr{X}\} \in \mathfrak{U}$$

for all $\mathscr{F} \in \mathbf{F}$. We may assume that the functionals $\varepsilon_{\mathscr{F}}$ are extended linearly to all of V. We then form the ultrapower

$$Q^* = Q^{\mathbf{F}}/\mathfrak{U},$$

and define

$$\varepsilon(A) = \langle \varepsilon_{\mathscr{F}}(A) : \mathscr{F} \in \mathbf{F}\rangle/\mathfrak{U}.$$

It is easy to check that ε has the desired properties. □

PROOF OF THEOREM 4.1. Without loss of generality we can assume $\mathscr{B} \neq \varnothing$, because if not, then there are two cases of whether $\mathscr{C} \nvdash \varnothing$ for some $\mathscr{C}$ or not. In the first case $\bigwedge \mathscr{C} \vdash \varnothing$ holds and we can conclude $\mathscr{A} \nvdash \mathscr{B}, \bigwedge \mathscr{C}$, which gives us a nonempty right-hand side. In the second case, the valuation v_0 which is constantly *true* is consistent with $\vdash$. This gives a trivial interpretation with $I = \{0\}$ and only one valuation.

---

[9] This method is related to the arguments of Scott [4] and replaces the idea in Chang [1] of extending an MV-algebra to a group and then using arguments about the first-order theory of ordered Abelian groups. It would seem to be a more direct approach.

[10] This is "classical"; cf. the reference given in [4, p. 235].

Let $C_0 = [\bigwedge \mathscr{A} \to \bigvee \mathscr{B}]$ so that $\not\vdash C_0$. Using Theorem 3.3 we know that $\not\vdash C_0$ also. Let $\vDash$ be a maximal consistent extension of $\vdash$ where $\not\Vdash C_0$. As we remarked in §3, we can suppose $\vdash$ extends $\vdash$; and so $\vDash$ does also. We define two relations between elements of $\mathscr{S}$ by means of $\vDash$:

(1) $\qquad\qquad\qquad A \gtreqqless B \quad$ iff $\quad \vDash A \to B$;

(2) $\qquad\qquad\qquad A \equiv B \quad$ iff $\quad A \gtreqqless B$ and $B \gtreqqless A$.

By virtue of the construction of $\vdash$, we find that $\gtreqqless$ is reflexive and transitive. In view of 3.2(iii) and the maximality of $\vDash$, the relation is also connected. Thus, since $\equiv$ is an equivalence relation, $\mathscr{S}/\equiv$ is linearly ordered. It is also easy to prove that $\equiv$ preserves the operations $\wedge$, $\vee$, $\to$ on $\mathscr{S}$; thus $\mathscr{S}/\equiv$ becomes an *algebra*, but we need no more than the congruence properties of $\equiv$ and will not compare this algebra with any other. (The *use* of one algebra is not enough justification for a *theory* of special algebras.)

We now take $\mathscr{S}$ to be the set of *basis* elements for a linear vector space V over Q. We may suppose for simplicity that $\mathscr{S} \subseteq V$. The sets $\mathscr{P}$ and $\mathscr{N}$ consist respectively of elements of the forms

(3) $\qquad\qquad A, \qquad\qquad$ where $\not\vdash A$; and

(4) $\qquad\qquad \pm(C + A - B), \quad$ where $C \equiv A \to B$ and $B \gtreqqless A$.

Suppose for a moment that we can show that $\mathscr{P}$ and $\mathscr{N}$ satisfy the hypothesis of Lemma 4.2. Let $\varepsilon: V \to Q^*$ be the linear functional that satisfies 4.2(i) and 4.2(ii). (We could have made ε bounded by the unit of Q^* if we had wanted to.)

Clearly ε is 0 on $\mathscr{N}$ because $\mathscr{N} = -\mathscr{N}$. If we put $A = B = C$ in (4) then $\varepsilon(C) = 0$ if $C \equiv C \to C$. Since, for all $C \in \mathscr{S}$, either $C \equiv C \to C$ or $\not\vdash C$, we see that ε is *nonnegative* on all of $\mathscr{S}$. We define valuations v_i for $i \in I = [0, \infty) \subseteq Q^*$ by the formula

(5) $\qquad\qquad\qquad v_i(A) = \mathbf{t} \quad$ iff $\quad \varepsilon(A) \leqq i$.

It then follows that

(6) $\qquad\qquad\qquad v_0(C_0) = \mathbf{f}.$

Note too that the system $\{v_i : i \in I\}$ satisfies (min) and condition (9) of §2.

To check (6) and (7) we use the fact that $\gtreqqless$ is a linear ordering. In case $A \gtreqqless B$, we note that

(7) $\qquad\qquad\qquad A \wedge B \equiv A,$

(8) $\qquad\qquad\qquad A \vee B \equiv B.$

We note too that

(9) $\qquad\qquad\qquad \varepsilon(C) = \varepsilon(D), \quad$ if $C \equiv D.$

Therefore, in general,

(10) $\varepsilon(A \wedge B) = \max(\varepsilon(A), \varepsilon(B))$,
(11) $\varepsilon(A \vee B) = \min(\varepsilon(A), \varepsilon(B))$.

Finally, by construction of $\mathcal{N}$,

$$\begin{aligned}
(12) \qquad \varepsilon(A \rightarrow B) &= 0 && \text{if } A \gneqq B, \\
&= \varepsilon(B) - \varepsilon(A) && \text{if } B \gneqq A.
\end{aligned}$$

From (5) above and (10), (11) and (12), the required (6), (7) and (8) of §2 now follow. It remains only to check the condition on $\mathcal{P}$ and $\mathcal{N}$.

Suppose we had an equality in V of the form

(13) $P_0 + \cdots + P_n = N_0 + \cdots + N_m$.

Since the elements of $\mathcal{S}$ are *independent* basis elements, this equation means that the *formal* combination on the right-hand side has to *cancel out* to the left-hand side. The form of the right-hand side would be best visualized in a rectangular array:

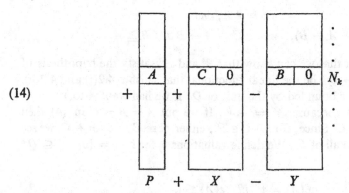

(14)

In each row we find the terms of a vector N_k. These have each three terms with two pluses and one minus or two minuses and one plus. One row has been illustrated: It has five spaces two of which have been set equal to zero. The zeros are placed so that the sum of the first column is $P = P_0 + \cdots + P_n$ (possibly with some zeros), whereas the sum of the remaining plus terms, call it X, is *formally cancelled* by the sum of the minus terms, called Y in (14). What we must show is that such a configuration is *impossible*.

If there were such a configuration, there would be one with a *minimum* number of terms in $\mathcal{P}$, i.e., with a minimum number of terms A where $\nleq A$. Inasmuch as there is *at least* one P-column term that must be nonzero and in $\mathcal{P}$, we may suppose it is the first in N_0. We must argue that if any *plus* term in an N_k is in $\mathcal{P}$, then at least one *minus* term is the negative of a $\mathcal{P}$-element. The possible cases are illustrated thus:

(15)
$$C + A - B \quad or$$
$$A + C - B \quad or$$
$$B - A - C$$

where $B \geqq A$ and $C \equiv A \to B$ and the *first* term in each of the three cases is the one in $\mathscr{P}$. Take them in order.

In the first instance, if $\vDash B$, then $\vDash A$ and $\vDash A \to B$, but we have $\nvDash C$. So $\nvDash B$. In the second, if $\vDash B$, then $\vDash A$. But $\nvDash A$, so $\nvDash B$. In the third, if $\vDash A$ and $\vDash C$, then $\vDash A \to B$; but then $\vDash B$, which is wrong. Thus $\nvDash A$ or $\nvDash C$ follows.

The trick now is to follow through the way the $\mathscr{P}$-terms cancel. A typical situation is illustrated as follows:

(16)

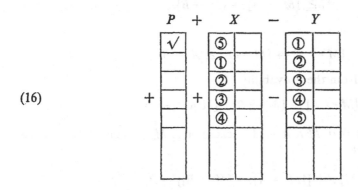

We began with a $\mathscr{P}$ in the P-column in the place marked $\sqrt{}$ in the upper left-hand corner. A minus term, marked ①, must also be a $\mathscr{P}$ and be in the same row. This ① must cancel *another* ① in the X-columns. (It has been moved to the second row.) This ① is also a $\mathscr{P}$, so there must be a minus term, marked ②, in the same row which is also a $\mathscr{P}$. And so on. A plus $\mathscr{P}$ *forces* a minus $\mathscr{P}$ which must cancel with a plus $\mathscr{P}$ in another row. Finally a *cycle* must be closed, since there are only finitely many rows. (The rows have been neatly ordered in (16) and the cycle closed at the *top*. But this was only an illustration: The cycle could have closed below the top. It is all the same; the term in the first column was only used to get the chain going—the only thing that is important is the existence of a cycle.)

Among all the terms in this cycle there must be one that is *minimal* in the sense of the linear ordering $\geqq$. Call this term M. It is in $\mathscr{P}$, and, in effect, we are going to *subtract* it from all the terms in the cycle. There are again several cases—four to be exact—but they reduce to two. We are going to replace each ⓚ in (16) by the term $(M \to$ ⓚ$)$. If we can show that the *rows* are still in $\mathscr{N}$, the new configuration will have the same P-property of the original (14) but it will have at least one *less* $\mathscr{P}$-term since $(M \to M) \notin \mathscr{P}$. This contradicts the assumed minimality of configuration (14).

The argument about the rows goes as follows: We wish to replace $\pm(C + A - B)$ by either

$$\pm(C + [M \to A] - [M \to B]) \quad \text{or} \quad \pm([M \to C] + A - [M \to B]),$$

where $B \geqq A$ and $C \equiv A \to B$. In the first instance $A \geqq M$; while in the second $C \geqq M$ (in any case $B \geqq C$). The question is whether the new vector is in $\mathcal{N}$. In the first instance we are asking whether $M \to B \geqq M \to A$ and $C \equiv [M \to A] \to [M \to B]$. Since it is provable that

(17) $B \to A \vdash [M \to B] \to [M \to A],$

the first inequality follows along with

(18) $C \geqq [M \to A] \to [M \to B].$

We must still show

(19) $[M \to A] \to [M \to B] \geqq A \to B.$

Statement (19) comes from the provability of

(20) $[M \to A] \to [M \to B] \vdash [M \wedge A] \to B$

together with the fact that $M \wedge A \equiv A$ because $A \geqq M$. Statement (20) comes from the provability of

(21) $M, A \vdash [[M \to A] \to [M \to B]] \to B,$

which in turn is equivalent to

(22) $M \to A \vdash M \to [[[M \to A] \to [M \to B]] \to B].$

This last is the same as

(23) $M \to A \vdash [[M \to A] \to [M \to B]] \to [M \to B],$

which indeed is provable.

Finally in the second instance we must show that $[M \to B] \geqq A$ and $[M \to C] \equiv A \to [M \to B]$. The equivalence is obvious because $A \to [M \to B] \equiv M \to [A \to B]$ and $C \equiv A \to B$. To prove the inequality remember that $A \to B \equiv C \geqq M$. We then find that $M \to B \geqq [A \to B] \to B \equiv A \vee B \equiv A$ because $B \geqq A$. The whole proof is now complete. $\square$

Theorem 4.1 has been given for "abstract" sets $\mathcal{S}$ in complete generality. If we considered syntactical *formulae* built up from sentential variables in the usual way, and if we "gave" the entailment relation $\vdash$ by means of *deductions* (the least relation satisfying the rules), then the argument stripped of the ultrapower refinement would show that an entailment is provable if and only if it is valid in all *finite* Łukasiewicz truth tables.

5. Conclusion. The proof is not entirely trivial, but the point of the presentation here is that all the *steps* are familiar—and that the axioms given in terms of $\vdash$ are easier to work with than those commonly used. I have also tried to show that if one started with such a problem, not knowing the answer, then the application of the general theorem on linear functionals (which is obviously very appropriate for this particular system) would *lead* one to the required axioms—since certain checks are necessary and certain properties are forced on our attention. (In fact I corrected an error in the original presentation in just this way.) We shall see whether other problems yield to this approach.

A problem that remains is to axiomatize the theory without the condition (min). This is interesting since the motivation of *degree of error* does not at all seem to demand (min). In particular in a predicate logic it would be more natural to define valuations that satisfied the condition

(1) $\qquad v_i(\exists x.A(x)) = t \quad$ iff $v_i(A(a)) = t$ for some $a \in U$.

In *infinite* universes U such quantified formulae would quite easily fail to satisfy (min). There is also an intriguing point that we might want to consider *variable* universes U_i in the style of the so-called Kripke models for intuitionistic logic. Before we do so, however, we ought to ask whether *degrees of error* form just a pleasant toy or whether their "logic" has serious applications.

BIBLIOGRAPHY

1. C. C. Chang, *A new proof of the completeness of the Łukasiewicz axioms*, Trans. Amer. Math. Soc. **93** (1959), 74–80. MR **23** #A58.

2. L. Henkin, J. D. Monk and A. Tarski, *Cylindric algebras*. Part I, North-Holland, Amsterdam, 1971.

3. N. Rescher, *Many-valued logic*, McGraw-Hill, New York, 1969.

4. D. Scott, *Measurement structures and linear inequalities*, J. Math. Psychology **1** (1964), 233–247.

5. ————, "Background to formalization. I," in *Truth, syntax and modality* (H. Leblanc, editor), North-Holland, Amsterdam, 1972.

6. A. Tarski, *Logic, semantics, metamathematics. Papers from 1923–1938*, Clarendon Press, Oxford, 1956. MR **17**, 1171.

7. A. Urquhart, *An interpretation of many-valued logic*, Z. Math. Logik Grundlagen Math. (to appear).

OXFORD UNIVERSITY, OXFORD, ENGLAND

D. Scott (✉)
Computer Science Department, Carnegie Mellon University, Pittsburgh, PA, USA
e-mail: dana.scott@cs.cmu.edu

© American Mathematical Society 1974. Reprinted, with kind permission, from: Scott, D., "Completeness and axiomatizability in many-valued logic". In: L. Henkin (ed), Proceedings of the Tarski Symposium, American Mathematical Society, Providence, pp. 411–435.

Part 13
J.A. Goguen and R.M. Burstall (1984)

Three Decades of Institution Theory

Răzvan Diaconescu

Abstract About three decades ago the paper 'Introducing Institutions' by Goguen and Burstall, included in this anthology, marked the beginning of the major trend in universal logic known as 'institution theory'. In this paper we survey some of the rather vast body of institution theory developments that have happened during this period of time, fuelled by computing science or by pure logic motivations. Our survey includes also some historical facts that are little known to the audience, but which may explain some of the mysteries surrounding this extraordinary research area.

Keywords Institution theory · Universal logic · Formal specification · Algebraic specification

Mathematics Subject Classification (2000) 00A30 · 03C95 · 18C50 · 68Q65

1 Introduction

30 years have passed since the introduction by Joseph Goguen and Rod Burstall of the concept of 'institution' (in [14] under the name 'language'). Since then institution theory has gradually developed from a simple and strikingly elegant general category theoretic formulation of the informal notion of logical system into an important trend of what is now called 'universal logic', with substantial applications and implications in both logic and computing science.

During this period of time many things happend. Very sadly, Joseph Goguen passed away in 2006, leaving behind an incredibly diverse scientific inheritance that will benefit many generations to come. It will perhaps take another several decades to fully understand the implications of his ideas. Rod Burstall has retired from the academia years ago. New young scientists from various fields of science continue to join and contribute to the growth of institution theory, for many of them this activity being an important component of their professional career. A worldwide distributed group of people working in this area is known under the name FLIRTS (http://www.informatik.uni-bremen.de/flirts/).

The aim of this presentation is to guide the reader through the development of institution theory from its seminal paper, included in this anthology, to its current status. We will recall important moments in this process, and discuss the most significant contributions of institution theory. Due to the rather big size of institution theory literature and also due to incompleteness in my knowledge, the omission of important works from this survey and from its references is inevitable. I apologize for all such omissions.

J.-Y. Béziau (ed.), *Universal Logic: An Anthology*, 309–321
Studies in Universal Logic, DOI 10.1007/978-3-0346-0145-0_25, © Springer Basel AG 2012

2 The Initial Decade

The Computing Science Origins   Institution theory may be the only important trend in universal logic that has emerged from within computing science, the others emerging from logic, and perhaps philosophy. This origin of institution theory may surprise many, since computing science is often blamed for its poor intellectual value. While this perception may be generally correct in average, there are very significant exceptions. In fact, what is now labelled as 'computing science' is hardly a science in the way mathematics or physics are. Some say it is still too young and there was not enough time to coagulate, however one can hear this since 50 years already and probably in the next 50 years too. It may be more realistic to view computing science as a playground were several actors, most notably mathematics, but also logic, engineering, philosophy, sociology, biology, play. Sometimes an extremely interesting play, which has not only brought significant changes and developments to the actors, but also has revolutionized our scientific thinking in many ways. Institution theory is such an example, the way we think of logic and model theory will never be the same as before.

As the paper included in this anthology shows, the birth of the institution concept came as a response to the population explosion of logical systems in use in specification theory and practice at the time. People felt that many of the theoretical developments (concepts, results, etc.), and even aspects of implementations, are in fact independent of the details of the actual logical systems, that especially in the area of structuring of specifications or programs, it would be possible to develop the things in a completely generic way. The benefit would be not only in uniformity, but also in clarity since for many aspects of specification theory the concrete details of actual logical systems may appear as irrelevant, with the only role being to suffocate the understanding. The first step to achieve this was to come up with a very general formal definition for the informal concept of logical system. Due to their generality, category theoretic concepts appeared as ideal tools. However there is something else which makes category theory so important for this aim: its deeply embedded non-substantialist thinking which gives prominence to the relationships (morphisms) between objects in the detriment of their internal structure. Moreover, category theory was at that time, and continues even now to be so, the mathematical field of the upmost importance for computing science. In fact, it was computing science which recovered the status of category theory, at the time much diminished in conventional mathematical areas. The essay [41] that Joseph Goguen wrote remains one of the most beautiful essays on the significance of category theory for computing science and not only.

The Name   In [14] the 'institutions' were called 'languages', but this did not last long. I think every newcomer to the area has wondered about the name 'institution' for the formal mathematical definition of the concept of logical system. It is not straightforward to see the connection between the meaning of this word in common languages and its meaning as an abstract mathematical structure. The fact that mathematics provides other notorious examples of this phenomenon, such as 'group' or even 'category', does not help much. I have heard the following explanation from Joseph Goguen in an Oxford café during my DPhil years. Apparently this name was given half-joke half-truth. At the time (but I think this is still true nowadays), computing scientists had a strong tendency to create social institutions around logical systems. They believed so much in their own logic, that they committed themselves to promoting it, building tools and systems implementing it, starting workshops and conferences devoted to that logic and its applications.

The ACM Paper   The paper [14] is perhaps the first properly published work introducing the concept of institution. Moreover, [14] also develops some basics of institution theory but disguised as properties of equational logic. However Goguen and Burstall mention clearly that this was done only to protect some of the audience from the hardship of abstract thinking, since all those properties can be presented at the level of abstract institutions. Although the first publication focused on the concept of institution is the one included in this anthology ([44] in our list of references), which many consider to be the first "true" institution theory publication, the academic community tends to cite [46] as the seminal paper of this area. This is so because the latter paper is a journal publication and consequently is more elaborated. What puzzles a bit is the rather late publication year of [46]. I remember when during my undergraduate years around mid 1980's, one of my computing science professors in Bucharest noticing how much I was in love with category theory, my passion for model theory, and also my lack of interest for other research trends available in our university at that time, gave me a draft of the institution paper included in this anthology, that at the time was circulating in the community. I can say that in sense that event completely shaped my professional future, I immediately realized that my interests fit perfectly the institution theoretic perspective. It is ironic that about six years later I found myself making some small contributions to the last version of its journal version [46], just before its printing. A prominent German scientist of the same generation also confessed to me that when he first read the same paper he was 'electrized', and one can see now how much institutions are part of his professional achievements. It took journal of ACM not less than 9 years to publish the paper. Why that long? Joseph Goguen explained to me that at some point, after the final acceptance, the chief editor of the journal of ACM was constantly delaying the actual printing of the paper, so strong was his emotion against the type of thinking promoted by that work. In a way what has happened since then proved him right... During my career I have encountered regularly such kind of emotional reactions, which can be explained by the fear induced by approaches going against the substantialist way of thinking characteristic to the classical western scientific culture.

Myriad of Logics as Institutions   An important activity of the initial decade of institution theory was to formalize various logics from computing science as institutions. This was mainly motivated by the wish to make use of the specification theoretic results and methods developed at the general level of abstract institutions to the respective logics. There was also another beneficial consequence, the process of formalizing a logic as institution has often led to a conceptual clarification of that logic. The paper included in this anthology presents various fragments of many-sorted first order logic as institutions. Other less conventional logics from computing science have been captured as institutions in a series of papers, some of them never properly published. These includes order sorted algebra in simple form [75] and with sort constraints [63, 86], unified algebras [69], lambda calculus [87], higher order logic with polymorphic types [70], multiple valued [1] and fuzzy logics [40], hidden sorted algebra [42] (and [13] for the order sorted extension), Edinburgh LF [74], or even a model theory for objects, XML, and databases [3].

The Beginnings of Institution-Independent Model Theory   The definition of the concept of institution provides an ideal meta-mathematical framework for the development of a true abstract model theory free of any commitment to a specific logical system. The main axiom of institutions, the so-called 'satisfaction condition', is inspired by the work of Barwise and others [10, 11], a trend known as 'abstract model theory'. However that trend was only concerned with extensions of conventional logic, hence one may say it is only 'half-abstract'. The true independence from actual logical systems has been achieved by institution theory through a full categorical abstraction of the main logical concepts of signature, sentence, model, and of the satisfaction relation between them. Other general categorical approaches to model theory such as works on sketches [38, 53, 85] or on satisfaction as cone injectivity [4–6, 56–58] are also unsatisfactory from the point of view of a true abstract model theory. While the former just develops another language for expressing (possibly infinitary) first order logic realities, the latter considers models as objects of abstract categories but it lacks the multi-signature aspect of institutions given by the signature morphism and the model reducts, which leads to severe methodological limitations. Moreover, in these categorical model theory frameworks the satisfaction of sentences by the models is usually defined rather than being axiomatized.

The first developments of an abstract model theory at the level of abstract institutions belong to Andrzej Tarlecki. Although they were motivated by model theoretic aspects in algebraic specification, they did not have a clear computing science flavor. These works include results about existence of free models of theories [80], axiomatizability of quasi-varieties [81], the initial formulation of the method of diagrams for abstract institutions [81], elements of internal logic such as Boolean connectives and quantifiers for abstract institutions [79]. The work [79] contained also the first formulation of the Craig interpolation property and of a very fundamental form of model amalgamation in abstract institutions.[1] In subsequent developments the latter property gained a crucial role since the majority of computing science or model theory results rely upon this form of model amalgamation. Conventional logic and model theory was unable to realize the importance of this form model amalgamation since most of the actual logics, in fact all of the conventional ones, have this property rather tacitly, and also because of the single-signature orientation in conventional logic and model theory.

3 The Computing Science Decade

In the nineties institution theory has witnessed very few model theoretic driven developments without computing science significance. It is mainly for this reason that we call this period the 'computing science decade'.

Foundations of Specification Languages   During this period institution theory achieved recognition as the most fundamental mathematical structure underlying formal specification, especially algebraic. It has thus become standard to base the definition of specification languages upon logic systems captured as institutions such that all the language constructs are reflected rigorously as mathematical entities in the respective institutions. Moreover, there was the awareness of the importance of a series of model theoretic properties of the underlying institution, as a guarantee for good semantic properties of the respective language, an important example being model amalgamation.

---

[1]Here we refer to a rather common form of model amalgamation across signature morphisms. This is very different from another form of model amalgamation much used in conventional model theory [54], which is across model homomorphisms, and which is much less common.

The nineties was the time of the development of the latest generation of algebraic specification languages. CafeOBJ [30, 31] and Maude [19] emerged as direct successors of the famous OBJ language [48], while CASL [8] was the result of a joint European effort to unify a series of specification frameworks into a new modern language. The definitions of both CafeOBJ and CASL have been strongly based upon institution theory. Due to some errors in the design of rewriting logic [60], Maude failed shortly from having an underlying institution.

Institution-Independent Specification and Programming   The effort to develop specification and programming theory at a generic level independent of any particular institution gave results especially in the area of modularization (structuring) of specification and programs, the so-called specification/programming in-the-large paradigm. These results showed that this paradigm is essentially institution-independent. The work [77] developed the semantics of a set of generic structuring operators at the level of arbitrary abstract institutions, concrete structuring constructs of actual specification languages being derived as combinations of these generic operators. A somehow parallel approach was that of the so-called 'module algebra' of [32], which developed an algebra for software modules applicable to any language rigorously based on an underlying logic captured as institution. The latter work revealed an intimate relationship between the semantical properties of the structuring mechanism and the interpolation properties of the underlying institution. A similar conclusion had emerged from studies on modularization [36, 37, 83] using the so-called 'π-institutions' of [39], an entailment theoretic abstraction of the concept of institution. Moreover, in [12] interpolation properties have shown to represent a crucial condition for a generic institution-independent lifting of complete proof calculi from the level of the basic specifications to that of the structured specifications built with the operators of [77]. At this moment it is important to remind the reader that all this series of institution theoretic developments has been effectively used for the design of languages such as CafeOBJ and CASL, and have also had a strong impact on their associated specification and verification methodologies.

Logic Translations   The study of translations between logical systems has an old tradition in logic, and it lies at the core of the universal logic approach since they are a concrete expression of a fundamental philosophical principle relevant for universal logic, that of the co-dependent origination, or interdependency, of logical systems. Therefore, it does not surprise that right from the beginning institution theory has developed concepts of maps between institutions and used them for various purposes such as expressing a logic into another, or for borrowing logical properties or even tools (such as theorem provers). These maps are defined such that they preserve the mathematical structure of the concept of institution. There are two main ways to define such structure preserving maps leading to two main kinds of homomorphisms between institutions: morphisms and comorphisms. Conceptually they are dual to each other, however their use differs a lot. While the former usually expresses a forgetful relationship between a more complex and a simpler institution, the latter is used to formalize embeddings of simpler logics into more complex ones, or to formalize encodings of more complex logics into simpler ones by means of the theories of the simpler logic. While the study of institution morphisms had started with the institution paper included in this anthology, the awareness about comorphisms has developed only gradually (an early reference being [59]). The work [47] is a survey on institution morphisms and comorphisms discussing both structural and methodological aspects related to these concepts of institution mappings. The duality between morphisms and comorphisms has been mathematically shown in [7] in the sense that under an adjunction between the categories of the signatures of the institutions $\mathcal{I}$ and $\mathcal{I}'$, morphisms $\mathcal{I}' \to \mathcal{I}$ and comorphisms $\mathcal{I} \to \mathcal{I}'$ bijectively determine each other. While comorphisms provide the main mathematical notion for developing a systematic theory of doing logic by translation in the sense of the transfer of properties (and even tools, from a more applied perspective) from one logical system to another, for this aim the community has also explored other less established notions for mappings between institutions [61]. Hence doing logic by translation has become a major trend within institution theory, an important pioneering work being [18]. On a more computing science note, institution mappings have been also used to relate formally between specification languages [63].

Logic Combination   There was an institution theoretic effort towards this notoriously difficult problem using the concept of *parchment* (see [68]). The so-called 'charters' and 'parchments' have been introduced by the fathers of institution theory, Goguen and Burstall, in [45] as generic technical devices to present institutions, the main axiom (i.e. the satisfaction condition) of institutions being derived at a very general level. While mathematically the charters represent a middle layer between parchments and institutions, the parchments appear as a rather useful concept in its own since they represent meta-level many sorted equational specifications of both the syntax and the semantics of actual logical systems. Later on, the parchment based work [45] inspired other efforts towards the problem of logic combination, such as [16] and [17]. A completely different approach to logic combination is to internalize the features of a specific logic $L1$ to abstract institutions. Then any actual logic $L2$ considered in the role of the abstract institution gives rise to a combination between $L1$ and $L2$. This idea has been realized for possible worlds semantics in [35].

4 The Model Theory Decade

In the third decade all the computing science inspired trends and applications mentioned above have continued. Moreover new applications of institution theory, outside formal specification or declarative programming, have emerged in areas such as ontologies and cognitive semantics [43], concurrency [67], or quantum computing [15]. But probably the most significant developments during this decade were the so-called 'Grothendieck institution' approach to multi-logic heterogeneous specification and the renaissance of a strong model theory activity within institution theory that was not primarily computing science motivated, and which continued at a much deeper level what has been started in the initial decade. Consequently, for the first time institution theoretic papers have been published by non-computing journals such as *J. Symbolic Logic*, *Studia Logica*, or *Logica Universalis*, and institution theory has emerged as an important actor for the universal logic programme. For this reason let us call this decade the 'model theory decade'. The recent monograph [29] includes most of the institution-independent model theory resulting from this activity.

Multi-logic Heterogeneous Specification   One of the important applications of the institution theoretic approach to logic translations is that of specification languages and frameworks based upon a system of logics rather than upon a single logic. This recent paradigm reflects the understanding that different applications might require different logics, that no single logical system is appropriate for a variety of applications which differ substantially in their nature. One of the earliest works in this direction is [82]. The first specification language that was designed as a multi-logic heterogeneous language was perhaps CafeOBJ [30, 31]. Its semantics was based upon a system of institutions, each of them reflecting a particular specification paradigm, and these were related by a network of embeddings defined formally as comorphisms. A serious problem had emerged: how to make use of the rich existing institution theoretic specification technology for such situation which is not based upon a single underlying institution. One solution, explored in [21], was to extend the institution-independent specification theory, including all basic concepts and results, from a single abstract institution to a system of institutions, i.e. a diagram of institutions, more precisely. However the thinking along these lines has led to a rather different and much more efficient solution, namely that of the 'flattening' of the respective diagram of institutions to a single institution by extending a corresponding construction from category theory [49] to institutions. The resulting concept of *Grothendieck institution* [22, 62] is emerging as the fundamental mathematical structure for the multi-logic heterogeneous specification paradigm. Apart of CafeOBJ, the heterogeneous specification framework with CASL extensions [64] is also based upon the theory of Grothendieck institutions. Quite surprisingly, Grothendieck institutions have been applied to pure model theory, such as for obtaining interpolation results [29].

Doing Model Theory Without Concrete Structure   The development of model theory at the very general level of abstract institutions is based upon the observation that the most important model theory methods are independent of the conventional first order logic context in which they have originally been developed. This means that all these methods can be formulated and developed at a much more abstract level independent of any particular logical structure. The breakthrough was given by the institution-independent method of ultraproducts [23], which was followed by a rather drastic reformulation in [24] of the institution-independent method of diagrams of [80, 81]. The development of institution-independent saturated model theory [29, 33] came a bit later. These have been used for developing general results about compactness [23], axiomatizability [29, 80, 81], elementary chains [51], interpolation [25, 52], definability [72], completeness [20, 71], generating a big array of novel concrete results in actual unconventional, or even in conventional well studied logics. Moreover, the institution-independent approach to model theory makes the access to highly difficult model theoretic results considerably easier, an example being the Keisler–Shelah isomorphism theorem [29, 33].

Illuminating Model Theoretic Phenomena   The institution-independent approach has lead to the redesign of important fundamental logic concepts and to the clarification of some causality relationships between model theoretic phenomena including the demounting of some deep theoretical preconceptions. One such example is that of interpolation which has been extended to sets of sentences instead of single sentences and to arbitrary commutative squares of signature morphisms instead of the traditional intersection-union squares of signatures. The first extension corrects a traditional misunderstanding about the lack of interpolation properties of logics such Horn clause logic or equational logic. It is the merit of [76] to have proved a Craig interpolation property for sets of sentences in equational logic based upon its Birkhoff-style axiomatizability property, thus revealing a previously unknown cause for interpolation. This idea has been generalized to abstract institutions in [25], thus leading to a myriad of new concrete interpolation results (for fragments of first order logic see also [73]). The second extension of the interpolation concept comes from the practice of algebraic specification which requires interpolation for arbitrary pushout squares of signature morphisms. When interpolation is considered in this way a significant difference between the single and the many sorted logics shows up. The interpolation problem for many sorted first order logic, which stayed for several years as a conjecture, had received a rather elegant solution in [52] as a particular concrete case of a general institution-independent interpolation result. The institution theoretic study of interpolation has also revealed that the Craig-Robinson form of interpolation [78], which stregthens the Craig formulation by adding to the set of the premises a set of 'secondary' premises from the second signature, is actually more appropriate than the traditional Craig formulation. This conclusion is motivated by applications such as definability [29, 72], translation of interpolation [27, 29], modularization of formal specifications [32, 37, 83], completeness of structured specifications proof calculi [12, 29]. A somehow similar situation happens with (Beth) definability, it can also be extended to arbitrary signature morphisms and formulated more properly in terms of sets of sentences [66, 72], and it can also be obtained as a consequence of Birkhoff-style axiomatizability properties [72]. Another example is given by completeness, which was discovered to have a 'layered' structure as explained below. Both Birkhoff and Gödel-Henkin forms of completeness have been developed at the generic level of abstract institutions in [20] and [50, 71], respectively, by a technique common to both of them, originally developed by [12], and which consists of separating the proof rules and the completeness phenomenon on several layers. In this approach the base layer consists of an institution with a given sound and complete proof system. Since this base layer refers usually to the 'atomic' sentences, its completeness is rather easy to establish in each particular case. The other layers are built on top of the base layer successively by considering more complex sentences and consequently adding new proof rules and meta-rules. This layered construction is done fully abstractly and the respective completeness results are proved fully generally relative to the completeness of the predecessor level, thus leading especially in the Birkhoff case to a multitude of concrete complete proof calculi for various logics, some of them rather unconventional. Many of these complete proof calculi are new, and quite surprising in that they appear rather remote from the original Birkhoff completeness.

Stratified Institutions   This is a recent refinement of the concept of institution which captures uniformly the concept of open formulæ and the concept of models with states (such as possible worlds semantics for modal logics) in a fully abstract setting. Stratified institutions have been developed in [2, 9], however a precursor can be found in [34]. They have already been used to develop a very general version of Tarski's elementary chain theorem applicable to both classical and non-classical (i.e. modal) logics. Stratified institutions also represent a big promise for logic combination, which is one of the great challenges in contemporary logic.

Proof Theoretic Developments   Although institution theory is primarily model theoretic approach, there have been a proof theory development within institution theory

[26, 29, 59, 66, 74] motivated primarily by the foundations for formal verifications. The main goal of the recent approach to proof theory of [26, 29, 66] is to liberate it from the Curry–Howard isomorphism dogma in order to achieve greater simplicity, generality, and harmony with the model theory. Another recent approach to extend institutions with proofs is proposed by [74], its most interesting feature being the conceptual symmetry between the model and the proof theory. Technically speaking, the proof theory of [26, 29, 66], as well as that of [74], follows the proofs-as-arrows idea of categorical logic [55], but it has a much broader range of applications than the latter. Moreover it treats concepts such as implication or quantifiers in a more realistic manner than in categorical logic (for example in categorical logic implication presupposes conjunctions).

Categorical Abstract Algebraic Logic   Although algebraic logic is not a model theoretic approach, we should also mention here the new trend called 'categorical abstract algebraic logic' that aims to develop algebraic logic at the level of abstract π-institutions. The paper [84] is a representative work from a long series of publications by the same author on this topic.

The UNILOG Connection   One of the consequences of the worldwide recent growth of universal logic activities is the UNILOG series of congresses and schools on the topic, one each two or three years. The theory of institutions has been an active UNILOG actor right from the beginning. Each UNILOG congress poses a research question and organizes a contest of papers on the respective topic. The institution theory paper came second in the first UNILOG contest (Switzerland 2005) for the question 'what is the identity of a logic' [66] and the same authors won the contest of the next congress (China 2007) for the question 'what is a logic translation' [65].

5 Looking to the Future

Future is hard to predict, especially in the current climate of scientific research in which theories are developing and trends are changing at an increased speed. Institution theory is already established as the most fundamental mathematical structure for logic based specification theory, and in this sense it will continue to play its foundational role. Moreover institution theoretic ideas will continue to spread in other areas of computing science, however it is difficult to see exactly in which of these and how. In the next period I think the interest for developing model theory at the very general level of abstract institutions, as part of the universal logic trend, will continue to grow. A related area of great interest consists of applying institution-independent model theory to provide a model theory for logical formalisms that do not have a proper one. The new developments such as stratified institutions and the institutional proof theory also represent a big research promise. In longer term I think the most important message given by institution theory is the non-substantialist way of thinking it promotes and its associated top-down methodologies (see [28] for a philosophical essay on this topic).

Acknowledgements   Thanks to Till Mossakowski, in general for his friendship and collaboration over the years, and in particular for reading a preliminary draft of this survey and for making a series of useful suggestions. Andrzej Tarlecki also helped to confirm some historical issues from the beginnings of institution theory.

References

1. Agusti-Cullel, J., Esteva, F., Garcia, P., Godo, L.: Formalizing multiple-valued logics as institutions. In: IPMU 1990. Lecture Notes in Computer Science, vol. 521, pp. 269–278 (1991)
2. Aiguier, M., Diaconescu, R.: Stratified institutions and elementary homomorphisms. Inf. Process. Lett. **103**(1), 5–13 (2007)
3. Alagi, S.: Institutions: Integrating objects, XML and databases. Inf. Softw. Technol. **44**, 207–216 (2002)
4. Andréka, H., Németi, I.: Łoś lemma holds in every category. Stud. Sci. Math. Hung. **13**, 361–376 (1978)
5. Andréka, H., Németi, I.: A general axiomatizability theorem formulated in terms of cone-injective subcategories. In: Csakany, B., Fried, E., Schmidt, E.T. (eds.) Universal Algebra, Colloquia Mathematics Societas János Bolyai, vol. 29, pp. 13–35. North-Holland, Amsterdam (1981)
6. Andréka, H., Németi, I.: Generalization of the Concept of Variety and Quasivariety to Partial Algebras Through Category Theory. Dissertationes Mathematicae, vol. CCIV (1983)
7. Arrais, M., Fiadeiro, J.L.: Unifying theories in different institutions. In: Haveraaen, M., Owe, O., Dahl, O.-J. (eds.) Recent Trends in Data Type Specification. Lecture Notes in Computer Science, vol. 1130, pp. 81–101. Springer, Berlin (1996)
8. Astesiano, E., Bidoit, M., Kirchner, H., Krieg-Brückner, B., Mosses, P., Sannella, D., Tarlecki, A.: CASL: The common algebraic specification language. Theor. Comp. Sci. **286**(2), 153–196 (2002)
9. Barbier, F.: Géneralisation et préservation au travers de la combinaison des logique des résultats de théorie des modèles standards liés à la structuration des spécifications algébriques. PhD thesis, Université Evry (2005)
10. Barwise, J.: Axioms for abstract model theory. Ann. Math. Log. **7**, 221–265 (1974)
11. Barwise, J., Feferman, S.: Model-Theoretic Logics. Springer, Berlin (1985)
12. Borzyszkowski, T.: Logical systems for structured specifications. Theor. Comput. Sci. **286**(2), 197–245 (2002)
13. Burstall, R., Diaconescu, R.: Hiding and behaviour: An institutional approach. In: Roscoe, A.W. (ed.) A Classical Mind: Essays in Honour of C.A.R. Hoare, pp. 75–92. Prentice Hall, New York (1994). Also in Technical Report ECS-LFCS-8892-253, Laboratory for Foundations of Computer Science, University of Edinburgh (1992)
14. Burstall, R., Goguen, J.: The semantics of Clear, a specification language. In: Bjorner, D. (ed.) 1979 Copenhagen Winter School on Abstract Software Specification. Lecture Notes in Computer Science, vol. 86, pp. 292–332. Springer, Berlin (1980)
15. Caleiro, C., Mateus, P., Sernadas, A., Sernadas, C.: Quantum institutions. In: Futatsugi, K., Jouannaud, J.-P., Meseguer, J. (eds.) Algebra, Meaning, and Computation. LNCS, vol. 4060, pp. 50–64. Springer, Berlin (2006)
16. Caleiro, C., Ramos, J.: Cryptomorphisms at work. In: Fiadeiro, J., Mosses, P., Orejas, F. (eds.) Recent Trends in Algebraic Development Techniques. Lecture Notes in Computer Science, vol. 3432, pp. 45–60. Springer, Berlin (2005)
17. Caleiro, C., Ramos, J.: From fibring to cryptofibring: A solution to the collapsing problem. Logica Universalis **1**(1), 71–92 (2007)
18. Cerioli, M., Meseguer, J.: May I borrow your logic? (transporting logical structures along maps). Theor. Comput. Sci. **173**, 311–347 (1997)
19. Clavel, M., Durán, F., Eker, S., Lincoln, P., Martí-Oliet, N., Meseguer, J., Talcott, C.: All About Maude – A High-Performance Logical Framework, Lecture Notes in Computer Science, vol. 4350. Springer, Berlin (2007)
20. Codescu, M., Găină, D.: Birkhoff completeness in institutions. Logica Universalis **2**(2), 277–309 (2008)
21. Diaconescu, R.: Extra theory morphisms for institutions: Logical semantics for multi-paradigm languages. Appl. Categ. Struct. **6**(4), 427–453 (1998). A preliminary version appeared as JAIST Technical Report IS-RR-97-0032F (1997)
22. Diaconescu, R.: Grothendieck institutions. Appl. Categ. Struct. **10**(4), 383–402 (2002). Preliminary version appeared as IMAR Preprint 2-2000, ISSN 250-3638 (February 2000)
23. Diaconescu, R.: Institution-independent ultraproducts. Fundam. Inform. **55**(3–4), 321–348 (2003)
24. Diaconescu, R.: Elementary diagrams in institutions. J. Log. Comput. **14**(5), 651–674 (2004)

25. Diaconescu, R.: An institution-independent proof of Craig Interpolation Theorem. Stud. Log. **77**(1), 59–79 (2004)
26. Diaconescu, R.: Proof systems for institutional logic. J. Log. Comput. **16**(3), 339–357 (2006)
27. Diaconescu, R.: Borrowing interpolation. J. Log. Comput. doi:10.1093/logcom/exr007
28. Diaconescu, R.: Institutions, Madhyamaka and universal model theory. In: Béziau, J.-Y., Costa-Leite, A. (eds.), Perspectives on Universal Logic, pp. 41–65. Polimetrica (2007)
29. Diaconescu, R.: Institution-Independent Model Theory. Birkhäuser, Basel (2008)
30. Diaconescu, R., Futatsugi, K.: CafeOBJ Report: The Language, Proof Techniques, and Methodologies for Object-Oriented Algebraic Specification, AMAST Series in Computing, vol. 6. World Scientific, Singapore (1998)
31. Diaconescu, R., Futatsugi, K.: Logical foundations of CafeOBJ. Theor. Comput. Sci. **285**, 289–318 (2002)
32. Diaconescu, R., Goguen, J., Stefaneas, P.: Logical support for modularisation. In: Huet, G., Plotkin, G. (eds.) Logical Environments, Proceedings of a Workshop, Edinburgh, Scotland, May 1991, pp. 83–130. Cambridge (1993)
33. Diaconescu, R., Petria, M.: Saturated models in institutions. Arch. Math. Log. **49**(6), 693–723 (2010)
34. Diaconescu, R., Stefaneas, P.: Modality in open institutions with concrete syntax. Bull. Greek Math. Soc. **49**, 91–101 (2004). Previously published as JAIST Tech Report IS-RR-97-0046 (1997)
35. Diaconescu, R., Stefaneas, P.: Ultraproducts and possible worlds semantics in institutions. Theor. Comput. Sci. **379**(1), 210–230 (2007)
36. Dimitrakos, T.: Formal Support for Specification Design and Implementation. PhD thesis, Imperial College (1998)
37. Dimitrakos, T., Maibaum, T.: On a generalized modularization theorem. Inf. Process. Lett. **74**, 65–71 (2000)
38. Ehresmann, C.: Esquisses et types des structures algébriques. Bul. Inst. Politeh. Iaşi **14**(18), 1–14 (1968)
39. Fiadeiro, J.L., Sernadas, A.: Structuring theories on consequence. In: Sannella, D., Tarlecki, A. (eds.) Recent Trends in Data Type Specification. Lecture Notes in Computer Science, vol. 332, pp. 44–72. Springer, Berlin (1988)
40. Godo, L., Esteva, F., Garcia, P., Agusti-Cullel, J.: A formal semantical approach to fuzzy logic. In: Proceedings of the 21st Intl. Symp. on Multiple Valued Logic, pp. 72–79 (1991)
41. Goguen, J.: A categorical manifesto. Math. Struct. Comput. Sci. **1**(1), 49–67 (March 1991). Also, Programming Research Group Technical Monograph PRG–72, Oxford University (March 1989)
42. Goguen, J.: Types as theories. In: Reed, G.M., Roscoe, A.W., Wachter, R.F. (eds.) Topology and Category Theory in Computer Science, Proceedings of a Conference held at Oxford, June 1989, pp. 357–390. Oxford (1991)
43. Goguen, J.: Data, schema, ontology and logic integration. J. IGPL **13**(6), 685–715 (2006)
44. Goguen, J., Burstall, R.: Introducing institutions. In: Clarke, E., Kozen, D. (eds.) Proceedings, Logics of Programming Workshop. Lecture Notes in Computer Science, vol. 164, pp. 221–256. Springer, Berlin (1984)
45. Goguen, J., Burstall, R.: A study in the foundations of programming methodology: Specifications, institutions, charters and parchments. In: Pitt, D., Abramsky, S., Poigné, A., Rydeheard, D. (eds.) Proceedings, Conference on Category Theory and Computer Programming. Lecture Notes in Computer Science, vol. 240, pp. 313–333. Springer, Berlin (1986)
46. Goguen, J., Burstall, R.: Institutions: Abstract model theory for specification and programming. J. Assoc. Comput. Mach. **39**(1), 95–146 (1992)
47. Goguen, J., Roşu, G.: Institution morphisms. Form. Asp. Comput. **13**, 274–307 (2002)
48. Goguen, J., Winkler, T., Meseguer, J., Futatsugi, K., Jouannaud, J.-P.: Introducing OBJ. In: Goguen, J., Malcolm, G. (eds.) Software Engineering with OBJ: Algebraic Specification in Action. Kluwer Academic, Dordrecht (2000)
49. Grothendieck, A.: Catégories fibrées et descente. In Revêtements étales et groupe fondamental, Séminaire de Géométrie Algébrique du Bois-Marie 1960/61, Exposé VI. Institut des Hautes Études Scientifiques, 1963. Reprinted in Lecture Notes in Mathematics, vol. 224, pp. 145–194. Springer (1971)
50. Găină, D., Petria, M.: Completeness by forcing. J. Log. Comput. **20**(6), 1165–1186 (2010)
51. Găină, D., Popescu, A.: An institution-independent generalization of Tarski's Elementary Chain Theorem. J. Log. Comput. **16**(6), 713–735 (2006)

52. Găină, D., Popescu, A.: An institution-independent proof of Robinson consistency theorem. Stud. Log. **85**(1), 41–73 (2007)
53. Guitart, R., Lair, C.: Calcul syntaxique des modèles et calcul des formules internes. Diagramme **4** (1980)
54. Hodges, W.: Model Theory. Cambridge University Press, Cambridge (1993)
55. Lambek, J., Scott, P.: Introduction to Higher Order Categorical Logic. Cambridge Studies in Advanced Mathematics, vol. 7. Cambridge (1986)
56. Makkai, M.: Ultraproducts and categorical logic. In: DiPrisco, C.A. (ed.) Methods in Mathematical Logic. Lecture Notes in Mathematics, vol. 1130, pp. 222–309. Springer, Berlin (1985)
57. Makkai, M., Gonzolo, R.: First order categorical logic: Model-theoretical methods in the theory of topoi and related categories. Lecture Notes in Mathematics, vol. 611, Springer, Berlin (1977)
58. Matthiessen, G.: Regular and strongly finitary structures over strongly algebroidal categories. Can. J. Math. **30**, 250–261 (1978)
59. Meseguer, J.: General logics. In: Ebbinghaus, H.-D., et al. (eds.) Proceedings, Logic Colloquium, 1987, pp. 275–329. North-Holland, Amsterdam (1989)
60. Meseguer, J.: Rewriting as a unified model of concurrency. In: Proceedings, Concur'90 Conference, Lecture Notes in Computer Science, vol. 458, pp. 384–400. Springer, Amsterdam (August 1990)
61. Mossakowski, T.: Different types of arrow between logical frameworks. In: auf der Heide F.M., Monien B. (eds.) Proc. ICALP 96. Lecture Notes in Computer Science, vol. 1099, pp. 158–169. Springer, Berlin (1996)
62. Mossakowski, T.: Comorphism-based Grothendieck logics. In: Diks, K., Rytter, W. (eds.) Mathematical Foundations of Computer Science. Lecture Notes in Computer Science, vol. 2420, pp. 593–604. Springer, Berlin (2002)
63. Mossakowski, T.: Relating CASL with other specification languages: The institution level. Theor. Comput. Sci. **286**, 367–475 (2002)
64. Mossakowski, T.: Heterogeneous specification and the heterogeneous tool set. Habilitation thesis, University of Bremen (2005)
65. Mossakowski, T., Diaconescu, R., Tarleck, A.: What is a logic translation? Logica Universalis **3**(1), 59–94 (2009)
66. Mossakowski, T., Goguen, J., Diaconescu, R., Tarleck, A.: What is a logic? In: Béziau, J.-Y. (ed.) Logica Universalis, pp. 113–133. Birkhäuser, Basel (2005)
67. Mossakowski, T., Roggenbach, M.: Structured CSP – A process algebra as an institution. In: Fiadeiro, J. (ed.) WADT 2006. Lecture Notes in Computer Science, vol. 4409, pp. 92–110. Springer, Heidelberg (2007)
68. Mossakowski, T., Tarlecki, A., Pawłowski, W.: Combining and representing logical systems using model-theoretic parchments. In: Parisi Presicce, F. (ed.) Recent Trends in Algebraic Development Techniques. Proc. 12th International Workshop. Lecture Notes in Computer Science, vol. 1376, pp. 349–364. Springer, Berlin (1998)
69. Mosses, P.: Unified algebras and institutions. In: Proceedings, Fourth Annual Conference on Logic in Computer Science, pp. 304–312. IEEE (1989)
70. Nielsen, M., Platet, U.: Polymorphism in an Institutional Framework. Technical University of Denmark (1986)
71. Petria, M.: An institutional version of Gödel Completeness Theorem. In: Algebra and Coalgebra in Computer Science, vol. 4624, pp. 409–424. Springer, Berlin (2007)
72. Petria, M., Diaconescu, R.: Abstract Beth definability in institutions. J. Symbol. Log. **71**(3), 1002–1028 (2006)
73. Popescu, A., Şerbănuţă, T., Roşu, G.: A semantic approach to interpolation. In: Foundations of Software Science and Computation Structures. Lecture Notes in Computer Science, vol. 3921, pp. 307–321. Springer, Berlin (2006)
74. Rabe, F.: Representing Logics and Logic Translations. PhD thesis, Jacobs University Bremen (2008)
75. Roşu, G.: The institution of order-sorted equational logic. Bull. EATCS **53**, 250–255 (1994)
76. Rodenburg, P.-H.: A simple algebraic proof of the equational interpolation theorem. Algebra Universalis **28**, 48–51 (1991)
77. Sannella, D., Tarlecki, A.: Specifications in an arbitrary institution. Inform. Control **76**, 165–210 (1988)
78. Shoenfield, J.: Mathematical Logic. Addison-Wesley, Reading (1967)

79. Tarlecki, A.: Bits and pieces of the theory of institution. In: Pitt, D., Abramsky, S., Poigné, A., Rydeheard, D. (eds.) Proceedings, Summer Workshop on Category Theory and Computer Programming. Lecture Notes in Computer Science, vol. 240, pp. 334–360. Springer, Berlin (1986)
80. Tarlecki, A.: On the existence of free models in abstract algebraic institutions. Theor. Comput. Sci. **37**, 269–304 (1986)
81. Tarlecki, A.: Quasi-varieties in abstract algebraic institutions. J. Comput. Syst. Sci. **33**(3), 333–360 (1986)
82. Tarlecki, A.: Moving between logical systems. In: Haveraaen, M., Owe, O., Dahl, O.-J. (eds.) Recent Trends in Data Type Specification. Lecture Notes in Computer Science, vol. 1130, pp. 478–502. Springer, Berlin (1996)
83. Veloso, P.: On pushout consistency, modularity and interpolation for logical specifications. Inf. Process. Lett. **60**(2), 59–66 (1996)
84. Voutsadakis, G.: Categorical abstract algebraic logic: Algebrizable institutions. Appl. Categ. Struct. **10**, 531–568 (2002)
85. Wells, C.F.: Sketches: outline with references. Unpublished draft
86. Yan, H.: Theory and Implementation of Sort Constraints for Order Sorted Algebra. PhD thesis, Programming Research Group, Oxford University, draft of 1993
87. Yukawa, K.: The Untyped Lambda Calculus as a Logical Programming Language. City University of New York, New York (1990)

R. Diaconescu (✉)

Institute of Mathematics "Simion Stoilow" of the Romanian Academy, Bucharest, Romania

e-mail: Razvan.Diaconescu@imar.ro

Introducing Institutions

J. A. Goguen[1] and R. M. Burstall

SRI International and the University of Edinburgh

Abstract

There is a population explosion among the logical systems being used in computer science. Examples include first order logic (with and without equality), equational logic, Horn clause logic, second order logic, higher order logic, infinitary logic, dynamic logic, process logic, temporal logic, and modal logic; moreover, there is a tendency for each theorem prover to have its own idiosyncratic logical system. Yet it is usual to give many of the same results and applications for each logical system; of course, this is natural in so far as there are basic results in computer science that are independent of the logical system in which they happen to be expressed. But we should not have to do the same things over and over again; instead, we should generalize, and do the essential things once and for all! Also, we should ask what are the relationships among all these different logical systems. This paper shows how some parts of computer science can be done in any suitable logical system, by introducing the notion of an **institution** as a precise generalization of the informal notion of a "logical system." A first main result shows that if an institution is such that interface declarations expressed in it can be glued together, then **theories** (which are just sets of sentences) in that institution can also be glued together. A second main result gives conditions under which a theorem prover for one institution can be validly used on theories from another; this uses the notion of an institution morphism. A third main result shows that institutions admiting free models can be extended to institutions whose theories may include, in addition to the original sentences, various kinds of constraints upon interpretations; such constraints are useful for defining abstract data types, and include so-called "data," "hierarchy," and "generating" constraints. Further results show how to define insitutions that mix sentences from one institution with constraints from another, and even mix sentences and (various kinds of) constraints from several different institutions. It is noted that general results about institutions apply to such "multiplex" institutions, including the result mentioned above about gluing together theories. Finally, this paper discusses some applications of these results to specification languages, showing that much of that subject is in fact independent of the institution used.

1 Introduction

Recent work in programming methodology has been based upon numerous different logical systems. Perhaps most popular are the many variants of first order logic found, for example, in the theorem provers used in various program verification projects. But also popular are equational logic, as used in the study of abstract data types, and first order Horn clause logic, as used in "logic programming." More exotic logical systems, such as temporal logic, infinitary logic, and continuous algebra have also been proposed to handle features such as concurrency and non-termination, and most systems exist in both one-sorted and many-sorted forms. However, it seems apparent that much of programming methodology is actually *completely*

---

[1]Research supported in part by Office of Naval Research contracts N00014-80-0296 and N00014-82-C-0333, and National Science Foundation Grant MCS8201380.

independent of what underlying logic is chosen. In particular, if [Burstall & Goguen 77] are correct that the essential purpose of a specification language is to say how to put (small and hopefully standard) theories together to make new (and possibly very large) specifications, then much of the syntax and semantics of specification does not depend upon the logical system in which the theories are expressed; the same holds for implementing a specification, and for verifying correctness. Because of the proliferation of logics of programming and the expense of theorem provers, it is useful to know when sentences in one logic can be translated into sentences in another logic in such a way that it is sound to apply a theorem prover for the second logic to the translated sentences.

This paper approaches these problems through the theory of institutions, where an **institution** is a logical system suitable for programming methodology; the paper also carries out some basic methodological work in an arbitrary institution. Informally, an institution consists of

- a collection of signatures (which are vocabularies for use in constructing sentences in a logical system) and signature morphisms, together with for each signature Σ
- a set of Σ-sentences,
- a set of Σ-models, and
- a Σ-satisfaction relation, of Σ-sentences by Σ-models,

such that when you change signatures (with a signature morphism), the satisfaction relation between sentences and models changes consistently. One main result in this paper is that any institution whose syntax is nice enough to support gluing together interface declarations (as given by signatures) will also support gluing together **theories** (which are collections of sentences) to form larger specifications. A second main result shows how a suitable **institution morphism** permits a theorem prover for one institution to be used on theories from another. A third main result is that any institution supporting free constructions extends to another institution whose sentences may be either the old sentences, or else any of several kinds of constraint on interpretations. Such constraints are useful, for example, in data type declarations, and we believe that they are valuable as a general formulations of the kinds of induction used in computer science. Again using the notion of institution morphism, it is shown how sentences from one institution can be combined with constraints from another in a "duplex" institution; more generally, sentences and constraints from several institutions can be mixed together in a "multiplex" institution. This gives a very rich and flexible framework for program specification and other areas of theoretical computer science.

The notion of institution was first introduced as part of our research on Clear, in [Burstall & Goguen 80] under the name "language." The present paper adds many new concepts and results, as well as an improved notation. The "abstract model theory" of [Barwise 74] resembles our work in its intention to generalize basic results in model theory and in its use of elementary category theory; it differs in being more concrete (for example, its syntactic structures are limited to the usual function, relation and logical symbols) and in the results that are generalized (these are results of classical logic, such as those of Lowenheim-Skolem and Hanf).

The first two subsections below try to explain the first two paragraphs of this introduction a little more gradually. The third indicates what we will assume the reader knows about category theory. Section 2 is a brief review of general algebra, emphasizing results used to show that equational logic is indeed an institution. Section 3 introduces the basic definitions and results for institutions and theories, and considers the equational, first order, first order with equality, Horn clause, and conditional equational institutions. Section 4 discusses the use of constraints to specify abstract data types. Section 5 considers the use of two or more institutions together. Section 6 discusses applications to programming methodology and shows how certain general specification concepts can be expressed in any suitable institution. All of the more difficult proofs and many of the more difficult definitions are omitted in this condensed version of the paper; these details will appear elsewhere at a later time.

1.1 Specifications and Logical Systems

Systematic program design requires careful specification of the problem to be solved. But recent experience in software engineering shows that there are major difficulties in producing consistent and rigorous specifications to reflect users' requirements for complex systems. We suggest that these difficulties can be ameliorated by making specifications as modular as possible, so that they are built from small, understandable and re-usable pieces. We suggest that this modularity may be useful not only for understanding and writing specifications, but also for proving theorems about them, and in particular, for proving that a given program actually satisfies its specification. Modern work in programming methodology supports the view that abstractions, and in particular data abstractions, are a useful way to obtain such modularity, and that parameterized (sometimes called *generic*) specifications dramatically enhance this utility. One way to achieve these advantages is to use a specification language that puts together parameterized abstractions.

It is important to distinguish between specifications that are written directly in some logical system, such as first order logic or the logic of some particular mechanical theorem prover, and specifications that are written in a genuine specification language, such as Special [Levitt, Robinson & Silverberg 79], Clear [Burstall & Goguen 77], OBJ [Goguen & Tardo 79], or Affirm [Gerhard, Musser et al. 79]. The essential purpose of a logical system is to provide a relationship of *satisfaction* between its *syntax* (i.e., its theories) and its *semantics* or models; this relationship is sometimes called a *model theory* and may appear in the form of a *Galois connection* (as in Section 3.2 below). A specification written directly in such a logical system is simply an unstructured, and possibly very large, set of sentences. On the other hand, the essential purpose of a specification language is to make it easy to write and to read specifications of particular systems, and especially of large systems. To this end, it should provide mechanisms for putting old and well-understood specifications together to form new specifications. In particular, it should provide for parameterized specifications and for constructions (like blocks) that define local environments in which specification variables may take on local values.

In order for a specification written in a given specification language to have a precise meaning, it is necessary for that language to have a precise semantics! Part of that semantics will be an underlying logic which must have certain properties in order to be useful for this task. These properties include a suitable notion of model; a satisfaction relationship between sentences and models; and a complete and reasonably simple proof theory. Examples include first order logic, equational logic, temporal logic, and higher order logic. In general, any mechanical theorem prover will have its own particular logical system (e.g., [Boyer & Moore 80], [Aubin 76], [Shostak, Schwartz & Melliar-Smith 81]) and the utility of the theorem prover will depend in part upon the appropriateness of that logical system to various application areas.

There is also the interesting possibility of using two (or even more) institutions together, as discussed in Section 5 below: then one can specify data structures in one institution and at the same time use the more powerful axioms available in other institutions; this has been used to advantage in some specifications in Clear [Burstall & Goguen 81] and also in the language

Ordinary [Goguen 82a, Goguen 82b]. This permits utilizing induction in institutions where it makes sense, and also writing sentences in other more expressive institutions where induction does not always make sense.

We wish to emphasize that a specification language is not a programming language. For example, the denotation of an Algol text is a function, but the denotation of a specification text is a **theory**, that is, a set of sentences *about* programs.

1.2 Parameterization over the Underlying Logic

Since a specification language is intended to provide mechanisms for structuring specifications, its definition and meaning should be as *independent* as possible of what underlying logical system is used. In fact, dependence on the logical system can be relegated to the level of the syntax of the sentences that occur in theories. This also serves to simplify the task of giving a semantics for a specification language.

In order to achieve the benefits listed above, we introduce the notion of an institution, which is an abstraction of the notion of a logical system [Burstall & Goguen 80]. We have designed our specification languages Clear and Ordinary in such a way that they can be used with *any* institution. This means, in particular, that they could be used in connection with any theorem prover whose underlying logic is actually an institution. Roughly speaking, an **institution** is a collection of signatures (which are vocabularies for constructing sentences) together with for each signature Σ: the set of all Σ-sentences; the category of all Σ-models; and a Σ-satisfaction relationship between sentences and models, such that when signatures are changed (with a signature morphism), satisfaction is preserved.

1.3 Prerequisites

Although relatively little category theory is required for most of this paper, we have not resisted the temptation to add some more arcane remarks for those who may be interested. We must assume that the reader is acquainted with the notions of category, functor and natural transformation. Occasional non-essential remarks use adjoint functors. There are several introductions to these ideas which the unfamiliar reader may consult, including [Arbib & Manes 75], [Goguen, Thatcher, Wagner & Wright 75a] and [Burstall & Goguen 82], and for the mathematically more sophisticated [MacLane 71] and [Goldblatt 79]. Familiarity with the initial algebra approach to abstract data types is helpful, but probably not necessary. Colimits are briefly explained in Section 3.4.

1.4 Acknowledgements

Thanks go to the Science Research Council of Great Britain for financial support and a Visiting Fellowship for JAG, to the National Science Foundation for travel money, to the Office of Naval Research and the National Science Foundation for research support, to Eleanor Kerse for typing some early drafts, and to Jose Meseguer for his extensive comments. We are grateful to many people for helpful conversations and suggestions, notably to our ADJ collaborators Jim Thatcher, Eric Wagner, and Jesse Wright, also to Peter Dybjer, Gordon Plotkin, David Rydeheard, Don Sanella, John Reynolds and Steve Zilles. Special thanks to Kathleen Goguen and Seija-Leena Burstall for extreme patience.

2 General Algebra

This section is quick review of many-sorted general algebra. This will provide our first example of the institution concept, and will also aid in working out the details of several other institutions. If you know all about general algebra, you can skip to Section 3, which defines the abstract notion of institution and gives some further examples. We will use the notational approach of [Goguen 74] (see also [Goguen, Thatcher & Wagner 78]) based on "indexed sets," in contrast to the more complex notations of [Higgins 63], [Benabou 68] and [Birkhoff & Lipson 70]. If I is a set (of "indices"), then an I-**indexed set** (or a family of sets indexed by I) A is just an assignment of a set A_i to each **index** i in I. If A and B are I-indexed sets, then a **mapping**, **map**, or **morphism** of I-indexed sets, f: A→B, is just an I-indexed family of functions f_i: $A_i \rightarrow B_i$ one for each i in I. There is an obvious composition[2] of I-indexed mappings, $(f;g)_i = f_i;g_i$ where f;g denotes composition of the functions f and g in the order given by $(f;g)(x) = g(f(x))$. This gives a category $\mathbf{Set}_I$ of I-indexed sets. We may use the notations $A = \langle A_i \mid i \text{ in } I \rangle$ for an I-indexed set with components A_i and $f = \langle f_i \mid i \text{ in } I \rangle$ for an I-indexed mapping A→B of I-indexed sets where f_i: $A_i \rightarrow B_i$. Notice that the basic concepts of set theory immediately extend component-wise to I-indexed sets. Thus, $A \subseteq B$ means that $A_i \subseteq B_i$ for each i in I, $A \cap B = \langle A_i \cap B_i \mid i \text{ in } I \rangle$, etc.

2.1 Equational Signatures

Intuitively speaking, an equational signature declares some sort symbols (to serve as names for the different kinds of data around) and some operator symbols (to serve as names for functions on these various kinds of data), where each operator declaration gives a tuple of input sorts, and one output sort. A morphism between signatures should map sorts to sorts, and operators to operators, so as to preserve their input and output sorts.

Definition 1: An **equational signature** is a pair $\langle S, \Sigma \rangle$, where S is a set (of **sort** names), and Σ is a family of sets (of **operator** names), indexed by $S^* \times S$; we will often write just Σ instead of $\langle S, \Sigma \rangle$. σ in Σ_{us} is said to have **arity** u, **sort** s, and **rank** u,s; we may write σ: u→s to indicate this. []

In the language of programming methodology, a signature declares the interface for a package, capsule, module, object, abstract machine, or abstract data type (unfortunately, there is no standard terminology for these concepts).

Definition 2: An **equational signature morphism** ϕ from a signature $\langle S, \Sigma \rangle$ to a signature $\langle S', \Sigma' \rangle$ is a pair $\langle f, g \rangle$ consisting of a map f: S→S' of sorts and an $S \times S^*$-indexed family of maps g_{us}: $\Sigma_{us} \rightarrow \Sigma'_{f^*(u)f(s)}$ of operator symbols, where f^*: $S^* \rightarrow S'^*$ is the extension of f to strings[3]. We will sometimes write $\phi(s)$ for f(s), $\phi(u)$ for $f^*(u)$, and $\phi(\sigma)$ or $\phi\sigma$ for $g_{us}(\sigma)$ when $\sigma \in \Sigma_{us}$. []

The signature morphism concept is useful for expressing the *binding* of an actual parameter to the formal parameter of a parameterized software module (see Section 6.2). As is standard in category theory, we can put the concepts of Definitions 1 and 2 together to get

Definition 3: The **category of equational signatures**, denoted **Sig**, has equational

---

[2]This paper uses ; for composition in any category.

[3]This extension is defined by: $f^*(\lambda) = \lambda$, where λ denotes the empty string; and $f^*(us) = f^*(u)f(s)$, for u in S^* and s in S.

signatures as its objects, and has equational signature morphisms as its morphisms. The identity morphism on $\langle S, \Sigma \rangle$ is the corresponding pair of identity maps, and the composition of morphisms is the composition of their corresponding components as maps. (This clearly forms a category.) []

2.2 Algebras

Intuitively, given a signature Σ, a Σ-algebra interprets each sort symbol as a set, and each operator symbol as a function. Algebras, in the intuition of programming methodology, correspond to concrete data types, i.e., to data representations in the sense of [Hoare 72].

Definition 4: Let $\langle S, \Sigma \rangle$ be a signature. Then a Σ-**algebra** A is an S-indexed family of sets $|A| = \langle A_s \mid s \text{ in } S \rangle$ called the **carriers** of A, together with an $S^* \times S$-indexed family α of maps α_{us}: $\Sigma_{us} \to [A_u \to A_s]$ for u in S^* and s in S, where $A_{s1...sn} = A_{s1} \times ... \times A_{sn}$ and $[A \to B]$ denotes the set of all functions from A to B. (We may sometimes write for A for $|A|$ and A_s for $|A|_s$.) For u$=$s1...sn, for σ in Σ_{us} and for $(a1,...,an)$ in A_u we will write $\sigma(a1,...,an)$ for $\alpha_{us}(\sigma)(a1,...,an)$ if there is no ambiguity. []

Definition 5: Given a signature $\langle S, \Sigma \rangle$ a Σ-**homomorphism** from a Σ-algebra $\langle A, \alpha \rangle$ to another $\langle A', \alpha' \rangle$, is an S-indexed map f: $A \to A'$ such that for all σ in Σ_{us} and all $a = (a1,...,an)$ in A_u the **homomorphism condition**

$$f_s(\alpha(\sigma)(a1,...,an)) = \alpha'(\sigma)(f_{s1}(a1),...,f_{sn}(an))$$

holds. []

Definition 6: The **category Alg$_\Sigma$** of Σ-**algebras** has Σ-algebras as objects and Σ-homomorphism as morphisms; composition and identity are composition and identity as maps. (This clearly forms a category). []

We can extend **Alg** to a functor on the category **Sig** of signatures: it associates with each signature Σ the category of all Σ-algebras, and it also defines the effect of signature morphisms on algebras. In the definition below, **Cat**^{op} denotes the opposite of the category **Cat** of all categories, i.e., **Cat**^{op} is **Cat** with its morphisms reversed.

Definition 7: The functor **Alg**: **Sig**$\to$**Cat**^{op} takes each signature Σ to the category of all Σ-algebras, and takes each signature morphism $\phi = \langle f: S \to S', g \rangle$: $\Sigma \to \Sigma'$ to the functor **Alg**(ϕ): **Alg**$_{\Sigma'} \to$ **Alg**$_\Sigma$ sending

1. a Σ'-algebra $\langle A', \alpha' \rangle$ to the Σ-algebra $\langle A, \alpha \rangle$ with $A_s = A'_{f(s)}$ and $\alpha = g; \alpha'$, and
2. sending a Σ'-homomorphism h': $A' \to B'$ to the Σ-homomorphism

$$\textbf{Alg}(\phi)(h') = h: \textbf{Alg}(\phi)(A') \to \textbf{Alg}(\phi)(B') \text{ defined by } h_s = h'_{f(s)}.$$

It is often convenient to write $\phi(A')$ or $\phi A'$ for **Alg**$(\phi)(A')$ and to write $\phi(h')$ for **Alg**$(\phi)(h')$. []

If S is the sort set of Σ, then there is a **forgetful functor** U: **Alg**$_\Sigma \to$**Set**$_S$ sending each algebra to its S-indexed family of carriers, and sending each Σ-homomorphism to its underlying S-indexed map.

For each S-indexed set X, there is a **free algebra** (also called a "term" or "word" algebra), denoted $T_\Sigma(X)$, with $|T_\Sigma(X)|_s$ consisting of all the Σ-terms of sort s using "variable" symbols from X; i.e., $(T_\Sigma(X))_s$ contains all the s-sorted terms with variables from X that can be constructed using operator symbols from Σ; moreover, the S-indexed set T_Σ forms a Σ-algebra in a natural way.

We begin our more precise discussion with a special case, defining $(T_\Sigma)_s$ to be the least set of strings of symbols such that

1. $\Sigma_{\lambda,s} \subseteq T_{\Sigma,s}$, and
2. σ in $\Sigma_{s1...sn,s}$ and ti in $T_{\Sigma,si}$ implies the string $\sigma(t1,...,tn)$ is in $T_{\Sigma,s}$.

Then the Σ-structure of T_Σ is given by α defined by:

1. for σ in $\Sigma_{\lambda,s}$ we let $\alpha(\sigma)$ be the string σ of length one in $T_{\Sigma,s}$; and
2. for σ in $\Sigma_{s1...sn,s}$ and ti in $T_{\Sigma,si}$ we let $\alpha(\sigma)(t1,...,tn)$ be the string $\sigma(t1,...,tn)$ in $T_{\Sigma,s}$.

Next, we define $\Sigma(X)$ to be the S-sorted signature with $(\Sigma(X))_{us} = \Sigma_{us} \cup X_s$ if $u=\lambda$ and $(\Sigma(X))_{us} = \Sigma_{us}$ if $u \neq \lambda$. Then $T_\Sigma(X)$ is just $T_{\Sigma(X)}$ regarded as a Σ-algebra rather than as a $\Sigma(X)$-algebra.

The freeness of $T_\Sigma(X)$ is expressed precisely by the following.

Theorem 8: Let $i_X: X \to U(T_\Sigma(X))$ denote the inclusion described above. Then the following "universal" property holds: for any Σ-algebra B, every (S-indexed) map f: $X \to B$, called an **assignment**, extends uniquely to a Σ-homomorphism $f^\#: T_\Sigma(X) \to B$ such that $i_X;U(f^\#)=U(f)$.
◻

We will often omit the U's in such equations, as in the following traditional diagram of S-indexed sets and mappings for the above equation:

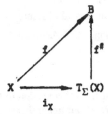

In particular, taking $X=\emptyset$, we see that there is a unique Σ-homomorphism from T_Σ to any other Σ-algebra; for this reason T_Σ is called the **initial** Σ-algebra.

2.3 Equations and Satisfaction

We now define the equations over a given signature, and what it means for an algebra to satisfy an equation.

Definition 9: A Σ-equation e is a triple $(X,\tau1,\tau2)$ where X is an S-indexed set (of variable symbols) and $\tau1$ and $\tau2$ in $|T_\Sigma(X)|_s$ are terms over X of the same sort s in S. Such an equation might be written

for all X, $\tau1=\tau2$

or

$(\forall X)\ \tau1=\tau2$.
◻

The necessity for explicitly including declarations for variables in equations (as in [Burstall & Goguen 77]) is shown in [Goguen & Meseguer 81]: without this one gets an unsound deductive system for many-sorted equational logic.

Definition 10: A Σ-algebra A **satisfies** a Σ-equation $(X,\tau 1,\tau 2)$ iff for all assignments f: $X \to |A|$ we have $f^{\#}(\tau 1) = f^{\#}(\tau 2)$. We will write A $\models$ e for "A satisfies e." []

We now define another functor, Eqn, on the category of signatures. In order to do so, we first define for each signature morphism ϕ: $\Sigma \to \Sigma'$ a function $\phi^{\sim}$ from Σ-terms to Σ'-terms. To this end, we first give a simpler definition for sort maps.

Definition 11: If X is an S-sorted set of variables and if f: $S \to S'$, then we define $f^{\sim}(X)$ to be the S'-indexed set X' such that

$$X'_{s'} = \cup \{X_s \mid f(s) = s'\} .$$

(Notice that without loss of generality, we can assume disjointness of the sets X_s for s in S: if the X_s were not disjoint, we could just take a disjoint union in the above formula.)

Now let ϕ: $\Sigma \to \Sigma'$ be the signature morphism $(f: S \to S', g)$. Let X be an S-indexed set (of variables) and let X' be $f^{\sim}(X)$. We will define an S-indexed map $\phi^{\sim}$: $|T_{\Sigma}(X)| \to |\phi(T_{\Sigma'}(X'))|$. First, note that[4] $X \subseteq |\phi(T_{\Sigma'}(X'))|$ since if x is in X_s then x is in $X'_{f(s)}$ and $X'_{f(s)} \subseteq |T_{\Sigma'}(X')|_{f(s)} = |\phi(T_{\Sigma'}(X'))|_s$; let j: $X \to |\phi(T_{\Sigma'}(X'))|$ denote this inclusion. Then j has a unique extension as a Σ-homomorphism $j^{\#}$: $T_{\Sigma}(X) \to \phi(T_{\Sigma'}(X'))$ by Theorem 8, and we simply define $\phi^{\sim} = |j^{\#}|$. []

Definition 12: The functor Eqn: **Sig** $\to$ **Set** takes each signature Σ to the set Eqn(Σ) of all Σ-equations, and takes each $\phi = (f,g)$: $\Sigma \to \Sigma'$ to the function Eqn(ϕ): Eqn(Σ) $\to$ Eqn(Σ') defined by

$$\text{Eqn}(\phi)((X,\tau 1,\tau 2)) = (f^{\sim}(X), \phi^{\sim}(\tau 1), \phi^{\sim}(\tau 2)) .$$

It is often convenient to write $\phi(e)$ or ϕe instead of Eqn(ϕ)(e). []

Proposition 13: Satisfaction Condition. If ϕ: $\Sigma \to \Sigma'$, if e is an Σ-equation, and if A' is a Σ'-algebra, then

$$A' \models \phi(e) \text{ iff } \phi(A') \models e . []$$

The not entirely trivial proof is omitted from this version of the paper. This concludes our review of general algebra. We now turn to our generalization.

3 Institutions

An institution consists of a category of signatures such that associated with each signature are sentences (e.g., equations), models (e.g., algebras), and a relationship of satisfaction that is, in a certain sense, invariant under change of signature. This is an abstraction of first order model theory. A different approach, axiomatizing the category of theories as well as that of signatures, is given in [Goguen & Burstall 78]; another uses the notion of monadic theory [Burstall & Rydeheard 80]. The generality of these approaches is useful in dealing with aspects of programming, such as errors and infinite data structures, that seem to require theories and models that are in some way more complicated than those of ordinary logic. Another motivation is the elegant way that data definitions can be handled (see Section 5).

---

[4]This means that $X_s \subseteq |\phi(T_{\Sigma'}(X'))|_s$ for each s in S.

3.1 Definition and Examples

The essence of the notion of institution is that when signatures are changed (with a signature morphism) then sentences and models change consistently. This consistency is expressed by the "Satisfaction Condition," which goes a step beyond the classical conception of "semantic truth" in [Tarski 44]. The wide range of consequences and the fact that Proposition 13 is not entirely trivial, suggest that this is not a trivial step. A philosophical argument can also be given: it is a familiar and basic fact that the truth of a sentence (in logic) is independent of the vocabulary chosen to represent the basic relationships that occur in it. It is also fundamental that sentences translate in the same direction as a change of symbols, while models translate in the opposite direction; the Satisfaction Condition is an elegant expression of the invariance of truth under the renaming of basic symbols taking account of the variance of sentences and the covariance of models. (It also generalizes a condition called the "Translation Axiom" in [Barwise 74].)

Definition 14: An **institution** I consists of

1. a category **Sign** of "signatures,"
2. a functor Sen: **Sign**→**Set** giving the set of **sentences** over a given signature,
3. a functor Mod: **Sign**→**Cat**^{op} giving the category (sometimes called the **variety**) of models of a given signature (the arrows in **Mod**(Σ) are called **model morphisms**), and
4. a **satisfaction relation** $\models$ $\subseteq$ $|$**Mod**(Σ)$|$ x Sen(Σ) for each Σ in **Sign**, sometimes denoted $\models_\Sigma$

such that for each morphism ϕ: Σ→Σ' in **Sign**, the **Satisfaction Condition**

$$m' \models \phi(e) \text{ iff } \phi(m') \models e$$

holds for each m' in $|$**Mod**(Σ')$|$ and each e in Sen(Σ). []

For some purposes this definition can be simplified by replacing **Mod**: **Sign**→**Cat**^{op} by Mod: **Sign**→**Set**^{op}. Then Mod(Σ) is the *set* of all Σ-models; the two versions of the definition are thus related by the equation Mod(Σ)=$|$**Mod**(Σ)$|$. Indeed, our original version [Burstall & Goguen 80] was the second. Some reasons for changing it are: first, it is more consistent with the categorical point of view to consider morphisms of models along with models; and secondly, we would like every liberal institution to be an institution, rather than just to determine one (liberal institutions will play an important role later in this paper).

There is a more categorical definition of the institution concept, replacing the rather ad hoc looking family of satisfaction relations by a functor into a category of "twisted relations." This will be given in the full version of this paper.

Example 1: Equational Logic. The work of Section 2 shows that (many sorted) equational logic is an institution, with **Sign** the category **Sig** of equational signatures (Definition 3), with **Mod**(Σ)=**Alg**(Σ) (Definition 6; see Definition 7 for the functor **Alg** which instantiates the functor **Mod**), with Sen(Σ) the set of all Σ-equations (Definition 8; see Definition 12 for the functor Eqn, which instantiates the functor Sen), and with satisfaction given in the usual way (Definition 10). The Satisfaction Condition holds by Proposition 13. Let us denote this institution $\mathcal{EQ}$. **End of Example 1.**

Example 2: First Order Logic. We follow the path of Section 2 again, but now showing that a more complicated logical system is an institution. Our work on equational logic will greatly aid with this task. Many details are omitted.

Definition 15: A **first order signature** Ω is a triple (S,Σ,Π), where

1. S is a set (of **sorts**),
2. Σ is an S*xS-indexed family of sets (of **operator** symbols, also called **function** symbols), and
3. Π is an S*-indexed family of sets (of **predicate symbols**).

A **morphism of first order signatures**, from Ω to Ω', is a triple $\langle\phi_1,\phi_2,\phi_3\rangle$, where

1. $\phi_1\colon S\to S'$ is a function,
2. $\phi_2\colon \Sigma\to\Sigma'$ is an S*xS-indexed family of functions $(\phi_2)_{us}\colon \Sigma_{us}\to\Sigma'_{\phi_1^*(u)\phi_1(s)}$ and
3. $\phi_3\colon \Pi\to\Pi'$ is an S*-indexed family of functions $(\phi_3)_u\colon \Pi_u\to\Pi'_{\phi_1^*(u)}$.

Let **FoSig** denote the category with first order signatures as its objects and with first order signature morphisms as its morphisms. []

Definition 16: For Ω a first order signature, an Ω-**model** (or Ω-**structure**) A consists of

1. an S-indexed family $|A|$ of non-empty sets $\langle A_s \mid s \text{ in } S\rangle$, where A_s is called the **carrier** of sort s,
2. an S*xS-indexed family α of functions $\alpha_{us}\colon \Sigma_{us}\to[A_u\to A_s]$ assigning a function to each function symbol, and
3. an S*-indexed family β of functions $\beta_u\colon \Pi_u\to\text{Pow}(A_u)$ assigning a relation to each predicate symbol, where $\text{Pow}(X)$ denotes the set of all subsets of a set X.

For π in Π_u with $u=s1...sn$ and ai in A_{si} for $i=1,...,n$, we say that "$\pi(a1,...,an)$ holds" iff $(a1,...,an)$ is in $\beta(\pi)$; and as usual, we may abbreviate this assertion by simply writing "$\pi(a1,...,an)$."

Next, we define a **first order Ω-homomorphism** $f\colon A\to A'$ of Ω-models A and A' to be an S-indexed family of functions $f_s\colon A_s\to A'_s$ such that the homomorphism condition holds for Σ (as in Definition 5) and such that for π in Π_u with $u=s1...sn$, and with ai in A_{si} for $i=1,...,n$,

$$\pi(a1,...,an) \text{ implies } \pi'(f_{s1}(a1),...,f_{sn}(an)) \,,$$

where π denotes $\beta(\pi)$ and π' denotes $\beta'(\pi)$.

Let **FoMod** denote the category with first order models as its objects and with first order morphisms as its morphisms. We now extend **FoMod** to a functor **FoSig**$\to$**Cat**^{op}. Given a first order signature morphism $\phi\colon \Omega\to\Omega'$, define the functor **FoMod**$(\Omega')\to$**FoMod**(Ω) to send: first of all, A' in **FoMod**(Ω') to $A=\phi A'$ defined by

1. $A_s=A'_{s'}$ for s in S with $s'=\phi_1(s)$,
2. $\alpha_{us}(\sigma)=\alpha'_{u's'}((\phi_2)_{us}(\sigma))$ for u in S*, s in S and σ in Σ_{us} where $u'=\phi_1^*(u)$ and $s'=\phi_1(s)$, and
3. $\beta_{us}(\pi)=\beta'_{u'}((\phi_3)_u(\pi))$ for u in S* and π in Π_u with u' as above;

and secondly, to send $f'\colon A'\to B'$ in **FoMod**(Ω') to $f=\phi f'\colon A\to B$ in **FoMod**(Ω), where $A=\phi A'$ and $B=\phi B'$, defined by $f_s=f'_{s'}$ where $s'=\phi_1(s)$. This construction extends that of Definition 7, and it is easy to see that it does indeed give a functor. []

The next step is to define the sentences over a first order signature Ω. We do this in the usual way, by first defining terms and formulas. Let X be an S-indexed set of variable symbols, with each X_s the infinite set $\{x_1^s, x_2^s, ...\}$; we may omit the superscript s if the sort is clear from context. Now define the S-indexed family TERM(Ω) of Ω-**terms** to be the carriers of $T_\Sigma(X)$, the free Σ-algebra with generators X, and define a(n S-indexed) function Free on TERM(Ω) inductively by

1. $\text{Free}_s(x) = \{x\}$ for x in X_s, and
2. $\text{Free}_s(\sigma(t1,...,tn)) = \cup_{i=1}^n \text{Free}_{si}(ti)$.

Definition 17: A (well formed) Ω-**formula** is an element of the carrier of the (one sorted) free algebra WWF(Ω) having the **atomic formulae** $\{ \pi(t1,...,tn) \mid \pi \in \Pi_u$ with u=s1...sn and ti $\in$ TERM(Ω)$_{si} \}$ as generators, and having its (one sorted) signature composed of

1. a constant true,
2. a unary prefix operator $\neg$,
3. a binary infix operator &, and
4. a unary prefix operator $(\forall x)$ for each x in X.

The functions Var and Free, giving the sets of **variables** and of **free variables** that are used, respectively, in Ω-formulae, can be defined in the usual way, inductively over the above logical connectives. We then define Bound(P)=Var(P)-Free(P), the set of **bound variables** of P.

We can now define the remaining logical connectives in terms of the basic ones given above in the usual way.

Finally, define an Ω-**sentence** to be a **closed** Ω-formula, that is, an Ω-formula P with Free(P)=$\emptyset$. Finally, let FoSen(Ω) denote the set of all Ω-sentences. []

We now define the effect of FoSen on first order signature morphisms, so that it becomes a functor **FoSig**$\rightarrow$**Set**. Given $\phi: \Omega \rightarrow \Omega'$, we will define FoSen($\phi$): FoSen($\Omega$)$\rightarrow$FoSen($\Omega'$) using the initiality of TERM(Ω) and WFF(Ω). Since $(\phi_1, \phi_2): \Sigma \rightarrow \Sigma'$ is a signature morphism, there is an induced morphism $\psi: T_\Sigma(X) \rightarrow T_{\Sigma'}(X)$ which then gives $\psi:$ TERM(Ω)$\rightarrow$TERM(Ω'). We can now define WFF(ϕ): WFF(Ω)$\rightarrow$WFF(Ω') by its effect on the generators of WFF(Ω), which are the atomic formulae, namely

$$\text{WFF}(\phi)(\pi(t1,...,tn)) = \phi_3(\pi)(\psi(t1),...,\psi(tn)) .$$

Finally, we define FoSen(ϕ) to be the restriction of WFF(ϕ) to FoSen(Ω) $\subseteq$ WFF(Ω). For this to work, it must be checked that WFF(ϕ) carries closed Ω-formulae to closed Ω'-formulae; but this is easy.

It remains to define satisfaction. This corresponds to the usual "semantic definition of truth" (originally due to [Tarski 44]) and is again defined inductively. If A is a first order model, let Asgn(A) denote the set of all **assignments** of values in A to variables in X, i.e, [X$\rightarrow$A], the set of all S-indexed functions f: X$\rightarrow$A.

Definition 18: Given a sentence P, define Asgn(A,P), the set of assignments in A for which P is true, inductively by

1. if P=$\pi(t1,...,tn)$ then f$\in$Asgn(A,P) iff $(f^\#(t1),...,f^\#(tn)) \in \beta(\pi)$, where $f^\#(t)$ denotes the evaluation of the Σ-term t in the Σ-algebra part of A, using the values of variables given by the assignment f,

2. Asgn(A,true)=Asgn(A),

3. Asgn(A,¬P)=Asgn(A)-Asgn(A,P),

4. Asgn(A,P&Q)=Asgn(A,P)∩Asgn(A,Q), and

5. Asgn(A,(∀x)P)={f | Asgn(A,f,x)⊆Asgn(A,P)}, where Asgn(A,f,x) is the set of all assignments f ' that agree with f except possibly on the variable x.

Then a model A **satisfies** a sentence P, written A |= P, iff Asgn(A,P)=Asgn(A). []

Finally, we must verify the satisfaction condition. This follows from an argument much like that used for the equational case, and is omitted here. Thus, first order logic is an institution; let us denote it $\mathcal{FO}$. **End of Example 2.**

Example 3: First Order Logic with Equality. This institution is closely related to that of Example 2. A signature for first order logic with equality is a first order signature Ω=(S,Σ,Π) that has a particular predicate symbol $\equiv_s$ in Π_{ss} for each s in S. A morphism of signatures for first order logic with equality must preserve these predicate symbols, i.e., $\phi_3(\equiv_s)=\equiv_{\phi_1(s)}$. This gives a category **FoSigEq** of signatures for first order logic with equality.

If Ω is a signature for first order logic with equality, then a model for it is just an Ω-model A in the usual first order sense satisfying the additional condition that for all s in S, and for all a,a' in A_s, $a\equiv_s a'$ iff a=a'.

A homomorphism of first order Ω-models with equality is just a first order Ω-homomorphism (in the sense of Definition 16), and we get a category **FoModEq**(Ω) of Ω-models for each signature Ω in |**FoSigEq**|, and **FoModEq** is a functor on **FoSigEq**. Ω-sentences are defined just as in Example 2, and so is satisfaction. We thus get a functor FoSenEq: **FoSigEq**→**Set**. The Satisfaction Condition follows immediately from that of first order logic without equality. Let us denote this institution by $\mathcal{FOEQ}$. **End of Example 3.**

Example 4: Horn Clause Logic with Equality. We now specialize the previous example by limiting the form that sentences can take, but without resticting either the predicate or operator symbols that may enter into them. In particular, we maintain the equality symbol with its fixed interpretation from Example 3; but we require that all sentences be of the form

(∀$\underline{x}$) A_1&A_2&...A_n ⇒ A ,

where each A_i is an atomic formula $\pi(t_1,...,t_m)$. In particular, we do not allow disjunction, negation or existential quantifiers. That this is an institution follows from the fact that first order logic with equality is an institution. Let us denote this institution $\mathcal{HORN}$. **End of Example 4.**

Example 5: Conditional Equational Logic. As a specialization of the first order Horn clause logic with equality, we can consider the case where equality is the only predicate symbol. This gives the institution often called conditional equational logic. **End of Example 5.**

Example 6: Horn Clause Logic without Equality. We can also restrict Example 4 by dropping equality with its fixed interpretation. This too is obviously an institution because it is just a restriction of ordinary first order logic. **End of Example 6.**

It seems clear that we can do many-sorted temporal or modal logic in much the same way, by adding the appropriate modal operators to the signature and defining their correct interpretation in all models; the models may be "Kripke" or "alternative world" structures. Higher order equational logic is also presumably an institution; the development should again follow that of equational logic, but using higher order sorts and operator symbols.[5] Quite possibly, the "inequational logic" of [Bloom 76], the order-sorted equational logic of [Goguen 78], and various kinds of infinitary equational logic, such as the logic of continuous algebras in [Goguen, Thatcher, Wagner & Wright 75b] and [Wright, Thatcher, Wagner & Goguen 76] are also institutions. We further conjecture that in general, mechanical theorem provers, such as those of [Boyer & Moore 80], [Aubin 76] and [Shostak, Schwartz & Melliar-Smith 81], are based on logical systems that are institutions (or if they are not, should be modified so that they are!). Clearly, it would be helpful to have some general results to help in establishing whether or not various logical systems are institutions.

There is a way of generating many other examples due to [Mahr & Makowsky 82a]: take as the sentences over a signature Σ all specifications using Σ written in some specification language (such as Special or Affirm); we might think of such a specification as a convenient abbreviation for a (possibly very large) conjunction of simpler sentences. However, this seems an inappropriate viewpoint in the light of Section 6 which argues that the real purpose of a specification language is to define the meanings of the basic symbols (in Σ), so that the signature should be constructed (in a highly structured manner) right along with the sentences, rather than being given in advance.

This and the next section assume that a fixed but arbitrary institution has been given; Section 5 discusses what can be done with more than one institution.

3.2 Theories and Theory Morphisms

A Σ-theory, often called just a "theory" in the following, consists of a signature Σ and a closed set of Σ-sentences. Thus, this notion differs from the [Lawvere 63] notion of "algebraic theory," which is independent of any choice of signature; the notion also simplifies the "signed theories" of [Burstall & Goguen 77]. The simpler notion is more appropriate for the purposes of this paper, as well as easier to deal with. In general, our theories contain an infinite number of sentences, but are defined by a finite presentation.

Definition 19:

1. A Σ-**theory presentation** is a pair $\langle \Sigma, E \rangle$, where Σ is a signature and E is a set of Σ-sentences.

2. A Σ-model A **satisfies** a theory presentation $\langle \Sigma, E \rangle$ if A satisfies each sentence in E; let us write $A \models E$.

3. If E is a set of Σ-sentences, let E^* be the set[6] of all Σ-models that satisfy each sentence in E.

4. If M is a set of Σ-models, let M^* be the set of all Σ-sentences that are satisfied by each model in M; we will hereafter also let M^* denote $\langle \Sigma, M^* \rangle$, the **theory of** M.

5. By the **closure** of a set E of Σ-sentences we mean the set E^{**}, written $E^\bullet$.

6. A set E of Σ-sentences is **closed** iff $E = E^\bullet$. Then a Σ-**theory** is a theory presentation $\langle \Sigma, E \rangle$ such that E is closed.

---

[5]The sorts involved will be the objects of the free Cartesian closed category on the basic sort set [Parsaye-Ghomi 82].

[6]In the terminology of axiomatic set theory, this will often turn out to be a **class** rather than a **set** of models, but this distinction will be ignored in this paper.

7. The Σ-theory **presented by** the presentation $\langle\Sigma,E\rangle$ is $\langle\Sigma,E^*\rangle$.

[]

Notice that we have given a model-theoretic definition of closure. For some institutions, a corresponding proof-theoretic notion can be given, because there is a complete set of inference rules. For the equational institution, these rules embody the equivalence properties of equality, including the substitution of equal terms into equal terms [Goguen & Meseguer 81].

We can also consider closed sets of models (in the equational institution these are usually called varieties). The **closure** of a set M of models is M^{**}, denoted M^*, and a full subcategory of models is called **closed** iff its objects are all the models of some set of sentences.

Definition 20: If T and T' are theories, say $\langle\Sigma,E\rangle$ and $\langle\Sigma',E'\rangle$, then a **theory morphism** from T to T' is a signature morphism F: $\Sigma\to\Sigma'$ such that $\phi(e)$ is in E' for each e in E; we will write F: $T\to T'$. The **category of theories** has theories as objects and theory morphisms as morphisms, with their composition and identities defined as for signature morphisms; let us denote it **Th**. (It is easy to see that this is a category.) []

Notice that there is a forgetful functor Sign: **Th**$\to$**Sign** sending $\langle\Sigma,E\rangle$ to Σ, and sending ϕ as a theory morphism to ϕ as a signature morphism.

Proposition 21: The two functions, *: sets of Σ-sentences $\to$ sets of Σ-models, and *: sets of Σ-models $\to$ sets of Σ-sentences, given in Definition 19, form what is known as a **Galois connection** [Cohn 65], in that they satisfy the following properties, for any sets E,E' of Σ-sentences and sets M,M' of Σ-models:

1. E $\subseteq$ E' implies $E'^* \subseteq E^*$.
2. M $\subseteq$ M' implies $M'^* \subseteq M^*$.
3. E $\subseteq E^{**}$.
4. M $\subseteq M^{**}$.

These imply the following properties:

5. $E^* = E^{***}$.
6. $M^* = M^{***}$.
7. There is a dual (i.e., inclusion reversing) isomorphism between the closed sets of sentences and the closed sets of models.
8. $(\cup_n E_n)^* = \cap E_n^*$.
9. $\phi(E^*) = (\phi E)^*$, for ϕ: $\Sigma\to\Sigma'$ a signature morphism.

[]

For a theory T with signature Σ, let **Mod**(T) and also T* denote the **full subcategory** of **Mod**(Σ) of all Σ-models that satisfy all the sentences in T. This is used in the following.

Definition 22: Given a theory morphism F: $T\to T'$, the **forgetful functor** F*: $T'^*\to T^*$ sends a T'-model m' to the T-model F(m'), and sends a T'-model morphism f: $m'\to n'$ to $F^*(f) = \textbf{Mod}(F)(f): F(m')\to F(n')$. This functor may also be denoted **Mod**(F). []

For this definition to make sense, we must show that if a given Σ'-model m' satisfies T', then $\phi^*(m')$ satisfies T. Let e be any sentence in T. Because ϕ is a theory morphism, $\phi(e)$ is a sentence of T', and therefore m' $\models \phi(e)$. The Satisfaction Condition now gives us that $\phi(m')$

$\models$ e, as desired. We also need that the morphism $\phi^*(f)$ lies in T*, but this follows because T* is a full subcategory of **Mod**(Σ), and the source and target objects of $\phi^*(f)$ lie in T*.

3.3 The Closure and Presentation Lemmas

Let us write $\phi(E)$ for $\{\phi(e) \mid e \text{ is in } E\}$ and $\phi(M)$ for $\{\phi(m) \mid m \text{ is in } M\}$. Let us also write $\phi^{-1}(M)$ for $\{m \mid \phi(m) \text{ is in } M\}$. Using this notation, we can more compactly write the Satisfaction Condition as

$$\phi^{-1}(E^*) = \phi(E)^* ,$$

and using this notation, we can derive

Lemma 23: Closure. $\phi(E^\bullet) \subseteq \phi(E)^\bullet$.

Proof: $\phi(E^{**})^* = \phi^{-1}(E^{***}) = \phi^{-1}(E^*) = \phi(E)^*$, using the Satisfaction Condition and 5. of Proposition 21. Therefore $\phi(E^\bullet) = \phi(E^{**}) \subseteq \phi(E^{**})^{**} = \phi(E)^{**} = \phi(E)^\bullet$, using 3. of Proposition 21 and the just proved equation. []

The following gives a necessary and sufficient condition for a signature morphism to be a theory morphism.

Lemma 24: Presentation. Let $\phi: \Sigma \to \Sigma'$ and suppose that $\langle \Sigma, E \rangle$ and $\langle \Sigma', E' \rangle$ are presentations. Then $\phi: \langle \Sigma, E^\bullet \rangle \to \langle \Sigma', E'^\bullet \rangle$ is a theory morphism iff $\phi(E) \subseteq E'^\bullet$.

Proof: By the Closure Lemma, $\phi(E^\bullet) \subseteq \phi(E)^\bullet$. By hypothesis, $\phi(E) \subseteq E'^\bullet$. Therefore, $\phi(E^\bullet) \subseteq \phi(E)^\bullet \subseteq E'^{\bullet\bullet} = E'^\bullet$, so ϕ is a theory morphism. Conversely, if ϕ is a theory morphism, then $\phi(E^\bullet) \subseteq E'^\bullet$. Therefore $\phi(E) \subseteq E'^\bullet$, since $E \subseteq E^\bullet$. []

If an institution has a complete set of rules of deduction, then the Presentation Lemma tells us that to check if ϕ is a theory morphism, we can apply ϕ to each sentence e of the source presentation E and see whether $\phi(e)$ can be proved from E'. There is no need to check all the sentences in $E^\bullet$.

3.4 Putting Theories Together

A useful general principle is to describe a large widget as the interconnection of a system of small widgets, using widget-morphisms to indicate the interfaces over which the interconnection is to be done. [Goguen 71] (see also [Goguen & Ginali 78]) gives a very general formulation in which such interconnections are calculated as colimits. One application is to construct large theories as colimits of small theories and theory morphisms [Burstall & Goguen 77]. In particular, the pushout construction for applying parameterized specifications that was implicit in [Burstall & Goguen 77] has been given explicitly by [Burstall & Goguen 78] and [Ehrich 78]. For this to make sense, the category of theories should have finite colimits. For the equational institution, [Goguen & Burstall 78] have proved that the intuitively correct syntactic pasting together of theory presentations exactly corresponds to theory colimits. Colimits have also been used for many other things in computer science, for example, pasting together graphs in the theory of graph grammars [Ehrig, Kreowski, Rosen & Winkowski 78]. Let us now review the necessary categorical concepts.

Definition 25: A **diagram** D in a category **C** consists of a graph G together with a labelling of each node n of G by an object D_n of **C**, and a labelling of each edge e (from node n to node n' in G) by a morphism $D(e)$ in **C** from D_n to $D_{n'}$; let us write D: G→**C**. Then a **cone** α in **C** over the diagram D consists of an object A of **C** and a family of morphisms $\alpha_n: D_n \to A$, one for each node n in G, such that for each edge e: n→n' in G, the diagram

338

D(e)

commutes in C. We call D the **base** of the cone α, A its **apex**, G its **shape**, and we write $\alpha: D \Rightarrow A$. If α and β are cones with base D and apexes A,B (respectively), then a **morphism of cones** $\alpha \to \beta$ is a morphism f: A$\to$B in C such that for each node n in G the diagram

commutes in C. Now let **Cone(D,C)** denote the resulting category of all cones over D in C. Then a **colimit** of D in C is an initial object in **Cone(D,C)**. []

The uniqueness of initial objects up to isomorphism implies the uniqueness of colimits up to cone isomorphism. The apex of a colimit cone α is called the **colimit object**, and the morphisms $\alpha(n)$ to the apex are called the **injections** of D(n) into the colimit. The colimit object is also unique up to isomorphism.

Definition 26: A category C is **finitely cocomplete** iff it has colimits of all finite diagrams, and is **cocomplete** iff it has colimits of all diagrams (whose base graphs are not proper classes). A functor F: C$\to$C$'$ **reflects colimits** iff whenever D is a diagram in C such that the diagram D;F in C$'$ has a colimit cone $\alpha': D \Rightarrow A'$ in C$'$, then there is a colimit cone $\alpha: D \Rightarrow A$ in C with $\alpha' = \alpha;F$ (i.e., $\alpha'_n = F(\alpha_n)$ for all nodes n in the base of D). []

Here is the general result about putting together theories in any institution with a suitable category of signatures. The proof is omitted.

Theorem 27: The forgetful functor Sign: **Th**$\to$**Sign** reflects colimits. []

It now follows, for example, that the category **Th** of theories in an institution is [finitely] cocomplete if its category **Sign** of signatures is [finitely] cocomplete.

The category of equational signatures is finitely cocomplete (see [Goguen & Burstall 78] for a simple proof using comma categories), so we conclude that the category of signed equational theories is cocomplete. Using similar techniques, we can show that the category of first order signatures is cocomplete, and thus without effort conclude that the category of first order theories is cocomplete (this might even be a new result).

4 Constraints

To avoid overspecifying problems, we often want *loose* specifications, i.e., specifications that have many acceptable models. On the other hand, there are also many specifications where one wants to use the natural numbers, truth values, or some other fixed data type. In such cases, one wants the subtheories that correspond to these data types to be given *standard* interpretations. Moreover, we sometimes want to consider parameterized standard data types, such as **Set**[X] and **List**[X]. Here, one wants sets and lists to be given standard interpretations, once a suitable interpretation has been given for X.

So far we have considered the category $T^* = \mathbf{Mod}(T)$ of *all* interpretations of a theory T; this section considers how to impose constraints on these interpretations. One kind of constraint requires that some parts of T have a "standard" interpretation relative to other parts; these constraints are the "data constraints" of [Burstall & Goguen 80], generalizing and relativizing the "initial algebra" approach to abstract data types introduced by [Goguen, Thatcher, Wagner & Wright 75b, Goguen, Thatcher & Wagner 78]. However, the first work in this direction seems to be that of [Kaphengst & Reichel 71], later generalized to "initially restricting algebraic theories" [Reichel 80]. Data constraints make sense for any "liberal" institution, and are more expressive even in the equational institution. Actually, our general results apply to many different notions of what it means for a model to satisfy a constraint. In particular, we will see that "generating constraints" and "hierarchy constraints" are special cases; these correspond to the conditions of "no junk" and "no confusion" that together give the more powerful notion of "data constraint." Section 5 generalizes this to consider "duplex constraints" that involve a different insitition than the one in which models are taken.

4.1 Free Interpretations

For example, suppose that we want to define the natural numbers in the equational institution. To this end, consider a theory N with one sort **nat**, and with a signature Σ containing one constant 0 and one unary operator inc; there are no equations in the presentation of this theory. Now this theory has many algebras, including some where inc(0)=0. But the natural numbers, no matter how represented, give an **initial** algebra in the category $\mathbf{Alg}_\Sigma$ of all algebras for this theory, in the sense that there is *exactly* one Σ-homomorphism from it to any other Σ-algebra. It is easy to prove that any two initial Σ-algebras are Σ-isomorphic; this means that the property "being initial" determines the natural numbers uniquely up to isomorphism. This characterization of the natural numbers is due to [Lawvere 64], and a proof that it is equivalent to Peano's axioms can be found in [MacLane & Birkhoff 67], pages 67-70.

The initial algebra approach to abstract data types [Goguen, Thatcher, Wagner & Wright 75b, Goguen, Thatcher & Wagner 78] has taken this "Lawvere-Peano" characterization of the natural numbers as a paradigm for defining other abstract data types; the method has been used for sets, lists, stacks, and many many other data types, and has even been used to specify database systems [Goguen & Tardo 79] and programming languages [Goguen & Parsaye-Ghomi 81]. The essential ideas here are that concrete data types are algebras, and that "abstract" in "abstract data type" means *exactly* the same thing as "abstract" in "abstract algebra," namely defined uniquely up to isomorphism. A number of less abstract equivalents of initiality for the equational institution, including a generalized Peano axiom system, are given in [Goguen & Meseguer 83].

Let us now consider the case of the parameterized abstract data type of sets of elements of a sort **s**. We add to **s** a new sort **set**, and operators[7]

$\emptyset$: **set**
$\{\_\}$: **s** → **set**
$\_ \cup \_$: **set,set** → **set** ,

subject to the following equations, where S, S' and S" are variables of sort **set**,

$\emptyset \cup S = S = S \cup \emptyset$
$S \cup (S' \cup S") = (S \cup S') \cup S"$
$S \cup S' = S' \cup S$

---

[7] We use "mixfix" declarations in this signature, in the style of OBJ [Goguen 77, Goguen & Tardo 79]: the underbars are placeholders for elements of a corresponding sort from the sort list following the colon.

340

$S \cup S = S$.

Although we want these operators to be interpreted "initially" in some sense, we do *not* want the initial algebra of the theory having sorts **s** and **set** and the operators above. Indeed, the initial algebra of this theory has the *empty* carrier for the sort **s** (since there are no operators to generate elements of **s**) and has only the element ∅ of sort **set**. What we want is to permit *any* interpretation for the parameter sort **s**, and then to require that the new sort **set** and its new operators are interpreted freely *relative* to the given interpretation of **s**.

Let us make this precise. Suppose that F: T→T' is a theory morphism. Then there is a forgetful functor from the category of T'-models to the category of T-models, F*: T'*→T* as in Definition 22. In the equational case T*=**Alg**(T) and a very general result of [Lawvere 63] says that there is a functor that we will denote by F$: T*→T'* called the **free functor** determined by F, characterized by the following "universal" property: given a T-model A, there is a T'-model F$(A) and a T-morphism η_A: A→F*(F$(A)), called the **universal** morphism, such that for any T'-model B' and any T-morphism f: A→F*(B'), there is a unique T'-morphism f#: F$(A)→B' such that the diagram

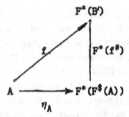

commutes in **Mod**(T). It can be shown [MacLane 71] that if for each A there is an object F$(A) with this universal property, then there is a unique way to define F$ on morphisms so as to get a functor.

For this to make sense in an arbitrary institution, we need the existence of free constructions over arbitrary theory morphisms, as expressed in the following.

Definition 28: An institution is **liberal** iff for every theory morphism F: T→T' and every T-model A, there is a T'-model F$(A) with a **universal** T-morphism η_A: A→F*(F$(A)) having the property that for each T'-model B' and T-morphism f: A→F*(B'), there is a unique T'-morphism f#: F$(A)→B' such that the above diagram commutes (in the category **Mod**(T)). []

As in the equational case, the existence of a universal morphism for each T-model A guarantees the existence of a unique functor F$ having the value F$(A) on the object A [MacLane 71]; this functor F$ is called a **free** functor, just as F* is called a **forgetful** functor. (T$ is left adjoint to F*, and is unique up to isomorphism of its value objects if it exists. Thus an institution is liberal iff the forgetful functors induced by theory morphisms always have left adjoints.) The equational institution is liberal, but the first order logic instituion is not. The Horn clause first order logic with equality discussed in Example 4 is another liberal institution, by deep results of [Gabriel & Ulmer 71]; so is the conditional equational institution of Example 5.

Returning to our set example, consider the theory morphism, Set, that is the inclusion of the trivial theory, **Triv** having just the sort **s**, into the theory of sets of **s**, let's call it **Set-of-Triv**, obtained by adding the sort **set** and the operators and equations given above. Then Set* takes a **Set-of-Triv**-algebra (which is just a set) and forgets the new sort **set** and the three new operators, giving an algebra that has just the set of **s**-sorted elements. The free functor Set$

takes a **Triv**-algebra A, and extends it freely to a **Set-of-Triv**-algebra, the new operators giving distinct results except where the equations of **Set-of-Triv** force equality.

Given a **Set-of-Triv**-algebra B, there is a natural way to check whether or not its **set** sort and operators are free over its parameter sort **s**: in the above diagram, let A=Set*(B) and let $f=\mathrm{id}_{\mathrm{Set}^*(B)}$; then $f^\#$: Set$^\$$(Set*(B))→B restricted to the parameter sort is the identity (because Set*$(f^\#)$=f), and $f^\#$ itself should be an isomorophism if the **set** part of B is to be free over its parameter part. In the general case, the morphism $(\mathrm{id}_{F^*(B)})^\#$: $F^\$(F^*(B))$→B is called the **counit** (of the adjunction) and denoted ϵ_B. The discussion of this example motivates the following.

Definition 29: Let F: T→T' be a theory morphism. Then a T'-model B is **F-free** iff the counit morphism $\epsilon_B=(\mathrm{id}_{F^*(B)})^\#$: $F^\$(F^*(B))$→B is an isomorphism. []

The notion of F-free for the equational case is due to [Thatcher, Wagner & Wright 79][8].

Results of [Mahr & Makowsky 82b] and [Mahr & Makowsky 82a] show that, in a sense, the most general sublanguage of first order logic that admits initial models is Horn clause logic with infinitary clauses; further, the most general finitary language uses finitary Horn clauses; the most general equational language uses (infinitary) conditional equations; and the most general finitary equational sublanguage consists of finitary conditional equations. These results apply to the existence of initial models, rather than to left adjoints of forgetful functors; it would be interesting to know whether or not they extend in this way. It is also interesting to note that they have made use of abstractions of the notion of "logical system" similar to that of an "institution."

4.2 Constraining Theories

It is very convenient in program specification to let theories include not just the sentences provided by an institution, but also declarations that certain subtheories should be interpreted freely relative others in the sense of Section 4.1. We call such sentences **data constraints** and we call theories that can include them **constraining theories**. Such a theory could require that subtheories defining data types such as natural numbers, sets, and strings be given their standard interpretations (uniquely up to isomorphism), while permiting a sort with a partial ordering to have any interpretation (that satisfies the partial order theory), and thus to be "loose," or to be parametric as are the "meta-sort" or "requirement" theories of Clear [Burstall & Goguen 77].

Returning to the Set example of the previous subsection, let us further enrich **Set-of-Triv** with some new sorts and operators to get a theory **Set-of-Triv-Etc**. For example, we might add an operator
        choice: **set** → **s**
that chooses an arbitrary element of any non-empty set; this is a good example of a loose specification. For example, the enriched theory might include the conditional equation
        choose(s) in s if s≠∅ .
Let Etc: **Set-of-Triv**→**Set-of-Triv-Etc** be the theory inclusion and let
        etc: Sign(**Set-of-Triv**)→Sign(**Set-of-Triv-Etc**)

---

[8] [Burstall & Goguen 80] defined B to be F-free if B≈$F^\$(F^*(B))$; however, there are examples in which these two objects are isomorphic, but not naturally so by the counit morphism η. Our thanks to Eric Wagner for this comment.

be the corresponding signature morphism. Then a **Set-of-Triv-Etc**-algebra A interprets sets as intended if etc*(A) satisfies **Set-of-Triv** and is Set-free. This motivates the following.

Definition 30: Let Σ be a signature. Then a Σ-**constraint** is a pair

$$\langle F: T'' \rightarrow T', \; \theta: \text{Sign}(T') \rightarrow \Sigma \rangle$$

consisting of a theory morphism and a signature morphism. (We may call a Σ-constraint a Σ-**data constraint** if it is used to define a data type.) A Σ-model A **satisfies** the Σ-constraint $c = \langle F: T'' \rightarrow T', \; \theta: \text{Sign}(T') \rightarrow \Sigma \rangle$ iff $\theta(A)$ satisfies T' and is F-free; we write $A \models_{\Sigma} c$ in this case.
[]

A picture of the general situation in this definition may help:

$$
\begin{array}{ccccc}
 & F & & \theta & \\
T'' & \longrightarrow & T' & \text{Sign}(T') \longrightarrow \Sigma \\
\end{array}
$$

$$
\begin{array}{ccccc}
 & F* & & \theta* & \\
T''* & \rightleftarrows & T'* & \text{Mod}(\Sigma') \longleftarrow \text{Mod}(\Sigma) \\
 & F\$ & & & \\
\end{array}
$$

In the Set-etc example, F: T''→T' is the theory inclusion Etc: **Triv→Set-of-Triv**, and θ: Sign(T')→Σ is the signature morphism underlying the theory inclusion Etc: **Set-of-Triv→Set-of-Triv-Etc**. For any Σ-algebra A, it makes sense to ask whether θA satisfies T'* and is F-free, as indeed Definition 30 does ask.

Our work on constraints dates from the Spring of 1979, and was influenced by a lecture of Reichel in Poland in 1978 and by the use of functors to handle parametric data types in [Thatcher, Wagner & Wright 79]. There are three main differences between our approach and the "initial restrictions" of [Reichel 80]: first, an initial restriction on an algebraic theory consists of a pair of subtheories, whereas we use a pair of theories with an arbitrary theory morphism between them, and a signature morphism from the target theory. The use of subtheories seems very natural, but we have been unable to prove that it gives rise to an institution, and conjecture that it does not do so; we also believe that the added generality may permit some interesting additional examples. The second difference is simply that we are doing our work over an arbitrary liberal institution. The third difference lies in the manner of adding constraints: whereas [Reichel 80] defines a "canon" to be an algebraic theory together with a set of initial restrictions on it, we will define a new institution whose sentences on a signature Σ include both the old Σ-sentences, and also Σ-constraints. This route has the technical advantage that both kinds of new sentence refer to the same signature. Historically the first work in this field seems to have been the largely unknown paper of [Kaphengst & Reichel 71], which apparently considered the case of a single chain of theory inclusions.

We now show that constraints behave like sentences even though they have a very different internal structure. Like sentences, they impose restrictions on the allowable models. Moreover, a signature morphism from Σ to Σ' determines a translation from Σ-constraints to Σ'-constraints just as it determines a translation from of Σ-sentences to Σ'-sentences.

Definition 31: Let ϕ: $\Sigma \rightarrow \Sigma'$ be a signature morphism and let $c = \langle F, \theta \rangle$ be a Σ-constraint. Then the **translation** of c by ϕ is the Σ'-constraint $\langle F, \theta;\phi \rangle$; we write this as $\phi(c)$. []

It is the need for the translation of a constraint to be a constraint that leads to constraints in which θ is not an inclusion. We now state the Satisfaction Condition for constraints.

Lemma 32: Constraint Satisfaction. If $\phi: \Sigma \to \Sigma'$ is a signature morphism, if c is a Σ-constraint, and if B is a Σ'-model then

$$B \models \phi(c) \text{ iff } \phi(B) \models c \text{ . } []$$

Given a liberal institution I, we can construct another institution having as its sentences both the sentences of I and also constraints.

Definition 33: Let I be an arbitrary liberal institution. Now construct the institution $C(I)$ whose theories contain both constraints and I-sentences as follows: the category of signatures of $C(I)$ is the category **Sign** of signatures of I; if Σ is an I-signature, then $\text{Sen}_{C(I)}(\Sigma)$ is the union of the set $\text{Sen}_I(\Sigma)$ of all Σ-sentences from I with the set of all Σ-constraints[9]; also $\text{Mod}(\Sigma)$ is the same for $C(I)$ as for I; we use the concept of constraint translation in Definition 31 to define $\text{Sen}(\phi)$ on constraints; finally, satisfaction for $C(I)$ is as in I for Σ-sentences from I, and is as in Definition 30 for Σ-constraints. []

We now have the following.

Proposition 34: If I is a liberal institution, then $C(I)$ is an institution. []

This means that all the concepts and results of Section 3 can be applied to $C(I)$-theories that include constraints as well as sentences. Let us call such theories **constraining theories**. Thus, we get notions of presentation and of closure, as well as of theory. In particular, the Presentation and Closure Lemmas hold and we therefore get the following important result:

Theorem 35: Given a liberal institution with a [finitely] cocomplete category of signatures, its category of constraining theories is also [finitely] cocomplete.

Proof: Immediate from Theorems 27 and 34. []

Let us consider what this means for the equational institution. While the proof theory for inferring that an equation is in the closure of a set of equations is familiar, we have no such proof theory for constraints. However, this should be obtainable because a constraint corresponds to an induction principle plus some inequalities [Burstall & Goguen 81, Goguen & Meseguer 83]; in particular, the constraint that sets are to be interpreted freely will give us all the consequences of the induction principle for sets. In more detail, this constraint for sets demands that all elements of sort **set** be generated by the operators $\emptyset$, $\{\_\}$ and $\_\cup\_$, and this can be expressed by a principle of structural induction over these generators (notice that this is not a first order axiom). The constraint also demands that two elements of sort **set** are *unequal* unless they can be proved equal using the given equations. We cannot express this distinctness with equations; one way to express it is by adding a new boolean-valued binary operator on the new sort, say $\equiv$, with some new equations such that $t \equiv t' = \text{false}$ iff $t \neq t'$, and also such that $\text{true} \neq \text{false}$ [Goguen 80].

Let now us consider an easier example, involving equational theories with the following three (unconstrained) presentations:

1. **E** -- the empty theory: no sorts, no operators, no equations.
2. **N** -- the theory with one sort **nat**, one constant 0, one unary operator inc, and no equations.

---

[9]There are some foundational difficulties with the size of the closure of a constraint theory that we will ignore here; they can be solved by limiting the size of the category of signatures used in the original liberal institution.

3. **NP** -- the theory with one sort **nat**, two constants 0 and 1, one unary operator inc, one binary infix operator +, and equations 0+n==n, inc(m)+n==inc(m+n), 1==inc(0).

Let Σ^N and Σ^{NP} be the signatures of **N** and **NP** respectively, and let F^N: **E**→**N** and F^{NP}: **N**→**NP** denote the inclusion morphisms. Now **NP** as it stands has many different interpretations, for example the integers modulo 10, or the truth values with 0==false, 1==true, inc(false)==false, inc(true)==true, false+n==false, true+n==true. In the latter model, + is not commutative. In order to get the *standard* model suggested by the notation, we need to impose the constraint $\langle F^N, F^{NP} \rangle$ on the theory **NP**. Then the only model (up to isomorphism) is the natural numbers with the usual addition. Note that the equation m+n==n+m is satisfied by this model and therefore appears in the equational closure of the presentation; it is a property of + provable by induction. There are also extra constraints in the closure, for example $\langle f', \phi \rangle$, where f' is the inclusion of the empty theory **E** in the theory with 0, 1 and + with identity, associativity and commutativity equations, and ϕ is its inclusion in **NP**. This constraint is satisfied in all models that satisfy the constraint $\langle F^N, F^{NP} \rangle$. In this sense, the constraint on 0, 1 and + gives a derived induction principle. Further examples can be found in [Burstall & Goguen 81].

In general, the closure of a constraining presentation to a "constraining theory" adds new equations derivable by induction principles corresponding to the constraints; and it also adds some new constraints that correspond to derived induction principles. The new equations are important in giving a precise semantics for programming methodology. For example, we may want to supply **NP** as an actual parameter theory to a parameterized theory whose meta-sort demands a commutative binary operator. The new constraints seem less essential. [Clark 78] and [McCarthy 80] discuss what may be a promising approach to the proof-theoretic aspect of constraining theories. Clark calls his scheme "predicate completion" and thinks of it as a method for infering new sentences under a "closed world" assumption (the common sense assumption that the information actually given about a predicate is all and only the relevant information about that predicate; McCarthy identifies this with Occam's famous razor). We, of course, identify that "closed world" with the initial model of the given theory. Clark's scheme is simply to infer the converse of any Horn clause. This is *sound* for the institutions of conditional equations and first order Horn clauses (in the sense that all the sentences thus obtained are true of the initial model); it is not clear that it is complete. McCarthy calls his scheme "circumscription" and is interested in its application in the context of full first order logic; he has shown that it is sound when there exist minimal models. It can be unsound when such models do not exist.

It is worth considering what happens if we add extra silly equations to **NP** constrained by $\langle F^N, \Sigma^{NP} \rangle$. For example, 1+n==n contradicts the constraint; in fact, if we add it, we simply get an inconsistent constraining theory (it has no algebras).

4.3 Other Kinds of Constraint

Because a number of variations on the notion of data constraint have recently been proposed, it seems worthwhile to examine their relationship to the very general machinery developed in this paper. In fact, very little of Section 4.2 or Section 5 depends upon "F-freeness" in the definition of constraint satisfaction. This suggests weakening that notion. Recall that a Σ-algebra A satisfies a data constraint c==$\langle$F: T→T', θ: Sign(T')→$\Sigma\rangle$ iff θA satisfies T' and is F-free, which means that $\epsilon_{\theta A}$: $F^{\$}(F^*(\theta A))$→A is an isomorphism. For the equational institution, the most obvious ways of weakening the F-free concept are to require that $\epsilon_{\theta A}$ is only injective or only surjective, rather than bijective as for F-free. For $\epsilon_{\theta A}$ to be injective corresponds to what has been called a "hierarchy constraint" [Broy, Dosch, Partsch, Pepper & Wirsing 79]; it means that no elements of θA are identified by the natural mapping from $F^\ast(F^\ast(\theta A))$; this condition generalizes the "no confusion" condition of [Burstall & Goguen 82]. For $\epsilon_{\theta A}$ to be surjective corresponds to what has been called a "generating constraint" [Ehrig, Wagner & Thatcher 82]; it means that all elements of θA are generated by elements of $F^\ast(\theta A)$; this condition generalizes the "no junk" condition of [Burstall & Goguen 82]. Notice that these two together give that θA is F-free. Now here is the general notion:

Definition 36: Let I be an institution, let M be a class of model morphisms from I, let $c = \langle F: T \to T', \theta: \mathrm{Sign}(T') \to \Sigma \rangle$ be a constraint from I, and let A be a Σ-model from I. Then A **M-satisfies** c iff θA satisfies T' and $\epsilon_{\theta A}$ lies in M. []

In particular, for a liberal institution I, modifying the construction of $C(I)$ to use the notion of M-satisfaction, gives an institution (generalizing Proposition 34), and theories in this institution will be as cocomplete as I is (generalizing Theorem 35). This means, for example, that we can glue together theories that use hierarchy constraints with the usual colimit constructions, and in particular, we can do the specification language constructions of Section 6.

5 Using More than One Institution

After the work of Section 4.2, we know how to express constraints in any liberal institution; in particular, we can use constraints in the equational institution to specify parameterized abstract data types. We can also give loose specifications in any institution. Liberality is a fairly serious restriction since non-liberal institutions can often be more expressive than liberal institutions. For example, if one adds negation to the equational institution, it ceases to be liberal. Thus, the ambitious specifier might want both the rich expressive power of first order logic and also the data structure definition power of the equational institution. This section shows how he can eat his cake and have it too. The basic idea is to precisely describe a relationship between two institutions, the second of which is liberal, in the form of an institution morphism, and then to permit constraints that use theories from the second institution as an additional kind of sentence in this "duplex" institution. There are also other uses for institution morphisms; in particular, we will use this concept in showing when a theorem prover for one institution is sound for theories from another institution.

5.1 Institution Morphisms

Let us consider the relationship between the institution of first order logic with equality, $\mathcal{FOEQ}$, and the equational institution, $\mathcal{EQ}$. First of all, any first order signature can be reduced to an equational signature just by forgetting all its predicate symbols. Secondly, any equation can be regarded as a first order sentence just by regarding the equal sign in the equation as the distinguished binary predicate symbol (the equal sign in the equations of the equational institution is *not* a predicate symbol, but just a punctuation symbol that separates the left and right sides). Thirdly, any first order model can be viewed as an algebra just by forgetting all its predicates. These three functions are the substance of an institution morphism $\mathcal{FOEQ} \to \mathcal{EQ}$; there are also some conditions that must be satisfied.

Definition 37: Let I and I' be institutions. Then an **institution morphism** $\Phi: I \to I'$ consists of

    1. a functor $\Phi: \mathbf{Sign} \to \mathbf{Sign'}$,

2. a natural transformation α: Φ;Sen'$\Rightarrow$Sen, that is, a natural family of functions
α_Σ: Sen'$(\Phi(\Sigma))\rightarrow$Sen$(\Sigma)$, and

3. a natural transformation β: Mod$\Rightarrow\Phi$;Mod', that is, a natural family of functors
β_Σ: Mod$(\Sigma)\rightarrow$Mod'$(\Phi(\Sigma))$,

such that the following **satisfaction condition** holds

$$A \models_\Sigma \alpha_\Sigma(e') \quad \text{iff} \quad \beta_\Sigma(A) \models'_{\Phi(\Sigma)} e'$$

for A a Σ-model from I and e' a $\Phi(\Sigma)$-sentence from I'. []

The reader may wish to verify that Φ: $\mathcal{FOEQ}\rightarrow\mathcal{EQ}$ as sketched in the first paragraph of this subsection really is an institution morphism.

Notice that an institution morphism Φ: $I\rightarrow I'$ induces a functor Φ: Th$_I\rightarrow$Th$_{I'}$ on the corresponding categories of theories by sending a Σ-theory T to the $\Phi(\Sigma)$-theory $\beta_\Sigma(T^*)^*$. We now have the following useful result, whose proof is omitted in this version of this paper:

Theorem 38: If Φ: $I\rightarrow I'$ is an institution morphism such that Φ: **Sign**$\rightarrow$**Sign'** is [finitely] cocontinuous, then Φ: Th$_I\rightarrow$Th$_{I'}$ is also [finitely] cocontinuous. []

Actually, a stronger result is true: Φ on theories preserves whatever colimits Φ preserves on signatures. By Theorem 38, if a large theory in the source institution is expressed as a colimit of smaller theories, the corresponding theory in the target institution can also be so expressed. Another reason for being interested in this result has to do with using theorem provers for one institution on theories in another (see below).

Definition 39: An institution morphism Φ: $I\rightarrow I'$ is **sound** iff for every signature Σ' and every Σ'-model A' from I', there are a signature Σ and a Σ-model A from I' such that A'$=\beta_\Sigma(A)$. []

This condition is clearly satisfied by the institution morphism from $\mathcal{FOEQ}$ to $\mathcal{EQ}$ discussed above.

Proposition 40: If Φ: $I\rightarrow I'$ is a sound institution morphism and if P is a set of Σ'-sentences from I', then a Σ'-sentence e' is in P$^\bullet$ iff αe is in $\alpha_\Sigma(P)^\bullet$, where Σ is a signature from I such that $\Sigma'=\Phi(\Sigma)$. []

We now know, for example, that the institution morphism from $\mathcal{FOEQ}$ to $\mathcal{EQ}$ permits using a theorem prover for first order logic with equality on equational theories.

5.2 Duplex Institutions

This subsection gives a construction for an institution whose theories can contain both sentences from a nonliberal institution I and constraints from a liberal institution I'. What is needed to make the construction work is an institution morphism Φ: $I\rightarrow I'$ expressing the relationship between the two institutions. This idea was introduced informally in [Goguen 82a] and [Burstall & Goguen 81]. Note that all the results of this section generalize to the notion of M-satisfaction given in Definition 36, although they are stated for the case of data constraints, i.e., the case where M is the class of isomorphisms.

Definition 41: Let I' be a liberal institution, let Φ: $I\rightarrow I'$ be an institution morphism, and let Σ be a signature from I. Then a Σ-**duplex constraint** is a pair

$$c=\langle F: T''\rightarrow T', \theta: \text{Sign}(T')\rightarrow\Phi(\Sigma)\rangle ,$$

where F is a theory morphism from I' and θ is signature morphism from I. Furthermore, a Σ-model A from I **satisfies** the duplex constraint c iff $\theta(\beta_\Sigma(A))$ satisfies T' and is F-free; as usual, write A $\models_\Sigma$ c. []

A picture of the general situation in this definition may help:

$$
\begin{array}{ccc}
& F & & \theta \\
T'' \longrightarrow T' & & \mathrm{Sign}(T') \longrightarrow \Phi(\Sigma)
\end{array}
$$

$$
\begin{array}{c}
F* \\
T''* \underset{F^\sharp}{\overset{}{\rightleftharpoons}} T'* \quad \mathrm{Mod}'(\mathrm{Sign}(T')) \xleftarrow{\;\;\theta*\;\;} \mathrm{Mod}(\Phi(\Sigma)) \xleftarrow{\;\;\beta_\Sigma\;\;} \mathrm{Mod}(\Sigma)
\end{array}
$$

It is worth remarking that in practice T'' and T' could be just presentations, as long as F is a theory morphism in I.

Definition 42: Let Φ: $I \to I'$ be an institution with I' liberal, let Σ be a signature from I, let $c = \langle F, \theta \rangle$ be a Σ-duplex constraint, and let ϕ: $\Sigma \to \Sigma'$ be a signature morphism from I. Then the **translation** of c by ϕ is the Σ'-duplex constraint $\phi c = \langle F, \theta; \Phi(\phi) \rangle$. []

Lemma 43: Satisfaction. Let Φ: $I \to I'$ be an institution morphism with I' liberal, let Σ be a signature from I, let $c = \langle F, \theta \rangle$ be a Σ-duplex constraint, let ϕ: $\Sigma \to \Sigma'$ be a signature morphism from I, and let B be a Σ'-model from I. Then

$\quad$ B $\models_{\Sigma'}$ ϕc iff ϕB $\models_\Sigma$ c . []

Definition 44: Let Φ: $I \to I'$ be an institution morphism with I' liberal and let Σ be a signature from I. Then the **duplex institution** over Φ, denoted $D(\Phi)$ has: its signatures those from I; its Σ-sentences the Σ-sentences from I plus the Σ-duplex constraints; its Σ-models the Σ-models from I; and satisfaction is as defined separately for the sentences from I and the duplex constraints. []

Theorem 45: If I' is a liberal institution and Φ: $I \to I'$ is an institution morphism, then $D(\Phi)$ is an institution. []

This result implies (by Theorem 27) that if the category of signatures of I is [finitely] cocomplete then so is the category of constraining theories that use as sentences both the sentences from I and the duplex constraints constructed using I', since this is the category $\mathrm{Th}_{D(\Phi)}$ of theories of the duplex institution $D(\Phi)$. Examples using this idea have been given in the specification languages Clear [Burstall & Goguen 81] and Ordinary [Goguen 82a, Goguen 82b].

There is another much simpler way to use an institution morphism Φ: $I \to I'$ to construct a new institution in which one simply permits sentences from either I or I'.

Definition 46: Let Φ: $I \to I'$ be an institution morphism. Then $T(\Phi)$ is the institution with: its signatures Σ those from I; its Σ-sentences either Σ-sentences from I, or else pairs of the form $c = \langle T, \theta$: $\mathrm{Sign}(T) \to \Phi(\Sigma) \rangle$, where T is a theory from I' and θ is a signature morphism; the Σ-models of $T(\Phi)$ are those of I; and c is **satisfied** by a Σ-model A means that $\theta(\beta_\Sigma(A))$ satisfies T. []

Proposition 47: $T(\Phi)$ is an institution. []

For example, it may be convenient to use already existing equational theories when construcing new first order theories.

5.3 Multiplex Institutions

Actually, we can use many different institutions all at once, some of them liberal institutions for various kinds of constraints, and some of them not necessarily liberal, for expressive power; all that is needed is a morphism to each from the basic institution. Let I be an institution and let $\Phi_i\colon I\to I_i$ be institution morphisms for $i=1,...,m+n$, where I_i are liberal for $i=1,...,n$. Then we define $P(\Phi_1,...,\Phi_n; \Phi_{n+1},...,\Phi_{n+m})$ to be the institution with: signatures those from I; sentences either sentences from I, or else constraints $\langle F\colon T''\to T', \theta\colon \mathrm{Sign}(T')\to\Phi_i(\Sigma)\rangle$ for θ, T'', T' from I_i for some $1\leq i\leq n$, or else pairs $\langle T, \theta\colon \mathrm{Sign}(T)\to\Phi_i(\Sigma)\rangle$ with θ, T from I_i for some $n+1\leq i\leq n+m$; with its models those of I; and with satisfaction as usual for the I-sentences and for the constraints, and as in Definition 46 for the others. Notice that the liberal insititutions can each use a different collection $\mathcal{M}$ of morphisms to define constraint satisfaction. One can even use the same liberal institution with different classes $\mathcal{M}$. In particular, this means that one can glue together (with colimits) first order theories that use combinations of generating constraints, hierarchy constraints, and data constraints in the equational institution, as well as loose first order axioms.

6 Specification

There is now enough technical machinery so that we can consider some operations on theories that are useful in constructing specifications. For more detailed explanations and examples, see the Clear language and its semantics [Burstall & Goguen 80, Burstall & Goguen 81]; similar ideas appear in several other specification languages.

A specification can be given directly by a presentation in the appropriate institution, taken to denote the theory which is the closure of the presentation. More complex theories can be built up from smaller ones by applying combination operations. A specification language, such as Clear, gives a notation for these operators on theories. An alternative approach is to consider operations on categories of models instead of operations on theories. This has been pursued by [Thatcher, Wagner & Wright 79] and should also fit into the institutional framework. Another approach favoured by some is to consider operations on the presentations; however, we prefer a more "semantic" approach.

6.1 Combining Theories

Suppose that we have separately specified a theory **Bool** of truth values and a theory **Nat** of natural numbers, each with appropriate operators. We would have to combine these to get, say, **Bool** + **Nat**, before defining an extra operator such as $\leq$: **nat x nat→bool**, whose definition involves sorts and operators taken from each theory. (We call adding such extra operators or sorts **enriching** a theory).

There are some issues about the most convenient definition of the combine operation. Suppose that we combine theories T and T′ in the equational institution. Should we take the union of their sorts, operators and equations? This would handle correctly cases where they both use sorts like **bool** which should be the same in each (such cases are quite common). Or should we take the disjoint union? This would correctly handle cases where there is an operator appearing in each with the same name, but inadvertently so; for example, the writers of the two specifications may have used the same name by chance (this is not unlikely with large specifications).

Clear allowed shared subtheories, while still attempting to avoid the problems associated with a chance reuse of the same name. Each theory keeps track of the theories from which it is constructed, so that, for example, it is known that T + T' and T' + T" both contain T. If these two theories are themselves combined, then the sorts and operators in T will not be duplicated. But if the same name is used by chance in T' and in T", then separate copies will be kept, disjoint union style. The mechanism is to use not simple theories but rather "based theories," which are cones in the category of theories. The tip of the cone is the current theory, and the diagram forming the base of the cone shows which theories were used to build it up (see [Burstall & Goguen 80] for details). The coproduct of two such cones gives the required combine operation, written T + T'.

This works, but it is rather sophisticated. Another approach is to ensure that each sort or operator gets a unique name when it is declared; then we can take a simple union of the theories. But union is a set theoretic notion, while the institutional framework uses an arbitrary category of signatures. The solution seems to be to use a subcategory of signature **inclusions**, that is, a full subcategory that is a poset and whose morphisms are called inclusions, and should all be monics. We then no longer need to use cones, and can simply take as T + T' their coproduct in the subcategory of inclusions. Enrichments will give rise to inclusions. (This proposed simpler semantics needs further study.)

6.2 Parameterised Specifications

We often want to write a specification using some "unknown" sorts and operators, that can later be instantiated (as with generic packages in Ada). Then we can develop a library of such parameterised specifications. Thinking of them as specification-building operations, or as specification-valued procedures, we can design a specification language as a functional language whose data elements are theories. An attractive mathematical framework for this uses the pushout construction to apply a parameterised specification to an argument, as we shall explain.

A parameterised specification may be taken to denote a theory, P, having a distinguished subtheory, R; that is, there is an inclusion morphism p: R→P. The subtheory R shows which part of P is the parameter part to be instantiated. For example, P might be a theory about ordered sequences of elements, and R might be the subtheory of the elements and their ordering. By choosing different instances of R, for example natural numbers with their usual ≤ ordering, or sets with the inclusion ordering, we get different instances of P, namely ordered sequences of numbers, or ordered sequences of sets.

What do we mean by saying that a theory A is an instance of a theory R? For equational theories we need to map the sorts and operators of R to sorts and operators of A, in such a way that equations of R are preserved. In general we need a theory morphism f: R→A. We may call R the **requirement**, f the **fitting morphism**, and A the **actual parameter**. Thus we have the diagram

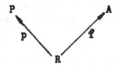

350

The fitting morphism serves to *bind* the material in the actual parameter to the "required" material in the requirement theory. Since the fitting morphism must be a theory morphism, just its existence implies that the actual parameter satisfies the axioms given in the requirement theory, once its syntax has been changed in accord with the bindings given by the fitting morphism.

What do we mean by applying the parameterised theory P to the argument theory A (or more accurately, by applying p to f)? We want a theory that includes what is in P, as well as the new material from A, suitably instantiated by identifying the R part of P with the corresponding elements in A (this correspondence being given by f). It turns out that the pushout of the above diagram is the required categorical construction; we can think of it as yielding the sum of P and A, amalgamating their R parts. (For the equational institution, it has been *proven* that theory really do behave this way [Goguen & Burstall 78].) Writing p[f] for the result of the application, we have the pushout diagram

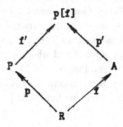

We would expect p and p' to both be inclusions in this diagram.

6.3 Enrichments

We can enrich a specification to get a larger specification. In the case of equational theories, this is done by adding new sorts, operators and equations; for example, adding to **Bool** + **Nat** the new operator $\leq$: **nat x nat**→**bool** with its defining equations, such as $0 \leq n$=true. Note that the new material does not itself constitute a theory, so we cannot use combine here.

However, enrichment is really just a special case of applying a parameterised specification. Suppose that we wish to enrich a theory T. The new operator declarations may refer to sorts in T, and the new equations to operators in T. So there is a minimal subtheory of T needed to support this new material, say T', with T'$\subseteq$T, that is, with an inclusion i: T'→T. T' together with the new material forms a theory T", with T'$\subseteq$T". We may think of the enrichment as an abbreviation for an inclusion j: T'→T". The result of the enrichment is just the pushout of i and j, or considering j as a parameterised specification and i as the fitting morphism, it is j[i].

6.4 An Analogy

This section has applied ideas of the preceding sections to problems of program specification, including theory combination, theory environments, and parameterized theories. The discussion follows an analogy [Burstall & Goguen 77] between specification languages and programming languages, in which theories are treated as values, parameterized theories correspond to procedures, based theories are environments, and requirement theories correspond to types. This last part of the analogy can be seen as an application of the "correspondence principle" of [Landin 66].

Just as declarations for variables and procedures create environments containing local values for identifiers in a programming language, so do declarations in a specification language create environments having local values. Whereas an environment in a programming language is just an assignment of values (and possibly types) to identifiers, environments in a specification language must also keep track of relationships between theories, so that there is not only an assignment of theories to identifiers, but also a collection of theory inclusions among the given theories, indicating that some theories are contained in others. The basic semantic insight is that to put theories together, whether for combining, enriching, or for applying a parameterized theory to an actual theory, one can simply take colimits of appropriate diagrams.

7 Summary

We have formalized the intuitive notion of a "logical system" or "abstract model theory" with the abstract notion of an institution, and have shown that institutions whose signatures have finite colimits are suitable for use in programming methodology. In a liberal institution, the forgetful functors induced by theory morphisms, have corresponding free functors, and data types can be defined by "constraints," which are abstract induction principles. Allowable kinds of constraint include data, generating, and hierarchy. We have introduced the notion of an institution morphism and shown how to use it in determining when a theorem prover for one institution can be soundly used on (translations of) theories from another institution. Institution morphisms were also used in defining duplex and multiplex institutions, which formalize the simultaneous use of more than one institution; this permits combining the greater expressive power of nonliberal institutions with the data type definition capability of liberal institutions. Finally, we showed how to formulate a number of basic specification language constructions in an arbitrary institution having finite colimits.

References

[Arbib & Manes 75]
        Arbib, M. A. and Manes, E.
        Arrows, Structures and Functors.
        Academic Press, 1975.

[Aubin 76]   Aubin, R.
        Mechanizing Structural Induction.
        PhD thesis, University of Edinburgh, 1976.

[Barwise 74]   Barwise, J.
        Axioms for Abstract Model Theory.
        Annals of Mathematical Logic 7:221-265, 1974.

[Benabou 68]   Benabou, J.
        Structures Algebriques dans les Categories.
        Cahiers de Topologie et Geometrie Differentiel 10:1-126, 1968.

[Birkhoff & Lipson 70]
        Birkhoff, G. and Lipson, J.
        Heterogeneous Algebras.
        Journal of Combinatorial Theory 8:115-133, 1970.

[Bloom 76]   Bloom, S. L.
        Varieties of Ordered Algebras.
        Journal of Computer and System Sciences 13:200-212, 1976.

352

[Boyer & Moore 80]
    Boyer, R. and Moore, J. S.
    A Computational Logic.
    Academic Press, 1980.

[Broy, Dosch, Partsch, Pepper & Wirsing 79]
    Broy, M., Dosch, N., Partsch, H., Pepper, P. and Wirsing, M.
    Existential Quantifiers in Abstract Data Types.
    In *Proceedings, 6th ICALP*, pages 73-87. Springer-Verlag, 1979.
    Lecture Notes in Computer Science, volume 71.

[Burstall & Goguen 77]
    Burstall, R. M. and Goguen, J. A.
    Putting Theories together to Make Specifications.
    Proceedings, Fifth International Joint Conference on Artificial Intelligence
      5:1045-1058, 1977.

[Burstall & Goguen 78]
    Burstall, R. M. and Goguen, J. A.
    Semantics of Clear.
    1978.
    Unpublished notes handed out at the Symposium on Algebra and Applications,
      Stefan Banach Center, Warszawa, Poland.

[Burstall & Goguen 80]
    Burstall, R. M., and Goguen, J. A.
    The Semantics of Clear, a Specification Language.
    In *Proceedings of the 1979 Copenhagen Winter School on Abstract Software*
      Specification, pages 292-332. Springer-Verlag, 1980.
    Lecture Notes in Computer Science, Volume 86.

[Burstall & Goguen 81]
    Burstall, R. M. and Goguen, J. A.
    An Informal Introduction to Specifications using Clear.
    In Boyer, R. and Moore, J (editor), *The Correctness Problem in Computer*
      Science, pages 185-213. Academic Press, 1981.

[Burstall & Goguen 82]
    Burstall, R. M. and Goguen, J. A.
    Algebras, Theories and Freeness: An Introduction for Computer Scientists.
    In *Proceedings, 1981 Marktoberdorf NATO Summer School*, . Reidel, 1982.

[Burstall & Rydeheard 80]
    Burstall, R. M. and Rydeheard, D. E.
    Signatures, Presentations and Theories: a Monad Approach.
    Technical Report, Computer Science Department, University of Edinburgh,
      1980.

[Clark 78]   Clark, K. L.
    Negation as Failure.
    In H. Gallaire and J. Minker (editor), *Logic in Data Bases*, pages 293-322.
      Plenum Press, 1978.

[Cohn 65]   Cohn, P. M.
Universal Algebra.
Harper and Row, 1965.
Revised edition 1980.

[Ehrich 78]   Ehrich, H.-D.
On the Theory of Specification, Implementation and Parameterization of Abstract Data Types.
Technical Report, Forschungsbericht, Dortmund, 1978.

[Ehrig, Kreowski, Rosen & Winkowski 78]
Ehrig, E., Kreowski, H.-J., Rosen, B. K. and Winkowski, J.
Deriving Structures from Structures.
In *Proceedings, Mathematical Foundations of Computer Science*, . Springer-Verlag, Zakopane, Poland, 1978.
Also appeared as technical report RC7046 from IBM Watson Research Center, Computer Sciences Dept.

[Ehrig, Wagner & Thatcher 82]
Ehrig, H., Wagner, E. and Thatcher, J.
Algebraic Specifications with Generating Constraints.
Technical Report, IBM Research Center, Yorktown Heights, New York, 1982.
Draft report.

[Gabriel & Ulmer 71]
Gabriel, P. and Ulmer, F.
Lokal Prasentierbare Kategorien.
Springer-Verlag, 1971.
Springer Lecture Notes in Mathematics, vol. 221.

[Gerhard, Musser et al. 79]
Gerhard, S. L., Musser, D. R., Thompson, D. H., Baker, D. A., Bates, R. W., Erickson, R. W., London, R. L., Taylor, D. G., and Wile, D. S.
An Overview of AFFIRM: A Specification and Verification System.
Technical Report, USC Information Sciences Institute, Marina del Rey, CA, 1979.

[Goguen 71]   Goguen, J.
Mathematical Foundations of Hierarchically Organized Systems.
In E. Attinger (editor), *Global Systems Dynamics*, pages 112-128. S. Karger, 1971.

[Goguen 74]   Goguen, J. A.
Semantics of Computation.
In *Proceedings, First International Symposium on Category Theory Applied to Computation and Control*, pages 234-249. University of Massachusetts at Amherst, 1974.
Also published in Lecture Notes in Computer Science, Vol. 25., Springer-Verlag, 1975, pp. 151-163.

[Goguen 77]   Goguen, J. A.
Abstract Errors for Abstract Data Types.
In *IFIP Working Conference on Formal Description of Programming Concepts*, . MIT, 1977.
Also published by North-Holland, 1979, edited by P. Neuhold.

354

[Goguen 78]   Goguen, J. A.
                 Order Sorted Algebra.
                 Technical Report, UCLA Computer Science Department, 1978.
                 Semantics and Theory of Computation Report No. 14; to appear in *Journal of
                     Computer and System Science.*

[Goguen 80]   Goguen, J. A.
                 How to Prove Algebraic Inductive Hypotheses without Induction: with
                     applications to the correctness of data type representations.
                 In W. Bibel and R. Kowalski (editor), *Proceedings, 5th Conference on
                     Automated Deduction,* pages 356-373. Springer-Verlag, Lecture Notes in
                     Computer Science, Volume 87, 1980.

[Goguen 82a]   Goguen, J. A.
                 Ordinary Specification of Some Construction in Plane Geometry.
                 In Staunstrup, J. (editor), *Proceedings, Workshop on Program Specification,*
                     pages 31-46. Springer-Verlag, 1982.
                 Lecture Notes in Computer Science, Volume 134.

[Goguen 82b]   Goguen, J. A.
                 Ordinary Specification of KWIC Index Generation.
                 In Staunstrup, J. (editor), *Proceedings, Aarhus Workshop on Specification,*
                     pages 114-117. Springer-Verlag, 1982.
                 Lecture Notes in Computer Science, Volume 134.

[Goguen & Burstall 78]
                 Goguen, J. A. and Burstall, R. M.
                 *Some Fundamental Properties of Algebraic Theories: a Tool for Semantics of
                     Computation.*
                 Technical Report, Dept. of Artificial Intelligence, University of Edinburgh,
                     1978.
                 DAI Research Report No. 5; to appear in *Theoretical Computer Science.*

[Goguen & Ginali 78]
                 Goguen, J. A. and Ginali, S.
                 A Categorical Approach to General Systems Theory.
                 In Klir, G. (editor), *Applied General Systems Research,* pages 257-270.
                     Plenum, 1978.

[Goguen & Meseguer 81]
                 Goguen, J. A. and Meseguer, J.
                 Completeness of Many-sorted Equational Logic.
                 SIGPLAN Notices 16(7):24-32, July, 1981.
                 Also appeared in *SIGPLAN Notices,* January 1982, vol. 17, no. 1, pages 9-17;
                     extended version as SRI Technical Report, 1982, and to be published in
                     Houston Journal of Mathematics.

[Goguen & Meseguer 83]
                 Goguen, J. A. and Meseguer, J.
                 An Initiality Primer.
                 In Nivat, M. and Reynolds, J. (editor), *Application of Algebra to Language
                     Definition and Compilation,* . North-Holland, 1983.
                 To appear.

[Goguen & Parsaye-Ghomi 81]
Goguen, J. A. and Parsaye-Ghomi, K.
Algebraic Denotational Semantics using Parameterized Abstract Modules.
In J. Diaz and I. Ramos (editor), *Formalizing Programming Concepts*, pages
292-309. Springer-Verlag, Peniscola, Spain, 1981.
Lecture Notes in Computer Science, volume 107.

[Goguen & Tardo 79]
Goguen, J. A. and Tardo, J.
An Introduction to OBJ: A Language for Writing and Testing Software
Specifications.
In *Specification of Reliable Software*, pages 170-189. IEEE, 1979.

[Goguen, Thatcher & Wagner 78]
Goguen, J. A., Thatcher, J. W. and Wagner, E.
An Initial Algebra Approach to the Specification, Correctness and
Implementation of Abstract Data Types.
In R. Yeh (editor), *Current Trends in Programming Methodology*, pages
80-149. Prentice-Hall, 1978.
Original version, IBM T. J. Watson Research Center Technical Report RC
6487, October 1976.

[Goguen, Thatcher, Wagner & Wright 75a]
Goguen, J. A., Thatcher, J. W., Wagner, E. G., and Wright, J. B.
An Introduction to Categories, Algebraic Theories and Algebras.
Technical Report, IBM T. J. Watson Research Center, Yorktown Heights,
N. Y., 1975.
Research Report RC 5369.

[Goguen, Thatcher, Wagner & Wright 75b]
Goguen, J. A., Thatcher, J. W., Wagner, E. and Wright, J. B.
Abstract Data Types as Initial Algebras and the Correctness of Data
Representations.
In *Computer Graphics, Pattern Recognition and Data Structure*, pages 89-93.
IEEE, 1975.

[Goldblatt 79]   Goldblatt, R.
Topoi, The Categorial Analysis of Logic.
North-Holland, 1979.

[Higgins 63]   Higgins, P. J.
Algebras with a Scheme of Operators.
Mathematische Nachrichten 27:115-132, 1963.

[Hoare 72]   Hoare, C. A. R.
Proof of Correctness of Data Representation.
Acta Informatica 1:271-281, 1972.

[Kaphengst & Reichel 71]
Kaphengst, H. and Reichel, H.
Algebraische Algorithemtheorie.
Technical Report WIB Nr. 1, VEB Robotron, Zentrum fur Forschung und
Technik, Dresden, 1971.
In German.

356

[Landin 66]   Landin, P. J.
The Next 700 Programming Languages.
Communications of the Association for Computing Machinery 9, 1966.

[Lawvere 63]   Lawvere, F. W.
Functorial Semantics of Algebraic Theories.
Proceedings, National Academy of Sciences 50, 1963.
Summary of Ph.D. Thesis, Columbia University.

[Lawvere 64]   Lawvere, F. W.
An Elementary Theory of the Category of Sets.
Proceedings, National Academy of Sciences, U.S.A. 52:1506-1511, 1964.

[Levitt, Robinson & Silverberg 79]
Levitt, K., Robinson, L. and Silverberg, B.
The HDM Handbook.
Technical Report, SRI, International, Computer Science Lab, 1979.
Volumes I, II, III.

[MacLane 71]   MacLane, S.
Categories for the Working Mathematician.
Springer-Verlag, 1971.

[MacLane & Birkhoff 67]
MacLane, S. and Birkhoff, G.
Algebra.
Macmillan, 1967.

[Mahr & Makowsky 82a]
Mahr, B. and Makowsky, J. A.
An Axiomatic Approach to Semantics of Specification Languages.
Technical Report, Technion, Israel Institute of Technology, 1982.
Extended Abstract.

[Mahr & Makowsky 82b]
Mahr, B. and Makowsky, J. A.
Characterizing Specification Languages which Admit Initial Semantics.
Technical Report, Technion, Israel Institute of Technology, February, 1982.
Technical Report #232.

[McCarthy 80]   McCarthy, J.
Circumscription - A Form of Non-Monotonic Reasoning.
Artificial Intelligence 13(1,2):27-39, 1980.

[Parsaye-Ghomi 82]
Parsaye-Ghomi, K.
Higher Order Data Types.
PhD thesis, UCLA, Computer Science Department, January, 1982.

[Reichel 80]   Reichel, H.
Initially Restricting Algebraic Theories.
Springer Lecture Notes in Computer Science 88:504-514, 1980.
Mathematical Foundations of Computer Science.

[Shostak, Schwartz & Melliar-Smith 81]
Shostak, R., Schwartz, R. & Melliar-Smith, M.
STP: A Mechanized Logic for Specification and Verification.
Technical Report, Computer Science Lab, SRI International, 1981.

[Tarski 44]
Tarski, A.
The Semantic Conception of Truth.
Philos. Phenomenological Research 4:13-47, 1944.

[Thatcher, Wagner & Wright 79]
Thatcher, J. W., Wagner, E. G. and Wright, J. B.
Data Type Specification: Paramerization and the Power of Specification Techniques.
In *Proceedings of 1979 POPL*, . ACM, 1979.

[Wright, Thatcher, Wagner & Goguen 76]
Wright, J. B., Thatcher, J. W., Wagner, E. G. and Goguen, J. A.
Rational Algebraic Theories and Fixed-Point Solutions.
Proceedings, 17th Foundations of Computing Symposium :147-158, 1976.
IEEE.

R.M. Burstall
School of Informatics, Edinburgh University, Scotland, UK

© Springer 1984. Reprinted, with kind permission, from: Goguen, J. & Burstall, R., "Introducing institutions". In: Clarke, E. & Kozen, D. (eds), Proceedings, Logics of Programming Workshop, Lecture Notes in Computer Science, Vol. 164, pp. 221–255

Part 14
Andrea Loparić and Newton
da Costa (1984)

Paranormal Logics and the Theory of Bivaluations

Jean-Yves Béziau

Abstract In the paper *Paraconsistency, paracompleteness and valuations* (1984), Loparić and da Costa use a general theory of bivaluations to develop a paranormal logic, i.e. a logic which is both paraconsistent and paracomplete. We explain how this theory works, in which sense it allows to claim that every logic is bivalent and permits to develop a general theory of logics. We also explain how the theory of bivaluations is used to give a better formulation of the principle of non-contradiction and the principle of excluded middle, and how we can develop some generalized truth-tables, useful to work with such kind of non-truth-functional bivalent semantics.

Keywords Paraconsistent logic · Paracomplete logic · Bivaluation · Truth-functionality · Abstract logic · Universal logic

Mathematics Subject Classification (2000) Primary 03B53 · Secondary 03B22

1 Introduction

Newton da Costa is best known for his work on paraconsistent logic, a field he has crucially contributed to promote, developing many paraconsistent logical systems and clarifying the subject both from a philosophical and a mathematical viewpoint. But he also has been working on many different subjects: other non-classical logics, foundations of logic, foundations of mathematics, foundations of physics and foundations of science in general. All these lines of research are related. Da Costa's work on paraconsistent logic is in particular related to the development of a general theory of logics, based on semantics, the theory of bivaluations. This is perfectly illustrated by the paper under consideration.

Roughly speaking paraconsistent logics are logics in which the principle of non-contradiction (PNC hereafter) does not hold. PNC is considered, since Aristotle, as a fundamental principle of logic. The predominance of the PNC has prevailed during all the development of occidental culture and civilization, from Aristotle to modern times. But there also have been people proposing a diametrically opposed viewpoint within this same culture. From Heraclitus to Marx via Hegel some people have sustained that contradiction is essential in reality and/or in thought. The positive occidental view on contradiction has mixed with oriental culture of contradiction in China, the climax being Maoism (see [42]). But the oriental view on contradiction does not reduce to the Maoist version of Marxism. It has its own flavor illustrated by the Yin-Yang symbol of Taoism. The physicist Niels

Bohr used to wear this symbol, he thought that Taoism was a good explanation for the recent advances of modern physics he was characterizing by the notion of complementarity.

How can we know who is right? Is the PNC a fundamental principle or not? Marxism and Maoism look like ideologies rather than scientific and rational theories and one can say the same about Taoism. The Yin-Yang sign remembers symbols such as the Christian cross, the swastika or the Mercedes logo. But to believe, like Aristotle and his followers, that the PNC is a fundamental principle, is this not the same as to believe in Contradiction or God? What are the differences between science, ideology and religion?

Da Costa's inquiry about the PNC touches upon such difficult question. By investigating how to develop logical systems in which the PNC does not hold, he has contributed to demystify the PNC. He was able to successfully construct such systems using mathematical logic, showing therefore that the PNC is not an absolute foundation of logic. So what is the basis of logic? Da Costa started to develop general tools to deal with logics, appearing more fundamental than the PNC, among them the theory of bivaluations.

The theory of bivaluations is a theory using bivaluations to describe and construct logical systems. According to this theory one can rely on two values, truth and falsity, to develop any kind of reasoning. In the present paper the authors claim that "every logical system whatever has a two-valued semantics of valuations" (p. 376). We will explain the meaning of this claim and see its import for the development of a general theory of logics[1].

2 Logical Calculus

The point of departure of da Costa in this paper (and other papers) is the notion of *logical calculus*, which is considered as a set of axioms Δ and a set of inference rules $\mathcal{R}$ on a given language $\mathcal{L}$. On the basis of such logical calculus a consequence relation $\vdash_{\mathcal{C}}$ is defined and it is also possible to define a logic $\langle \mathcal{L}; \vdash_{\mathcal{C}} \rangle$. Since Loparić and da Costa put here the emphasis on logical calculus rather than on the logical structure corresponding to it, we don't find explicitly in this paper the logical structure but rather the logical calculus $\mathcal{C} = \langle \Delta; \mathcal{R} \rangle$.

Such an approach is a basis for a general theory of logics, close to Tarski's consequence operator. As in Tarski's approach the set of formulas is not really specified and it can be many things, although in this paper the authors have in mind the usual language of classical propositional logic. There is a kind of vagueness leaving space for many interpretations and applications.

The notion of rule is presented as follows: "a rule relates a new formula (the conclusion), to a given set of formulas (the premises). For a given rule, the number of premises is always finite and fixed" (p. 375). This is typically the definition of a Hilbertian style rule of inference.

The way how the consequence relation $\vdash_{\mathcal{C}}$ is then defined is not given, but implicitly this is in the usual way as for Hilbertian systems. This means that the consequence relation

---

[1]Da Costa generally uses the expression "theory of valuation", but it seems better to use the expression "theory of bivaluations" which is more explicit. This theory was developed by da Costa and its collaborators, mainly Andréa Loparić, Elias Alves and the present author. Nicola Grana has published a booklet in Italian which is an elementary presentation of this theory. References are in the bibliography.

has necessarily the properties corresponding to the three axioms of a consequence operator: reflexivity, monotonicity and transitivity. As we have suggested [15], Tarski originally chose these axioms because he had Hilbertian systems in view. As we said, Tarski was axiomatizing axiomatic systems—Hilbertian proof systems are known as *axiomatic systems* because they generally have a lot of axioms and few rules.

In this kind of logical calculus since the rules are finitary the consequence relation also is, that is to say, if $\Gamma \vdash A$ then there is a finite subset Γ_0 of Γ such that $\Gamma_0 \vdash A$. This means that the consequence relation obeys also the further Tarskian axiom of finiteness. Finiteness justifies here the name *calculus*: we have something which is computable.

One may wonder if starting with a logical calculus is not too restrictive. In fact it is not more restrictive than dealing with a finitary consequence operator since any such logical calculus defines such an operator and vice-versa. The advantage of such a logical calculus is that the notions of rules and proofs appear explicitly, the inconvenient is that we cannot modulate the basic properties, sometimes called *structural*, they are necessarily valid.

3 Bivaluations

3.1 Non-truth-functional Bivalent Semantics

In this paper two notions of bivaluations are considered, the notion of *evaluation* and the notion of *valuation*, both are functions from the set of formulas $\mathcal{L}$ to $\{0, 1\}$.

The most important feature of these bivaluations is that contrarily to the case of bivaluations of bivalent semantics for classical propositional logic, they are not homomorphisms between the set of formulas to an algebra of similar type on $\{0, 1\}$. In fact no structure is considered on the set $\{0, 1\}$. This generates semantics that can be called *non-truth-functional bivalent semantics*. It is in fact proved in this paper (Theorem 7, p. 383) that the system π presented cannot be characterized by any finite matrices, an analog of Dugundji's theorem. An important difference between truth-functional and non-truth-functional bivalent semantics is that the latter cannot be generated by distributions on atomic formulas.

Evaluations are bivaluations that respect axioms and rules. Valuations are characteristic functions of saturated sets. What are exactly these sets and why considering them? The reply to these questions is connected to the central result of this paper claiming that every logical system has a two-valued semantics.

First it is important to note that considering bivaluations and theories (sets of formulas) is the same if we identify a theory with its characteristic function. Giving any class of theories *THE* of a given language $\mathcal{L}$ one can wonder if it is sound and complete for a logic $\langle \mathcal{L}; \vdash_C \rangle$ generated by a calculus C. Using *THE* (considering theories as bivaluations), one can define a consequence relation using the usual definition of semantical consequence:

$\Gamma \models_{THE} A$ iff for any bivaluation $\beta \in THE$, if $\beta(X) = 1$ for every element X of Γ, then $\beta(A) = 1$

We have then the following definitions of soundness and completeness: *THE* is *sound* for $\langle \mathcal{L}; \vdash_C \rangle$ iff if $\Gamma \vdash_C A$ then $\Gamma \models_{THE} A$ and *THE* is *complete* iff if $\Gamma \models_{THE} A$ then $\Gamma \vdash_C A$.

3.2 Evaluation and Soundness

An *evaluation e* is a bivaluation that respects axioms and rules. This means that it gives the value 1 to all axioms and that if it gives the value 1 to all the premises of a rule, it gives the value 1 to the conclusion. Loparić and da Costa probably victims of the fear of trivialization add that an evaluation should give the value 0 to at least one formula. The class of theories *EVA* corresponding to evaluations is trivially sound for (the logic generated by) the calculus, as pointed out by property i) of p. 375.[2]

It is easy to prove that evaluations are characteristic functions of closed theories. A theory is said to be *closed* iff if $\Gamma \vdash_C A$ then $A \in \Gamma$. Some people, for example in the Polish school, used to call theories only closed theories and this is what Loparić and da Costa also do here.

It is possible to prove that a class of theories is sound for a calculus iff it is included in *CLO* (the class of closed theories). Therefore if a semantics is sound and complete for a calculus it should be included in *CLO*. *CLO* is also complete for the logic generated by the calculus. This result was proved by Gentzen in his first paper [28], which was about Hertz's *Satzsysteme* [31].

3.3 Saturated Set and Completeness

The question now is which kind of smaller semantics one may consider for (a logic generated by) a calculus. In this paper the authors deal with the semantics *SAT* of saturated theories. *SAT* is a sound and complete semantics for the logic generated by a calculus. Elements for the proof are given in this paper but the proof is not explicitly presented. The proof relies in fact on a version of Lindenbaum lemma presented as item iv) of p. 376, whose proof is not given. This version of Lindenbaum lemma is known as Lindenbaum–Asser lemma.

This version of the lemma can be found in a book by Günter Asser [1] where we can also found the concept of saturated theory, but under another terminology: relatively maximal theory.[3] This terminology makes sense establishing a clear connection between saturated and maximal theories. A maximal theory Υ is a theory such that there is a formula A with $\Upsilon \not\vdash A$ (this means that Υ is *non-trivial*) and such that any strict extension of it is trivial. A relatively maximal (or saturated) theory Σ is a theory such that there is a formula A with $\Sigma \not\vdash A$ and such that A is a consequence of any strict extension of Σ. We say that Σ is relatively maximal with respect to A. One can prove that a theory is maximal iff it is relatively maximal with respect to all formulas not belonging to it. This shows that the set of maximal theories *MAX* is included in the set of relatively maximal theories *REM* (i.e. *SAT*).

---

[2]Loparić and da Costa use the notation $\mathcal{E}$ for the set of evaluations.

[3]The question is open to know if G. Asser was the first to introduce the notion of *relatively maximal theory* and the first to state *Lindenbaum–Asser lemma*. Such name was given to this lemma, but this is not a proof: for example as we know the notion of *Lindenbaum algebra* is due to Tarski and not to Lindenbaum (see [30, p. 169, footnote 2]). Note however that Lindenbaum lemma is due to Lindenbaum.

REM is included in *CLO* as stated by property iii) p. 376, so *REM* is a sound semantics. And *REM* is complete due to Lindenbaum–Asser lemma. *MAX* is included in *REM* and in *CLO*, *MAX* also is sound. One can wonder why considering *REM* and not *MAX*—after all the semantics of classical propositional logic is *MAX*. A clear answer was given in [10]: *REM* is a minimal complete semantics, this means that in case when *MAX* is strictly included in *REM*, *MAX* is not complete. But in which logics is *MAX* strictly included in *REM*? This is the case of positive intuitionistic logic. But this not the case of the system π presented in this paper. In the system π there is a classical negation and a classical implication and it has been proved in [10] that in such situations *REM* = *MAX*.

The concept of relatively maximal theory is known in the Polish tradition (see [38, 41]). In Brazil it was first used by Loparić [33], dealing with the paraconsistent logic C_ω which is a logic, like positive intuitionistic logic, which has no classical negation and no classical implication. In the present paper it would have been sufficient to consider the semantics *MAX* to prove the completeness of the system π. But the introduction of the concept of relatively maximal theory, named here saturated set, makes sense in this paper in connection with the statement of a general completeness theorem.

4 Every Logic is Bivalent?

Loparić and da Costa claim in the introduction of their paper: "every logical system whatever has a two-valued semantics of valuations". This claim is connected with Theorem 1 (p. 376) of their paper stating that *REM* is a sound and complete semantics for any logic generated by a calculus. But in fact we can make a similar claim if we consider that *CLO* is a sound and complete semantics for any logic generated by a calculus. As we said, this result can be attributed to Gentzen [28], but he didn't make a claim about bivalency because if he was considering theories rather than bivaluations.

One does not need therefore to prove that *REM* is a sound and complete semantics for any logic generated by a calculus to claim that every such logic is bivalent. Moreover to prove that *REM* is sound and complete one needs more requisites than to prove that *CLO* is sound and complete. The fact that *REM* is sound and complete is a direct consequence of Lindenbaum–Asser lemma requiring the consequence relation to be finitary.

But *REM* is interesting for another reason related to Gentzen's work. In [6, 12], the following theorem has been proved:

Birmseq Theorem *Relatively maximal theories respect sequent rules of structurally standard systems of sequents.*

A structurally standard systems of sequents, SSSS for short, is a system of sequents with all the standard structural rules and in which moreover the sequents have the usual structure: multiplicity of formulas on both side and no context restrictions. It is possible to immediately translate sequent rules into conditions for bivaluations by considering Gentzen's original idea according to which a sequent $A_1, \ldots, A_n \to B_1, \ldots, B_m$ means $A_1 \wedge \cdots \wedge A_n \supset B_1 \vee \cdots \vee B_m$ in classical logic. Given a set of sequent rules, we can define the set *SRT* of bivaluations generated by translations of sequent rules into conditions for bivaluations.

If we have a SSSS the completeness of *SRT* is a corollary of the Birmseq Theorem: a relatively maximal theory respects the rules, therefore it obeys the conditions given by translations of sequent rules, *REM* is therefore included in *SRT*, and it is easy to show that any set of bivaluations which includes *REM* is complete in case we have a finitary consequence relation. Using this fact we can prove the completeness of bivalent semantics for sequent calculi of many logics.

This methodology has been applied to Łukasiewicz logic *L*3 [9] and De Morgan logic [16]. The claim that "every logical system whatever has a two-valued semantics of valuations" is certainly a bit provocative and is not completely true. It is nevertheless true that it is possible to provide a bivalent semantics for *L*3 and the four-valued logic of Dunn and Belnap. The apparent paradox is cleared by making the distinction between truth-functional and non-truth functional bivalent semantics.

The soundness and completeness of *REM* and *CLO* depend on the fact that we have a consequence relation which is reflexive, transitive, monotonic, and for *REM*, moreover finitary. If we drop one of these conditions, we don't have anymore these results and we cannot rely on them to claim bivalency. A way to claim bivalency for all logical systems is to sustain that logical systems have to obey these properties, that they have to be calculi or something similar: Tarskian consequence relations. But one may wonder if this really makes sense, because it is difficult to believe in the absolute validity of such properties, as shown for example for the axiom of monotonicity, with the rise of non-monotonic logics.

The idea of da Costa was to develop a general theory of logics using the theory of bivaluations. In view of the limitation of such an approach, I started to develop universal logic, considered as a general theory of logics with logical structures obeying no axioms (see [5, 6, 17]) similarly to universal algebra developed by Garrett Birkhoff (see [18, 19]).

Although the expression *universal logic* is analogous to the expression *universal algebra*, this does not mean that logic reduces to algebra, since logical structures are not the same as algebraic structures. Da Costa himself has insisted on this point by defending the idea that logics which are not algebraizable, like the paraconsistent logic *C*1, can still be considered as logics (for a discussion about this, see [7]). The system π presented here by Loparić and da Costa also is not algebraizable.

5 Paraconsistency and Paracompleteness

Paraconsistent logics are logics in which the principle of non-contradiction (PC) does not hold and paracomplete logics are logics in which the principle of excluded middle (EM) does not hold. There is a duality between paraconsistency and paracompleteness well expressed using the theory of bivaluations. In this paper a system of logic π which is both paraconsistent and paracomplete is presented. Such systems have been called *non-alethic logics* or alternatively *paranormal logics*.[4]

---

[4]The terminology "paranormal" was devised by myself and used for example in [26] where another paranormal logic is presented—about paranormal logics see also: [37, 39]. The prefix "para" has the advantage to provide uniformity with "paraconsistent" and "paracomplete", terminology introduced by Quesada and da Costa. The terminology "non-alethic" is due to Francisco Miró Quesada. It is ambiguous because "alethic" is used in a different way in the context of modal logic. Moreover if bivalent semantics based on truth and falsity can be constructed for a logic which is both paraconsistent and paracomplete,

In developing paraconsistent logic Newton da Costa was led to the theory of valuations, a general technique to study and build logical systems. This paper is an application of this technique showing how it is possible to construct logics derogating not only PC but also EC, paranormal logics infringing two basic principles of good old classical logic. Let us see how this can be done. First one needs to clearly formulate these principles. The theory of bivaluations is a way to do so, avoiding some common confusions. Some people claim that the principle of bivalence (PB) is the conjunction of the principle of non-contradiction with the principle of excluded middle, so that we have the following equation:

$$PB = PC + EM.$$

But what is the principle of bivalence? It can be expressed as follows: A proposition is either true or false (PB). This is the conjunction of two principles: A proposition cannot be true and false (PB1) and A proposition cannot be neither true, nor false (PB2). A bivalent semantics of the theory of bivaluations clearly respects the principle of bivalence. This is not the case for example of the three-valued semantics of Łukasiewicz where a proposition can have a third value, called "possible" or "indeterminate". Such a three-valued semantics infringes PB2.

In order to get the equation $PB = PC + EM$ some people will argue that PC $=$PB1 and EM $=$ PB2. In this case it is true that Łukasiewicz three-valued semantics infringes EM, and the fact that in the logic $L3$ the law of excluded middle $A \lor \neg A$ is not valid. could comfort the one equating EM with PB2. But it is not difficult to see how one can construct a logic in which $A \lor \neg A$ is a tautology based on a semantics infringing PB2. Should we still say then that such logic is infringing EM? This is dubious and the fact that on the other hand it is possible with a bivalent semantics to develop a logic (obeying then PB2) in which $A \lor \neg A$ is not a tautology should definitely lead to the rejection of such equation. The equation PC $=$ PB1 is similarly false and at the end we see that the equation $PB = PC + EM$ makes no sense.

We can argue that PC and EM are principles about *negation*. If one has a bivalent semantics in which A is true iff $\neg A$ is false, condition expressed by the table

A	$\neg A$
0	1
1	0

then "A and $\neg A$ cannot be true together" (PC) is equivalent to "A cannot be true and false" (PB1). But the identity between PC and PB1 can be false if we have a non-truth-

---

there is no good reason to say that it is "against truth"—as a literal transcription of the Greek expression "non-alethic" would suggest. The expression "paranormal" expresses the fact that we are beyond the standard norms PC and EM of classical logic.

functional bivalent semantics in which we have the following table:

A	$\neg A$
0	1
1	0
1	1

In this case PB1 is true but PC is false.

PC and EM can be stated as follows

> PC: A and $\neg A$ cannot be true together

> EM: A and $\neg A$ cannot be false together

This is close to the theory of the square of opposition. The conjunction of PC and EM does not lead to PB but to the notion of contradiction. Using the theory of the square of opposition, PC is the principle of (non) subcontrariety, EM the principle of (non) contrariety and PC+EM the principle of (non) contradiction. Using the theory of bivaluations, PC and EM are formulated as follows:

> PC: if $\beta(A) = 1$ then $\beta(\neg A) = 0$

> EM: if $\beta(A) = 0$ then $\beta(\neg A) = 1$

for any bivaluation β of a bivalent semantics for the given logic.

The system π presented in this paper has a bivalent semantics which does not obey neither PC nor EM. But one can argue that the negation of this system is still a negation, because it has some interesting properties that the authors get by additional features compatible with the rejection of PC and EM, which are expressed in particular by items 4, 5, 6, 7 on pp. 379–380. Let us explain how this works.

In a given bivaluation β, we can say that a formula A obeys PC and EM iff $\beta(A) \neq \beta(\neg A)$. Having this in mind, we can interpret the conditions 4 to 7 (read contrapositively) as saying:

(4) If A obeys PC and EM, then $\neg A$ obeys PC and EM
(5) For a binary connective c, if A or B obeys PC and EM, then $A c B$ obeys PC and EM
(6) $A \wedge \neg A$ obeys PC and EM
(7) $A \vee \neg A$ obeys PC and EM

How this precisely works can be understood with the help of truth-tables. This is what we will see now.

6 Generalization of the Concept of Truth-Table

To have a better understanding of the system π, it is useful to draw some truth-tables. Da Costa in developing the theory of bivaluations, with Andréa Loparić and Elias Alves, has generalized the idea of truth-table. We already saw a first specimen in the previous section, let us now see what it is in more details. We consider the following truth-table:

p	$\neg p$
0	0
0	1
1	0
1	1

This table clearly shows that when we know the value of an atomic formula p, we don't know the value of its negation, since when p is false $\neg p$ can be false or can be true and similarly when p is true $\neg p$ can be false or can be true. So one may say that we have here a non-deterministic semantics, in the sense that the value of $\neg p$ is not determined by the value of p. Nevertheless we have decidability, since we are able to describe all possible solutions with an algorithm which is precisely this truth-table. So in this context non-determinism does not imply undecidability.[5]

Let us see more precisely how this algorithm works. First let us emphasize that the above table is not a truth-table defining the semantics of the negation of the system π. It is a truth-table for negation of atomic formulas of the system π. The following truth-table:

$A \wedge \neg A$	$\neg(A \wedge \neg A)$
0	1
1	0

corresponds to condition 6) p. 380 of the bivaluation semantics for the system π. We can say that it is a truth-table for negation of contradictions, corresponding to "$A \wedge \neg A$ obeys PC and EM." The two above tables put together give rise to the following table

p	$\neg p$	$p \wedge \neg p$	$\neg(p \wedge \neg p)$
0	0	0	1
0	1	0	1
1	0	0	1
1	1	1	0

showing that $\neg(p \wedge \neg p)$ is not a tautology in π. But this compound table is still not a truth-table describing the full semantics of negation in π. There is in fact not a unique truth-table for negation in π. Conditions 4–7 of pp. 379–380 can be described by a series of truth-tables expressing the behavior of negation in π.

Given a negative formula in π, i.e. a formula of type $\neg A$, it is possible to draw a table which permits to determine all its possible truth-values. To do so we have to examine the value not only of its proper subformulas but also the value of some negations of its

---

[5]Non-deterministic semantics have also been developed for many-valued semantics, see [2, 14, 20].

proper subformulas. For example here is a table describing all possible truth-values of the formula $\neg(p \wedge q)$.

p	q	$\neg p$	$\neg q$	$p \wedge q$	$\neg(p \wedge q)$
0	0	0	0	0	0
0	0	0	0	0	1
0	0	0	1	0	1
0	0	1	0	0	1
0	0	1	1	0	1
0	1	0	0	0	1
0	1	0	1	0	0
0	1	0	1	0	1
0	1	1	0	0	1
0	1	1	1	0	1
1	0	0	0	0	1
1	0	0	1	0	1
1	0	1	0	0	0
1	0	1	0	0	1
1	0	1	1	0	1
1	1	0	0	1	0
1	1	0	1	1	0
1	1	1	0	1	0
1	1	1	1	1	0
1	1	1	1	1	1

The lines 1, 7, 13 and 20 express the fact that if $p \wedge q$ does not obey PC or EM then p and q do not obey PC or EM. Let us now see the general definition of these truth-tables.

A *quasi truth-table*[6] is a table with a finite number of columns and lines such that on the first line of each column we have a formula and on the other lines (proper lines) in each column we have 0 or 1 obeying the following conditions:

- each proper line of the table can be extended into a bivaluation β from $\mathcal{L}$ to $\{0, 1\}$
- for any bivaluation β, there is a proper line of the table such that $\beta(x) = y$, for any x, x being a formula given by the first line, and y being 0 or 1 according to the given line.

The above table has some additional features, corresponding to the following definition: a truth-table is a *complete quasi truth-table* iff it is a *quasi truth-table* and the set of formulas of the first line is closed by the subnegformula property, i.e. it contains all proper subformulas and negations of proper subformulas of this set of formulas. For any formula it is possible to construct a complete truth-table having on the first line the formula and the set of its subnegformulas. The method for constructing complete truth-tables for the system π is similar to the one for constructing truth-tables for the paraconsistent logic C1, see e.g. [4].

---

[6]da Costa and his school like to use the word *quasi*, other examples of such compound expressions are: *quasi truth* and *quasi set*. For this reason we can say that da Costa is not only a *paralogician* but also a *quasi logician*.

This generalized truth-table method is very useful to deal with logics built with non-truth-functional bivalent semantics. The following table for example shows that in a paracomplete logic where negation is defined just with half of the condition for negation (PC), the formula $p \to \neg p$ has the behavior of a classical negation for p, using the standard semantic definition of implication.

p	$\neg p$	$p \to \neg p$
0	0	1
0	1	1
1	0	0

We can therefore translate classical logic in a very simple paracomplete logic strictly included into classical logic (for more details about this logic see [8]). This is the translation paradox [32], an important phenomenon to analyze for developing a general theory of relations between logics.

References

1. Asser, G.: Einfürhung in die mathematische Logik, Teil 1: Aussagenkalkül. Teubner, Leipzig (1959)
2. Avron, A.: Non-deterministic matrices and modular semantics of rules. In: Beziau, J.-Y. (ed.) Logica Universalis – Towards a General Theory of Logic, pp. 155–174. Birkhäuser, Basel (2005)
3. Beziau, J.-Y.: Calcul des séquents pour logique non-alèthique. Log. Anal., **125–126**, 143–155 (1989)
4. Beziau, J.-Y.: Nouveaux résultats et nouveau regard sur la logique paraconsistante C1. Log. Anal., **141–142**, 45–58 (1993)
5. Beziau, J.-Y.: Universal logic. In: Childers, T., Majer, O. (eds.) Proceedings of the 8th International Colloquium – Logica'94, pp. 73–93. Czech Academy of Sciences, Prague (2004)
6. Beziau, J.-Y.: Recherches sur la Logique Universelle. PhD Thesis, Department of Mathematics, University Denis Diderot, Paris (1995)
7. Beziau, J.-Y.: Logic may be simple (logic, congruence and algebra). Log. Log. Philos. **5**, 129–147 (1997)
8. Beziau, J.-Y.: Classical negation can be expressed by one of its halves. Log. J. IGPL **7**, 145–151 (1999)
9. Beziau, J.-Y.: A sequent calculus for Łukasiewicz's three-valued logic based on Suszko's bivalent semantics. Bull. Sect. Log. **28**, 89–97 (1999)
10. Beziau, J.-Y.: La véritable portée du théorème de Lindenbaum–Asser. Log. Anal., **167–168**, 341–359 (1999)
11. Beziau, J.-Y.: From paraconsistent logic to universal logic. Sorites **12**, 5–32 (2001)
12. Beziau, J.-Y.: Sequents and bivaluations. Log. Anal. **176**, 373–394 (2001)
13. Beziau, J.-Y.: Bivalence, excluded middle and non contradiction. In: Behounek, L. (ed.) The Logica Yearbook 2003, pp. 73–84. Academy of Sciences, Prague (2003)
14. Beziau, J.-Y.: Non truth-functional many-valued semantics, In: Beziau, J.-Y., Costa-Leite, A., Facchini, A. (eds.) Aspects of Universal Logic, pp. 199–218. University of Neuchâtel, Neuchâtel (2004)
15. Beziau, J.-Y.: Les axiomes de Tarski. In: Rebuschi, M., Pouivet, R. (eds.) La logique en Pologne, pp. 135–149. Vrin, Paris (2006)
16. Beziau, J.-Y.: Bivalent semantics for De Morgan logic (the uselessness of four-valuedness). In: Carnielli, W.A., Coniglio, M.E., D'Ottaviano, I.M.L. (eds.) The Many Sides of Logic, pp. 391–402. King's College, London (2009)
17. Beziau, J.-Y.: What is a logic? Towards axiomatic emptiness. Log. Investig. **16**, 272–279 (2010)
18. Birkhoff, G.: Universal Algebra. Comptes Rendus du Premier Congrès Canadien de Mathématiques pp. 310–326. University of Toronto Press, Toronto (1946)
19. Birkhoff, G.: Universal algebra. In: Rota, G.-C. Oliveira, J.S. (eds.) Selected Papers on Algebra and Topology by Garrett Birkhoff. Birkhäuser, Basel (1987)

20. Carnielli, W., Lima-Marques, M.: Society semantics for multiple-valued logics. In: Carnielli, W.A., D'Ottaviano, I.M.L. (eds.) Advances in Contemporary Logic and Computer Science, pp. 33–52. American Mathematical Society, Providence (1999)
21. da Costa, N.C.A.: Calculs propositionnels pour les systèmes formels inconsistants. C. R. Acad. Sci. Paris **257**, 3790–3793 (1963)
22. da Costa, N.C.A., Alves, E.H.: Une sémantique pour le calcul C1. C. R. Acad. Sci. Paris **283**, 729–731 (1976)
23. da Costa, N.C.A., Alves, E.H.: A semantical analysis of the calculi C_n. Notre Dame J. Form. Log. **18**, 621–630 (1977)
24. da Costa, N.C.A., Beziau, J.-Y.: Théorie de la valuation. Log. Anal., **145–146**, 95–117 (1994)
25. da Costa, N.C.A., Beziau, J.-Y.: La théorie de la valuation en question. In: Proceedings of the XI Latin American Symposium on Mathematical Logic (Part 2), pp. 95–104. Universidad Nacional del Sur, Bahia Blanca (1994)
26. da Costa, N.C.A., Beziau, J.-Y.: Overclassical logic. Log. Anal. **157**, 31–44 (1997)
27. da Costa, N.C.A., Beziau, J.-Y., Bueno, O.A.S.: Malinowski and Suszko on many-valued logics: On the reduction of many-valuedness to two-valuedness. Mod. Log. **6**, 272–299 (1996)
28. Gentzen, G.: Über die Existenz unabhängiger Axiomensysteme zu unendlichen Satzsystemen. Math. Ann. **107**, 329–350 (1932)
29. Grana, N.: Sulla teoria delle valuationi di N.C.A. da Costa, Ligori, Napoli (1990)
30. Henkin, L., Monk, J.D., Tarski, A.: Cylindric Algebras. Part 1. North-Holland, Amsterdam (1971)
31. Hertz, P.: Über Axiomensysteme für beliebige Satzsysteme. Math. Ann. **101**, 457–514 (1929)
32. Humberstone, L.: Beziau's translation paradox. Theoria **71**, 138–181 (2005)
33. Loparić, A.: Une étude sémantique de quelques calculs propositionnels. C. R. Acad. Sci. Paris **284a**, 835–838 (1977)
34. Loparić, A.: A semantical study of some propositional calculi. J. Non-Class. Log. **3**, 74–95 (1986)
35. Loparić, A., Alves, E.H.: The semantics of systems Cn of da Costa. In: Arruda, A.I., da Costa, N.C.A., Sette, A.M. (eds.) Proceedings of the Third Brazilian Conference on Mathematical Logic pp. 161–172. Sociedade Brasileira de Lógica, São Paulo (1980)
36. Łukasiewicz, J.: O logice trójwartosciowej. Ruch Filoz. **5**, 170–171 (1920)
37. Marcos, J.: Nearly every normal modal logic is paranormal. Log. Anal. **48**, 279–300 (2005)
38. Pogorzelski, W.A., Wojtylak, P.: Completeness Theory for Propositional Logics. Birkhäuser, Basel (2008)
39. Popov, V.: On one four-place paranormal logic. Log. Stud. **10** (2003)
40. Tarski, A.: Remarques sur les notions fondamentales de la méthodologie des mathématiques. Ann. Soc. Pol. Math. **6**, 270–271 (1928)
41. Wójcicki, R.: Theory of Logical Calculi. Kluwer, Dordrecht (1988)
42. Zedong, M.: *On Contradiction*, Yenan (1937)

J.-Y. Béziau (✉)
Institute of Philosophy, Federal University of Rio de Janeiro, Rio de Janeiro, Brazil
e-mail: jean-yves.beziau@logica-universalis.org
url: http://www.jyb-logic.org

PARACONSISTENCY, PARACOMPLETENESS, AND VALUATIONS

Andréa LOPARIĆ and Newton C. A. DA COSTA

1. Introduction

A theory $\mathscr{T}$ is called inconsistent if among its theorems there are at least two, one of which is the negation of the other. When this is not the case; $\mathscr{T}$ is said to be consistent. We call $\mathscr{T}$ trivial when all formulas (or all closed formulas) of its language are also theorems of $\mathscr{T}$. If there is at least one formula (or closed formula) of the language of $\mathscr{T}$ that is not a theorem of $\mathscr{T}$, then $\mathscr{T}$ is said to be nontrivial.

A paraconsistent logic is a logic (a logical calculus or simply a calculus) which can be used in the systematization of inconsistent but nontrivial theories. These 'paraconsistent' theories, therefore, may contain inconsistencies (contradictions), i.e., pairs of theorems such that one is the negation of the other, without being trivial. Obviously, most of the extant systems of logic, such as the systems of classical logic, are not paraconsistent(in relation to paraconsistent logic, its applications, and its philosophy, see [1] and [3]).

Every logical system whatever has a two-valued semantics of valuations, which constitutes a generalization of the standard semantics (cf. [8] and the next section of this paper). Taking this fact into account, we can define precisely the notion of paraconsistent logic; a logic is paraconsistent if it can be the underlying logic of theories containing contradictory theorems which are both true. Such theories we call paraconsistent.

Similarly, we define the concept of paracomplete logic: a logical system is paracomplete if it can function as the underlying logic of theories in which there are (closed) formulas such that these formulas and their negations are simultaneously false. We call such theories paracomplete.

As a consequence, paraconsistent theories do not satisfy the principle of contradiction (or of non-contradiction), which can be

J.-Y. Béziau (ed.), *Universal Logic: An Anthology*, 373–385
Studies in Universal Logic, DOI 10.1007/978-3-0346-0145-0_28, © Springer Basel AG 2012

stated as follows: from two contradictory propositions, i.e., one of which is the negation of the other, one must be false. Moreover, paracomplete theories do not satisfy the principle of the excluded middle, formulated in the following form: from two contradictory propositions, one must be true.

The objective of the present paper is twofold: 1) to develop a propositional system of logic at the same time paraconsistent and paracomplete, which, in a certain sense, contains the classical propositional logic; 2) to illustrate how the method of valuations is convenient for the better understanding of a logical system (some applications of the semantics of valuations may be found in [2], [4], [10] and [16]).

The system studied in this paper may be seen as a kind of logic of vagueness (in the sense of [6] and perhaps also in the sense of [7]; it also constitutes an alternative to the dialectical logic in the sense of the logic DL of da Costa and Wolf (see [5] and [6]), which formalizes some views of McGill and Parry ([12]). Moreover, our system presents some connections with the dialectical logic DK of Routley and Meyer (cfr. [13]). One of the consequences of these parallels is that the main ideas of dialectics seem to be really vague, not being susceptible to unique characterization by a formal system.

Nonetheless, we do not intend here to explore the possible applications of our system, from the philosophical point of view or otherwise. Our aim is merely a technical one: to emphasize the relevance of the semantics of valuations, showing that with the help of such semantics one can, in various cases, obtain decision methods which really constitute a natural extension of the common two-valued truth-table decision method of the classical propositional calculus. In fact, we conjecture that any calculus that is decidable at all, is decidable by the corresponding semantics of valuations, i.e., more precisely, by the device which we shall name 'valuation tableaux' (for these tableaux see, for example, [4], [9], [10] and [11]).

2. *The semantics of valuations*

A (logical) calculus $\mathscr{C}$ is an ordered pair $<\Delta, \mathscr{R}>$, in which Δ is a set of formulas of a given language $\mathscr{L}$, and $\mathscr{R}$ is a collection of inference rules; Δ and $\mathscr{R}$ are supposed to be nonempty, and Δ is called the set of axioms of $\mathscr{C}$. In this paper, $\mathscr{L}$ will a propositional language. In general, we assume that $\mathscr{L}$ contains propositional variables (normally a denumerable infinite set of such variables), parentheses and connectives. The connectives are supposed to have finite ranks (although this point is not essential to our discussion). Usually, the language contains connectives for implication ($\rightarrow$), conjunction (&), disjunction ($\vee$) and negation ($\neg$); there may also be intensional ones, like the modal and deontic connectives. The concept of formula is introduced as usual. Capitan Roman letters will stand for formulas; sets of formulas will be denoted by capital Greek letters. The notion of inference rule could be made precise, but this is not essential to our objectives here. We recall only that a rule relates a new formula (the conclusion) to a given set of formulas (the premisses). For a given rule, the number of premisses is always finite and fixed.

In a calculus $\mathscr{C}$, it is easy to define when a formula A is a syntactical consequence of a set Γ of formulas. When this is done, we write $\Gamma \vdash_{\mathscr{C}} A$. If $\Gamma = \Phi$, A is said to be a thesis or a theorem of $\mathscr{C}$, and this fact is symbolized as follows: $\vdash_{\mathscr{C}} A$. The symbol '$\vdash_{\mathscr{C}}$' has all the expected properties. We write simply '$\vdash$', instead of '$\vdash_{\mathscr{C}}$', when there is no doubt about the calculus we are considering.

Definition 1. – Let $\mathscr{C} = \langle \Delta, \mathscr{R} \rangle$ be a calculus and e a function from the set of formulas of the language L of $\mathscr{C}$ into $\{0, 1\}$. We say that e is a (two-valued) evaluation associated with $\mathscr{C}$ if we have:

1) If $A \in \Delta$, then $e(A) = 1$;

2) If all premisses of an application of a rule belonging to $\mathscr{R}$ assume the value 1 under e, then the corresponding conclusion also assumes the value 1;

3) There exist at least one formula A such that $e(A) = 0$.

Let $\mathscr{E}$ be the set of evaluations of a calculus $\mathscr{C}$, $e \in \mathscr{E}$, and Γ a set formulas of the language of $\mathscr{C}$. We say that e satisfies Γ if, for every $A \in \Gamma$, $e(A) = 1$. We can easily prove the following properties.

i) If $\Gamma \vdash A$, then, for every $e \in \mathscr{E}$, if e satisfies Γ, then $e(A) = 1$;

ii) $\mathscr{C}$ is trivial if, and only if, $\mathscr{E} = \Phi$ ($\mathscr{C}$ is said to be trivial if, for every formula A of its language, we have that $\vdash A$).

Let $\Sigma \cup \{A\}$ be a set of formulas of a nontrivial calculus $\mathscr{C}$. Σ is called A-saturated if $\Sigma \nvdash A$ and, for every $B \notin \Sigma$, $\Sigma \cup \{B\} \vdash A$. We clearly have (see [8] and [9]):

   iii) If Σ is A-saturated, then $\Sigma \vdash B$ if, and only if, $B \in \Sigma$;

   iv) If $\Gamma \nvdash A$, then there exists an A-saturated set Σ such that $\Gamma \subset \Sigma$.

   v) The characteristic function of an A-saturated set is an evaluation.

An evaluation which is the characteristic function of an A-saturated set is called a *valuation*. Let $\mathscr{V}$ be the set of all valuations associated with $\mathscr{C}$. Using ii, iii, iv, and v, we can easily show that:

   vi) $\mathscr{E} = \Phi$ if, and only if, $\mathscr{V} = \Phi$.

The notion of semantical consequence, with respect to a non-trivial calculus $\mathscr{C}$, is introduced without difficulty. We say that A is a semantical consequence of Γ if, for every valuation v which satisfies Γ, $v(A) = 1$ (in which case we write $\Gamma \vDash_{\mathscr{C}} A$, or simply $\Gamma \vDash A$). If $\Gamma = \Phi$, we say that A is valid in $\mathscr{C}$ (and we write $\vDash_{\mathscr{C}} A$ or $\vDash A$).

Any nontrivial calculus (logical system) has a two-valued semantics, in the sense of the following theorem:

Theorem 1. – $\Gamma \vdash A$ if, and only if, $\Gamma \vDash A$.
   Proof. – Follows from i-v above.

The soundness and completeness of the classical propositional calculus are special cases of the preceeding theorem. The same is true of several other calculi (see, for example, [4] and [10].

A valuation v, such that $v(A) = 1$ for every A belonging to a collection Γ of formulas, is called a model of Γ.

A theory based on a calculus $\mathscr{C}$ is any set $\mathscr{T}$ of formulas of $\mathscr{L}$, the language of $\mathscr{C}$, such that if $\mathscr{T} \vdash A$, then $A \in \mathscr{T}$. When $\mathscr{T} = \{A: \mathscr{X} \vdash A\}$, $\mathscr{X}$ is called a set of axioms for $\mathscr{T}$. A model of $\mathscr{T}$ is any valuation v of $\mathscr{C}$ such that $v(A) = 1$ for every $A \in \mathscr{T}$. The theorems (or theses) of $\mathscr{T}$ are the formulas that belong to $\mathscr{T}$.

$\mathscr{C}$ is paraconsistent if, and only if, there are theories, based on $\mathscr{C}$, having models v for which $v(A) = v(\neg A) = 1$, for some $A \in \mathscr{L}$. Similarly, $\mathscr{C}$ is paracomplete if for some theory $\mathscr{T}$, based on $\mathscr{C}$, $\mathscr{T}$ has a model v such that, for some formula A, $v(A) = v(\neg A) = 0$, and conversely.

Remark. – Sometimes, in order to define the relation $\Gamma \vdash A$ with respect to a given calculus $\mathscr{C} = \langle \Delta, \mathscr{R} \rangle$, it may occur that some rules of $\mathscr{R}$ have to undergo certain restrictions, which take into account the cases where $\Gamma \neq \Phi$ (as occurs in modal logic with the rule of necessitation and in predicate logic with the rule of generalization). However, Theorem 1 remains valid under convenient, clear adaptations.

It seems worthwhile to observe that the semantics of valuations satisfies Tarski's conditions of formal correctness and of material adequacy (see [14] and [15]). In particular, Tarski's criterion $\mathscr{C}$ remains valid. These aspects of the semantics of valuations would become more evident if we were to consider predicate logic instead of propositional logic (cf. [2]).

A large part of our results and comments apply to first-order logic, with or without identity, and even to higher-order logic; of course, profound adaptations are required, since, in particular, we have to analyse the behaviour of the quantifiers.

3. The system π

We introduce now a paraconsistent and paracomplete calculus, π, with a very weak primitive negation, but where a kind of classical negation is definable. The language of π contains the set $\{\rightarrow, \&, \vee, \neg\}$ of primitive connectives, and an axiomatic basis for it is given by the following postulates (where $A^{\circ} \underset{\text{Def}}{=} \neg(A \& \neg A) \& (A \vee \neg A)$:

1) $A \rightarrow (B \rightarrow A)$
2) $(A \rightarrow B) \rightarrow ((A \rightarrow (B \rightarrow C)) \rightarrow (A \rightarrow C))$
3) $A, A \rightarrow B / B$
4) $(A \& B) \rightarrow A$
5) $(A \& B) \rightarrow B$
6) $A \rightarrow (B \rightarrow (A \& B))$
7) $A \rightarrow (A \vee B)$
8) $B \rightarrow (A \vee B)$

9) $(A \to C) \to ((B \to C) \to ((A \lor B) \to C))$

10) $A^{\circ} \lor (A \,\&\, \neg A) \lor \neg (A \lor \neg A)$

11) $\neg(A \lor \neg A) \to \neg(A \,\&\, \neg A)$

12) $\neg(A \,\&\, \neg A) \to ((A \,\&\, \neg A) \to B)$

13) $\neg(A \lor \neg A) \to ((A \lor \neg A) \to B)$

14) $(A^{\circ} \,\&\, B^{\circ}) \to ((A \,\&\, B)^{\circ} \,\&\, (A \lor B)^{\circ} \,\&\, (A \to B)^{\circ} \,\&\, (\neg A)^{\circ})$

Theorem 2. – The following schemes are not valid in π;

1) $\neg\neg A \to A$

2) $A \to \neg\neg A$

3) $\neg(A \,\&\, \neg A)$

4) $(A \,\&\, \neg A) \to B$

5) $A \lor \neg A$

6) $(A \lor \neg A) \to B$

7) $(A \to \neg A) \to \neg A$

8) $\neg A \lor \neg\neg A$

9) $(A \to B) \to ((A \to \neg B) \to \neg A)$

10) $(\neg B \to \neg A) \to (A \to B)$

11) $(A \to B) \to (\neg B \to \neg A)$

12) $\neg A \to (A \to B)$

13) $\neg(A \to A) \to B$

14) $\neg(A \lor B) \to (\neg A \,\&\, \neg B)$

15) $\neg(A \,\&\, B) \to (\neg A \lor \neg B)$

16) $(\neg A \lor \neg B) \to \neg(A \,\&\, B)$

17) $(\neg A \,\&\, \neg B) \to \neg(A \lor B)$

18) $(A \to B) \to \neg(A \,\&\, \neg B)$

19) $\neg(A \,\&\, \neg B) \to (A \to B)$

20) $\neg\neg(A \lor \neg A)$

Theorem 3. – The following schemes are provable in π:

T1) $\neg(A \lor \neg A) \to (A \to B)$

T2) $(\neg A \,\&\, \neg(A \,\&\, \neg A)) \to (A \to B)$

T3) $A \lor (A \to B)$

T4) $(A \,\&\, \neg A) \lor \neg(A \,\&\, \neg A)$

T5) $((A \,\&\, \neg A) \,\&\, \neg(A \,\&\, \neg A)) \to B$

T6) $\neg((A \,\&\, \neg A) \,\&\, \neg(A \,\&\, \neg A))$

T7) $(A \lor \neg A) \lor \neg(A \lor \neg A)$

T8) $\neg(A \vee \neg A) \rightarrow ((A \,\&\, \neg A) \rightarrow B)$

T9) $A \vee \neg(A \,\&\, \neg A)$

T10) $\neg A \vee \neg(A \,\&\, \neg A)$

T11) $\neg((A \vee \neg A) \,\&\, \neg(A \vee \neg A))$

T12) $(A \,\&\, \neg A)^{\circ}$

T13) $(\neg(A \,\&\, \neg A))^{\circ}$

T14) $(A \vee \neg A)^{\circ}$

T15) $A^{\circ\circ}$

T16) $(A \rightarrow B) \rightarrow ((A \rightarrow (B \rightarrow \neg(B \vee \neg B))) \rightarrow (A \rightarrow \neg(A \vee \neg A)))$

T17) $((A \rightarrow \neg(A \vee \neg A)) \rightarrow \neg((A \rightarrow \neg(A \vee \neg A)) \vee$
$\neg(A \rightarrow \neg(A \vee \neg A)))) \rightarrow A$

T18) $(\neg(A \,\&\, \neg A) \,\&\, (B \vee \neg B)) \rightarrow ((\neg B \rightarrow \neg A) \rightarrow (A \rightarrow B))$

T19) $(\neg(B \,\&\, \neg B) \,\&\, (A \vee \neg A)) \rightarrow ((A \rightarrow B) \rightarrow (\neg B \rightarrow \neg A))$

T20) $A^{\circ} \rightarrow (A \rightarrow \neg\neg A)$

T21) $A^{\circ} \rightarrow (\neg\neg A \rightarrow A)$

Definition 2. – $\sim A \underset{\text{Def}}{=} A \rightarrow \neg(A \vee \neg A)$. ('$\sim$' is the *strong* or *classical negation* of π).

Theorem 4. – '$\sim$' has all properties of the classical negation.

Proof. – In fact, we have in π:

$\vdash (A \rightarrow B) \rightarrow (\sim B \rightarrow \sim A)$ (by T16),

and

$\vdash \sim\sim A \rightarrow A$ (by T17).

Thus, since postulates 1-9 above, together with the schemes $(A \rightarrow B) \rightarrow (\sim B \rightarrow \sim A)$ and $\sim\sim A \rightarrow A$, constitute an axiomatic basis for the classical propositional calculus (in which '$\rightarrow$', '&', '$\vee$' and '$\sim$' are the primitive connectives), π contains, in an obvious sense, that calculus.

4. *Valuation semantics and decision method for* π

Let V_{π} be the set of functions from the set of formulas of π into $\{0, 1\}$, such that, for every $v \in V_{\pi}$, we have:

1) $v(A \,\&\, B) = 1$ if, and only if, $v(A) = 1$ and $v(B) = 1$;

2) $v(A \vee B) = 1$ if, and only if, $v(A) = 1$ or $v(B) = 1$;

3) $v(A \rightarrow B) = 1$ if, and only if, $v(A) = 0$ or $v(B) = 1$;

4) If $v(\neg\neg A) = v(\neg A)$, then $v(\neg A) = v(A)$;

5) If $v(A \S B) = v(\neg(A \S B))$, then $v(A) = v(\neg A)$ and
   $v(B) = v(\neg B)$, for $\S \in \{\rightarrow, \wedge, \vee\}$;
6) $v(\neg(A \& \neg A)) \neq v(A \& \neg A)$;
7) $v(\neg(A \vee \neg A)) \neq v(A \vee \neg A)$.

Theorem 5. – The set $\mathscr{V}_\pi$ is the set of valuations associated with π.

Proof. – It is easy to demonstrate that $v \in \mathscr{V}_\pi$ if, and only if, there exists a set of formulas $\Sigma \cup \{A\}$, such that Σ is A-saturated and v is its characteristic function.

Remark. – Since every theorem of π is classically valid, it is easy to see that every classical valuation belongs to $\mathscr{V}_\pi$; thus, there are valuations v for π such that, for some A, $v(A) \neq v(\neg A)$. But there are also valuations v of $\mathscr{V}_\pi$ such that, for some formula A, $v(A) = v(\neg A)$; in this case, we say that A is not 'semantically well-behaved'. Otherwise, A is 'semantically well-behaved'. Our system π has an interesting property: the semantical behaviour of any given formula A can be expressed, in a certain sense, by a formula. For example, A & A° *means* that A is true and well behaved, in this sense: for every $v \in \mathscr{V}_\pi$, if $v(A \& A°) = 1$, then $v(A) = 1$ and $v(\neg A) = 0$. In the same way: A & $\neg$A *means* 'A is true but not well-behaved'; $\neg$A & A° *means* that A is false and well-behaved; finally, $\neg$A & $\neg$(A $\vee$ $\neg$A) *means* that A is false and not well-behaved.

With the help of $\mathscr{V}_\pi$, we can obtain a decision method for the calculus π.

Let $\{A_1, A_2, \ldots, A_n\}$ be a set of formulas of the language π. We say that $A_1, A_2, \ldots, A_n$ constitutes a π-sequence if, for $1 \leqslant i \leqslant n$, we have:

a) If B is a subformula of A_i, then, for some $j \leqslant i$, B = A_j;
b) If A_i is $\neg(A \S B)$, where $\S \in \{\rightarrow, \&, \vee\}$, then there are i, k < i such that A_j is $\neg A$ and A_k is $\neg B$;
c) For $1 \leqslant j < i$, $A_j \neq A_i$.

Definition 3. – Suppose that $A_1, A_2, \ldots, A_n$ is a π-sequence. The tableaux for $A_1, A_2, \ldots, A_n$, denoted by $t_n(A_1, A_2, \ldots, A_n)$ or simply by t_n, when the formulas $A_1, A_2, \ldots, A_n$ are manifest, is a function from $I_n \times J(A_1, A_2, \ldots, A_n)$ into $\{0, 1\}$, where $I_n = \{1, 2, \ldots, n\}$ and

1) $J(A_1) = \{1, 2\}$, $t_1(1, 1) = 1$, and $t_1(1, 2) = 0$;

2) $J(A_1, A_2, \ldots, A_{n-1}) \subseteq J(A_1, A_2, \ldots, A_n)$ and, for $J(A_1, A_2, \ldots, A_{n-1}) = \{1, 2, \ldots, m\}$, one has:

2.1) For $1 \leq i \leq n$, $1 \leq j \leq m$, $t_n(i,j) = t_{n-1}(i,j)$;

2.2) a) If A_n is a propositional variable, $J(A_1, A_2, \ldots, A_n) = \{1, \ldots, 2m\}$, and, for $j \leq m$, $t_n(n,j) = 1$, and for $j > m$, $t_n(n,j) = 0$;

b) If A_n is a $A_k \& A_l$, $J(A_1, A_2, \ldots, A_n) = \{1, 2, \ldots, m\}$; for $1 \leq j \leq m$, $t_n(n,j) = 1$ if, and only if, $t_n(k,j) = 1$ or $t_{n(l, j)} = 1$;

c) If A_n is $A_k \vee A_l$, $J(A_1, A_2, \ldots, A_n) = \{1, 2, \ldots, m\}$; for $1 \leq j \leq m$, $t_n(n,j) = 1$ if, and only if, $t_n(k,j) = 1$ or $t_n(l,j) = 1$;

d) If A_n is $A_k \to A_l$, $J(A_1, A_2, \ldots, A_n) = \{1, 2, \ldots, m\}$; for $1 \leq j \leq m$, $t_n(n,j) = 1$ if, and only if, $t_n(k,j) = 0$ or $t_n(l,j) = 1$;

e) If A_n is $\neg A_k$, then:

e_1) If A_k is a propositional variable, $J(A_1, A_2, \ldots, A_n) = \{1, 2, \ldots, 2m\}$ and, for $j \leq m$, $t_n(n,j) \neq t_n(k,j)$ for $j > m$, $t_n(n,j) = t_n(k,j)$;

e_2) If a_k is $A_p \& \neg A_p$ or A_k is $A_p \vee \neg A_p$, then $J(A_1, A_2, \ldots, A_n) = \{1, 2, \ldots, m\}$; for $1 \leq i \leq m$, $t_n(n, j) \neq t_n(k, j)$;

e_3) If A_k is $A_p \& A_q$ or A_k is $A_p \vee A_q$, where A_q is different from $\neg A_p$, then let $\bar{p}$ and $\bar{q}$ be such that $A_{\bar{p}}$ is $\neg A_p$ and $A_{\bar{q}}$ is $\neg A_q$, and let $\bar{J}(A_1, A_2, \ldots, A_n) = \{j \in J(A_1, A_2, \ldots, A_{n-1}): t_n(p, j) = t_n(\bar{p}, j)$ or $t_n(q, j) = t_n(\bar{q}, j)\}$; if $J(A_1, A_2, \ldots, A_n) = \{m_1, m_2, \ldots, m_r\}$, then $J(A_1, A_2, \ldots, A_n) = \{1, 2, \ldots, m + r\}$; for $1 \leq i \leq n - 1$, $1 \leq s \leq r$, $t_n(i, m + s)$; for $j \leq m$, $t_n(n, j) \neq t_n(k, j)$; for $j > n$, $t_n(n, j) = t_n(k, j)$;

e_4) If A_k is $A_p \to A_q$, everything goes as in case e_3) above.

e_5) If A_k is $\neg A_p$, let $\bar{J}(A_1, A_2, \ldots, A_n) = \{j \in J(A_1, A_2, \ldots, A_{n-1}: t_n(p,j) = t_n(k,j)\}$; if $\bar{J}(A_1, A_2, \ldots, A_n) = \{m_1, \ldots, m_r\}$. then, $J(A_1, A_2, \ldots, A_n) = \{1, 2, \ldots, m + r\}$; for $1 \leq i \leq n - 1$, $1 \leq s \leq r$, $t_n(i, m + s) = t_n(i, m_s)$; for $j \leq m$, $t_n(n, j) \neq t_n(k, j)$; for $j \leq m$, $t_n(n, j) = t_n(k, j)$.

Let $t_n(A_1, A_2, \ldots, A_n)$ be the tableau for a π-sequence $A_1, A_2, \ldots, A_n$. We have:

Lemma 1. – For every j ∈J(A$_1$, A$_2$, . . . , A$_n$) there is a ν ∈𝒱$_π$ such that, for 1 ≤ i ≤ n, ν(A$_i$) = t$_n$(i, j).

Proof. – Construct the set α as follows:

1) If A ∈ {A$_1$, A$_2$, . . . , A$_n$}, then A ∈α if, and only if, t$_n$(i, j) = 0:
2) If A ∉ {A$_1$, A$_2$, . . . , A$_n$}, then A ∈α if, and only if:
   a) A is B & C and B ∈α or C ∈α;
   b) A is B ∨ C and A, B ∈α;
   c) A is B →C and A ∈α and B ∉α;
   d) A is ⌐(B & ⌐B) and B & ⌐B ∈α;
   e) A is ⌐(B ∨ ⌐B) and B ∨ ⌐B ∈α;
   f) A is ⌐⌐B and B ∈α and ⌐B ∈α;
   g) A is ⌐(B § C), B § C ∉α and B ∈α if and only if ⌐B ∉α and C ≠α if and only if ⌐C ∉α, for § ∈{→, &, ∨}

Now, it is not difficult to show that the complement of α, with respect to the set of all formulas of π, is A-saturated relative to every formula A of the form B & ⌐B & ⌐(B & ⌐B). Thus, its characteristic function ν belongs to 𝒱$_π$ and, by construction, for any A$_i$∈{A$_1$, A$_2$, . . . , A$_n$}, ν(A$_i$) = t$_n$(i, j).

Lemma 2. – For every ν ∈𝒱$_π$, there is a j ∈J(A$_1$, A$_2$, . . . , A$_n$), such that, for 1≤ i ≤ n, t$_n$(i, j) = ν(A$_i$).

Proof. – By induction on n.

Proof. – ⊢ A$_i$ if, and only if, ⊨ A$_i$ by Theorem 1. Now, by lemmas 1 and 2, ⊨ A$_i$ if, and only if, for every j ∈J(A$_1$, A$_2$, . . . , A$_n$), t$_n$(i, j) = 1.

On the other hand, we can easily stipulate a canonical way of associating, to any formula A, a π-sequence A$_1$, A$_2$, . . . , A$_n$ such that A$_n$ is A.

Therefore, π is decidable by our *tableaux*, based on the valuation semantics.

Example: The graphic below presents the tableau for the schematic π-sequence 'A, ⌐A, A ∨ ⌐A, ⌐(A ∨ ⌐A), A & ⌐A, ⌐(A & ⌐A), ⌐(A ∨ ⌐A) → ⌐(A & ⌐A)' and shows the validity of the scheme' ⌐(A ∨ ⌐A) → ⌐(A & ⌐A)' as well as the non-validity of 'A ∨ ⌐A' and '⌐(A & ⌐A)'.

A	$\neg$A	A $\vee$ $\neg$A	$\neg$(A $\vee$ $\neg$A)	A & $\neg$A	$\neg$(A & $\neg$A)	$\neg$(A $\vee$ $\neg$A) $\rightarrow$ $\neg$(A & $\neg$A)
1	0	1	0	0	1	1
0	1	1	0	0	1	1
1	1	1	0	1	0	1
0	0	0	1	0	1	1

Using valuation-tableaux we can prove an interesting result: π is not a finite many-valued logic. The proof runs as follows:

Lemma 3. – Let $A_1, A_2, \ldots, A_n$ be the π-sequence where A_1 is the propositional variable 'p' and, for $1 < i \le n$, $A_i = \neg A_{i-1}$. Then, a) $J(A_1, A_2, \ldots, A_n) = \{1, 2, \ldots, 2_n\}$; b) for $1 \le i \le n$, $t_n(i, 2n-1) = 1$, $t_n(i, 2_n) = 0$; c) for $n \ge 2$, $t_n(n, 2n-3) = 0$ and, for $1 \le i \le n-1$, $t_n(i, 2n-3) = 1$; d) for $n \ge 2$, $t_n(n, 2n-2) = 1$ and, for $1 \le i \le n-1$, $t_n(i, 2n-2) = 0$.

Proof. – By induction on n.

Corollary. – a) $\vdash \neg_k p \rightarrow \neg_{k+m} p$, for $k \ge 0$, $m > 1$;

b) $\nvdash \neg_{k+m} p \rightarrow \neg_k p$, for $k \ge 0$, $m > 1$, (where '$\neg_n p$' is the formula obtained by putting n occurrences of '$\neg$' before 'p').

Proof. – Take the π-sequence 'p, $\neg$p, $\ldots$, $\neg_k p, \ldots, \neg_{k+m} p$'. By Lemma 3, $t_{k+m+1}(k+m+1, 2(k+m+1)-3) = 0$ and $t_{k+m+1}(k+1, 2(k+m+1)-3) = 1$; thus, by Lemma 1, there is some $v \in \mathcal{V}_\pi$ such that $v(\neg_{k+m} p) = 0$ and $v(\neg_k p) = 1$, hence $v(\neg_k p \rightarrow \neg_{k+m} p) = 0$ and, by the Theorem 6, $\nvdash \neg_k p \rightarrow \neg_{k+m} p$; on the other hand, by Lemma 3, $t_{k+m+1}(k+m+1, 2(k+m+1)-2) = 1$ and $t_{n+m+1}(k+1, 2(k+m+1)-2) = 0$; therefore, by lemma 1, there is some $v \in \mathcal{V}_\pi$ such that $v(\neg_{k+m} p) = 1$ and $v(\neg_k p) = 0$, thus $v(\neg_{k+m} p \rightarrow \neg_k p) = 0$ and, by Theorem 6, $\nvdash \neg_{k+m} p \rightarrow \neg_k p$.

Theorem 7. – π has not a finite characteristic matrix; thus, π is not a finite many-valued logic.

Proof. – In any finite matrix there is some $k \ge 0$ and some $m > 1$ such that, for every value x of the matrix, $\neg_k(x) = \neg_{k+m}(x)$. Since $A \rightarrow A$ is π-valid, in any adequate matrix for π, and for every value y,

we should have a distinguished value for → (y, y); thus both
'$\daleth_k p \to \daleth_{k+m} p$' and '$\daleth_{k+m} p \to \daleth_k p$' would be valid in the matrix.

Universidade Estadual de Campinas   Andréa LOPARIĆ

Newton C. A. DA COSTA

Universidade Estadual de Campinas
Caixa Postal 1170
131.000 Campinas SP Brasil
Telex (019) 1150

REFERENCES

[1] Arruda, A. I., 'A survey of paraconsistent logic', in *Mathematical Logic in Latin America*, edited by A. I. Arruda, R. Chuaqui and N. C. A. da Costa, North-Holland, 1980, pp. 1-41.

[2] Arruda, A. I. and N. C. A. da Costa, 'Une semantique pour le calcul $C_1^=$', *C.R. Acad. Sc. Paris*, 284 (1977), 279-282.

[3] Da Costa, N. C. A., 'The philosophical import of paraconsistent logic', *The Journal of Non Classical Logic*, 1 (1982), 1-19.

[4] Da Costa, N. C. A. and Alves, E. H., 'Une sémantique pour le calcul C_1', *C.R. Acad. Sc. Paris*, 283 (1976), 729-731.

[5] Da Costa, N. C. A. and R. G. Wolf, 'Studies in paraconsistent logic I: the dialectical principle of the unity of opposites', *Philosophia*, 9 (1980), 189-217.

[6] Da Costa, N. C. A. and R. G. Wolf, 'Studies in paraconsistent logic II: quantifiers and the unity of opposites', to appear.

[7] Korner, S., *The Philosophy of Mathematics*, Hutchinson, 1960.

[8] Kotas, I. and N. C. A. da Costa, 'Some problems on logical matrices and valorizations', in *Proceedings of the Third Brazilian Conference on Mathematical Logic*, edited by A. I. Arruda, N. C. A. da Costa and A. M. Sette, Soc. Brasileira de Lógica, 1980, pp. 131-146.

[9] Loparić, A., 'Une étude sémantique de quelques calculs propositionnels'.

[10] Loparić, A., 'The method of valuations in modal logic', in *Mathematical Logic: Proceedings of the First Brazilian Conference*, edited by A. I. Arruda, N. C. A. da Costa and R. Chuaqui, Marcel Dekker, 1978, pp. 141-157.

[11] Loparić, A. and Alves, E. H., 'The semantics of the system C_n of da Costa', in *Proceedings of the Third Brazilian Conference on Mathematical Logic*, edited by A. I. Arruda, N. C. A. da Costa and A. M. Sette, Soc. Brasilieira de Lógica, 1980, pp. 161-172.

[12] McGill, V. J. and W. T. Barry, 'The unity of opposites: a dialectical principle'. *Science and Society*, 12 (1948), 418-444.

[13] Routley, R. and R. K. Meyer, 'Dialectical logic, classical logic, and the consistency of the world', Studies in Soviet Thought, 16 (1978), 1-25.

[14] Tarski, A., 'The Concept of truth in formalized languages', in *Logic, Semantics, Metamathematics, by A. Tarski, Clarendon Press, 1956, pp. 152-278.

[15] Tarski, A., 'The semantic conception of truth', *Philosophy and Phenomenological Research*, 4 (1944), 13-47.

[16] Wójcicki, R., *Theory of Propositional Calculi*, Polish Academy of Sciences, 1981.

A. Loparić (✉)
Department of Philosophy, University of São Paulo, São Paulo, Brazil
e-mail: aloparic@gmail.com

N.C.A. da Costa
Department of Philosophy, Federal University of Santa Catarina, Santa Catarina, Brazil
e-mail: ncacosta@terra.com.br

Part 15
Dov Gabbay (1996)

Fibring Logics

C. Caleiro and A. Sernadas

Logics of a combined nature were abundant in the literature when, in the mid 1990s, the study of general mechanisms for combining logics developed into a well-posed research area [5, 6]. Several motivations, both theoretical and applied, concur to justify such a new line of research, but the increasing number of ever more complex logics and logical features appearing in application areas ranging from software engineering to linguistics was certainly amongst the most important. The idea of combining logics had been cooking in a low flame for more than a decade, namely in the particular context of modal logic [29, 32, 50, 63, 64, 74], and within the theory of institutions, with emphasis on equational logic [43, 56]. However, Dov Gabbay's 1996 article *Fibred Semantics and the Weaving of Logics – Part I: Modal and Intuitionistic Logics* [34], included in this anthology, brought the whole enterprise to a new era. There are two principal reasons to justify the impact of the paper. First, among various application examples, the paper put forward the notions underlying the general mechanism for combining logics known as *fibring*. Reinterpreting the original phrasing, the general problem of fibring two logics $\mathcal{L}_1$ and $\mathcal{L}_2$ is:

(P0) Characterize the logics $\mathcal{L}$ built over the combined language that conservatively extend the two, and in particular the minimal such logic $\mathcal{L}_1 * \mathcal{L}_2$.

Perhaps even more importantly, the paper also clearly outlined for the first time the main objectives and subproblems that should guide a systematic study of the general problem, namely:

(P1) Characterize the notion of a logical system.
(P2) Present methodologies for combining any two logics.
(P3) Investigate transfer properties.
(P4) Compare the combined logics obtained with known logics.
(P5) Study possible interactions between the logics being combined.

The specialized study of fibred modal logics, from fusions to products, has followed its own course [30, 37–39, 41, 42, 46, 49, 52, 55, 77–79], and we do not go through it here.

This work was partially supported by FCT and EU FEDER, namely via the KLog project PTDC/MAT/68723/2006 of SQIG-IT. The authors are indebted to many colleagues with whom they have collaborated, over the years, on the topic of combining logics.

J.-Y. Béziau (ed.), *Universal Logic: An Anthology*, 389–396
Studies in Universal Logic, DOI 10.1007/978-3-0346-0145-0_29, © Springer Basel AG 2012

Instead, we concentrate on the development of the general theory of combining logics, as entailed by the idea of *universal logic* [4], and which subsumes the work on combining institutions [8, 11, 14, 26, 57, 58]. At the light of problem (P0), we briefly describe the work in each of the five directions mentioned. According to our view of the field, we first analyze subproblems (P1) and (P2), then we look into subproblems (P3) and (P5), and last we tackle subproblem (P4). We conclude with a short appreciation of the paths pursued and those that lay ahead.

1 Methodology

The statement of (P0) is relatively ambiguous, its precise meaning lying in the possible answers to subproblems (P1) and (P2). For fibring to be rigorously defined, one must first agree on what is a logical system, including the structure of its syntax, in order to make precise what it means to write combined formulas by mixing constructors from different logics. Moreover, one needs to settle what it means for a logic to extend another, namely in a conservative way. The task suggests taking an algebraic, or even categorial, approach to the fibring operation [16, 17, 68]: identifying the class of mathematical objects one wants to combine, and then defining and studying the operation that combines two such objects into a third one.

What is a Logical System?   There are many conceptions of what a logic system may be [33], a question prone to many philosophical wanderings. Even adopting a modern view of logic based on consequence operators, after Tarski or Scott, there is still room for large variations depending on the richness of the underlying language (e.g. sortedness and structurality), or on the underlying way of presenting them, namely semantically (e.g. using appropriate notions of model and formula satisfaction), or deductively (e.g. via Hilbert style axiomatizations, sequents or tableaux). Fortunately, a definitive answer is not necessary here. As long as one rigorously characterizes a collection of objects that one is willing to classify as logical systems, combining them makes perfect sense. Different notions of logical system, some very abstract, others quite particular, have been addressed and combined.

Combination Mechanisms   From the beginning there was an informal but robust notion of combining deductive calculi, namely axiomatizations, but only a mechanical definition of fibred semantics stemming from the work on modal-based logics, as introduced in [34]. Fibred Kripke-like semantics was then rigorously defined in [68, 80], and later extended to general logical matrices and abstract structural consequence operators in [7, 36], still over unsorted (propositional based) languages. The scope of the fibring operation was then extended to logical systems with more general many-sorted languages [61, 70, 80], as well as to other useful formalisms, including labelled calculi and tableaux [3, 25, 62], logics represented as theories in a meta-logic [15, 19, 22] and higher-order logic [23]. An important variation of the general fibring construction occurs when one takes in consideration the possibility that the logics being combined may share some of the syntactic constructors, leading to notions like *constrained fibring* [68] or *dovetailing* [34]. As hinted above, it is relatively easy to deal with these issues in deductive settings. However, fibring semantic-based logical systems is much harder, and these questions raise very interesting and deep problems in connection with the meta-properties of the fibring operation, that we will discuss below.

Besides fibring, in its full generality, some attention has also been given to interesting and useful restrictions of the main notion. Simpler forms of combining logics, with less ambitious syntactic aims and restricted forms of mixing syntactic constructors, were studied in detail, namely *temporalization* [29], *synchronization* [66, 67], *parameterization* [13], *exogenous enrichment* [54], and *approximation* [31].

2 Metatheorems

Once the methodological questions are dealt with and fibring has been rigorously defined over a given class of logical systems, studying the properties of the fibring operation seems to be the obvious next step. Which properties transfer from $\mathcal{L}_1$ and $\mathcal{L}_2$ to $\mathcal{L}_1 * \mathcal{L}_2$? Or better, under which general sufficient conditions on the logics $\mathcal{L}_1$ and $\mathcal{L}_2$ can one guarantee that $\mathcal{L}_1 * \mathcal{L}_2$ will enjoy a certain property? Indeed, subproblem (P3) is deeply related to subproblem (P5), as preservation results depend strongly on the amount of interaction allowed (or required) between the logical systems being combined. In fact, ultimately, these investigations lead to another fundamental question. How does the combined logic $\mathcal{L}_1 * \mathcal{L}_2$ relate, in general, to the logics one started with?

Preservation Results Already mingled in ancestor papers like [50], as well as in [34], one can find several transference results and non-preservation counterexamples about several interesting meta-logical properties, although just in the context of fibred modal logics. With the development of the general theory of combined logics, a great effort has been put in proving general transference results of meaningful properties, including, axiomatizability [68], algebraizability [27, 28, 47], and interpolation [20]. As explained above, the problems of preserving properties by fibring are particularly interesting but also substantially harder when one considers logical systems with rich multi-sorted languages with shared constructors. Special attention has been dedicated to the fundamental difficulties posed by semantical considerations, and their connection to the deductive ones, and therefore to finding general sufficient conditions for the preservation of soundness and completeness [7, 69, 80].

Conservativeness A fundamental case of preservation problem concerns logical derivations. To what extent are the inferences allowed in each of the ingredient logics $\mathcal{L}_1$ and $\mathcal{L}_2$ preserved to the fibred logic? Is $\mathcal{L}_1 * \mathcal{L}_2$ a conservative extension of the original systems? This desideratum is explicitly mentioned in the statement of the main problem (P0). However, as long as one allows the ingredient logics to share syntactic constructors, it becomes clear that it is not reasonable nor desirable to demand conservativeness to hold, in general. The fibring construction must, however, provide a combined logic which is "as conservative as possible". Such concerns are of prime importance in dealing with a particular instance of failure of conservativeness that became known as the *collapsing problem* [34, 35]. As mentioned above, this study motivated important developments and generalizations of the original definition of fibring for semantic-based logical systems, most notably using logical matrices, including *modulated fibring* [73] and *cryptofibring* [10].

3 Applications

The whole abstract idea of combining logics in an algebraic way is very appealing but, even if it is very well motivated, the question posed by subproblem (P4) is of utmost importance. Of course, having a hammer in hand and knowing how it works one should of course try it on a few nails. Still, this subproblem should be interpreted in a much deeper way. Are the general methods and results applicable and useful when one needs to build and study a logical system suitably combining features coming from several simpler logics? How do these features, even if thoroughly understood in isolation, interact with each other when combined? These are actually very difficult and challenging questions. Of course modal logics are an excellent playground for experimentation. However, there, specific results have been obtained that go much farther than the ones obtained in general for fibring. Anyway, things seem to be quite ok if one considers deductive formalisms, or logics represented in more general logical frameworks, as the typical tasks of putting together axioms and inference rules, even when there are interactions, are omnipresent in modern logic. However, in semantic-based formalisms, there is still a long way to go. There are promising steps forward, like [44], but collapsing phenomena have highlighted defects that can only be circumvented by further developing the theory of modulated fibring and cryptofibring, where one can aim at obtaining a better grip on the semantics of combined logics, while also opening room for less demanding general preservation results [8–10, 73].

Promising applications have been successfully addressed using simpler restricted combination mechanisms, namely temporalizations [2, 48] and exogenous enrichments [53]. The field of software engineering (where it is frequently necessary to work simultaneously with specifications written in different logics) was and remains the main target of application. However, most of the theoretical results on fibring fall short of this goal, namely because most were obtained for propositional-based logics, assume the homogeneous scenario for fibring (combining calculi of the same kind) and deal mainly with well behaved, structural logics. Artificial intelligence is another promising field of application [45, 51].

4 Outlook and Final Remarks

With the applications in mind, further work is needed towards expanding the theory of fibring towards more complex, possibly non-structural, logics, preferably within heterogeneous scenarios. The latter are considered in [24]. A promising approach to encompassing more general logics relies on graph-theoretic presentations and uses an abstraction map for relating models with the language [71, 72], thus avoiding the traditional homomorphic semantics underlying most of the previous work on fibring.

On another front, only a few transference results have been obtained so far. Much work is needed concerning the preservation by fibring of other interesting properties of logics, namely decidability, for instance via the preservation of the finite model property.

Given the recent progress in the related theory and applications of combining first-order theories [1, 59, 60, 75, 76], it seems worthwhile to try to establish a bridge between the two fields and thus open the door for a possible interplay between abstract model theory and the theory of fibring.

In order to contain the size of the list of bibliographic references, we refer instead the reader to recent surveys [12, 16, 18, 65] and books [21, 40]. As ever, Gabbay's book [36] is also mandatory reading to this end.

References

1. Baader, F., Ghilardi, S.: Connecting many-sorted theories. J. Symb. Log. **72**(2), 535–583 (2007)
2. Baltazar, P., Mateus, P.: Temporalization of probabilistic propositional logic. In: Logic Foundations of Computer Science 2009. Lecture Notes in Computer Science, vol. 5407, pp. 46–60. Springer, Berlin (2009)
3. Beckert, B., Gabbay, D.: Fibring semantic tableaux. In: Automated Reasoning with Analytic Tableaux and Related Methods. Lecture Notes in Computer Science, vol. 1397, pp. 77–92. Springer, Berlin (1998)
4. Béziau, J.-Y.: From consequence operator to universal logic: A survey of general abstract logic. In: Logica Universalis: Towards a General Theory of Logic, pp. 3–17. Birkhäuser, Basel (2005)
5. Blackburn, P., de Rijke, M.: Why combine logics? Stud. Log. **59**(1), 5–27 (1997)
6. Blackburn, P., de Rijke, M.: Zooming in, zooming out. J. Logic, Lang. Inf. **6**(1), 5–31 (1997)
7. Caleiro, C.: Combining logics. PhD thesis, IST – TU Lisbon, Portugal (2000)
8. Caleiro, C., Ramos, J.: Cryptomorphisms at work. In: Recent Trends in Algebraic Development Techniques – Selected Papers. Lecture Notes in Computer Science, vol. 3423, pp. 45–60. Springer, Berlin (2005)
9. Caleiro, C., Ramos, J.: Combining classical and intuitionistic implications. In: Frontiers of Combining Systems 07. Lecture Notes in Artificial Intelligence, pp. 118–132. Springer, Berlin (2007)
10. Caleiro, C., Ramos, J.: From fibring to cryptofibring: A solution to the collapsing problem. Logica Universalis **1**(1), 71–92 (2007)
11. Caleiro, C., Gouveia, P., Ramos, J.: Completeness results for fibred parchments: Beyond the propositional base. In: Recent Trends in Algebraic Development Techniques – Selected Papers. Lecture Notes in Computer Science, vol. 2755, pp. 185–200. Springer, Berlin (2003)
12. Caleiro, C., Sernadas, A., Sernadas, C.: Fibring logics: Past, present and future. In: We Will Show Them: Essays in Honour of Dov Gabbay, vol. 1, pp. 363–388. King's College Publications (2005)
13. Caleiro, C., Sernadas, C., Sernadas, A.: Parameterisation of logics. In: Recent Trends in Algebraic Development Techniques. Lecture Notes in Computer Science, vol. 1589, pp. 48–62. Springer, Berlin (1999)
14. Caleiro, C., Mateus, P., Ramos, J., Sernadas, A.: Combining logics: Parchments revisited. In: Recent Trends in Algebraic Development Techniques – Selected Papers. Lecture Notes in Computer Science, vol. 2267, pp. 48–70. Springer, Berlin (2001)
15. Caleiro, C., Carnielli, W., Coniglio, M., Sernadas, A., Sernadas, C.: Fibring non-truth-functional logics: Completeness preservation. J. Logic, Lang. Inf. **12**(2), 183–211 (2003)
16. Caleiro, C., Carnielli, W., Rasga, J., Sernadas, C.: Fibring of logics as a universal construction. In: Handbook of Philosophical Logic vol. 13, 2nd edn., pp. 123–187. Springer, Berlin (2005)
17. Carnielli, W., Coniglio, M.: A categorial approach to the combination of logics. Manuscrito **22**(2), 69–94 (1999)
18. Carnielli, W., Coniglio, M.: Combining Logics. Stanford Encyclopedia of Philosophy (2007)
19. Carnielli, W., Coniglio, M., D'Ottaviano, I.: New dimensions on translations between logics. Logica Universalis **3**(1), 1–18 (2009)
20. Carnielli, W., Rasga, J., Sernadas, C.: Preservation of interpolation features by fibring. J. Log. Comput. **18**(1), 123–151 (2008)
21. Carnielli, W., Coniglio, M., Gabbay, D., Gouveia, P., Sernadas, C.: Analysis and Synthesis of Logics – How to Cut and Paste Reasoning Systems. Applied Logic, vol. 35. Springer, Berlin (2008). http://www.springer.com/west/home/math?SGWID=4-10042-22-173762910-0
22. Coniglio, M., Carnielli, W.: Transfers between logics and their applications. Stud. Log. **72**(3), 367–400 (2002)
23. Coniglio, M., Sernadas, A., Sernadas, C.: Fibring logics with topos semantics. J. Log. Comput. **13**(4), 595–624 (2003)

24. Cruz-Filipe, L., Sernadas, A., Sernadas, C.: Heterogeneous fibring of deductive systems via abstract proof systems. Log. J. IGPL **16**, 121–153 (2008)
25. D'Agostino, M., Gabbay, D.: Fibred tableaux for multi-implication logics. In: Theorem Proving with Analytic Tableaux and Related Methods. Lecture Notes in Computer Science, vol. 1071, pp. 16–35. Springer, Berlin (1996)
26. Diaconescu, R.: Three decades of institution theory. In: This Anthology (2009)
27. Fernández, V.: Fibring logics in the Leibniz hierarchy. PhD thesis, State University of Campinas (UNICAMP), Brazil (2005). In Portuguese
28. Fernández, V., Coniglio, M.: Fibring in the Leibniz hierarchy. Log. J. IGPL **15**(5–6), 475–501 (2007)
29. Finger, M., Gabbay, D.: Adding a temporal dimension to a logic system. J. Logic, Lang. Inf. **1**(3), 203–233 (1992)
30. Finger, M., Gabbay, D.: Combining temporal logic systems. Notre Dame J. Form. Log. **37**(2), 204–232 (1996)
31. Finger, M., Wassermann, R.: Anytime approximations of classical logic from above. J. Log. Comput. **17**(1), 53–82 (2007)
32. Fitting, M.: Logics with several modal operators. Theoria **35**, 259–266 (1969)
33. Gabbay, D. (ed.): What Is a Logical System? Studies in Logic and Computation, vol. 4. Clarendon Press (1994)
34. Gabbay, D.: Fibred semantics and the weaving of logics: Part 1. J. Symb. Log. **61**(4), 1057–1120 (1996)
35. Gabbay, D.: An overview of fibred semantics and the combination of logics. In: Frontiers of Combining Systems. Applied Logic Series, vol. 3, pp. 1–55. Kluwer Academic, Dordrecht (1996)
36. Gabbay, D.: Fibring Logics. Oxford University Press, Oxford (1999)
37. Gabbay, D., Shehtman, V.: Products of modal logics. I. Log. J. IGPL **6**(1), 73–146 (1998)
38. Gabbay, D., Shehtman, V.: Products of modal logics. II. Relativised quantifiers in classical logic. Log. J. IGPL **8**(2), 165–210 (2000)
39. Gabbay, D., Shehtman, V.: Products of modal logics. III. Products of modal and temporal logics. Stud. Log. **72**(2), 157–183 (2002)
40. Gabbay, D., Kurucz, A., Wolter, F., Zakharyaschev, M.: Many-dimensional modal logics: Theory and applications. Studies in Logic and the Foundations of Mathematics, vol. 148. Elsevier, North-Holland (2003)
41. Gabelaia, D., Kontchakov, R., Kurucz, A., Wolter, F., Zakharyaschev, M.: Combining spatial and temporal logics: expressiveness vs. complexity. J. Artif. Intell. Res. **23**, 167–243 (2005)
42. Gabelaia, D., Kurucz, A., Wolter, F., Zakharyaschev, M.: Products of transitive modal logics. J. Symb. Log. **70**(3), 993–1021 (2005)
43. Goguen, J., Burstall, R.: A study in the foundations of programming methodology: Specifications, institutions, charters and parchments. In: Category Theory and Computer Programming. Lecture Notes in Computer Science, vol. 240, pp. 313–333. Springer, Berlin (1986)
44. Governatori, G., Padmanabhan, V., Sattar, A.: On fibring semantics for BDI logics. In: Logics in computer science – JELIA. Lecture Notes in Artificial Intelligence, vol. 2424, pp. 198–210. Springer, Berlin (2002)
45. Halpern, J., Pucella, R.: Characterizing and reasoning about probabilistic and non-probabilistic expectation. J. ACM **54**(3), 15–49 (2007)
46. Halpern, J., van der Meyden, R., Vardi, M.: Complete axiomatizations for reasoning about knowledge and time. SIAM J. Comput. **33**(3), 674–703 (2004)
47. Jánossy, A., Kurucz, A., Eiben, A.: Combining algebraizable logics. Notre Dame J. Form. Log. **37**(2), 366–380 (1996)
48. Kontchakov, R., Lutz, C., Wolter, F., Zakharyaschev, M.: Temporalising tableaux. Stud. Log. **76**(1), 91–134 (2004)
49. Kontchakov, R., Kurucz, A., Wolter, F., Zakharyaschev, M.: Spatial logic + Temporal logic = ?. In: Handbook of Spatial Logic, pp. 497–564. Springer, Berlin (2007)
50. Kracht, M., Wolter, F.: Properties of independently axiomatizable bimodal logics. J. Symb. Log. **56**(4), 1469–1485 (1991)
51. Liu, C., Ozols, M., Orgun, M.: A fibred belief logic for multi-agent systems. In: AI 2005: Advances in Artificial Intelligence. Lecture Notes in Computer Science, vol. 3809, pp. 29–38. Springer, Berlin (2005)

52. Marx, M.: Complexity of products of modal logics. J. Log. Comput. **9**(2), 197–214 (1999)
53. Mateus, P., Sernadas, A.: Weakly complete axiomatization of exogenous quantum propositional logic. Inf. Comput. **204**(5), 771–794 (2006)
54. Mateus, P., Sernadas, A., Sernadas, C.: Exogenous semantics approach to enriching logics. In: Essays on the Foundations of Mathematics and Logic, vol. 1, pp. 165–194. Polimetrica (2005)
55. Metcalfe, G., Olivetti, N., Gabbay, D.: Analytic calculi for product logics. Arch. Math. Log. **43**(7), 859–889 (2004)
56. Mossakowski, T.: Using limits of parchments to systematically construct institutions of partial algebras. In: Recent Trends in Data Type Specifications. Lecture Notes in Computer Science, vol. 1130, pp. 379–393. Springer, Berlin (1996)
57. Mossakowski, T., Tarlecki, A., Pawlowski, W.: Combining and representing logical systems. In: Category Theory and Computer Science. Lecture Notes in Computer Science, vol. 1290, pp. 177–196. Springer, Berlin (1997)
58. Mossakowski, T., Tarlecki, A., Pawlowski, W.: Combining and representing logical systems using model-theoretic parchments. In: Recent Trends in Algebraic Development Techniques. Lecture Notes in Computer Science, vol. 1376, pp. 349–364. Springer, Berlin (1998)
59. Nelson, G., Oppen, D.: Simplification by cooperating decision procedures. ACM Trans. Program. Lang. Syst. **1**(2), 245–257 (1979). doi:10.1145/357073.357079
60. Oppen, D.: Complexity, convexity and combinations of theories. Theor. Comp. Sci. **12**(3), 291–302 (1980)
61. Rasga, J.: Fibring labelled first-order based logics. PhD thesis, IST – TU Lisbon, Portugal (2003)
62. Rasga, J., Sernadas, A., Sernadas, C., Viganò, L.: Fibring labelled deduction systems. J. Log. Comput. **12**(3), 443–473 (2002)
63. Segerberg, K.: Two-dimensional modal logic. J. Philos. Log. **2**(1), 77–96 (1973)
64. Šehtman, V.: Two-dimensional modal logics. Akad. Nauk SSSR Mat. Zametki **23**(5), 759–772 (1978)
65. Sernadas, A., Sernadas, C.: Combining logic systems: Why, how, what for? CIM Bull. **15**, 9–14 (2003)
66. Sernadas, A., Sernadas, C., Caleiro, C.: Synchronization of logics. Stud. Log. **59**(2), 217–247 (1997)
67. Sernadas, A., Sernadas, C., Caleiro, C.: Synchronization of logics with mixed rules: Completeness preservation. In: Algebraic Methodology and Software Technology. Lecture Notes in Computer Science, vol. 1349, pp. 465–478. Springer, Berlin (1997)
68. Sernadas, A., Sernadas, C., Caleiro, C.: Fibring of logics as a categorial construction. J. Log. Comput. **9**(2), 149–179 (1999)
69. Sernadas, A., Sernadas, C., Zanardo, A.: Fibring modal first-order logics: Completeness preservation. Log. J. IGPL **10**(4), 413–451 (2002)
70. Sernadas, A., Sernadas, C., Caleiro, C., Mossakowski, T.: Categorial Fibring of Logics with Terms and Binding Operators. Studies in Logic and Computation, vol. 7, pp. 295–316. Research Studies Press Ltd. (2000)
71. Sernadas, A., Sernadas, C., Rasga, J., Coniglio, M.: A graph-theoretic account of logics. J. Log. Comput. doi:http://dx.doi.org/10.1093/logcom/exp023
72. Sernadas, A., Sernadas, C., Rasga, J., Coniglio, M.: On graph-theoretic fibring of logics. J. Log. Comput. doi:http://dx.doi.org/10.1093/logcom/exp024
73. Sernadas, C., Rasga, J., Carnielli, W.: Modulated fibring and the collapsing problem. J. Symbol. Log. **67**(4), 1541–1569 (2002)
74. Thomason, R.: Combinations of tense and modality. In: Handbook of Philosophical Logic, vol. II. Synthese Library, vol. 165, pp. 135–165. Reidel, Dordrecht (1984)
75. Tinelli, C., Ringeissen, C.: Unions of non-disjoint theories and combinations of satisfiability procedures. Theor. Comput. Sci. **290**(1), 291–353 (2003)
76. Tinelli, C., Zarba, C.: Combining nonstably infinite theories. J. Autom. Reason. **34**(3), 209–238 (2005)
77. van Benthem, J., Bezhanishvili, G., ten Cate, B., Sarenac, D.: Multimodal logics of products of topologies. Stud. Log. **84**(3), 369–392 (2006)
78. Wolter, F.: Fusions of modal logics revisited. In: Advances in Modal Logic, vol. 1, pp. 361–379. CSLI Publications (1998)

79. Wolter, F., Zakharyaschev, M.: Temporalizing description logics. In: Frontiers of Combining Systems, 2 vol. 7, pp. 379–401. Research Studies Press (2000)
80. Zanardo, A., Sernadas, A., Sernadas, C.: Fibring: Completeness preservation. J. Symb. Log. **66**(1), 414–439 (2001)

C. Caleiro (✉) · A. Sernadas
SQIG – Instituto de Telecomunicações, Department of Mathematics – Instituto Superior Técnico,
TU Lisbon, Lisbon, Portugal
e-mail: caleiro@gmail.com
e-mail: acs@math.ist.utl.pt

THE JOURNAL OF SYMBOLIC LOGIC
Volume 61, Number 4, Dec. 1996

FIBRED SEMANTICS AND THE WEAVING OF LOGICS
PART 1: MODAL AND INTUITIONISTIC LOGICS

D. M. GABBAY

Abstract. This is Part 1 of a paper on fibred semantics and combination of logics. It aims to present a methodology for combining arbitrary logical systems L_i, $i \in I$, to form a new system L_I. The methodology 'fibres' the semantics $\mathscr{K}_i$ of L_i into a semantics for L_I, and 'weaves' the proof theory (axiomatics) of L_i into a proof system of L_I. There are various ways of doing this, we distinguish by different names such as 'fibring', 'dovetailing' etc, yielding different systems, denoted by L_I^F, L_I^D etc. Once the logics are 'weaved', further 'interaction' axioms can be geometrically motivated and added, and then systematically studied. The methodology is general and is applied to modal and intuitionistic logics as well as to general algebraic logics. We obtain general results on bulk, in the sense that we develop standard combining techniques and refinements which can be applied to any family of initial logics to obtain further combined logics.

The main results of this paper is a construction for combining arbitrary, (possibly not normal) modal or intermediate logics, each complete for a class of (not necessarily frame) Kripke models. We show transfer of recursive axiomatisability, decidability and finite model property.

Some results on combining logics (normal modal extensions of **K**) have recently been introduced by Kracht and Wolter, Goranko and Passy and by Fine and Schurz as well as a multitude of special combined systems existing in the literature of the past 20–30 years. We hope our methodology will help organise the field systematically.

Contents

Received July 8, 1992; revised March 8, 1995.

SERC Senior Research Fellow under grant GR/H01014. Research supported by SERC project GR/G46671, *Syntactical Foundations for Nonmonotonic Reasoning*. I am grateful to Hans Kamp, Maarten de Rijke and the referees for valuable criticism.

J.-Y. Béziau (ed.), *Universal Logic: An Anthology*, 397–410
Studies in Universal Logic, DOI 10.1007/978-3-0346-0145-0_30, © Springer Basel AG 2012

§1. Overview and motivation. The aim of this paper is to present a methodological framework for the mathematical presentation (formulation) of, and possible solutions to, the following general weaving of logics problem:

Master Problem: General Weaving of Logics Problem. Let L be a logic with some connectives. Let $L_i, i \in I$ be a family of logics which are conservative extensions of L. Assume $L_i \cap L_j = L$, for $i \neq j$. Assume each L_i has, among others, the set $\mathscr{E}_i = \{\#_1^i, \ldots, \#_{k(i)}^i\}$ of additional connectives to those of L. Then the general weaving of logics problem is to characterise the set of all logics $\{L_I^\alpha \mid \alpha$ is a name for a way of combining $L_i, i \in I\}$ which are built on the connectives of L and $\bigcup_{i=1}^n \mathscr{E}_i$ and which are conservative extensions of each $L_i, i \in I$.

In particular we seek to characterise the minimal such logic L_I^{min}, as well as some distinguished other special logics L_I^δ, which possibly represent some special ways of combining $L_i, i \in I$. We use the notation $L_I^\delta = \otimes_{i \in I}^\delta L_i$.

The above general problem requires the formulation and solution of several secondary problems.

PROBLEM 1. Characterise the notion of a logical system (i.e., define the 'entities' (logics) involved in the weaving of logics 'problem').

PROBLEM 2. Present methodologies for combining any two logics, independent of how they are formulated (e.g., one may be formulated as a Hilbert system, the other through a semantical interpretation, the third may be an algebraic logic, etc.).

PROBLEM 3. Investigate transfer properties. If all $L_i, i \in I$ share some property (decidability, finite model property, Hilbert axioms, interpolation, etc.) do they transfer to L_I?

PROBLEM 4. Compare the combined logics thus obtained with existing known combinations of logics, which are abundant in the literature.

PROBLEM 5. Study possible natural interactions between the logics $L_i, i \in I$ during the combination process, possibly leading to non-conservative combinations. Try to identify generic special ways of non-conservatively combining logics, which are meaningful (e.g., combinations leading to temporal logics, n-dimensional cross products, combining a logic with its 'metalevel' consequence relation leading to conditional logics, making a logic fuzzy and so on).

Some attempts at a methodical solution to some of these problems for the case of normal modal logics can be found in [35, 14, 31] and for intuitionistic modal logics in [1]. In addition there are many papers presenting particular combinations of particular logics to suit particular applications (see the Conclusion and Future Research section below). These range from introducing modality into relevance logic to the fuzzification of an arbitrary logic. It is time to seek a general methodology for combining logics (which we refer to as 'weaving' of logics) and attempt to classify existing combinations within such a methodology. This is the task of this series of papers.

Upon reflection, it seems that the easiest way to combine logics is when they are presented semantically. The semantics characterises each connective individually and so any choice of connectives from a variety of systems can in principle be combined. The only problem that might arise is when the semantics are completely different in flavour. Can we do something in this case? Our best bet is to try and formulate a general semantics for general logics.

Our main tool in this paper is the notion of fibred semantics. We combine logics by fibring their semantics. Fibred semantics is a new concept of possible world semantics. It arises naturally in response to the needs of several independent areas of logic and its applications. In each of these areas, the simple, down to earth analysis of the local needs give rise essentially to the same solution—namely fibred semantics.

Our research plan is as follows:

• The present paper—*Fibred semantics and the weaving of logics part 1*—deals with modal and intuitionistic logics. Section 1 is a general introduction, Section 2 shows how to combine modal logics, Section 3 shows how to combine modal logics with an intuitionistic or intermediate logic and Section 4 compares our result with the literature and we conclude in Section 5 with a discussion of future research.

• Part 2 of this series, *Fibring non monotonic logics*, published in **Logic Colloquium 92** [29], addresses the question of how to fibre two general logics presented as non monotonic consequence relation. It first shows how to give such a logic a semantics and then uses the semantics to fibre the logics.

• Part 3 is entitled *How to make your logic fuzzy*, [28] shows that by fibring a logic with a many valued logic one can make it fuzzy. In fact many of the fuzzy logics in the literature are such fibrings.

• A related paper entitled *Conditional implications and non monotonic consequence* [27] shows, for a given consequence relation $\vdash\!\!\!\sim$, how to bring it into the object language as a conditional $>$, by fibring the semantics with itself.

For more details about related papers see the conclusion section.

We now proceed with a motivating example, to get our ideas into focus.

EXAMPLE 1.1 (Motivating example for fibred semantics). This example introduces the idea of fibred semantics. Imagine we want to form a combined logic with the intuitionistic arrow $\Rightarrow$ and the **K** modality $\Box$. We know the Kripke semantics for $\Rightarrow$ and the Kripke semantics for $\Box$. We also know the syntax of the combined logic, namely all wffs built up from the atoms using any of $\{\Rightarrow, \Box\}$ as connectives. What we do not have is any axioms or semantics for a combined system **L**.

Let us proceed nevertheless and take an arbitrary formula of the combined language, say $B = (q \Rightarrow \Box(p \Rightarrow r))$. We do not have models for the combined language, but we do notice that the main connective of B is $\Rightarrow$ and that $B = (q \Rightarrow A)$ with $A = \Box(p \Rightarrow r)$. We do have intuitionistic models for $\Rightarrow$ so let us take such a model of the form $\mathbf{m} = (S, \leq, a, h)$. S is the set of possible worlds, $\leq$ is a reflexive and transitive relation on S, $a \in S$ is the actual world and h is an assignment to the atoms, satisfying

(*) $$t \in h(q) \text{ and } t \leq s \text{ imply } s \in h(q),$$

where $h(q) \subseteq S$.

We can evaluate, or try to evaluate, B in $\mathbf{m}$, since its main connective is $\Rightarrow$. For B to hold in $\mathbf{m}$ we must have that it holds in the actual world a. Thus we continue:

$a \vDash_{\mathbf{m}} B$ iff $a \vDash_{\mathbf{m}} (q \Rightarrow A)$ iff for all t such that $a \leq t$ and $t \vDash_{\mathbf{m}} q$ we have $t \vDash_{\mathbf{m}} A$.

We know how to evaluate $t \vDash_{\mathbf{m}} q$, because q is atomic:

$$t \vDash_{\mathbf{m}} q \text{ iff (def) } t \in h(q).$$

The problem is to evaluate $t \vDash_{\mathbf{m}} A$, which is $t \vDash_{\mathbf{m}} \Box(p \Rightarrow r)$. $\Box$ is not in the intuitionistic language and so we do not have a recursive evaluation clause for it. Nor is A atomic so we cannot use the assignment h. So what shall we do?

We notice that in order to complete the evaluation of $a \vDash_{\mathbf{m}} B$ all we need is to have an answer, for each $t \in S$ such that $a \leq t$, to the question of whether $t \vDash_{\mathbf{m}} A$. Any answer will do! It is at this stage that we can introduce the basic idea of fibring! We notice that $A = \Box(p \Rightarrow r)$ is a formula beginning with the modality $\Box$. So if we take any Kripke model for $\Box$ (and we do have a semantics for $\Box$) then we can evaluate A there and get a value, and this value we give to $t \vDash_{\mathbf{m}} A$. So let us associate with each $t \in S$, a modal Kripke model

$$\mathbf{n}_t = \mathbf{F}(\mathbf{m}, t) = (T^t, R^t, a^t, h^t)$$

where $\mathbf{n}_t$ depends on $\mathbf{m}$ and on $t \in S$ and where T^t is the set of possible worlds of $\mathbf{n}_t$, R^t is the $\mathbf{K}$ accessibility relation. a^t is the actual world and h^t is the assignment.

We stipulate:

$$t \vDash_{\mathbf{m}} \Box A \text{ iff (definition) } \mathbf{n}_t \vDash \Box A.$$

The latter is the same as $a^t \vDash_{\mathbf{n}_t} \Box A$.

Since $\mathbf{n}_t$ is a modal model, a value can be found.

We continue the evaluation: $a^t \vDash \Box A$ iff for all $x \in T^t, a^t R^t x$ implies $x \vDash A$.

If A is in the modal language we can continue to evaluate. If A is atomic we get our value from h^t. What if A contains the intuitionistic $\Rightarrow$? I.e., $A = (p \Rightarrow r)$? Then we need to evaluate $x \vDash_{\mathbf{n}_t} (p \Rightarrow r)$.

Since $\mathbf{n}_t$ is a modal model, we do not know how to evaluate the above. However, all we need is a value. If we associate with x an intuitionistic model

$$\mathbf{m}_x = \mathbf{F}(\mathbf{n}_t, x) = (S^x, \leq^x, 0^x, h^x)$$

then we can continue our evaluation. $\mathbf{m}_x$ is functionally dependent on x and $\mathbf{n}_t$ through the function $\mathbf{F}$. We can represent the dependence as $\mathbf{F}(t, x)$.

We stipulate:

$$x \vDash_{\mathbf{n}_t} p \Rightarrow r \text{ iff (definition) } \mathbf{m}_x \vDash (p \Rightarrow r)$$

which is evaluated at the actual world of $\mathbf{m}_x$.

Thus we continue:

$$\mathbf{m}_x \vDash p \Rightarrow r \text{ iff } 0^x \vDash (p \Rightarrow r) \text{ iff for all } y \in S^x; 0^x \leq y \text{ and } y \vDash p \text{ implies } y \vDash r.$$

This process can be continued as many times as we need. To carry out the evaluation of arbitrary mixed formulas of $\mathbf{L}$, we need Kripke models of each logic and functions associating models of the other logic to each possible world, and infinitum, as illustrated in Figure 1.

These fibred models are not properly defined at the moment but are intuitively defined by the needs of combining logics. Note that we cannot put arbitrary modal

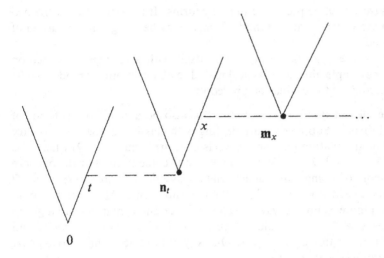

FIGURE 1

models $\mathbf{n}_t$, for points $t \in S$, because they must satisfy the intuitionistic persistence conditions

(*) $\qquad t \leq s$ and $\mathbf{n}_t \vDash \Box A$ imply $\mathbf{n}_s \vDash \Box A$

for any $\Box A$ of the modal language.

We need to secure the above condition (for arbitrary A):

$$\text{Let} \quad \mathbf{n}_t = (T^t, R^t, a^t, h^t)$$
$$\mathbf{n}_s = (T^s, R^s, a^s, h^s).$$

We can assume without loss of generality that our Kripke models $\mathbf{n}_t$ are pairwise disjoint and that each model $\mathbf{n} = (T, R, a, h)$ satisfies the condition:

$$x \in T \text{ iff } (\exists m \geq 0) a R^m x$$

where $x R^0 y$ iff $x = y$ and $x R^{m+1} y$ iff $\exists z (x R z \wedge z R^m y)$.

We will see in Section 3 that without loss of generality we can conservatively ensure that condition (*) holds for arbitrary A by asking that we have:

(**) $\qquad t \leq s$ implies $T^t \supseteq T^s$ and $R^t \supseteq R^s$, and $h^s \supseteq h^t \upharpoonright T^s$.

To summarize, we needed to have associated (*or fibred*) with each intuitionistic world a modal model and with each modal world an intuitionistic model. When we were forced to evaluate a formula D in a world belonging to the other semantics, we continued the evaluation in the associated model.

The basic fibring can be done by a two place function $\mathbf{F}$. If $\mathbf{m}$ is a model and t is a possible world in $\mathbf{m}$ then $\mathbf{F}(\mathbf{m}, t)$ is a Kripke model of the other semantics. We have $t \vDash_\mathbf{m} A$ iff $\mathbf{F}(\mathbf{m}, t) \vDash A$, iff $a^{\mathbf{F}(\mathbf{m}, t)} \vDash A$ where A is a formula with main connective not in the logic of the model $\mathbf{m}$ and $a^{\mathbf{F}(\mathbf{m}, t)}$ is the actual world of the model $\mathbf{F}(\mathbf{m}, t)$.

The model $F(\mathbf{m}, t)$ can be presented in many forms. If the semantics of the two logics involved are very similar, the model $F(\mathbf{m}, t)$ may be a slight modification of the model $\mathbf{m}$ itself.

It may even be the case that $F(\mathbf{m}, t)$ is $\mathbf{m}$ itself with a change in evaluation procedures. For example fibring a classical model to an intuitionistic model can be achieved by a change of the evaluation procedure.

The intuitive idea of the fibred semantics can also be understood in terms of classical model theory. Suppose we are dealing with classical models of a binary relation R and unary relation Q. These models have the form (S, R, Q) where S is a set and $R \subseteq S \times S$ and $Q \subseteq S$. Where do R and Q come from? For the purpose of the model theory of a binary and unary relations, we do not care where R and Q come from. All we need is a subset of $S \times S$ and a subset of S. However, we could get R and Q in a more elaborate way. We could, for example, map S onto a group G using $f : S \to G$ and let xRy hold in S iff $f(x) \cdot f(y) = f(y) \cdot f(x)$ in G and let $x \in Q$ hold if for example $f(x)^2 = 1$. Our way of looking at this is to say that we are fibring group structure onto S.

In computer science terms this can be looked upon as opening a window or what is called 'delegation' in object oriented programming. To compute whether xRy holds, we open a window and place $f(x), f(y)$ in the window (i.e., go to the associated group) and then compute something in the window (e.g., commutativity) and then come back. The group G itself may be fibred further, for example, into a field F. We need a function $g : G \to F$ with some translation of the group operations into field operations. For example $x \cdot y = z$ holds in G iff $g(x) \cdot g(y) = g(z)$ holds in F. (This means that G is viewed as a multiplicative group of a field.)

The above is not exactly fibring, it is more like representation. Fibring occurs when we double back into S. For example we can map G into S using a mapping $h : G \to S$ and require that for some formula $\Psi(u_1, u_2, u_3)$ of S the following holds: $x \cdot y = z$ in G iff $\Psi(h(x), h(y), h(z))$ holds in S.

1.1. Basic definitions and examples. The next series of definitions and examples will formally present the concept of fibring of this paper.

DEFINITION 1.2 (Logics and Semantics). (1) A propositional logical language $\mathbf{L}$ is comprised of an infinite set of atoms and a set of connectives $\{\sharp_1, \ldots, \sharp_k\}$. The connective $\sharp_j$ has n_j places.

(2) The wffs of $\mathbf{L}$ are defined by induction.
- An atom q is a wff.
- If $A_1, \ldots, A_{n_j}$ are wffs then $\sharp_j(A_1, \ldots, A_{n_j})$ is a wff with main connective $\sharp_j$.

(3) A consequence relation $\vdash\!\sim$ in the language $\mathbf{L}$ is a binary relation on wffs of $\mathbf{L}$ of the form $A\vdash\!\sim B$ satisfying:
- Identity

$$A\vdash\!\sim A.$$

- Transitivity

$$A\vdash\!\sim B \text{ and } B\vdash\!\sim C \Rightarrow A\vdash\!\sim C.$$

(4) A classical monadic logic language $\mathbf{CL}$ for a possible world semantics for a logic $\mathbf{L}$ with connectives $\{\sharp_1, \ldots, \sharp_k\}$ has the non logical predicates $\varphi_1, \ldots, \varphi_k, R_1,$

..., R_k where R_i is $n_i + 1$ place predicate (corresponding to the n_i place connective $\sharp_i$) and φ_i is m_i place predicate supporting (or connected to) the connective $\sharp_i$, ($m_i \leq n_i + 1$).

(5) An interpretation for the connectives $\sharp_i$ is a classical first-order formula $\Psi_i(t, Q_1, \ldots, Q_{n_i})$ of the language **CL** where t is the only free element variable in Ψ_i and Q_j are monadic variables (subset variables).[1]

(6) A possible world structure for the logic **L** in the language **CL** with interpretation $\{\Psi_i\}$ is any classical structure $\mathbf{m} = (S^{\mathbf{m}}, \varphi_j^{\mathbf{m}}, R_j^{\mathbf{m}}, \mathbf{e}^{\mathbf{m}}, h^{\mathbf{m}})$, where $\mathbf{e} \in S, \varphi_j$ and $R_j, \ldots$ are relations as in (4) above, and h is an assignment giving a subset $h(q) \subseteq S$, to each atom q of **L**.

(7) Given a structure $\mathbf{m}$, the function h can be extended to an arbitrary wff A of **L** by structural induction as follows

$$h(\sharp_j(A_1, \ldots, A_{n_j})) = \{t \mid \mathbf{m} \vDash \Psi_j(t, h(A_1), \ldots, h(A_{n_j}))\}.$$

We also write $t \vDash_h A$ when $t \in h(A)$.

We are abusing notation here. '$\vDash$' is classical satisfiability. $\Psi_j(t, Q_1, \ldots, Q_{n_j})$ is a formula with free variable t and free subset variables $Q_1, \ldots, Q_{n_j}$. By substituting $h(A_i)$ for Q_i we get a wff which we can evaluate at $\mathbf{m}$ as a classical model.

(8) $\mathbf{m}$ is said to satisfy $A, \mathbf{m} \vDash A$, if $\mathbf{e} \in h(A)$.

(9) We say that a class of model, $\mathscr{K}$ characterises a consequence relation $\vdash$ (or $\vdash$ is complete for $\mathscr{K}$) if we have for all A, B:

$$A \vdash B \text{ iff for all } \mathbf{m} \in \mathscr{K} \text{ (if } \mathbf{m} \vDash A \text{ then } \mathbf{m} \vDash B).$$

DEFINITION 1.3 (Fibred semantics). Let $\mathbf{L}_i$ be logic languages based on the same set of atomic propositions, with connectives $\{\sharp_1^i, \ldots, \sharp_{k_i}^i\}$, each of n_j^i places and $\mathbf{CL}_i$ be the corresponding monadic classical logic languages with interpretations Ψ_j^i for the respective connectives. Let **L** be a language containing (and based on) the disjoint union of the connectives of $\mathbf{L}_i$. Let **CL** be a monadic classical language based on the disjoint union of the predicates of $\mathbf{CL}_i$. We define a special class of structures for **CL**, obtained by 'fibring' the structures of $\mathbf{CL}_i$.

Let $\mathscr{K}_i$ be a class of structures for $\mathbf{L}_i$ of the form $\mathbf{m} = (S^{\mathbf{m}}, \varphi_j^{\mathbf{m}}, R_j^{\mathbf{m}}, \mathbf{e}^{\mathbf{m}}, h^{\mathbf{m}})$. We can assume that all sets of possible worlds $S^{\mathbf{m}}$ of all structures in $\cup_i \mathscr{K}_i$ are all pairwise disjoint. We now construct the fibred structures for **L**.

Let $W = \bigcup_{\mathbf{m} \in \cup_i \mathscr{K}_i} S^{\mathbf{m}}$. Since all $S^{\mathbf{m}}$ are pairwise disjoint, for each $w \in W$ there exists a unique $\mathbf{m}, i$ such that $w \in S^{\mathbf{m}}$ and $\mathbf{m} \in \mathscr{K}_i$. Let $\tau(w)$ be the function giving this unique i. Let $\mathbf{a}(w) = \mathbf{e}^{\mathbf{m}}$, i.e., $\mathbf{a}$ gives the actual world of the model $\mathbf{m}$ to which w belongs. Let $\mathbf{F}(i, w), \mathbf{F} : I \times W \mapsto W$ be an arbitrary function giving for each i and each w a value $w' = \mathbf{e}^{\mathbf{m}}, \mathbf{m} \in \mathscr{K}_i$. The function $\mathbf{F}$ is called a fibring function, associating with each possible world w and each semantic class $\mathscr{K}_i$ a structure $\mathbf{m} \in \mathscr{K}_i$. We also write $\mathbf{F}_i(w)$ when convenient. Let h be an assignment into W defined by

$$w \in h(q) \text{ iff } w \in h^{\mathbf{m}}(q), \text{ for the unique } \mathbf{m} \text{ such that } w \in S^{\mathbf{m}}.$$

We call any $\mathbf{n} = (W, \mathbf{a}, w_0, h, \mathbf{F}, \tau)$ a fibred structure, where $w_0 \in W$ is the 'actual' world of the structure.

---

[1] This clause is one dimensional. Fibring can be done in any dimension.

Let $\mathscr{K}$ be the class of all fibred structures, over the languages and semantics as above. Let A be a wff of $\mathbf{L}$. We define the notion of satisfaction as follows:

- $w \vDash_h q$ iff $w \in h(q)$, for q atomic.
- $w \vDash_h \sharp_j^i(A_1, \ldots, A_{n_j^i})$ iff

  (1) $\tau(w) = i$ and for the unique $\mathbf{m} \in \mathscr{K}_i$, such that $w \in S^{\mathbf{m}}$ we have
  $\mathbf{m} \vDash \Psi_j^i(h(A_1) \cap S^{\mathbf{m}}, \ldots, h(A_{n_j^i}) \cap S^{\mathbf{m}})$, or

  (2) $\tau(w) \neq i$ and $e^{\mathbf{m}} = \mathbf{F}(i, j)$ and $\mathbf{m} \vDash \sharp_j^i(A_1, \ldots, A_{n_j^i})$.

- We say $\mathbf{n} \vDash A$ iff $w_0 \vDash A$.
- We say $\mathscr{K} \vDash A$ iff for every $\mathbf{n}$ in $\mathscr{K}$, $\mathbf{n} \vDash A$.

Fibring the semantics of a modal logic to itself can be considered as a special case of two dimensional modal logics.

EXAMPLE 1.4 (Fibring modal Kripke semantics). Let us see what the fibred semantics looks like in the case of Kripke models for modal logics.

Let $\mathbf{L}_i$ be a modal language with connectives $\square_i, \lozenge_i$ whose semantics involves models of the form (S, R, a, h) where S is the set of possible worlds, $a \in S$ is the actual world and $R \subseteq S^2$ is the accessibility relation. h is the assignment function, which we present for convenience as a binary function, giving a value $h(t, q) \in \{0, 1\}$ for any $t \in S$ and atomic q. Assume the usual truth definitions for the $\square$ and $\lozenge$ of the language, namely

- $t \vDash \square A$ iff $\forall s(tRs \rightarrow s \vDash A)$.
- $t \vDash \lozenge A$ iff $\exists s(tRs \wedge s \vDash A)$.
- Satisfaction in the model is defined as satisfaction at a.

Assume $\mathbf{L}_i$ is a language with either a modality $\square_i$ or $\lozenge_i$ or both. Let $\mathscr{K}_i$ be a set of Kripke models which defines a logic $\vdash_i$ for this language. For example $\square_1$ may be **S4** modality, $\square_2$ may be the Löb modality and $\square_3$ may be some non normal modal extension of **K4**. In general, the logic $\vdash_i$ is defined by the class of models $\mathscr{K}_i$. $\mathscr{K}_i$ need not be a frame class or anything in particular, just a class of models of the form (S, R, a, h).

We can assume that the models satisfy the following condition:

- $S = \{x \mid \exists n \, aR^n x\}$.

This assumption does not affect satisfaction in models because points not accessible from a by any power R^n of R do not affect truth values at a.

We can assume that all sets of possible worlds in any $\mathscr{K}_i$ are all pairwise disjoint, and that there are infinitely many isomorphic (but disjoint) copies of each model in $\mathscr{K}_i$. We use the notation $\mathbf{m}$ for a model and present it as $\mathbf{m} = (S^{\mathbf{m}}, R^{\mathbf{m}}, a^{\mathbf{m}}, h^{\mathbf{m}})$ and write $\mathbf{m} \in \mathscr{K}_i$, when the model $\mathbf{m}$ is in the semantics $\mathscr{K}_i$. Thus our assumption boils down to

- $\mathbf{m} \neq \mathbf{n} \Rightarrow S^{\mathbf{m}} \cap S^{\mathbf{n}} = \varnothing$.

In fact a model can be identified by its actual world, i.e.,

- $\mathbf{m} = \mathbf{n}$ iff $a^{\mathbf{m}} = a^{\mathbf{n}}$.

We can now present the fibred semantics arising from Kripke models, in a much simplified form. Let

- $W = \bigcup_{\mathbf{m} \in \cup_i \mathscr{K}_i} S^{\mathbf{m}}$.

- $W_i = \{a^{\mathbf{m}} \mid \mathbf{m} \in \mathcal{H}_i\}$.
- $W_a = \bigcup_i W_i$.
- $R = \bigcup_{\mathbf{m} \in \bigcup_i \mathcal{H}_i} R^{\mathbf{m}}$.
- $h(t, q) = h^{\mathbf{m}}(t, q)$, for the unique $\mathbf{m}$ such that $t \in S^{\mathbf{m}}$.

Note that $R \upharpoonright S^{\mathbf{m}} = R^{\mathbf{m}}$ so that $R^{\mathbf{m}}$ are all retrievable from R. In fact $S^{\mathbf{m}}$ is retrievable from R and $a^{\mathbf{m}}$ as follows

$$S^{\mathbf{m}} = \{x \mid \exists n\, a^{\mathbf{m}} R^n x\}.$$

Notice that the master model (W, W_i, W_a, R, h), $W_a \subseteq W$ satisfies the following properties

(1) $\forall x \in W \exists y \in W_a \exists n (y R^n x)$.
(2) For any $t, s \in W_a$ let $S^t = \{x \mid t R^n x, \text{ for some } n\}$ then $t \neq s \to S^t \cap S^s = \varnothing$.

The semantics $\mathcal{H}_i$ is retrievable from the above master model by letting

$$\mathcal{H}_i = \{(S^t, R \upharpoonright S^t \times S^t, t, h \upharpoonright S^t) \mid t \in W_i\}.$$

We can thus view the fibring function $\mathbf{F}$ as a function giving for each i and each $w \in W$ another point (actual world) in W_i as follows:

$$\mathbf{F}_i(w) = \begin{cases} w & \text{if } w \in S^{\mathbf{m}} \text{ and } \mathbf{m} \in \mathcal{H}_i \\ \text{a value in } W_i, & \text{otherwise.} \end{cases}$$

The model $(W, W_i, W_a, R, w_0, h, \mathbf{F})$ with $w_0 \in W_a$ is referred to as a *master simplified fibred semantics model*. We have

$$t \vDash \Box_i A \text{ iff } \begin{cases} t \in \mathbf{m} \text{ and } \mathbf{m} \in \mathcal{H}_i \text{ and } \forall s (t R s \to s \vDash A\} \\ \text{or} \\ t \in \mathbf{m} \text{ and } \mathbf{m} \notin \mathcal{H}_i \text{ and } \mathbf{F}_i(t) \vDash \Box_i A. \end{cases}$$

It is obvious from the definition that the truth definition of $t \vDash \Box A$ is compatible with the truth condition:

$$t \vDash \Box_i A \text{ iff } \mathbf{F}_i(t) \vDash \Box_i A.$$

Similarly we define

$$t \vDash \Diamond_i A \text{ iff } \begin{cases} t \in \mathbf{m} \text{ and } \mathbf{m} \in \mathcal{H}_i \text{ and } \exists s (t R s \wedge s \vDash A\} \\ \text{or} \\ t \in \mathbf{m} \text{ and } \mathbf{m} \notin \mathcal{H}_i \text{ and } \mathbf{F}_i(t) \vDash \Diamond_i A. \end{cases}$$

We say the model satisfies A iff $w_0 \vDash A$.

There is a further simplifying assumption on $\mathbf{F}$ which we can require. Let $x, y, x \neq y$ be two points in W. We know there exist unique t, s such that $x \in S^t$ and $y \in S^s$. Consider $\mathbf{F}_i(x)$ and $\mathbf{F}_i(y)$.

$$\text{If } t \in W_i \text{ then } \mathbf{F}_i(x) = x.$$
$$\text{If } s \in W_i \text{ then } \mathbf{F}_i(y) = y.$$
$$\text{If } t \notin W_i \text{ then } \mathbf{F}_i(x) \in W_i.$$
$$\text{If } s \notin W_i \text{ then } \mathbf{F}_i(y) \in W_l.$$

The above allows for the possibility that $F_i(x) = F_i(y)$, in case $t, s \notin W_i$. We want to exclude this possibility, so we add the further assumption that

- $x \neq y \rightarrow F_i(x) \neq F_i(y)$.

This assumption can be made without loss of generality because we assumed that each Kripke model in each $\mathcal{K}_i$ has infinitely many isomorphic but pairwise disjoint copies of itself in $\mathcal{K}_i$.

We can now give an independent definition of what it means to be a *simplified fibred master model* (SFM-model) of a family of logics $L_i, i \in I$, as follows:

DEFINITION 1.5 (SFM-model). An SFM-model has the form $(W, W_i, W_a, R, w_0, h, F_i)$, where

(1) W is a set of worlds for some $i_0, w_0 \in W_{i_0}$.

(2) $W_i \subseteq W, i \in I$ are pairwise disjoint and nonempty, with $W_a = \bigcup_i W_i$.

(3) For $t, s \in W_i$ let $S^t = \{x \mid \exists n \; tR^n x\}$.

Then

- $t \neq s \rightarrow S^t \cap S^s = \varnothing$.
- $W = \bigcup_{t \in W_a} S^t$.

(4) F_i is a function satisfying

- $x \in S^t$ and $t \in W_i \rightarrow F_i(x) = x$.
- $x \in S^t$ and $t \notin W_i \rightarrow F_i(x) \in W_i$.
- $x \neq y \rightarrow F_i(x) \neq F_i(y)$.

(5) For each $t \in W_i$ the model $\mathbf{m}_t = (S^t, R \upharpoonright S^t \times S^t, t, h \upharpoonright S^t)$ is in the semantics $\mathcal{K}_i$ of L_i.

In case L_i is complete for the models $\{\mathbf{m}_t \mid t \in W_i\}$ we say the model is a *universal model*.

(6) We can also assume that F *generates* W as follows.

$$\text{Let } W^0 = S^{w_0}$$
$$W^{n+1} = W^n \cup \bigcup_{y \in W^n, i \text{ arbitrary}} S^{F_i(y)}.$$
$$\text{Then } W = \bigcup_n W^n.$$

Condition 6 can be assumed because evaluation at a model is evaluation at w_0. Any point not in $\bigcup_n W^n$ is not reachable from w_0 by embedded modalities so cannot affect truth values.

We now take a new look at the fibring function F. We can simplify the fibred semantics even further. Let us introduce new unary connectives of the form $J_i A$ (J_i for a 'jump' operator) and modal operators $\square$ and $\Diamond$ with the table:

- $t \vDash J_i A$ iff $F_i(t) \vDash A$.
- $t \vDash \Diamond A$ iff for some $s \, (tRs$ and $s \vDash A)$.
- $t \vDash \square A$ iff for all $s \, (tRs \rightarrow s \vDash A)$.

According to this definition we can let $\square_i = J_i \square, \Diamond_i = J_i \Diamond$. We are referring to J_i as 'jump' operators because their truth table 'jumps' the evaluation from a world t to the world $F(i, t)$.

Thus fibring several modalities together is like adding several unary jump operators to the modalities of the fibred semantics.

Conversely, assume we are given a modal logic with modalities $\square$ and $\Diamond$ and several J_i and a class $\mathcal{K}$ of *SMF*-models of the form (W, W_i, R, w_0, h, F_i), as defined above.

We can define modal semantics classes $\mathscr{K}_i$ by letting $\mathscr{K}_i$ be the set of all models of the form $(S^t, R \restriction S^t \times S^t, t, h \restriction S^t), t \in W_i$.

Note that the jump operators are slightly more expressive. For q atomic $\mathbb{J}q$ changes the index of evaluation of q from t to $\mathbf{F}_i(t)$. This cannot be done using modalities.

EXAMPLE 1.6 (Dovetailing modal Kripke semantics). Continuing the previous example, the case of dovetailing is when we require that for all atomic q

$$h(t, q) = h(\mathbf{F}_i(t), q)$$

for all i. This means the actual world of the model fibred at t can be identified with t. We must be careful however. For each t, we must first introduce the notation $tR_i^t y$ iff (definition) $\mathbf{F}_i(t)R^t y$. Then we can identify the point $\mathbf{F}_i(t)$ with t and the dovetailed model can be equivalently represented as (S, R_i^t, a, h) with

$$t \vDash \Box_i A \quad \text{iff} \quad \forall y(tR_i^t y \to y \vDash A)$$
$$t \vDash \Diamond_i A \quad \text{iff} \quad \exists y(tR_i^t y \wedge y \vDash A).$$

The function $\mathbf{F}$ is no longer needed, since we identified t with $\mathbf{F}_i(t)$.

Dovetailing looks particularly simple if we consider the simplified fibred semantics model $(W, R, w_0, h, \mathbf{F})$ of Example 1.4, then the corresponding relations R_i are not dependent on t.

It is interesting to see the role of the jump operators in the simplified dovetailed semantics.

We have in the fibred model $(W, R, w_0, h, \mathbf{F}_i)$

$$t \vDash \mathbb{J}_i \Box A \quad \text{iff} \quad \mathbf{F}_i(t) \vDash \Box A$$
$$\text{iff} \quad \forall x(\mathbf{F}_i(t)Rx \text{ implies } x \vDash A).$$

This becomes

$$t \vDash \mathbb{J}_i \Box A \text{ iff } \forall x(tR_i x \text{ implies } x \vDash A).$$

$\mathbb{J}_i$ becomes a mode shifting operator, changing the mode of evaluation, namely of which R_i to use.

We can consider the two clauses for a satisfaction $\vDash_i$ dependent on mode i.

- $t \vDash_i \Box A$ iff $\forall y(tR_i y$ implies $y \vDash_i A)$.
- $t \vDash_i \mathbb{J}_j A$ iff $t \vDash_j A$.

1.2. Ways of combining.

This paper discusses relationships between the consequence relations of $\mathbf{L}_i$ and the consequence relation of $\mathbf{L}$, for various conditions on $\mathbf{F}$ and h. Let $\delta(\mathbf{F}, h)$ be a condition on the combination (i.e., on $\mathbf{F}$ and h). We denote by $\mathbf{CL}^\delta$ the class of fibred structures obtained and by $\mathbf{L}^\delta$ the resulting logic (consequence relation). For some δ's, $\mathbf{L}^\delta$ is a conservative extension of each $\mathbf{L}_i$. For other δ's, we get a specific combined logic. The general problem to study is: given a δ and $\mathbf{L}_i$, can we characterise $\mathbf{L}^\delta$? If we have syntactical axiomatisations of $\mathbf{L}_i$, can we put them together to obtain an axiomatisation of $\mathbf{L}^\delta$? Can we identify known combined logics from the literature and present them as the result of some special conditions δ on the fibring?

Our method is to show how we can fibre logics $\mathbf{L}_i$ into a logic $\mathbf{L}^F$ through the most general fibring construction. Once we get semantics and axiomatisation for this general logic L^F, any combined logic existing in the literature can, in principle, be given a semantics which is related to the L^F semantics. The situation is similar to that of modal logic **K**. Any modal logic stronger than **K** can, in principle, be given semantics related to that of **K**. So, by studying the most general combination L^F of logics L_i, any existing axiomatic combined logic might, in principle, be given semantics. Conversely, we can try and axiomatise any restriction on the L^F semantics.[2]

For example, if we restrict our choice of fibred structures and require that for any atomic q and any w and i

$$w \vDash q \text{ iff } \mathbf{F}(i, w) \vDash q$$

we get a construction which we call dovetailing. Does this condition correspond to an axiom on **L**? We shall see that in modal logic this requirement is very natural.

EXAMPLE 1.7 (Sample Problem). Devise a logic with the classical connective '$\wedge$', the intuitionistic implication '$\Rightarrow$', and the **S4** modality '$\square$'. In other words, we want a logic **L** with connectives $\{\wedge, \Rightarrow, \square\}$ where these connectives behave in **L** respectively as they behave in their original logics. (Actually '$\wedge$' is the same in all three logics.)

To solve the above problem, we adopt an intuitive way of thinking. Given two logics, L_1 and L_2, how do we combine them? Never mind the question of whether we get a (minimal) conservative combination; we just want to put them together in some natural way.

Well, whether we can do anything depends on how the the logics are presented to us and what we know about them. Here are some possibilities:

(1) If L_1 and L_2 are presented in the same methodology, say proof theoretically as Gentzen systems, or say as Hilbert systems, then we can let **L** be the union of the languages and axioms and rules and allow substitution in **L** in both sets of axioms and rules. Thus if L_1 is a modal **K4** for $\square_1$ and L_2 is modal **K4** for $\square_2$ and the common language is classical logic then **L** will have both modalities with **K4** axioms and rules for each. There is no interaction between the two modalities. In fact, if we add the interactive axioms

$$A \rightarrow \square_2 \Diamond_1 A$$
$$A \rightarrow \square_1 \Diamond_2 A$$

we get the temporal logic $\mathbf{K}_t$ (with $\square_1$ future and $\square_2$ past).

(2) It may be that L_1 is presented as a Hilbert system and L_2 as a Gentzen system, how do we combine them?

(3) Worse still, suppose L_1 is presented proof theoretically and L_2 semantically, how do we combine them?

(4) If both L_1 and L_2 are faithfully translatable into a third logic **M** then we can combine the two logics in **M** and translate back. Let τ_1 and τ_2 translate (respectively) L_1 and L_2 into **M**. Assume that the translation is done by translating each connective $\sharp$ of L_i of the form $\sharp(x_1, \ldots, x_n)$ into a formula $\varphi_\sharp(x_1, \ldots, x_n) = \tau_i(\sharp)(x_1, \ldots, x_n)$ of **M**. Thus any formula of $L_{1,2}$ can be translated into **M** by applying recursively

---

[2]We agree that different modal logics can be based on the same semantics and that a given modal logic may have different semantics. However, once we have any reasonable semantics for a logic, we can hope to find some semantics for its extensions.

either τ_1 or τ_2 depending on the connective. Let τ be this 'combined' translation. Thus

- $\tau(\sharp(A_1, \ldots, A_n)) = \tau_i(\sharp)(\tau(A_1), \ldots, \tau(A_n)), \sharp \in \mathbf{L}_i.$
- $\tau(x) = x$, for x atomic.

Define $\mathbf{L}_{1,2}$ by

$$\mathbf{L}_{1,2} \vdash A \text{ iff (def) } \mathbf{M} \vdash \tau(A).$$

For example both $\square_1$ and $\square_2$ are translatable into classical logic via transitive possible world relations R_1 and R_2. Combining them in $\mathbf{M}$ gives a translation into (essentially Kripke semantics) $\mathbf{M}$ using both relations. In semantical terms we get Kripke semantics of the form (S, R_1, R_2). In this semantics $t \vDash \square_i A$ iff for all $s, tR_i s$ implies $s \vDash A$. Do we get the same system as when we combine their Hilbert Style formulation? In this case the answer is yes.

Going back to the example with $\{\wedge, \square\}$ of modal logic **S4** and $\{\Rightarrow\}$ of intuitionistic logic. We note that there are several ways of combining them. Are they identical?

- Combine through the translation into classical logic via their respective Kripke semantics.
- Translate modal logic via the Kripke semantics into classical logic $\mathbf{M}$. Consider the translation as a syntactical translation into $\mathbf{M}$ as a Hilbert system. Weaken $\mathbf{M}$ into $\mathbf{I}$ (intuitionistic predicate logic). Let $\mathbf{L}$ of $\{\wedge, \square, \Rightarrow\}$ be defined via the translation into $\mathbf{I}$.
- Combine $\wedge, \square$ and $\Rightarrow$ via the translation of intuitionistic logic into modal **S4**.
- Combine $\{\wedge, \square\}$ and $\{\Rightarrow\}$ as Hilbert systems. Take the union of languages and all instances of axioms and rules involving the connectives. Alternatively, take a natural deduction or Gentzen formulation of the two systems and combine the two proof theories.
- Combine the two systems by fibring their semantics as proposed in the present paper.

In this series of papers we develop several case studies of combining logics:

- Combining two modalities;
- intuitionistic modal logic.

The first case study, combining two modalities, is the simplest. This should be straightforward, as we are combining two logics of the same kind which, in general, have no special requirement on the assignments to the atoms.

The second case combines two slightly different logics; namely modal logic and intuitionistic logic. The novelty in this case is that intuitionistic logic has a restriction on the assignment to atoms which has to be satisfied. This slight complication has to be addressed.

We will introduce several methods of combining logics. Each method will have its own name. So for example we shall define what it means to 'fibre' two logics or to 'dovetail' two logics or to 'fuzzle' (make fuzzy) a logic by another or to 'join' two logics. Our purpose now is to give an intuitive idea of what these methodologies are. We explain the concepts by taking a simple example. Suppose we want to combine two modal logics $\mathbf{L}_1$ and $\mathbf{L}_2$. Let $\mathscr{K}_1, \mathscr{K}_2$ be the respective Kripke semantics of the logic. Let **m** be a model in $\mathscr{H}_1$ and let t be a possible world of **m**. The semantic construction which combines the logics associates a model **n** with t. The different methodologies of combination differ on the kind of model **n** that we use.

(1) For *fibring* logics, we require that **n** be any model in $\mathscr{H}_2$.

(2) For *dovetailing*, we require that **n** be a model of $\mathscr{H}_2$ such that for any *atomic q*

$$t \vDash q \text{ iff } \mathbf{n} \vDash q$$

(i.e., the fibred model must agree with the values t gives to atoms).

(3) For *joining* we form a multi-dimensional system out of the participating logics.

(4) For *fuzzling*, we simply assign at t values which are elements of the algebraic logic L_2 instead of just $\{0, 1\}$ values (e.g., boolean valued models are an example of 'making fuzzy' by weakening the 'crisp' $\{0,1\}$ values into boolean values).

D.M. Gabbay (✉)
Department of Informatics, King's College London, London, UK
e-mail: dov.gabbay@kcl.ac.uk